Digital Signal Processing

Digital Signal Processing

An Introduction

R. Anand, Ph.D.

Mercury Learning and Information
Dulles, Virginia
Boston, Massachusetts
New Delhi

Publisher: David Pallai
Mercury Learning and Information
22841 Quicksilver Drive
Dulles, VA 20166
info@merclearning.com
www.merclearning.com
1-800-232-0223

R. Anand. *Digital Signal Processing: An Introduction.*
ISBN: 978-1-68392-802-7

Library of Congress Control Number: 2022932248

222324321 Printed on acid-free paper in the United States of America.

CONTENTS

PREFACE

The great advancements in the design of microchips, digital systems, and computer hardware over the past 40 years have given birth to digital signal processing (DSP) which has grown over the years into a ubiquitous, multifaceted, and indispensable subject of study. As such, DSP has been applied in most disciplines ranging from engineering to economics and from astronomy to molecular biology. Consequently, it would take a multivolume encyclopedia to cover all the facets, aspects, and ramifications of DSP, and such a treatise would require many authors. This book focuses instead on the fundamentals of DSP, namely, on the representation of signals by mathematical models and on the processing of signals by discrete-time systems. Various types of processing are possible for signals, but the processing of interest in this volume is almost always linear. It typically involves reshaping, transforming, or manipulating the frequency spectrum of the signal of interest.

The author considers the processing of continuous- and discrete-time signals to be different facets of the same subject of study without a clear demarcation where the processing of continuous-time signals by analog systems ends, and the processing of discrete-time signals by digital systems begins. Discrete-time signals sometimes exist as distinct entities that are not derived from or related to corresponding continuous-time signals. The processing of such a signal would result in a transformed discrete-time signal, which would be, presumably, an enhanced, or in some way, more desirable version of the original signal. Obviously, reference to an underlying continuous time signal would be irrelevant in such a case. However, more often than not discrete-time signals are derived from corresponding continuous-time signals and, as a result, they inherit the spectral characteristics of the latter.

Discrete-time signals of this type are often processed by digital systems, and after that, they are converted back to continuous-time signals. A case in point can be found in the recording industry where music is first sampled to generate a discrete-time signal, which is then recorded on a disc. When the disc is played back, the discrete-time signal is converted into a continuous-time signal. In order to preserve the spectrum of the underlying continuous-time signal, e.g., that delightful piece of music, through this series of signal manipulations, special attention must be paid to the spectral relationships that exist between continuous- and discrete-time signals.

In the past, signal processing appeared in various concepts in more traditional courses like telecommunications, control, circuit theory, and in instrumentation. The signal processing done was analog, and discrete components were used to achieve the various objectives. However, in the later part of the 20th century we saw the introduction of computers and their fast and tremendous growth. In the late 1960s and early 1970s, a number of researchers resorted to modeling and simulation of various concepts in their research endeavors, using digital computers, in order to determine performance and optimize their designs. It is these endeavors that led to the development of many digital signal processing algorithms which we know today. With the rapid growth of computing power in terms of speed and memory capacity, a number of researchers wanted to obtain their results from near real-time to real time. This saw the development of processors and I/O devices that were dedicated to real-time data processing; though initially at lower speeds, they are currently capable of processing high speed data including video signals. The many algorithms that were developed in the research activities, combined with software and hardware that was developed for processing by industry, ushered in a new course into the university curriculum – *Digital Signal Processing*.

For many years, the course titled Digital Signal Processing was offered as a postgraduate course with students required to have a background in telecommunications (spectral analysis), circuit theory and of course mathematics. The course provided the foundation to do more advanced research in the field. Though this was useful, it did not provide all the necessary background that many industries required; to write efficient programs and to develop applications. In many institutions a simplified version of the postgraduate course has filtered into the undergraduate programs. This book is an attempt to bridge the gap. It can serve as a text for undergraduate or graduate courses and various scenarios are possible depending on the background preparation of the class and the curriculum of the institution.

CHAPTER 1

Introduction to Digital Signal Processing (DSP)

1.1 INTRODUCTION

Digital signal processing (DSP) is an area of science and technology that has developed rapidly over the past few decades. The techniques and applications of DSP are as old as Newton and Gauss and as new as digital computers and integrated circuits (ICs). The rapid development of DSP is a result of the significant advances in digital computer technology and IC fabrication.

DSP is concerned with the representation of signals by sequences of numbers or symbols and the processing of these sequences. Processing means the modification of sequences into a form that is in some sense more desirable.

In another words, DSP is a mathematical manipulation of discrete-time signals to get more desirable properties of the signal, such as less noise or distortion.

The classical numerical analysis formulas such as those used for interpolation, differentiation, and integration are also DSP algorithms.

DSP finds application in various fields such as speech communication, data communication, image processing, radar engineering, seismology, sonar engineering, biomedical engineering, acoustics, nuclear science, and many others.

DSP can be applied to one-dimensional signals as well as multidimensional signals. Example of the one-dimensional signal is speech and an example of the two-dimensional signal is an image. Many picture processing applications require the use of two-dimensional signal processing techniques.

Two-dimensional signal processing includes X-ray enhancement, analysis of aerial photographs (these photographs are necessary for detection of a forest fire or crop damage), analysis of satellite weather photographs, etc. Analysis of seismic data is required in oil exploration, earthquake measurements, and monitoring of nuclear tests. These utilize multidimensional signal processing techniques. The impact of DSP techniques will undoubtedly promote revolutionary advances in many fields of application. A notable example is telephony where digital techniques dramatically increased economy and flexibility in implementing switching and transmission systems.

1.2 APPLICATIONS OF DIGITAL SIGNAL PROCESSING

There are a variety of application areas of DSP because of the availability of high-resolution spectral analysis. It requires high-speed processor to implement the Fast Fourier transform (FFT). Some of these areas are

1. Speech processing,
2. Image processing,
3. Radar signal processing,
4. Digital communications,
5. Spectral analysis, and
6. Sonar signal processing.

Many of the above applications are discussed in Chapter 13.
Some of the other applications of DSP are in

a. Transmission lines,
b. Advanced optical fiber communication,
c. Analysis of sound and vibration signals,
d. Implementation of speech recognition algorithms,
e. Very Large-Scale Integration technology,
f. Telecommunication networks,
g. Microprocessor systems,
h. Satellite communications,

i. Telephony transmission,

j. Aviation,

k. Astronomy,

l. Industrial noise control, and

m. New DSP algorithms and many more.

Speech Processing: Speech is a one-dimensional signal. Digital processing of speech is applied to a wide range of speech problems such as speech spectrum analysis and channel vocoders (voice coders). DSP is applied to speech coding, speech enhancement, speech analysis and synthesis, speech recognition, and speaker recognition.

Image Processing: Any two-dimensional pattern is called an image. Digital processing of images requires two-dimensional DSP tools such as discrete Fourier transform, fast Fourier transform (FFT) algorithms, and z-transforms. Processing of electrical signals extracted from images by digital techniques includes image formation and recording, image compression, image restoration, image reconstruction, and image enhancement.

Radar Signal Processing: Radar stands for "radio detection and ranging." Improvement in signal processing is possible by digital technology. The development of DSP has led to greater sophistication of radar tracking algorithms. Radar systems consist of transmitting–receiving antenna, digital processing system, and control unit.

Digital Communications: Application of DSP in digital communication especially telecommunications comprises digital transmission using PCM, digital switching using time-division multiplexing, echo control, and digital tape recorders. DSP in telecommunication systems is found to be cost-effective due to the availability of medium- and large-scale digital ICs. These ICs have desirable properties such as small size, low cost, low power, immunity to noise, and reliability.

Spectral Analysis: Frequency-domain analysis is easily and effectively possible in DSP using fast Fourier transform (FFT) algorithms. These algorithms reduce computational complexity and also reduce the computational time.

Sonar Signal Processing: Sonar stands for "sound navigation and ranging." Sonar is used to determine the range, velocity, and direction of targets that are remote from the observer. Sonar uses sound waves at lower frequencies to detect objects underwater.

DSP can be used to process sonar signals, for the purpose of navigation and ranging.

1.3 SIGNALS

A signal can be defined as a function of one or more independent variable(s) which conveys information. Independent variables may be time, space, etc., and depend on the type of signals.

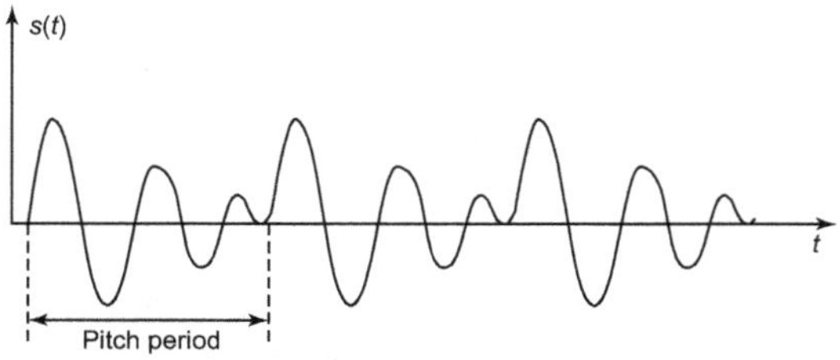

FIGURE 1.1 Speech signals.

Examples of signals are speech signals, pictures, electrocardiogram (ECG) signals, etc. A speech signal is represented mathematically as a function of time and a picture signal is represented as a brightness function of two spatial variables.

1.4 CLASSIFICATION OF SIGNALS

Any investigation in signal processing is started with a classification of signals involved in the specific application. Signals can be classified in the following classes:

1. Multichannel and multidimensional signals,
2. Continuous-time and discrete-time signals,
3. Analog and digital signals,
4. Deterministic and random signals,
5. Energy and power signals, and
6. Periodic and non-periodic signals.

Now, we will discuss these in detail in subsequent sections.

1.4.1 Multichannel and Multidimensional Signals

Multichannel Signals: Signals which are generated by multiple sources or multiple sensors are called multichannel signals. These signals are represented by a vector:

$$s(t) = \begin{bmatrix} S_1(t) \\ S_2(t) \\ S_3(t) \end{bmatrix}.$$

The signal represents a 3-channel signal. In electrocardiography, 3-lead and 12-lead electrocardiographs are often used in practice, which results in 3-channel and 12-channel signals, respectively.

Multidimensional Signal: A signal is called a multidimensional signal if it is a function of *M* independent variables. For example, *Speech signal* is a one-dimensional signal because the amplitude of the signal depends upon a single independent variable, namely, time. *TV Picture Signal:* A B/W picture signal is an example of a two-dimensional signal because the brightness of the signal at each point is a function of two spatial independent variables, namely, x and y. Variables x and y are width and height of the picture element.

A colored picture signal is an example of three-dimensional signal because brightness of the signal at each point is a function of three independent variables, namely, x, y, and time (t).

1.4.2 Continuous-time and Discrete-time Signals

Continuous-time Signals: A signal that varies continuously with time is called a *continuous-time signal.* These are defined for every value of the independent variable, namely, time. For example, *speech signal and temperature* of the room are continuous-time signals. The continuous-time signal is shown in Figure 1.2.

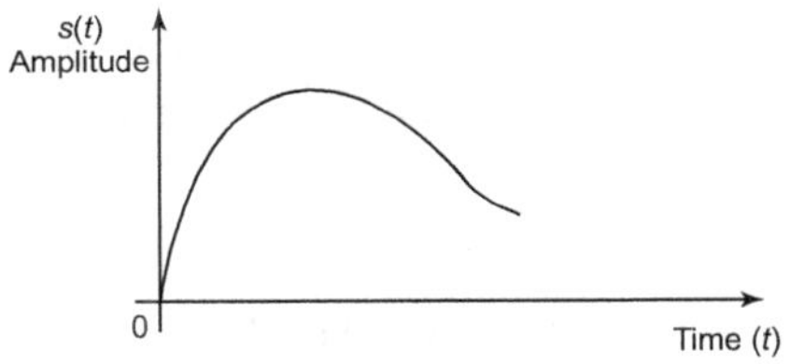

FIGURE 1.2 Continuous-time signal.

Discrete-time Signal: Discrete-time signals are signals which are defined at discrete times (Figure 1.3). These are represented by sequences of numbers. For example, the rail traffic signal is a discrete-time signal.

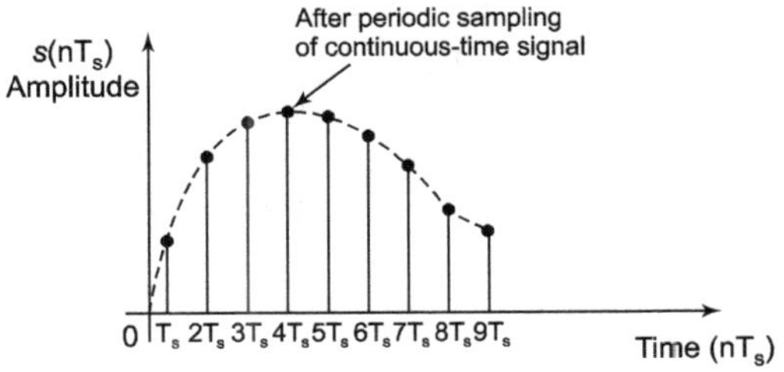

FIGURE 1.3 Discrete-time signal.

Discrete-time signals can be recovered by periodic sampling of continuous-time signals. Figure 1.3 illustrates the discrete-time signal.

1.4.3 Analog and Digital Signals

Analog Signals: Analog signals are signals of which both the dependent variable and the independent variable(s) are continuous in nature. Analog signals arise when a physical waveform is converted into an electrical signal. This conversion is performed by means of a transducer. For example, telephone speech signals, TV signals, etc., are very common types of the analog signal.

Telephone Speech Signals. A telephone message comprises speech sounds having vowels and consonants. These sounds produce an audio signal. These sound waves are converted into analog electrical signals by means of a transducer (microphone). The transducer is a device that converts non-electrical quantities into electrical signals, for example, a microphone. Continuous-amplitude, continuous-time signals are called analog signals. The Analog signal is shown in Figure 1.1.

Digital Signals: Digital signals are signals of which both the dependent variable and the independent variables are discrete in nature. Digital signals comprise pulses occurring at discrete intervals of time. Telegraph and teleprinter signals are the examples of digital signals. Figure 1.4 illustrates a telegraph signal.

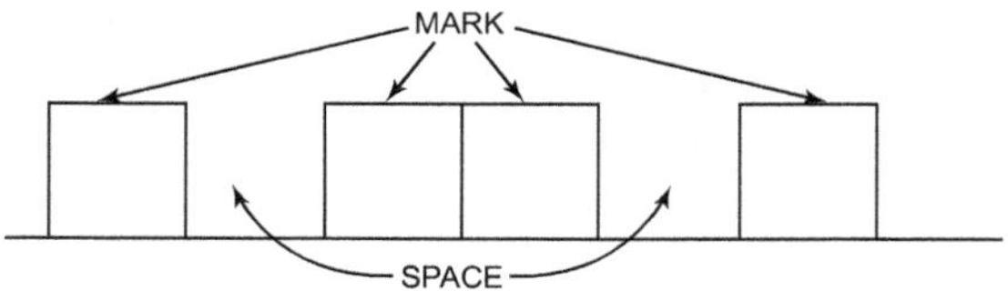

FIGURE 1.4 Telegraph signal (Digital signal).

1.4.4 Deterministic and Random Signals

Deterministic Signals. A deterministic signal is one that has no uncertainty with respect to its value at any value of an independent variable, namely, time. For example, the rectangular pulse given by Eq. (1.1) is a deterministic signal. Figures 1.5 and. 1.6 illustrate rectangular pulse and cosine signal, respectively; both are an example of the deterministic signal.

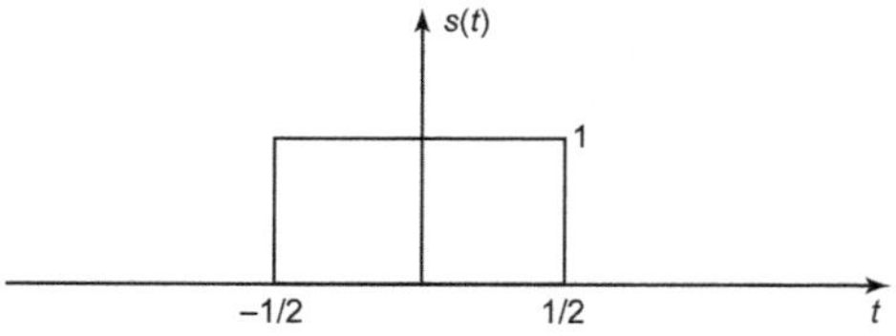

FIGURE 1.5 Rectangular pulse.

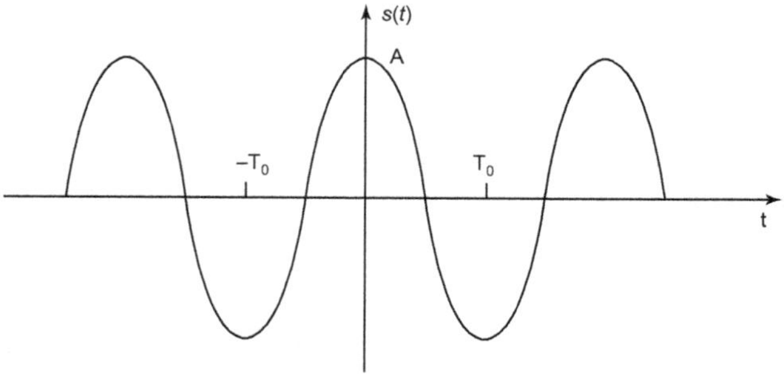

FIGURE 1.6 Cosine signal.

$$s(t) = \begin{cases} 1, |t| < \dfrac{1}{2} \\ 0, \text{otherwise} \end{cases}. \tag{1.1}$$

Another example of the deterministic signal is sinusoidal signals such as sine waves and cosine waves as given in Eq. (1.2):

$$s(t) = A \cos wt, -\infty < t < \infty. \tag{1.2}$$

Random signal: A random signal is a signal which has some degree of uncertainty with respect to its value at any value of independent variable namely, time. For example, thermal agitation noise in conductors is a random signal.

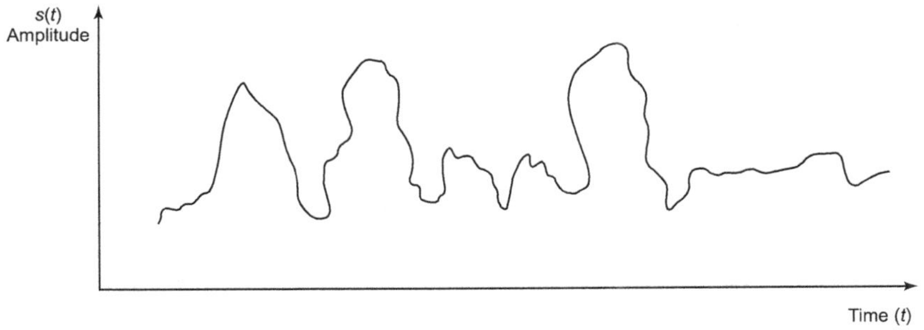

FIGURE 1.7 Random signal.

1.4.5 Energy and Power Signals

Energy signal: A signal is called an energy signal if and only if its total energy is finite. For example, the rectangular pulse is an energy signal.

Power signal: A signal is called a power signal if and only if its average power is finite. For example, sinusoidal waves are power signals.

The energy signals have zero average power and power signals have infinite energy. It means that both signals are mutually exclusive.

1.4.6 Periodic and Non-periodic Signals

Periodic Signal: A signal which repeats its waveform after a fixed period of time is called as a *periodic signal.* This fixed time is called *Time period* (T_0).

In other words, a signal which satisfies the condition $s(t) = s(t + T_0)$ for all t is called a periodic signal. For example, sinusoidal signals are the example of a periodic signal.

Non-periodic Signal: A signal which does not satisfy the above condition is called *non-periodic signal.*

Unit rectangular pulse is an example of a non-periodic signal.

Usually, periodic signals and random signals are power signals and deterministic signals, and non-periodic signals are energy signals.

1.5 SIGNAL PROCESSING SYSTEMS

A system responds to particular signals by producing other signals having some desired behavior.

Signal processing systems are of two types depending on the type of signal to be processed.

1. Continuous-time systems.
2. Discrete-time systems.

1.5.1 Continuous-time Systems

Continuous-time systems are the systems for which both input and output are continuous-time signals. *H(s)* is the transfer function of a continuous-time system. Figure 1.8 illustrates the block diagram of a continuous-time system.

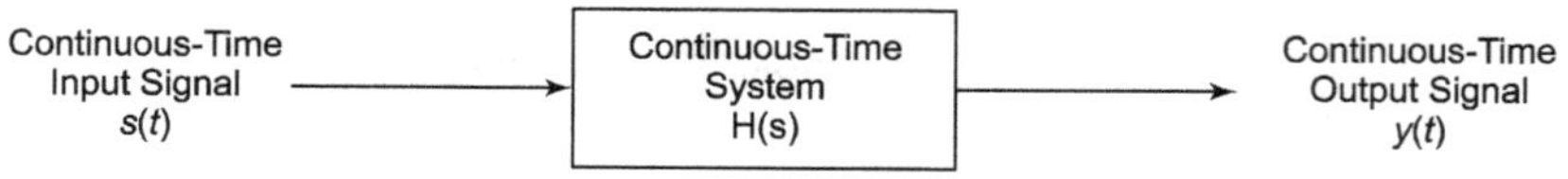

FIGURE 1.8 Block diagram of continuous-time system.

An example of continuous-time system is an analog filter which is used to reduce the noise corrupting a message signal.

1.5.2 Discrete-time Systems

Discrete-time systems are systems for which both the input and output are discrete-time signals. *H(z)* is the transfer function of a discrete-time system. Figure 1.9 illustrates the block diagram of a discrete-time system.

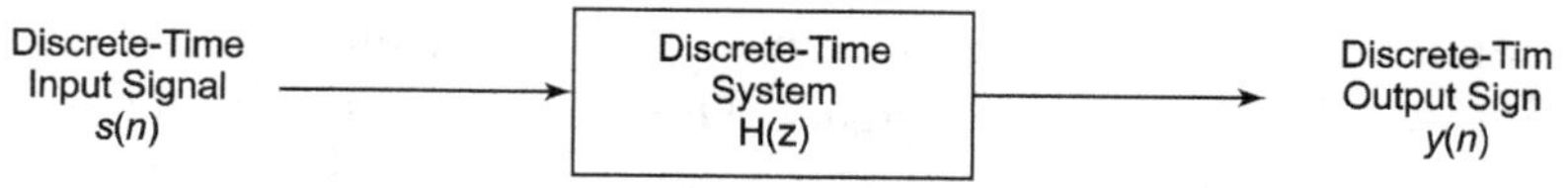

FIGURE 1.9 Block diagram of discrete-time system.

An example of a discrete-time system is a digital computer.

1.6 SIGNAL PROCESSING

Changing the basic nature of signal to obtain the desired shaping of the input signal is called *signal processing*. Signal processing is concerned with the representation, transformation, and manipulation of signals and the information they contain.

Signal processing is of two types depending upon the type of signal to be processed.

1. Analog signal processing (ASP), and
2. Digital signal processing (DSP).

1.6.1 Analog Signal Processing

In ASP, continuous-amplitude continuous-time signals are processed. Various types of analog signals are processed through low-pass filters, high-pass filters, band-pass filters, and band-reject filters to obtain the desired shaping of the input signal. Another example of ASP is the production of the modulated carrier using a high-frequency oscillator, and the modulating audio signal and a modulator. Figure 1.10 illustrates the block diagram of an ASP system.

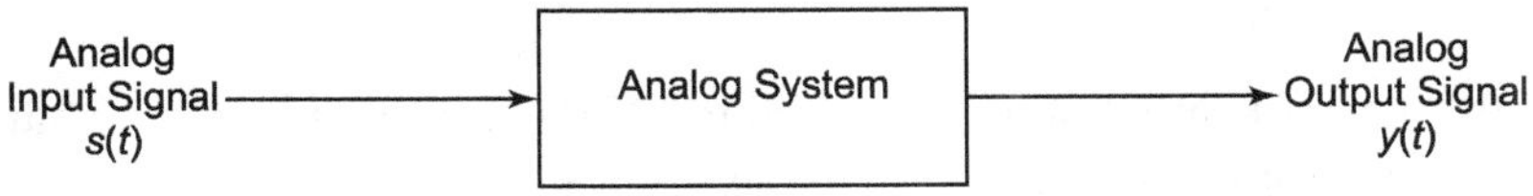

FIGURE 1.10 Block diagram of ASP system.

1.6.2 Digital Signal Processing

Digital signal processing (DSP) is a numerical processing of signals on a digital computer or some other data processing machine. Figure 1.11 illustrates the block diagram of DSP system.

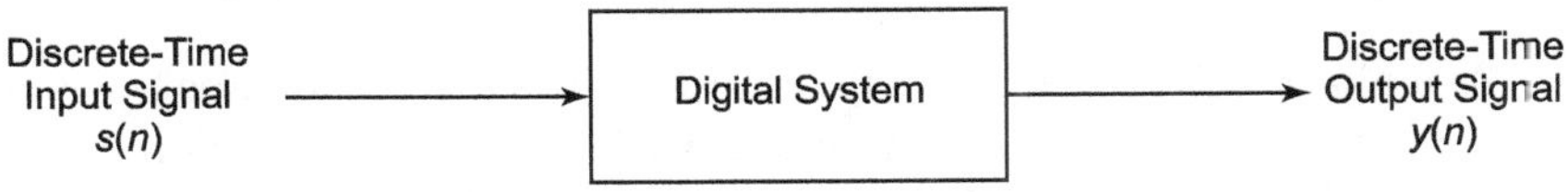

FIGURE 1.11 Block diagram of DSP system.

A digital system such as digital computer takes input signal in discrete-time sequence form and converts it in discrete-time output sequence.

1.7 ADVANTAGES OF DIGITAL SIGNAL PROCESSING OVER ANALOG SIGNAL PROCESSING

Digital signal processing has the following advantages:

1. Digital signal processing operations can be changed by changing the program in a digital programmable system, that is, these are flexible systems.
2. Better control of accuracy in digital systems is compared to analog systems.
3. Digital signals are easily stored on magnetic media such as magnetic tape without loss of quality of reproduction of the signal.
4. Digital signals can be processed offline, that is, these are easily transported.
5. Sophisticated signal processing algorithms can be implemented by DSP method.
6. Digital circuits are less sensitive to tolerances of component values.
7. Digital systems are independent of temperature, aging, and other external parameters.
8. Digital circuits can be reproduced easily in large quantities at a comparatively lower cost.
9. Cost of processing per signal in DSP is reduced by time-sharing of given processor among a number of signals.
10. Processor characteristics during processing, as in adaptive filters can be easily adjusted in digital implementation.
11. Digital system can be cascaded without any loading problems.

1.8 ELEMENTS OF DIGITAL SIGNAL PROCESSING SYSTEM

A majority of the signals encountered in science are analog in nature. In analog signals, both the dependent variable and independent variable(s) are continuous. Such signals may be processed directly by analog systems (i.e., analog filters) for the purpose of changing their characteristics or extracting some desired information.

Analog signals can also be processed digitally using DSP techniques. To process analog signals digitally, an interface between the analog signal and digital processor is needed. This interface is termed an analog-to-digital converter. The output of the analog-to-digital converter is a digital signal. This digital signal is appropriate for the digital processor.

The digital signal processor may be a large programmable digital computer or a small microprocessor.

In some applications such as in speech communication, we require digital signal in analog form at the receiver end. Here, we need another interface, called digital-to-analog converter. Figure 1.12 illustrates the block diagram of a DSP system.

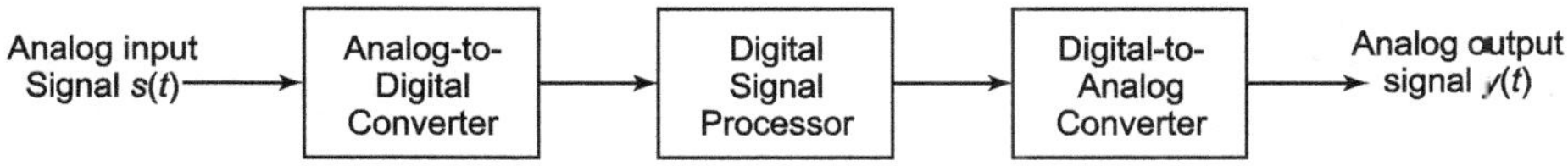

FIGURE 1.12 Block diagram of a digital signal processing system.

EXERCISES

1. Define a signal. Give some examples of signals.
2. Give the classification of signals.
3. What is signal processing? Differentiate between ASP and DSP.
4. What are the basic elements of the DSP system?
5. What are the advantages of DSP over ASP?
6. Differentiate multichannel and multidimensional signals. Give some examples of these signals.
7. What is the importance of DSP in various fields of engineering and technology? Give a brief account of its applications.

CHAPTER 2

Review of Discrete—Time Signals and Systems

2.1 INTRODUCTION

In Chapter 1, we have introduced the concept of digital signal processing. In this chapter, we will study discrete-time signals and systems. Discrete-time signals are obtained either by periodical sampling of continuous-time signals or by a recursion formula. Discrete-time signals are represented by discrete-time sequences.

If both input and output for a system are discrete, then this system is termed a discrete-time system. An example of a discrete-time system is a digital computer.

In this chapter, we first study discrete-time signals: various ways of representing discrete-time signals, different methods of obtaining discrete-time signals, elementary discrete-time signals, and manipulation of discrete-time signals.

After studying discrete-time signals, we will study discrete-time systems and their classification. In this chapter, we will also study LTI discrete-time systems, convolution and correlation operations for LTI discrete-time systems, inverse systems, and deconvolution operations.

Finally, we will study sampling of continuous-time signals, Nyquist rate, sampling theorem, aliasing, and reconstruction of the sampled version of continuous-time signals.

2.2 DISCRETE-TIME SIGNALS

Discrete-time signals are defined for discrete values of an independent variable (time). Discrete-time signal is not defined at instants between two successive samples.

Discrete-time signals are represented in two ways:

$$s(n),\ N_1 \le n \le N_2 \tag{2.1}$$

where N_1 and N_2 are the first and the last sample points, respectively, in a given discrete-time signal.

It represents non-uniformly spaced samples, and these are shown in Figure 2.1(*a*):

$$s(nT_s),\ N_1 \le n \le N_2 \tag{2.2}$$

It represents uniformly spaced samples, and these are shown in Figure 2.1(*b*).

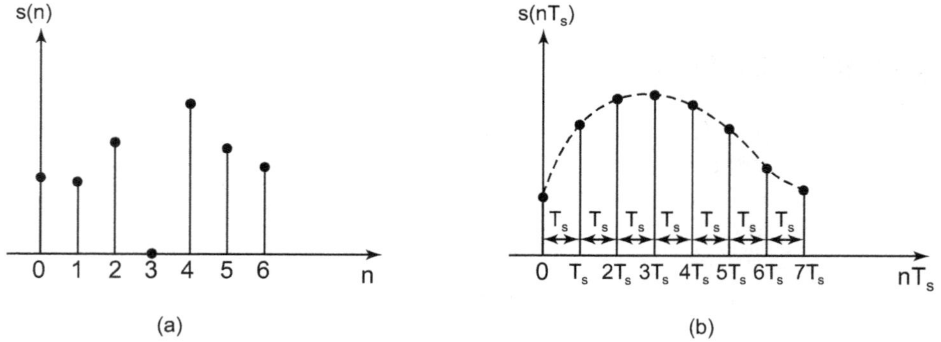

FIGURE 2.1 **(a)** Discrete-time signal showing non-uniformly spaced samples (there is no sampling period T_s) and **(b)** Discrete-time signal showing uniformly spaced samples.

2.2.1 Representation of Discrete-Time Signals

Discrete-time signal sequences can be represented in the following four ways:

1. Graphical Representation
2. Functional Representation
3. Tabular Representation
4. Sequence Representation.

Graphical Representation: Discrete-time signals can be represented by a graph when the signal is defined for every integer value of n for $-\infty < n < \infty$. This is illustrated in Figure 2.2.

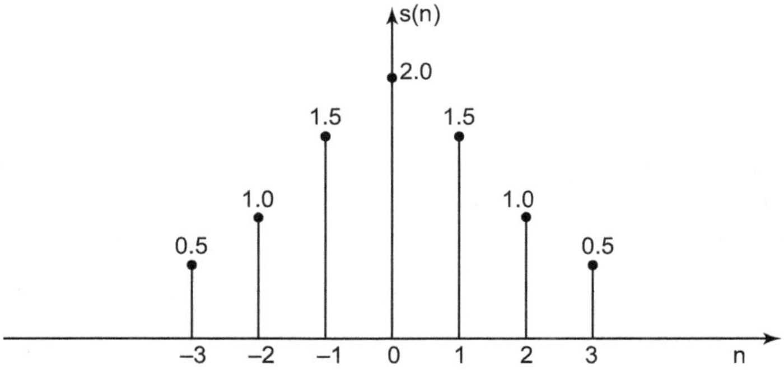

FIGURE 2.2 Graphical representation of a discrete-time signal.

Functional Representation: Discrete-time signals can be represented functionally as given below:

$$s(n)=\begin{cases}2, \text{ for } n=1,3\\ 4, \text{ for } n=2\\ 0, \text{ elsewhere}\end{cases} \tag{2.3}$$

Tabular Representation: Discrete-time signals can also be represented by a table as follows:

n	.....	–3	–2	–1	0	1	2	3	4	5	.....
s(n)		0	0	0	1	2	1	0	0	0	

Sequence Representation: An infinite-duration ($-\infty \le n \le \infty$) signal with the time as origin ($n = 0$) and indicated by the symbol ↑.

$$s(n) = \ldots 0,0,0,\underset{\uparrow}{1},3,1,0,0\ldots \tag{2.4}$$

2.2.2 Methods of Obtaining a Signal Sequence

There are three methods of obtaining a sequence:

1. To generate a set of numbers and order them into sequence form,

   Example: $s(n) = n,\ 0 \le n \le N - 1$ (2.5)

2. A sequence is generated by some recursion relation:

   Example: $s(n) = \dfrac{1}{s}s(n-1)$ (2.6)

   with initial condition $s(0) = 1$

generates a sequence

$$s(n) = \left(\frac{1}{2}\right)^n, \ 0 \le n \le \infty \tag{2.7}$$

3. A sequence is also obtained by periodic sampling of continuous-time signals. Periodic measurement of continuous-time signals is called periodic sampling.

   Discrete-time sequence, $s(nT_s) = s(t)\big|_{t=nT_s} \quad -\infty < n < \infty$ (2.8)

   where T_s is the sampling interval and $s(t)$ is a continuous-time signal.

2.2.3 Some Elementary Discrete-Time Signals

There are some basic signals which play an important role in the study of discrete-time signals and systems.

These signals are given as follows:

1. unit-sample (impulse) Sequence, $\delta(n)$,
2. unit-step sequence, $u(n)$,
3. unit-ramp sequence, $r(n)$,
4. exponential sequence, and
5. sinusoidal sequence.

Unit-Sample Sequence: Figure 2.3 shows a unit-sample sequence, it is denoted by $\delta(n)$ and is defined as follows:

$$\delta(n) = \begin{cases} 1, & n = 0 \\ 0, & n \neq 0 \end{cases} \tag{2.9}$$

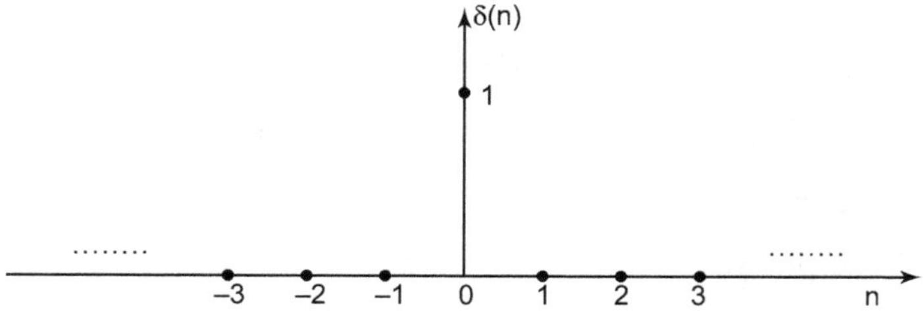

FIGURE 2.3 Graphical representation $\delta(n)$.

Unit-Step Sequence: It is denoted by $u(n)$ and is defined as follows:

$$u(n) = \begin{cases} 1, & n \ge 0 \\ 0, & n < 0 \end{cases} \tag{2.10}$$

Figure 2.4 illustrates the graphical representation of the unit-step sequence.

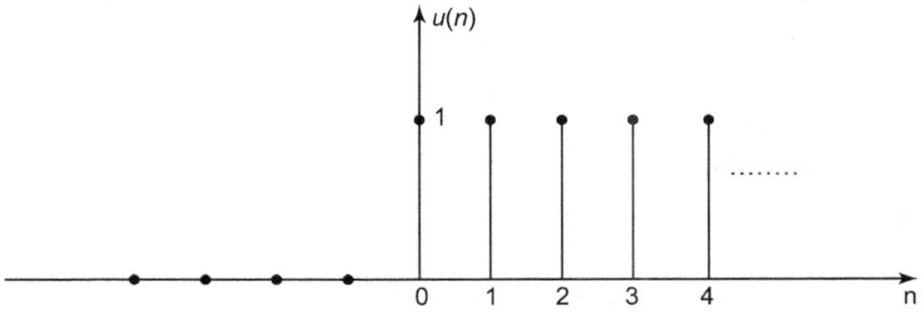

FIGURE 2.4 Graphical representation of $u(n)$.

Unit-Ramp Sequence: It is denoted by $r(n)$ and is defined as follows:

$$r(n)=\begin{cases}1, & \text{for } n\geq 0\\ 0, & \text{for } n<0\end{cases} \qquad (2.11)$$

Figure 2.5 shows the graphical representation of the unit-ramp sequence.

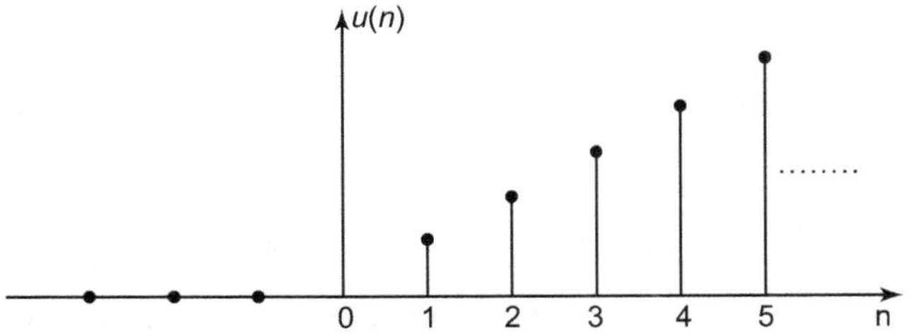

FIGURE 2.5 Graphical representation of $r(n)$.

Exponential Sequence: It is defined as

$$s(n) = (A)^n \text{ for all values of } n \qquad (2.12)$$

If parameter A is real, then $s(n)$ is a real sequence. Figure 2.6 illustrates a graphical representation of the exponential sequence.

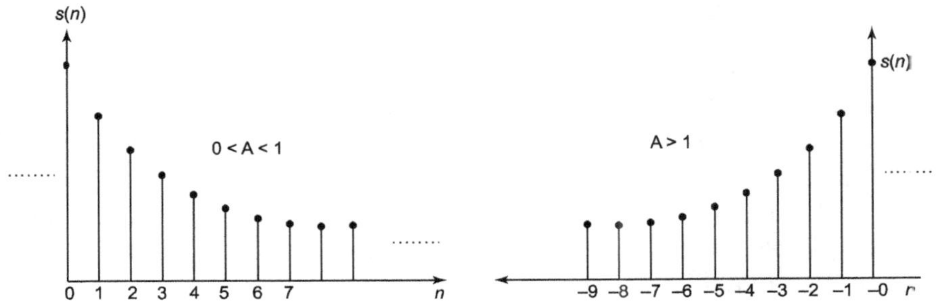

FIGURE 2.6 Graphical representation of exponential sequences.

Sinusoidal Sequences: There are two types of sinusoidal sequences, one is called the sine sequence and the other is called the cosine sequence.

Sine sequence is defined as follows:

$$s(n) = \sin \omega_0 n, \text{ for all } n$$

and cosine sequence is defined as follows:

$$s(n) = \cos \omega_0 n, \text{ for all } n$$

Figure 2.7 illustrates the graphical representation of cosine type sinusoidal sequence.

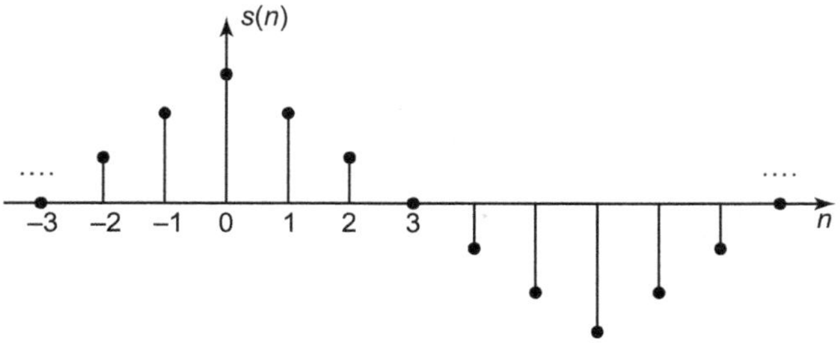

FIGURE 2.7 Graphical representation of cosine type sinusoidal sequence.

2.2.4 Manipulation of Discrete-Time Signals

Here, we will study some simple modifications in independent variable (time) and dependent variable (amplitude of signal). Such modification is required in DSP techniques.

Transformation of the Independent Variable (time): Modification of time can be done in three ways:

1. time shifting,
2. folding, and
3. time scaling.

Time Shifting: A signal can be shifted in time by replacing n by $n - k$, where k is integer and n is a discrete-time index.

If k is a positive integer, the result of time shifting is a delay of signal by k units of time.

If k is a negative integer, the result of time shifting is the advance of signal by $|k|$ units of time. Figure 2.8 illustrates a graphical representation of time shifting of a discrete-time sequence $s(n)$.

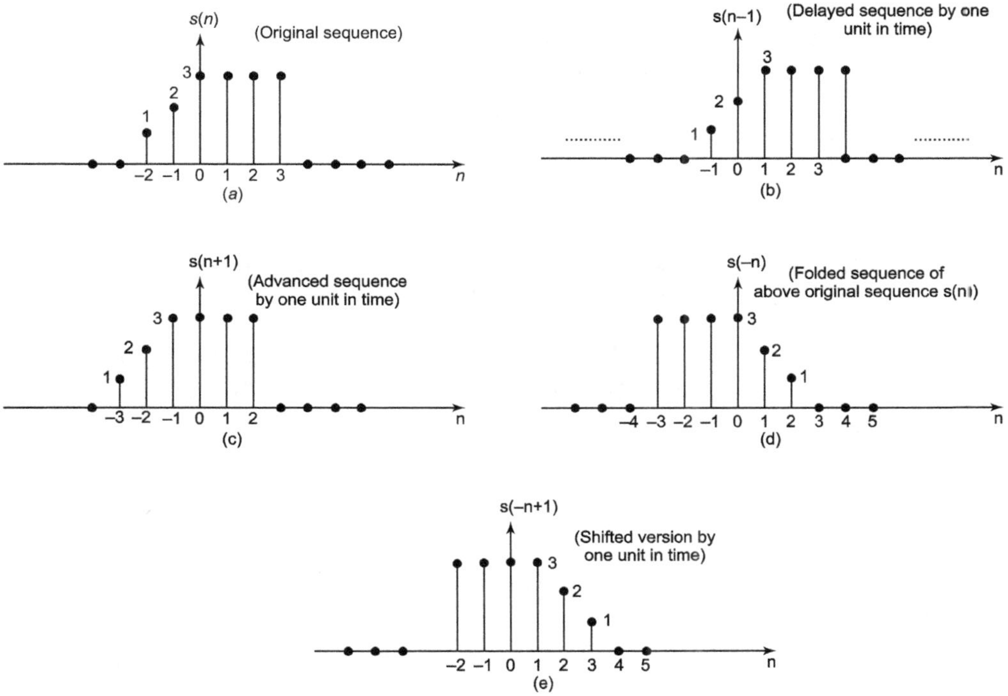

FIGURE 2.8 Graphical representation of (*a*) original sequence, $s(n)$ (*b*) delayed sequence by one unit in time, $s(n-1)$ (c) advanced sequence by one unit in time, $s(n+1)$ (d) folded sequence of above orig nal signal, $s(n)$, $s(-n)$ (*e*) shifted version of folded sequence $s(-n)$, $s(-n+1)$.

Folding: If independent variable (time) n is replaced by $-n$, then signal folding (mirror image) about the time origin ($n = 0$) will take place.

Operations of folding and time delaying (or advancing) a signal are not commutative. Figure 2.8(*d*) illustrates the graphical representation of folding operation of original sequence $s(n)$.

Time scaling: Time scaling is performed by replacing independent variable n by mn, where m is an integer. Time scaling is also called *down sampling*.

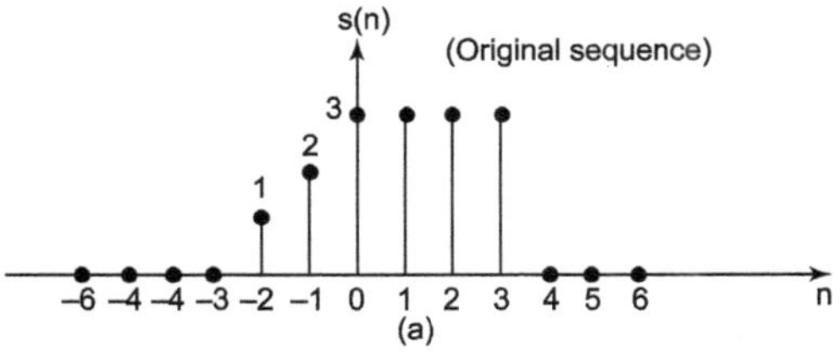

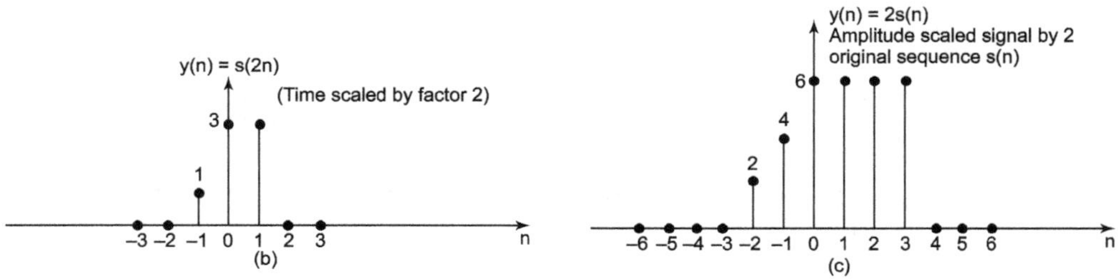

FIGURE 2.9 Graphical representation of (*a*) original sequence, *s*(*n*) (*b*) time-scaled version of *s*(*n*) by factor 2, *y*(*n*) = *s*(2*n*) (*c*) Amplitude-scaled version *s*(*n*) by factor 2, *y*(*n*) = 2*s*(*n*).

Transformation of the Dependent Variable (Signal Amplitude): Modification of signal amplitude can be done in three ways:

1. Addition of Sequences.
2. Multiplication of Sequences.
3. Amplitude Scaling of Sequence.

Addition of Sequences: The sum of two discrete-time sequences is given by

$$y(n) = s_1(n) + s_2(n), -\infty < n < \infty \tag{2.13}$$

Addition of two sequences is shown in Figure 2.10(*a*).

Multiplication of Sequences: The product of two discrete-time sequences is given by

$$y(n) = s_1\ (n), s_2(n), -\infty < n < \infty \tag{2.14}$$

Multiplication of two sequences is shown in Figure 2.10(*b*).

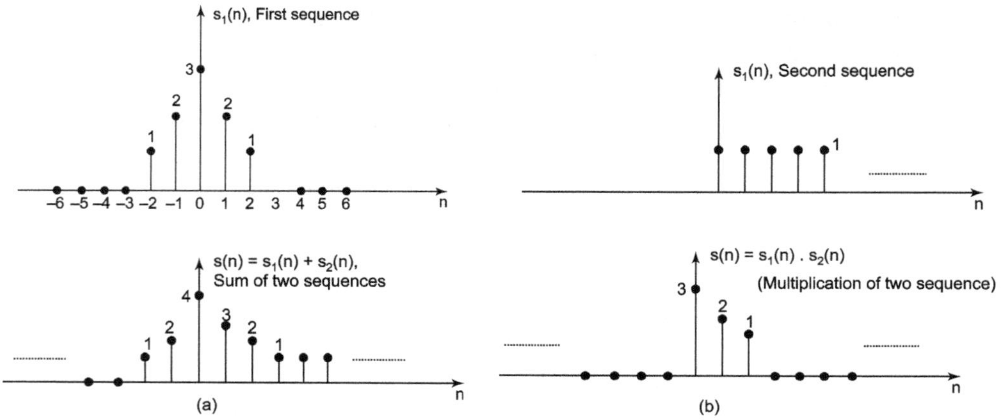

FIGURE 2.10 Graphical representation of (*a*) sum of two sequences, $s(n) = s_1(n) + s_2(n)$, (*b*) multiplication of two sequences, $s(n) \cdot s_1(n) \cdot s_2(n)$.

Amplitude Scaling of Sequence: Amplitude scaling of a signal by a constant B is accomplished by multiplying the value of every signal sample by B.

$$y(n) = Bs(n),\ -\infty < n < \infty \tag{2.15}$$

where B is real constant quantity.

2.3 DISCRETE-TIME SYSTEMS

A discrete-time system is a device or an algorithm in which both the input and the output are discrete-time signals. A block diagram representation of a discrete-time system is shown in Figure 2.11.

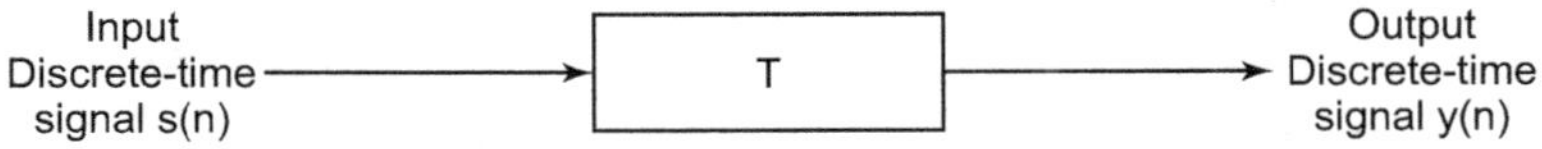

FIGURE 2.11 Block diagram representation of discrete-time system.

Output of a discrete-time system is given by

$$y(n) = T_s(n), \text{ where } T \text{ is an operator}$$

Examples of discrete-time system are

$$y(n) = s(n). \text{ This is an identity system.}$$

$$y(n) = \frac{1}{3}[s(n-1) + s(n) + s(n+1)]$$

This is a three sample averager.

Basic Building Blocks of a Discrete-time System: A discrete-time system (digital filter) consists of an interconnection of three simple building blocks or elements: *Adders, multipliers, and delay elements.* Adder is also called as a summing element. It performs addition of two or more discrete-time signals. The multiplier performs multiplication of a discrete-time signal with a scalar quantity. The adder and multiplier are conceptually simple components which are readily implemented in the arithmetic logic unit of a computer. Delay elements allow access to future and past values in the discrete-time signal. Delays are two types: positive and negative. A positive delay is simply called delay and it is implemented by a memory register which stores the current values of a discrete-time signal for one sample interval. These stored samples are available for future calculations. Positive delay is indicated by $\boxed{z^{-1}}$.

A negative delay is also called as *advance.* It is used to look ahead to the next value in the discrete-time signal and is indicated by $\boxed{z^{+1}}$.

Typical advances are used for non-real-time applications such as image processing. Advances in discrete-time signals simplify the analysis of discrete-time systems (digital filters). In real-time applications, advances are not permitted.

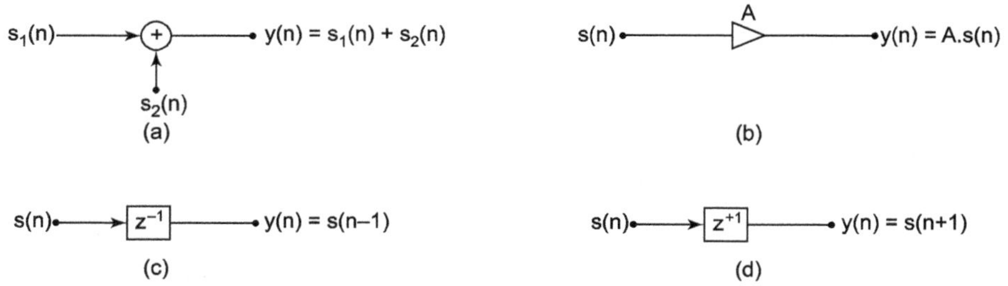

FIGURE 2.12 Illustration of basic building blocks or elements of a discrete-time system (digital filter) **(a)** adder or summing element **(b)** multiplier or multiplication element **(c)** positive delay element or simply "delay" **(d)** negative delay element or "advance."

A discrete-time system involves selecting and interconnecting a finite number of building block and determination of mutiplier coefficients. These three building blocks are shown in Figure 2.12.

We can easily understand the meaning of these building blocks (or elements) of a discrete-time systems or digital filters by an example.

Three sample averager is an example of a digital filter and its input–output relationship is given by the following difference equation:

$$y(n) = \frac{1}{3}\,[s(n-1) + s(n) + s(n+1)]$$

It means output of a three-sample averager is equal to the average the previous, present, and future input values. It is a non-recursive digital filter. The advance serves to access the next value of discrete-time sequence while the delay stores the previous value.

Its network structure is shown in Figure 2.13.

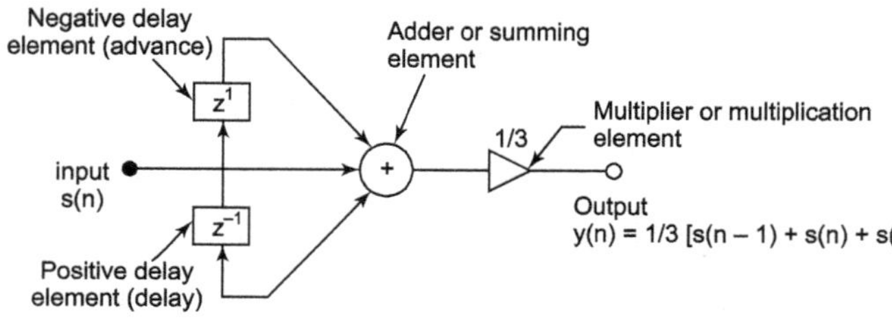

FIGURE 2.13 Three-sample averager. (It is an example of non-recursive digital filter).

2.3.1 Classification of Discrete-time Systems

Discrete-time systems can be classified in five groups:

1. Memoryless systems and systems with memory.
2. Time-invariant and time-varying systems.
3. Linear systems and non-linear systems.
4. Casual systems and non-casual systems.
5. Stable system and unstable systems.

We now discuss these groups one by one.

Memoryless Systems and Systems with Memory: Memoryless systems are also called *static systems.* A discrete-time system is called memoryless system if its output at any instant n depends at most on the input at the same instant, but not on past or future values of input samples.

$$\left.\begin{aligned} y(n) &= A.s(n) \\ y((n) &= ns(n) + Bs^2(n) \end{aligned}\right\} \text{ Both systems are static systems}$$

On the other hand, output of a system which depends on past or future samples of the input signal is called system with memory. It is also called dynamic system. These systems require memory for storage for future and past samples of input signal. For example, three-sample averager system,

$$y(n) = \frac{1}{3}\,[s(n-1) + s(n) + s(n+1)]$$

is a dynamic system.

Time-Invariant and Time-varying Systems: A system is called time invariant if its input–output characteristics do not change with time.

If the response to a delayed input and the delayed response are equal, then the system is called *time-invariant system.*

The response to a delayed input is denoted by $y(n, k)$ and the delayed response is denoted as $y(n - k)$. If both responses, $y(n, k)$ and $y(n - k)$ are equal, then the system is called *time-invariant system.* If both responses are not equal, then the system is called *time-varying system.* For example,

Differentiator

$y(n) = s(n) - s(n-1)$ is a time-invariant system.

Time multiplier

$y(n) = ns(n)$ is a time-varying system.

EXAMPLE 2.1

Check the system for time invariance which is characterized by the difference equation:

$$y(n) = ns(n).$$

Solution: The response to a delayed input is

$$y(n, k) = ns(n - k)$$

The delayed response is

$$y(n - k) = (n - k)\, s(n - k)$$

Both responses are not equal:

$$y(n, k) \neq y(n - k)$$

Therefore, the given discrete-time system $y(n) = ns(n)$ is not time invariant. It is a time-varying system.

EXAMPLE 2.2

Check the system for time invariance which is characterized by difference equation:

$$y(n) = s(n) - s(n - 1).$$

SOLUTION:

The response to a delayed input is

$$y(n, k) = s(n - k) - s(n - k - 1)$$

The delayed response is

$$y(n - k) = s(n - k) - s(n - k - 1)$$

Both responses are equal:

$$y(n, k) = y(n - k)$$

Above system is time invariant.

Linear Systems and Non-Linear Systems: A system which satisfies superposition principle is called a *linear system.* A system which does not satisfy superposition principle is termed as a *non-linear system.*

Superposition principle is stated as follows:

> *Response of the system to a weighted sum of input signals is equal to the corresponding weighted sum of responses of the system to each of the individual input signals.*

A system is linear if and only if

$$T\,[As_1(n) + Bs_2(n)] = AT\,[s_1(n)] + BT\,[s_2(n)]$$

where $s_1(n)$ and $s_2(n)$ are arbitrary input sequences and A and B are arbitrary constants.

$$y(n) = ns(n) \text{ is a linear system.}$$

$$y(n) = \alpha s(n) + \beta, \text{ where } \alpha \text{ and } \beta \text{ are constants.}$$

This is a non-linear system. This will be further clarified in Ex. 2.3.

EXAMPLE 2.3

Check the systems for linearity

i. $y(n) = ns(n)$
ii. $y(n) = \alpha s(n) + \beta$, where α and β are constants.

Solution:

i. The corresponding outputs for two discrete-time sequences $s_1(n)$ and $s_2(n)$ are

$$y_1(n) = ns_1(n)$$
$$y_2(n) = ns_2(n)$$

A linear combination of two input sequences results in the output

$$y_3(n) = T\,[s_3(n)] = T\,[As_1(n) + Bs_2(n)]$$
$$= n[As_1(n) + Bs_2(n)] = Ans_1(n) + Bns_2(n) \quad (1)$$

A linear combination of the two outputs results in the output

$$Ay_1(n) + By_2(n) = Ans_1(n) + Bns_2(n) \quad (2)$$

Since both outputs are equal, the system is linear.

ii. The corresponding outputs for two discrete-time sequences $s_1(n)$ and $s_2(n)$ are

$$y_1(n) = \alpha s_1(n) + \beta$$
$$y_2(n) = \alpha s_2(n) + \beta$$

A linear combination of $s_1(n)$ and $s_2(n)$ results in the output

$$y_3(n) = T\,[As_1(n) + Bs_2(n)] = \alpha[As_1(n) + Bs_2(n)] + \beta$$
$$= \alpha As_1(n) + \alpha Bs_2(n) + \beta \quad (3)$$

Linear combination of the two outputs results in the output

$$Ay_1(n) + By_2(n) = A\,[\alpha s_1(n) + \beta] + B\,[\alpha s_2(n) + \beta]$$
$$= \alpha A s_1(n) + A\beta + \alpha B s_2(n) + B\beta \qquad (4)$$

Here, $y_3(n) \neq Ay_1(n) + By_2(n)$, both outputs are not equal.

The system is non-linear.

Causal Systems and Non-Causal Systems: A system in which present output depends only on present and past inputs is called *causal system.*

A system is called *non-causal system* if its present output depends on future values of the input. Most of the real-time physical systems are causal systems, and processing of images and geophysical signals are the examples of non-causal systems.

Differentiator, $y(n) = s(n) - s(n-1)$ is a causal system because its present output $y(n)$ depends only on present input $s(n)$ and past input $s(n-1)$, and system $y(n) = s(n) + 3s(n+1)$ is a non-causal system because its present output $y(n)$ also depends on future input $s(n+1)$.

Stable Systems and Unstable Systems: An initially relaxed system is said to be bounded input bounded output (BIBO) stable if and only if every bounded input produces a bounded output.

Bounded input

$$|s(n)| \geq M_s < \infty \Rightarrow \text{Bounded Input}$$
$$|s(n)| \geq M_y < \infty \Rightarrow \text{Bounded Output}$$

2.3.2 Linear-Time-Invariant (LTI) Systems

A system which satisfies the condition of linearity and time invariance is called a *linear time-invariant (LTI) system.*

Response of a discrete time LTI system is computed by convolution sum. Discrete time LTI systems are described by constant coefficient difference equations.

In discrete time LTI systems category, we will study the following topics:

1. FIR and IIR discrete-time LTI systems.
2. Recursive and Non-recursive discrete-time systems.
3. Causal LTI systems.
4. Impulse response of LTI systems.
5. Stability of LTI systems.
6. Discrete-time systems described by difference equations.

FIR and IIR LTI Discrete-time Systems: LTI systems are classified in two groups on the basis of number of samples taken in computing its unit-sample (impulse) response.

a. Finite-Duration Impulse Response (FIR) LTI systems.

b. Infinite-Duration Impulse Response (IIR) LTI systems.

If impulse response of a LTI system is computed for a finite number of sample points [Finite duration], then such systems are called FIR systems.

FIR LTI discrete-time systems can be realized either recursively or non-recursively.

On the other hand, if impulse response of a LTI system is computed for an infinite number of sample points [Infinite duration], then the system is called IIR system.

IIR systems can only be realized by recursive method.

Recursive and Non-Recursive Discrete Time Systems: A system of which present output $y(n)$ at time n depends on any number of past output values $y(n-1)$, $y(n-2)$ ... is called a *recursive discrete-time system.*

The output of a causal recursive system is given by

$$y(n) = f\,[y(n-1), y(n-2), \ldots y(n-N), s(n), s(n-1) \ldots] \qquad (2.16)$$

where $y(n)$, $y(n-1)$ ... are outputs and $s(n)$, $s(n-1)$ ... are inputs.

A first-order system:

$y(n) = \alpha y(n-1) + s(n)$ is an example of recursive system.

A system of which present output $y(n)$ at time n depends only on present and past values of input signal, $s(n)$, $s(n-1)$, $s(n-2)$, ... is called *non-recursive system.*

The output of a causal non-recursive system is given by

$$y(n) = f\,[s(n), s(n-1), s(s-2), \ldots] \qquad (2.17)$$

A differentiator, $y(n) - s(n) - s(n-1)$, is an example of non-recursive system.

Causal LTI System: Causal system is a system of which output depends only on present and past inputs but does not depend on future input sample values.

Causality of LTI systems can be translated into a condition on the impulse responses, $h(n)$, or in other words, causality can be determined in terms of impulse response $h(n)$.

If impulse response is zero for negative values of n, then the system is called a causal LTI system. The convolution sum formula for causal LTI system may be modified and given as follows:

$$y(n) = \sum_{k=0}^{\infty} s(k)h(n-k) \qquad (2.18)$$

Response of a causal system to a causal input sequence is also a causal sequence.

If a sequence is zero for a negative value of n, then the sequence is called as *causal sequence.*

EXAMPLE 2.4

Determine the unit-step response of the LTI system with unit-sample response, $h(n) = A^n u(n)$, $|A| < 1$.

Solution: For computing the unit-step response, we put input sequence, $s(n)$ equal to unit-step sequence, $u(n)$.

$u(n)$ is a causal sequence, and the system is also causal.

From convolution sum formula. We learn more about convolution given in Art. 2.3.

$$y(n) = \sum_{k=-\infty}^{\infty} h(k)s(n-k) = \sum_{k=-\infty}^{\infty} A^k u(k)u(n-k)$$

Since

$$u\,(k) = \begin{cases} 1, & k \geq 0 \\ 0, & k < 0 \end{cases} \text{ and } u(n-k) \begin{cases} 1, & n-k \geq 0 \\ 0, & n-k < 0 \end{cases} = \begin{cases} 1, & k \leq 0 \\ 0, & k > 0 \end{cases}$$

$$= \sum_{k=0}^{n} A^k = \frac{1-A^{n+1}}{1-A}, n \geq 0$$

$$y(n) = 0; n < 0$$

Impulse Response of LTI Systems: The value of the response or output of a LTI system when the input is equal to unit-sample (impulse) sequence, d(n) is called *impulse response, h(n).* It is also called *unit-sample response.*

LTI systems are completely characterized by their impulse response.

Stability of LTI Systems: Stability is an important property for the practical implementation of a system.

A LTI system is bounded input bounded output (BIBO) stable if its impulse response, $h(n)$, is absolutely summable. Absolutely summable means summation of sequence $h(n)$ is possible and sequence $h(n)$ is a converging sequence.

Stability of a LTI system in terms of impulse response, $h(n)$, *is* given by

$$S_h \equiv \sum_{h=-\infty}^{\infty} |h(k)| < \infty \tag{2.19}$$

EXAMPLE 2.5

Check the stability of a LTI system with unit-sample (impulse) response,

$$h(n) = A^n u(n), \text{ where } A \text{ is a constant.}$$

Solution: Condition for stability for LTI system is given by

$$Sh \equiv \sum_{k=-\infty}^{\infty} |h(k)| = \sum_{k=0}^{\infty} \left|A^k\right| \text{ since } \left(u(k) = \begin{cases} 1, & k \geq 0 \\ 0, & k < 0 \end{cases} \right)$$

$$= \sum_{k=0}^{\infty} |A|^k = 1 + |A| + |A|^2 + \ldots$$

$$= \frac{1}{1-|A|}, \text{ for } |A| < 1,$$

This series coverages otherwise it diverges.

Therefore, the system is stable for $|A| < 1$.

Discrete-Time Systems described by Difference Equations: We are already familiar that continuous-time systems are described by differential equations. But discrete-time systems are described by *difference equations.*

For example, a differentiator for discrete-time systems are described by its difference equation

$$y(n) = s(n) - s(n-1)$$

A three-sample averager is also described by its difference equation

$$y(n) = \frac{1}{3} [s(n) + s(n-1) + s(n+1)]$$

The input–output relationship of discrete-time system is also described by its difference equation.

There are two methods by which difference equations can be solved.

1. **Direct Method.** This method is directly applicable in the time domain. We are not discussing this method in this book.
2. **Indirect Method.** It is also called z-transform method. This method will be discussed in Chapter 3.

2.4 CONVOLUTION OF TWO DISCRETE-TIME SIGNALS

Convolution sum is used to compute the response of Linear-time-invariant (LTI) discrete-time systems. LTI systems are completely characterized by its unit-sample (impulse) response, $h(n)$. This system is shown in Figure 2.14.

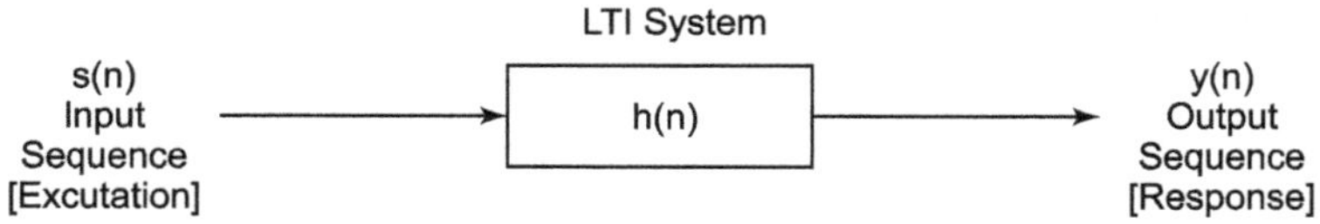

FIGURE 2.14 LTI system.

Convolution sum for a LTI discrete-time system is defined as follows:

$$y(n) = \sum_{h=-\infty}^{\infty} s(k)h(n-k) = \sum_{k=-\infty}^{\infty} h(k)s(n-k)$$

2.4.1 Procedure for Computing Convolution Sum

Convolution sum between $s(n)$ and $h(n)$ involves the following four steps:

1. **Folding:** Take the mirror image of $h(k)$ about $k = 0$ to obtain $h(-k)$.
2. **Shifting:** Shift $h(-k)$ by n_0 to the right (left) if n_0 is +ve ($-ve$) to obtain $h(n_0 - k)$.
3. **Multiplication:** Multiply $s(k)$ by $h(n_0 - k)$ to obtain the product sequence $Pn_0(k) = s(k)\, h(n_0 - k)$.
4. **Summation:** Sum all the values of the product sequences $Pn_0(k)$ to obtain the value of the output at the time $n = n_0$.

Above procedure results in the response of the system at a single time instant, $n = n_0$.

If we are interested in evaluating the response of the system over all time instants, $-\infty < n < \infty$, then repetition of steps 2 to 4 is necessary till the response at all time instants is obtained.

Note: *If one sequence has M points and second sequence has N points then convolution of these sequences will have M + N – 1 points.*

EXAMPLE 2.6

Determine the response of a discrete-time system to input signal $s(n) = \{2, 1, 3, 1\}$. Also given unit-sample (impulse) response

$$h(n) = \{1, 2, 2, -1\}.$$

Solution: Convolution sum is defined as follows:

$$y(n) = \sum_{K=-\infty}^{\infty} s(k)h(n-k)$$

$n = 0,$

$$y(0) = \sum_{K=-\infty}^{\infty} s(k)h(-k)$$

$$s(k) = \underset{\uparrow}{2}\ 1\ 3\ 1$$

$$h(k) = \underset{\uparrow}{1},\ 2,\ 2,\ -1$$

$$\boxed{\begin{array}{l} s(k) = 2,1,3,1 \\ \qquad\quad \uparrow \\ h(-k) = -1,2,2,1 \end{array}}$$

$$y(0) = \sum_{k=-\infty}^{\infty} s(k)h(-k) = 2\times 2 + 1\times 1 = 4 + 1 = 5$$

$n = 1,$ $\qquad y(1) = \sum_{k=-\infty}^{\infty} s(k)h(1-k)$

$$\boxed{\begin{array}{l} s(k) = 2,1,3,1 \\ \qquad\quad \uparrow \\ h(1-k) = -1,2,2,1 \end{array}}$$

$$y(1) = \sum_{k=-\infty}^{\infty} s(k)h(1-k) = 2\times 2 + 1\times 2 + 3\times 1$$
$$= 4 + 2 + 3 = 9$$

$n = 2,$ $\qquad y(2) = \sum_{h=-\infty}^{\infty} s(k)h(2-k)$

$$\boxed{\begin{array}{l} s(k) = 2,\ 1,\ 3,\ 1 \\ \qquad\quad \uparrow \\ h(z-k) = -1,\ 2,\ 2,\ 1 \end{array}}$$

$$y(2) = \sum_{k=-\infty}^{\infty} s(k)h(2-k)$$
$$= 2 \times (-1) + 1 \times 2 + 3 \times 2 + 1 \times 1 = -2 + 2 + 6 + 1 = 7$$

$n = 3,$ $\qquad y(3) = \sum_{k=-\infty}^{\infty} s(k)h(3-k)$

$$\boxed{\begin{array}{l} s(k) = 2,\ 1,\ 3,\ 1 \\ \qquad\quad \uparrow \\ h(3-k) = -1,\ 2,\ 2,\ 1 \end{array}}$$

$$y(3) = \sum_{k=-\infty}^{\infty} s(k)h(3-k) = 1\times(-1) + 3\times 2 + 1\times 2$$

$$= -1 + 6 + 2 = 7$$

$n = 4,$ $$y(4) = \sum_{k=-\infty}^{\infty} s(k)h(4-k)$$

$s(k) = 2,\ 1,\ 3,\ 1$	
$\uparrow$	
$h(4-k) = -1,\ 2,\ 2,\ 1$	

$$y(4) = \sum_{k=-\infty}^{\infty} s(k)h(4-k)$$

$$= 3 \times (-1) + 1 \times 2 = -3 + 2 = -1$$

$n = 5,$ $$y(5) = \sum_{k=-\infty}^{\infty} s(k)h(5-k)$$

$s(k) = 2,\ 1,\ 3,\ 1$	
$\uparrow$	
$h(5-k) = -1,\ 2,\ 2,\ 1$	

$$y(5) = \sum_{k=-\infty}^{\infty} s(k)h(5-k) = 1 \times (-1) = -1$$

$n = 6, y(6) = 0$
$n = 7, y(7) = 0$

. .

. .

If sequences $s(n)$ and $h(n)$ have M sample points and N sample points, respectively, then convolution of these sequences will have $M + N - 1$ sample points. In this example, sequence $s(n)$ has 4 points, and sequence $h(n)$ has 4 points.

Then convolution of these sequences will have $4 + 4 - 1 = 7$ points

$n = -1,$ $$y(-1) = \sum_{k=-\infty}^{\infty} s(k)(-1-k)$$

$s(k)$	$=$	$2,\ 1,\ 3,\ 1$
		$\uparrow$
$h(-1-k)$	$=$	$-1,\ 2,\ 2,\ 1$

$$y\ (-1) = \sum_{k=-\infty}^{\infty} s(k)h(-1-k) = 2 \times 1 = 2$$

Resultant of convolution sum of $s(n)$ and $h(n)$ is $y(n)$ and is given as follows:

$$\begin{aligned} y(n) &= \{y(-1), y(0), y(1), y(2), y(3), y(4), y(5)\} \\ &= \{2, \underset{\uparrow}{5}, 9, 7, 7, -1, -1\} \end{aligned}$$

EXAMPLE 2.7

Compute the convolution sum $y(n) = s(n) * h(n)$ of the pair of signals given by

$$\begin{aligned} s(n) &= A^n\, u(n) \\ h(n) &= B^n\, u(n) \text{ for both } A = B \text{ and } A \neq B. \end{aligned}$$

Solution: Since both $s(n)$ and $h(n)$ are causal sequences. These sequences are causal because both sequences are multiplier of $u(n)$, and $u(n)$ is defined as follows:

$$u(n) = \begin{cases} 1, & n \geq 0 \\ 0, & n < 0 \end{cases}$$

$u(n)$ is a causal sequence and its value is 1 for positive time instants.

Convolution sum is defined as follows:

$$y(n) = \sum_{k=-\infty}^{n} s(k)h(n-k)$$

$$= \sum_{k=-\infty}^{n} A^k u(k).B^{n-k} u(n-k), n = 0,1,2,...$$

Now, $u(k) = 1$ for $k = 0$ and $u(n-k) = 1$, for $k = n$

$$y(n) = \sum_{k=0}^{n} A^k B^{n-k} = B^n \sum_{k=0}^{n} \left(\frac{A}{B}\right)^k$$

for all integer values of k ranging from 0 to n.

Case I: If $A = B$

then,
$$y(n) = B^n \sum_{k=0}^{n} \left(\frac{A}{A}\right)^k = B^n \sum_{k=0}^{n} (1)^k$$

$$= B^n\,(n+1), n = 0, 1, ...$$

Case II: If $A \neq B$

then,
$$y(n) = B^n \sum_{k=0}^{n} \left(\frac{A}{B}\right)^k = B^n \frac{\left\{1 - \left(\frac{A}{B}\right)^{n+1}\right\}}{1 - \frac{A}{B}}$$

$$= \frac{B^{n+1} - A^{n+1}}{B - A}, n = 0,1,2,....$$

2.4.2 Linear Convolution

Linear convolution of two discrete-time sequences can be performed by graphical method. In this method, both discrete-time sequences are represented on graphs individually. We can understand linear convolution by graphical method with the help of following examples.

EXAMPLE 2.8

Determine the linear convolution $y(n) = s(n) * h(n)$ of the following two signals:

$$s(n) = A^n .u(n), 0 < A < 1$$
$$h(n) = u(n).$$

Solution: Linear convolution is defined as follows:

$$y(n) = \sum_{k=-\infty}^{\infty} h(n-k)s(k)$$

Figure 2.15 illustrates the computation of linear convolution of two discrete-time sequences using graphs.

For $n \le 0$, $h(n - k)\ s(k)$ is given by

$$s(k)\ h(n-k) = A^k, 0 \le 1 \le n$$
$$= 0, \text{ otherwise}$$

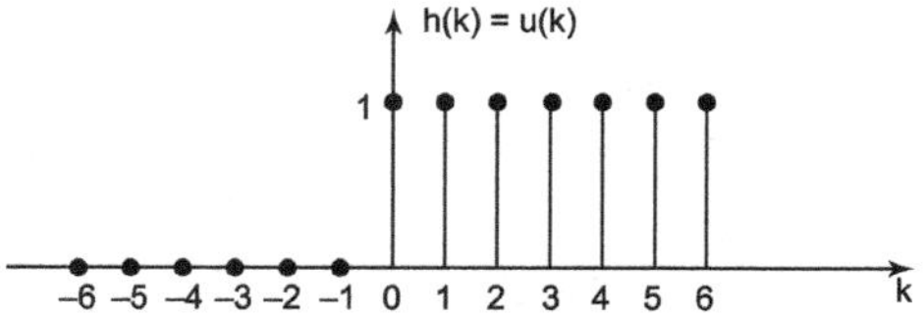

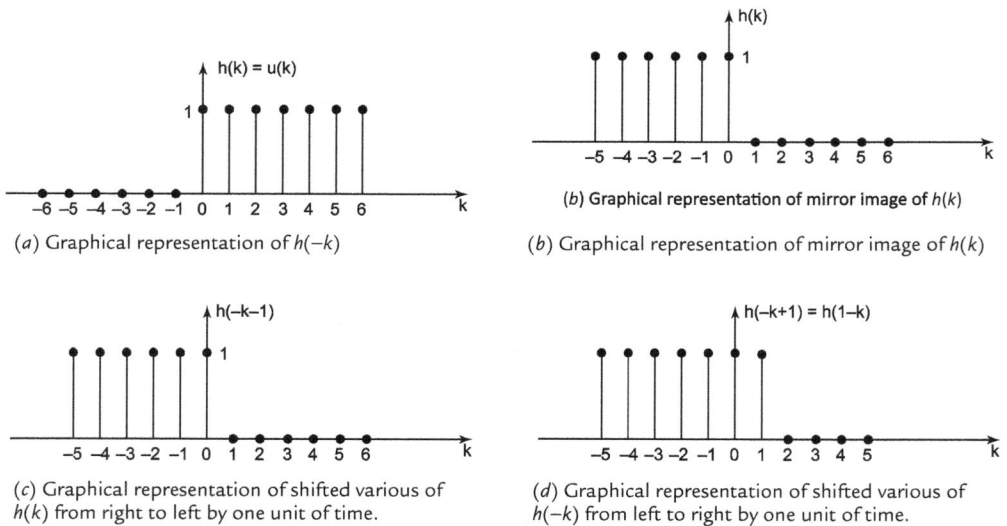

(a) Graphical representation of $h(-k)$

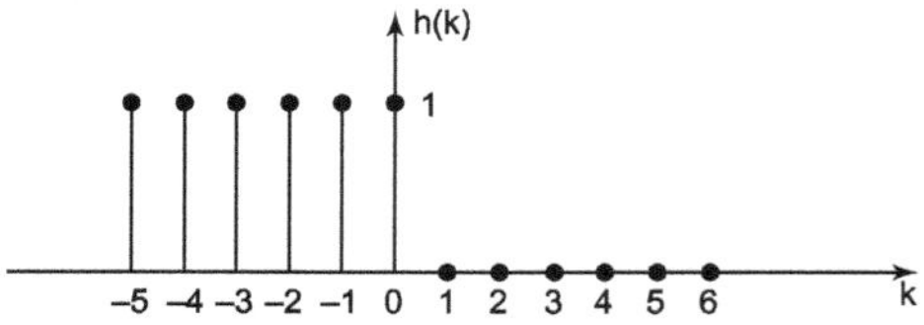

(b) Graphical representation of mirror image of $h(k)$

(b) Graphical representation of mirror image of $h(k)$

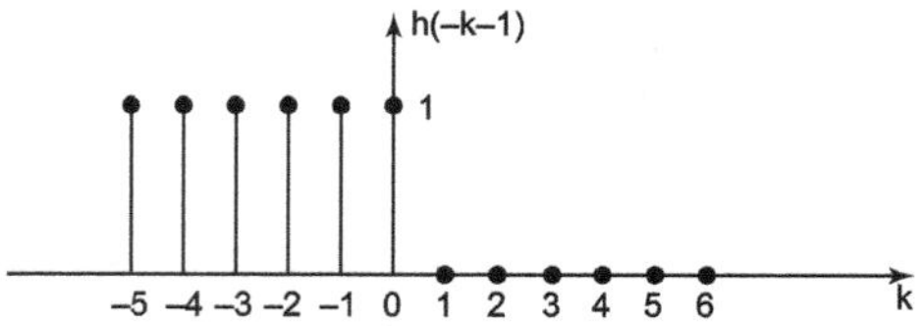

(c) Graphical representation of shifted various of $h(k)$ from right to left by one unit of time.

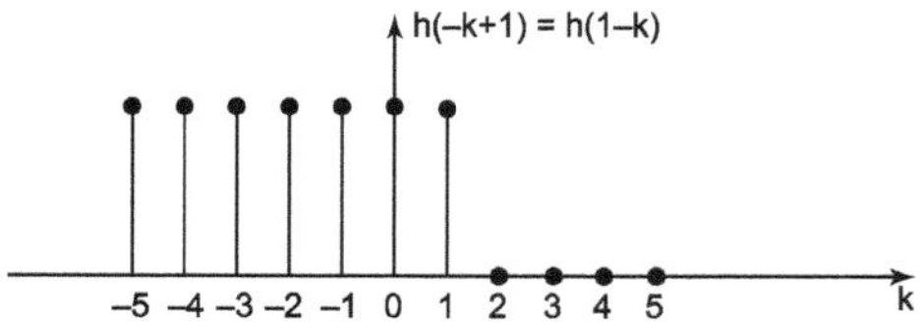

(d) Graphical representation of shifted various of $h(-k)$ from left to right by one unit of time.

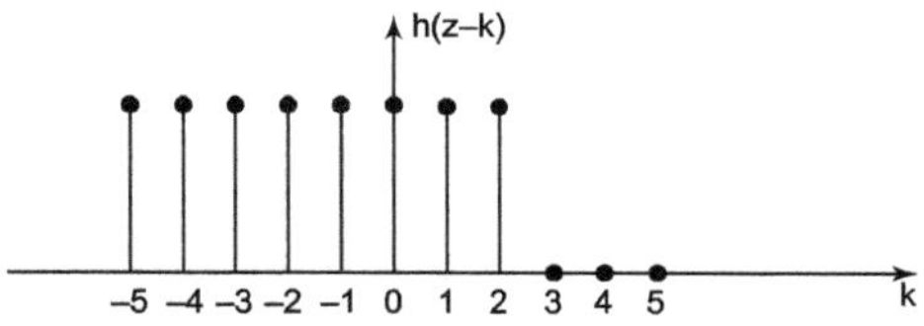

(*e*) Graphical representation of shifted various of $h(-k)$ from left to right by two unit of time.

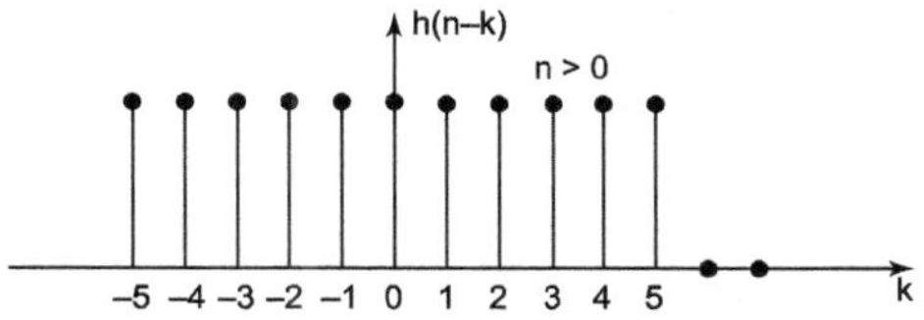

(*f*) Graphical representation of shifted various of $h(-k)$ from left to right by n unit of time when $n > 0$.

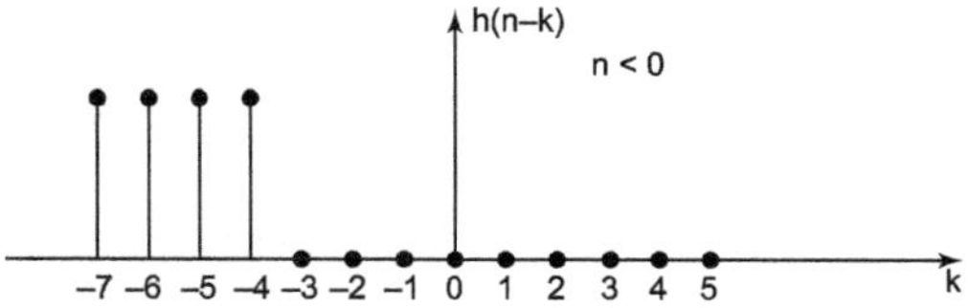

(*g*) Graphical representation of shifted various of $h(-K)$ from right to left by n unit of time when $n < 0$.

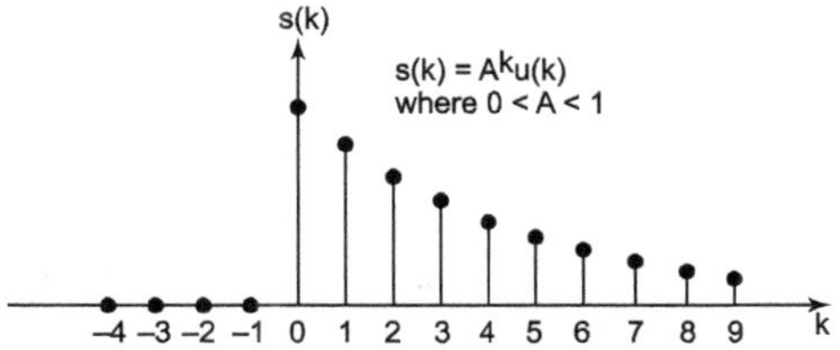

(*h*) Graphical representation of shifted various of $s(k) = Aku(k)$ where $0 < A < 1$.

FIGURE 2.15

Thus, for $n \ge 0$, $$y(n) = \sum_{k=0}^{n} A^k = \frac{1-A^{n+1}}{1-A} A\ A$$

For $n < 0$, $h(n-k)\, s(k) = 0$, *i.e.*,

$$y(n) = 0$$

Therefore, for all n, $y(n)$ is given by

$$y(n) = \left(\frac{1-(A)^{n+1}}{1-A}\right) u(n)$$

This is the resultant of convolution of $h(k)$ and $s(k)$ by graphical method

$$y(n) = \left[\frac{1-(A)^{n+1}}{1-A}\right] u(n)$$

can be sketched as shown in Figure 2.15(i).

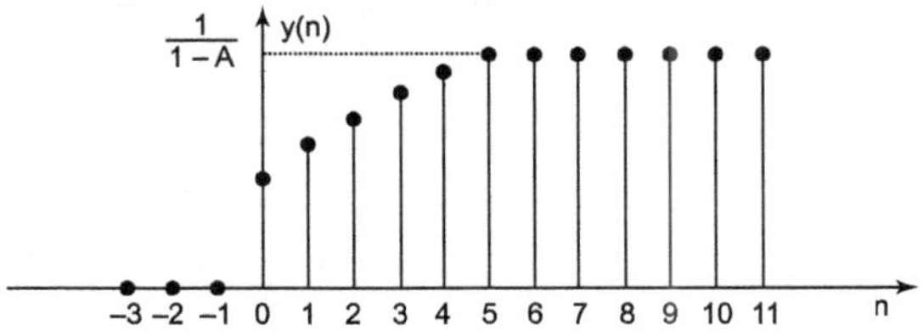

(*i*) Graphical representation of $y(n)$.

EXAMPLE 2.9

Determine linear convolution of the sequences

$$s(n) = 2^n u(-n) \text{ and } h(n) = u(n)$$

Solution: Linear convolution of two discrete-time sequences is given by

$$y(n) = \sum_{k=-\infty}^{\infty} s(k)h(n-k)$$

The sequences $s(k)$ and $h(n - k)$ are sketched as a function of k in Figure 2.16(a). Here, $s(k)$ is zero for $k > 0$ and $h(n - k)$ is zero for $k > n$.

Here, we observe that $s(k)\ h(n - k)$ is always non-zero samples along the k-axis.

Where $n \geq 0$, $s(k)\ h(n - k)$ has non-zero samples for $k \leq 0$. There will be two cases.

Case I: For $n \geq 0$,

$$y(n) = \sum_{k=-\infty}^{\infty} s(k)h(n-k) = \sum_{k=-\infty}^{0} (2)^k = \sum_{k=0}^{\infty} (2)^{-k}$$

$$= \sum_{k=-0}^{\infty} \left(2^{-1}\right)^k = \frac{1}{1-2^{-1}} = \frac{2}{2-1} = 2$$

Thus $y(n) = 2$, for $n \geq 0$. It is a constant value.

Case II: For $n < 0$, $s(k)\ h(n - k)$ has non-zero sample for $k \leq n$.

$$y(n) = \sum_{k=-\infty}^{n} s(k)h(n-k) = \sum_{k=-\infty}^{n} 2^k = \sum_{k=-n}^{\infty} (2)^{-k}$$

$$= \sum_{k=-n}^{\infty} \left(\frac{1}{2}\right)^k$$

Substituting $m = k + n$

$$= \sum_{k=0}^{\infty} \left(\frac{1}{2}\right)^{m-n} = \left(\frac{1}{2}\right)^{-n} \sum_{m=0}^{\infty} \left(\frac{1}{2}\right)^m$$

$$= (2)^n \left[\frac{1}{1-1/2}\right] = 2^{n+1}$$

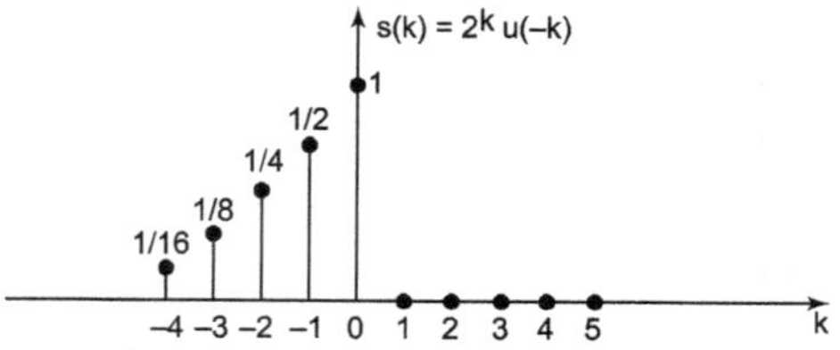

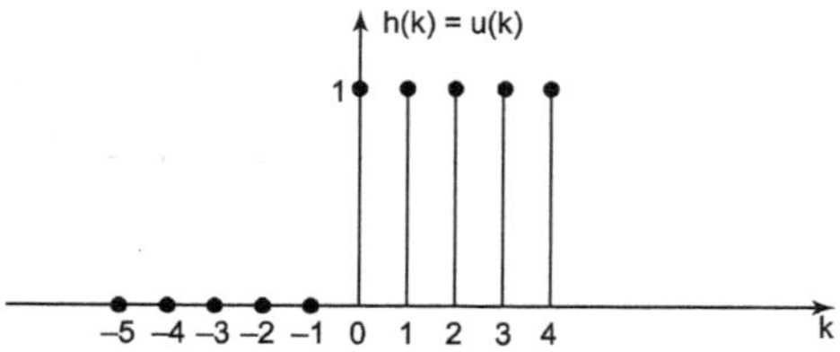

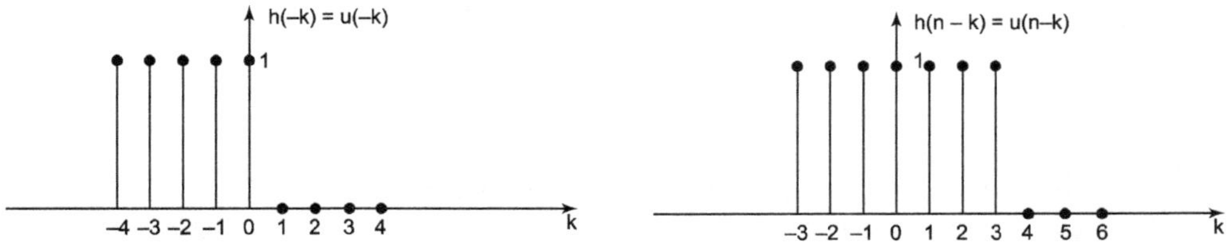

FIGURE 2.16 (a)

$$y(n) = 2^{n+1}, \text{ for } n < 0$$

$$y(n) = \begin{cases} 2, n \geq 0 \\ 2^{n+1}, n < 0 \end{cases}$$

This complex sequence $y(n)$ can be visualized by the graph given in Figure 2.16(b).

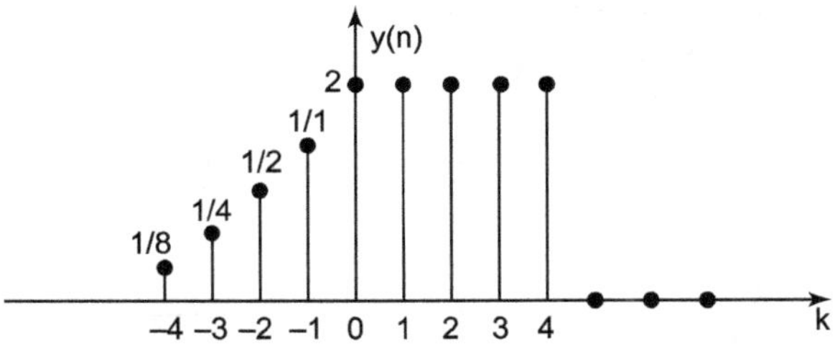

FIGURE 2.16 (b)

2.4.3 Properties of Convolution Sum

Convolution is a mathematical operation between two signal sequences $s(n)$ and $h(n)$. This operation satisfies following properties:

1. Commutative law
2. Associative law
3. Distributive law.

Commutative Law: Commutation sum satisfies commutative law. According to commutative law for a system shown in Figure 2.17,

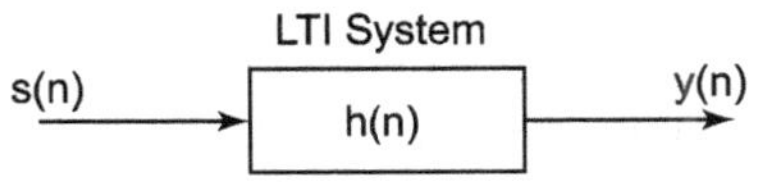

FIGURE 2.17 LTI system

$$s(n) * h(n) = h(n) * s(n)$$

or
$$\sum_{k=-\infty}^{\infty} s(k)h(n-k) = \sum_{k=-\infty}^{\infty} h(k)s(n-k)$$

This is true only for LTI discrete-time systems.
Associative Law: Convolution sum also satisfies the associative law. According to associative law for the systems shown in Figure 2.18,

$$[s(n) * h_1(n)] * h_2(n) = s(n) * [h_1(n) * h_2(n)]$$

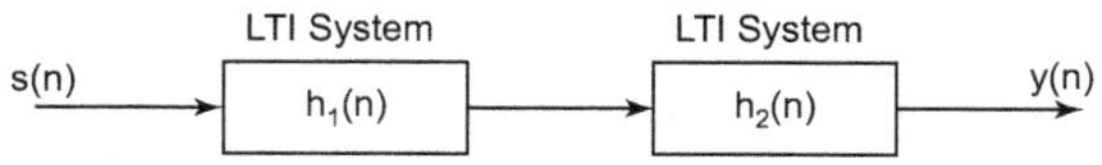

FIGURE 2.18 Cascading of two discrete-time LTI systems.

Distributive Law: This law is also satisfied by convolution sum of two discrete-time LTI systems. According to the distribution law for the systems shown in Figure 2.19,

$$s(n) * [h_1(n) + h_2(n)] = s(n) * h_1(n) + s(n) * h_2(n)$$

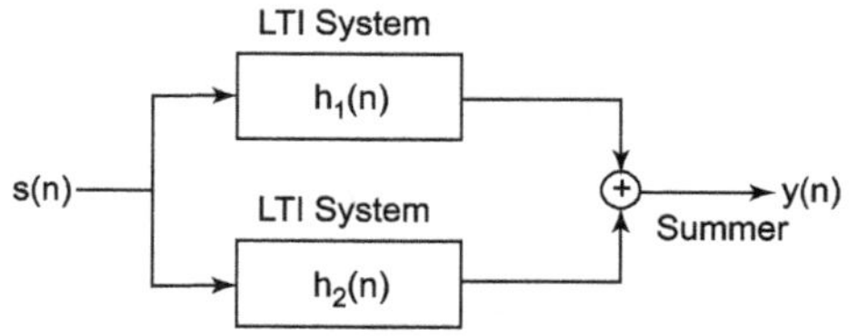

FIGURE 2.19 Two discrete-time LTI systems in parallel.

2.5 INVERSE SYSTEMS

Convolution is used to determine output $y(n)$ for any arbitrary input $s(n)$ and unit-sample response $h(n)$ of the LTI discrete-time system. There are some practical applications where we have an output signal $y(n)$ from a system of which characteristics are unknown and we are required to determine the input signal $s(n)$.

For example: In high-speed data (digital information) transmission through telephone channels, the channel distorts the signal and causes Inter Symbol Intereference (ISI) among the data symbols. ISI causes errors in the data recovered from the channel. In such circumstances, we require a corrective system which, when cascaded with the channel will produce a reciprocal signal of the desired transmitted signal. This system is called as an equalizer in digital communications. Figure 2.20 illustrates the block diagram of a system and an inverse system (both are cascaded).

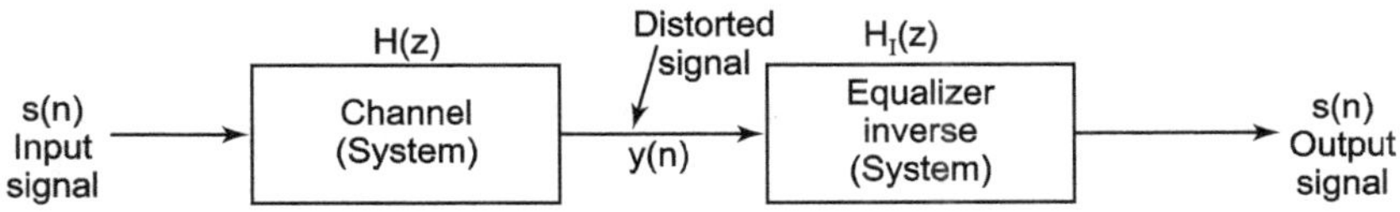

FIGURE 2.20 Block diagram showing system (channel) and inverse system (equalizer).

Inverse System: The frequency response of corrective system is basically reciprocal of the frequency response of the system which causes distortion (in telephony for digital transmission it is called ISI).

$$H(z) = \text{Transfer function of direct system.}$$
$$H_I(z) = \text{Transfer function of inverse system.}$$

For direct system–inverse system cascading

$$H(z)\,.\,H_I(z) = 1$$

Deconvolution: The distorted system produces an output $y(n)$ which is the convolution of the input with unit-sample response $h(n)$. Inverse system produces $s(n)$ by taking $y(n)$ as input and this operation is called *deconvolution.*

Inverse system can be designed by comparing the received signal with the transmitted signal. The process of determining the characteristics of an unknown system, $h(n)$ or $H(\omega)$ by a set of measurements performed on the system is termed as *system identification.*

Deconvolution is often used in seismic signal processing.

A system is said to be invertible if there is a one-to-one correspondence between its input and output signals. The cascading of a direct system and its inverse system is equivalent to the identity system, *i.e.*, $H(z).H_I(z) = 1$ or $h(n) * h_I(n) = \delta(n)$. Inverse systems are applicable in geophysics and digital communications. Figure 2.21 illustrates the block diagram showing cascading of direct and inverse systems.

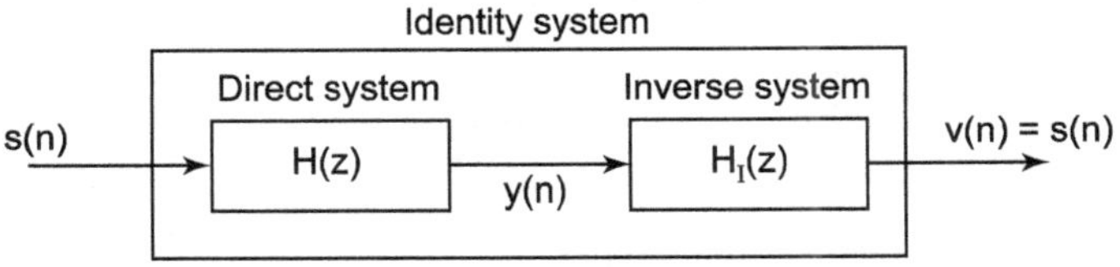

FIGURE 2.21 Block diagram showing cascading of direct system and inverse system.

From Figure 2.21,

$$v(n) = h(n) * h_1(n) * s(n) = s(n) \tag{2.20}$$

For identity system,

$$h(n) * h_I(n) = \delta(n) \tag{2.21}$$

Taking the z-transform * of Eqn. (2.21), we get

$$H(z)\, H_I(z) = 1 \tag{2.22}$$

Therefore, transfer function for inverse system will be

$$H_I(z) = \frac{1}{H(z)} \tag{2.23}$$

If transfer function of direct system is rational, then

$$H(z) = \frac{P(z)}{Q(z)} \tag{2.24}$$

The transfer function of an inverse system is given by

$$H_I(z) = \frac{Q(z)}{P(z)} \tag{2.25}$$

Thus, the zeros of $H(z)$ become the poles of $H_I(z)$ and poles of $H(z)$ become the zeros of $H_I\,(z)$. If $H(z)$ is an all-zero system (FIR system), then $H_I(z)$ is an all-pole system. If $H(z)$ is an all-pole system, then $H_I(z)$ will be all-zero system (FIR system).

EXAMPLE 2.10

Find out the inverse of the system with unit-sample response $h(n) = (1/3)^n\, u(n)$.

Solution:

Given $\quad h(n) = (1/3)n\, u(n)$

$$H(z) = z\text{-transform of } h(n)$$

$$= \frac{1}{1 - 1/3z^{-1}} \quad \text{ROC: } |z| > \frac{1}{3}$$

This is a causal, stable, and an all-pole system.

Its inverse system will be all-zero system (FIR system). It is given by

$$H_I(z) = \frac{1}{H(z)} = 1 - \frac{1}{3}z^{-1}$$

*z-transform and inverse z-transform are discussed in detail in Chapter 3.

$$h_I(n) = \text{inverse } z\text{-transform of } H_I(z)$$
$$= \mathcal{Z}^{-1}\left[1 - \frac{1}{3}z^{-1}\right]$$
$$= \delta(n) - \frac{1}{3}\delta(n-1)$$

$h_I(n)$ is the unit-sample response of inverse system of $H(z)$.

EXAMPLE 2.11

Find out the inverse of the system with unit-sample response:

$$h(n) = \delta(n) - \frac{1}{3}\delta(n-1).$$

Solution:

$$H(z) = \mathcal{Z}\left[\delta(n) - \frac{1}{3}\delta(n-1)\right]$$
$$= 1 - \frac{1}{3}z^{-1}, \text{ ROC: } |z| > 0$$

This is an all-zero system (FIR system).
Transfer function of the inverse system

$$H_I(z) = \frac{1}{H(z)} = \frac{1}{1 - \frac{1}{3}z^{-1}} = \frac{z}{z - \frac{1}{3}}$$

This inverse system has a zero at $z = 0$ (*i.e.*, origin) and a pole at $z = 1/3$.

In this case, there are two possible ROCs, and hence, there will be two possible inverse systems as shown in Figure 2.22.

Case I: ROC of $H_I(z)$: $|z| > \frac{1}{3}$

$$h_I(z) = \mathcal{Z}^{-1}[H_I(z)]$$
$$= \left(\frac{1}{3}\right)^n u(n)$$

This is the unit-sample response of a causal and stable system.
Case II: ROC of $H_I(z)$: $|z| < 1$

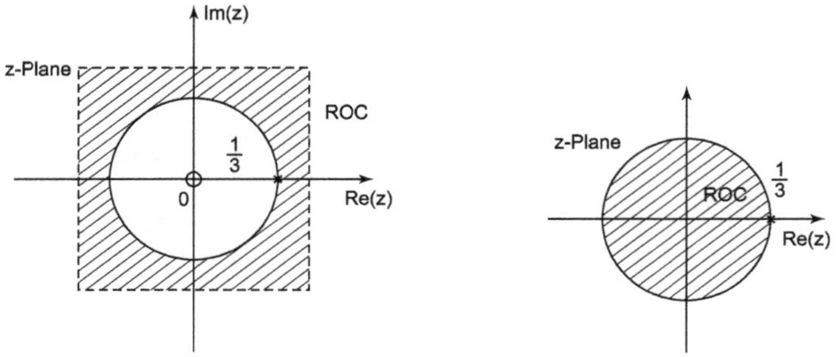

FIGURE 2.22 Two possible ROCs for $H(z) = \frac{3}{z - \frac{1}{2}}$

$$h_I(z) = \mathcal{Z}^{-1}[H_I(z)]$$

$$= \left(\frac{1}{3}\right)^n u(n-1)$$

This system is the unstable and anticausal.

2.6 CORRELATION OF TWO DISCRETE-TIME SIGNALS

A mathematical operation that has close resemblance with convolution is called *correlation*. Correlation operation also requires two discrete-time sequences just as convolution.

The objective in computing the correlation between two signals is to measure the degree of similarity of two signals. By measuring the degree of correlation, we can extract some information that depends on the application. Here, application means the type of system where correlation operation is used for extracting some information. It is required in radar, sonar, digital communications, and other areas of engineering and technology. Resultant of correlation operation of two discrete-time sequences is a discrete-time sequence.

If the two sequences are identical, then the resultant of correlation of two discrete-time sequences is called *auto-correlation sequence.*

If the two sequences are different, then the resultant of correlation of two sequences is called *cross-correlation sequence.*

Digital communication is one of the areas where correlation operation is often used.

2.6.1 Cross-correlation and Auto-correlation Sequences

Suppose $s(n)$ and $y(n)$ are two real signal sequences which have finite energy. It means that these are some energy sequences.

The cross-correlation of $s(n)$ and $y(n)$ is a sequence $R_{SY}(l)$ which is defined as follows:

$$R_{SY}(l) = \sum_{n=-\infty}^{\infty} s(n)y(n-1) \tag{2.26}$$

or,

$$R_{SY}(l) = \sum_{n=-\infty}^{\infty} y(n+l)s(n) \tag{2.27}$$

$$(l = 0,\ \pm 1,\ \pm 2,\ \ldots.)$$

Comparing equations (2.26) and (2.27), we get

$$R_{SY}(l) = R_{YS}(-l) \tag{2.28}$$

where $R_{SY}(l)$ is a folded version of $R_{YS}(l)$. Here, folding is done with respect to $l = 0$.

EXAMPLE 2.12

Determine the cross-correlation sequence $R_{SY}(l)$ of the sequences

$$s(n) = \{2,\ \underset{\uparrow}{1},\ 3\}$$

$$y(n) = \{1,\ \underset{\uparrow}{2},\ 2\}.$$

Solution: Number of sample points in resultant of correlation of two discrete-time sequences $= 3 + 3 - 1 = 5$.

Cross-correlation sequence is defined as follows:

$$R_{SY}(l) = \sum_{n=-\infty}^{\infty} s(n)y(n-l)\,,\ 1 = 0,\ \pm 1,\ \pm 2, \ldots$$

For $l = 0$, $\quad R_{SY}(0) = \sum_{n=-\infty}^{\infty} s(n)y(n)$

$$s(n) = 2, \underset{\uparrow}{1}, 3$$

$$y(n) = 1, \underset{\uparrow}{2}, 2$$

$$R_{SY}(0) = \sum_{n=-\infty}^{\infty} s(n)y(n) = 2 \times 1 + 1 \times 2 + 3 \times 2 = 2 + 2 + 6 = 10$$

For $l = 1$, $\quad R_{SY}(1) = \sum_{n=-\infty}^{\infty} s(n)y(n-1)$

$$\boxed{\begin{array}{l} s(n) = 2,1,3 \\ \qquad\quad \uparrow \\ y(n-1) = 1,2,2 \end{array}}$$

$$R_{SY}(1) = \sum_{n=-\infty}^{\infty} s(n)y(n-1) = 1 \times 1 + 3 \times 2 = 1 + 6 = 7$$

For $l = 2$, $$R_{SY}(2) = \sum_{n=-\infty}^{\infty} s(n)y(n-2)$$

$$\boxed{\begin{array}{l} s(n) = 2,1,3 \\ \qquad\quad \uparrow \\ y(n-2) = 1,2,2 \end{array}}$$

$$R_{SY}(2) = \sum_{n=-\infty}^{\infty} s(n)y(n-2) = 3 \times 1 = 3$$

$$R_{SY}(3) = 0$$
$$R_{SY}(4) = 0$$
$$R_{SY}(5) = 0$$

.
.
.

For $l = -1$, $$R_{SY}(-1) = \sum_{n=-\infty}^{\infty} s(n)y(n+1)$$

$$\boxed{\begin{array}{l} s(n) = 2,1,3 \\ \quad \uparrow \\ y(n+1) = 1,2,2 \end{array}}$$

$$R_{SY}(-1) = \sum_{n=-\infty}^{\infty} s(n)y(n+1) = 2 \times 2 + 1 \times 2 = 6$$

For $l = -z$, $R_{SY}(-2) = \sum_{n=-\infty}^{\infty} s(n)y(n+2)$

$$\begin{array}{r} s\ (n) = 2,\ 1,\ 3 \\ y\ (n+1) = 1,\ 2,\ 2 \qquad\quad \end{array}$$

$$R_{SY}(-2) = \sum_{n=-\infty}^{\infty} s(n)y(n+2) = 2 \times 2 = 4$$

$R_{SY}(-3) = 0$’

$$R_{SY}(-4) = 0$$
$$R_{SY}(-5) = 0$$

The resultant cross-correlation sequence

$$R_{SY}(l) = \{R_{SY}(-2), R_{SY}(-1), R_{SY}(0), R_{SY}(1), R_{SY}(2)\}$$
$$= \{4, 6, \underset{\uparrow}{10}, 7, 3\}$$

Note: *If $s(n)$ has M sample points and $y(n)$ has N sample points in their sequences, then its resultant cross-correlation sequence will have $M + N - 1$ sample points just like linear convolution of two discrete-time sequences will have.*

EXAMPLE 2.13

Compute the auto-correlation of the signal

$$s(n) = A^n u(n),\ 0 < A < 1.$$

Solution: Since $s(n)$ is an infinite-duration signal and its auto-correlation will also have infinite duration.

There will be two cases:

Case I: If $l \geq 0$

$$Rss(l) = \sum_{n=-\infty}^{\infty} s(n)s(n-l) = \sum_{n=-\infty}^{\infty} A^n u(n).A^{n-1}u(n-l)$$
$$= \sum_{n=l}^{\infty} A^n \cdot A^{n-1}$$
$$= \sum_{n=l}^{\infty} A^n.A^n.A^{-l} = A^{-1}\sum_{n=l}^{\infty}\left[A^2\right]^n$$

since $A < 1$, infinite series coverages

$$= A^{-l}\left[\frac{A^{2l}}{1-A^2}\right] = \frac{A^l}{1-A^2},\ l \geq 0 \qquad (1)$$

Case II: For $l < 0$

$$R_{ss}(l) = \sum_{n=0}^{\infty} s(n)s(n-l) = \sum_{n=0}^{\infty} A^n.A^{n-1} = A^{-l}\sum_{n=0}^{\infty}\left[A^2\right]^n$$
$$= A^{-l}\cdot\left[\frac{1}{1-A^2}\right] = \frac{A^{-l}}{1-A^2},\ l < 0 \qquad (2)$$

From Eqns. (1) and (2), we get

$$\left.\begin{aligned} R_{ss}(l) &= \frac{A^l}{1-A^2},\ l \geq 0 \\ R_{ss}(l) &= \frac{A^{-l}}{1-A^2},\ l < 0 \end{aligned}\right\}\ \text{Auto-correlation sequences}$$

Hence, auto-correlation of the signal $s(n) = A^n u(n), 0 < A < l$ is given as follows:

$$R_{ss}(l) = \frac{A^{|l|}}{1 - A^2}, -\infty < l < \infty \qquad (3)$$

2.7 SIGNALS AND VECTORS

There is a perfect analogy between signals and vectors. Signals are not just like vectors. Signals are vectors. A vector can be represented as a sum of its components in a variety of ways, depending on the choice of coordinate system. A signal can also be represented as a sum of its components in a variety of ways.

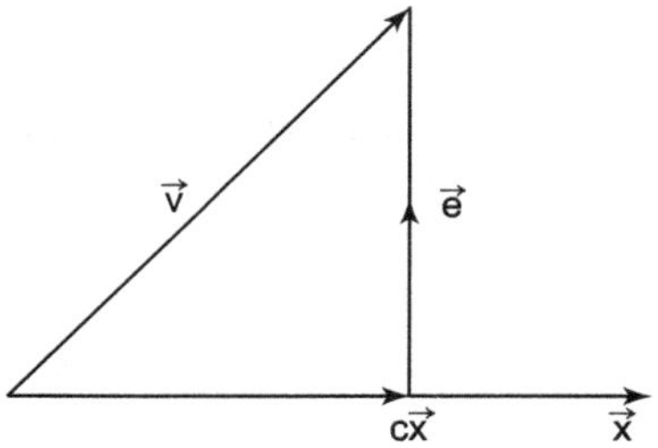

FIGURE 2.23 Illustration of two vectors *v* and *x*

2.7.1 Component of a Vector

A vector is specified by its magnitude and its direction. Here, vectors are represented by an alphabet over which an arrow is shown. For example, $\vec{x}$ is a vector with magnitude or length $|\vec{x}|$. Consider two vectors $\vec{v}$ and $\vec{x}$, as shown in Figure 2.23. Let the component of $\vec{v}$ and $\vec{x}$ be $c\vec{x}$. Geometrically the component of $\vec{v}$ along $\vec{x}$ is the projection of $\vec{v}$ on $\vec{x}$. The component of $\vec{v}$ along $\vec{x}$ is obtained by drawing a perpendicular from the tip of $\vec{v}$ on the vector $\vec{x}$. It is shown in Figure 2.23.

Vector $\vec{v}$ can be expressed in terms of $\vec{x}$ as follows:

$$\vec{v} = c\vec{x} + \vec{e} \qquad (2.29)$$

However, this is not the only way to express vector $\vec{v}$ in terms of vector $\vec{x}$. Figure 2.24 shows two of the infinite other possibilities.

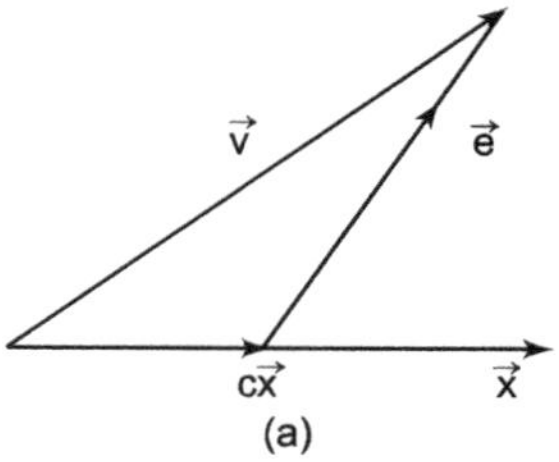

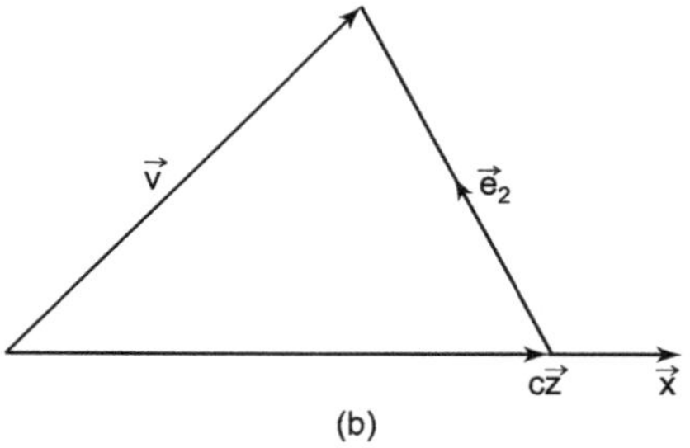

FIGURE 2.24 Illustration of approximation of a vector in terms of another vector.

From Figs. 2.24(a) and 2.24(b), we have

$$\vec{v} = c_1\vec{x} + \vec{e}_1$$
$$= c_2\vec{x} + \vec{e}_2 \tag{2.30}$$

In each of these three representations of Figs. 2.23, 2.24(a), and 2.24(b), vector $\vec{v}$ can be represented in terms of $\vec{x}$ plus another vector (called the error vector).

If we approximate $\vec{v}$ by $c\vec{x}$ in Figure 2.23,

$$\vec{v} \simeq c\vec{x} \tag{2.31}$$

The error in this approximation is the vector $\vec{e} = \vec{v} - c\vec{x}$. Similarly, the errors in the approximations in Figs. 2.24(a) and 2.24(b) are $\vec{e}$ and $\vec{e}_2$. But the error vector e is the smallest.

Now, we can define mathematically the component of a vector $\vec{v}$ along $\vec{x}$ to be $c\vec{x}$, where c is chosen to minimize the length of the error vector $\vec{e} = \vec{v} - c\vec{x}$.

For convenience, we can define the scalar or dot or inner product of two vectors $\vec{v}$ and $\vec{x}$ as follows:

$$\vec{v} \cdot \vec{x} = |\vec{v}||\vec{u}| \cos\theta \tag{2.32}$$

where θ = angle between vectors $\vec{v}$ and $\vec{x}$.

By using above definition, we can express magnitude of vector $\vec{x}$, *i.e.*, $|\vec{x}|$ as follows:

$$|\vec{x}|^2 = \vec{x} \cdot \vec{x} \tag{2.33}$$

Magnitude of a vector is also called length of the vector.

Now, the length of the component of $\vec{v}$ along $\vec{x}$ is $|\vec{v}|\cos\theta$, but it is also equal to $c|\vec{x}|$.

Therefore,
$$c|\vec{x}| = |\vec{v}| \cos\theta \tag{2.34}$$

On multiplying both sides of Eqn. (2.34) by $|\vec{x}|$, we get

$$c|\vec{x}|\,|\vec{x}| = |\vec{v}||\vec{x}| \cos\theta$$

or
$$c|\vec{x}|^2 = |\vec{v}||\vec{x}| \cos\theta$$
$$= \vec{v} \cdot \vec{x} \tag{2.35}$$

From Eqns. (2.33) and (2.35), we get

$$c = \frac{\vec{v} \cdot \vec{x}}{\vec{x} \cdot \vec{x}} = \frac{1}{|\vec{x}|^2}(\vec{v} \cdot \vec{x}) \tag{2.36}$$

since $|\vec{x}|^2 = \vec{x} \cdot \vec{x}$.

From Figure 2.23, it is apparent that when vectors $\vec{v}$ and $\vec{x}$ are perpendicular, or orthogonal, then vector $\vec{v}$ has a zero component along $\vec{x}$; consequently, $c = 0$.

Now, we can conclude from Eqn. (2.36) that if vectors $\vec{v}$ and $\vec{x}$ are to be orthogonal then their dot or scalar product must be zero, *i.e.*,

$$\vec{v} \cdot \vec{x} = 0.$$

2.7.2 Component of a Signal

Now, we can extend the concepts of vector component and orthogonality to signals. We consider the problem of approximating a real signal $v(t)$ in terms of another real signal $x(t)$ over an interval $[t_1, t_2]$:

$$v(t) \simeq c\,x(t),\ t_1 \le t \le t_2 \qquad (2.37)$$

The error $e(t)$ in this approximation is given by

$$e(t) = \begin{cases} v(t) - cx(t), t_1 \le t \le t_2 \\ 0, \text{otherwise} \end{cases} \qquad (2.38)$$

We now select some criterion for the "best approximation." We know that the signal energy is one possible measure of a signal size. For best approximation, we need to minimize the error signal $e(t)$. This error signal minimizes its size which is its energy E_e over the interval $[t_1, t_2]$.

Energy of error signal is given by

$$Ee = \overset{t_2}{\underset{t_1}{\equiv}} e^2(t)\,dt \qquad (2.39)$$

Substituting Eqn. (2.38) in Eqn. (2.39), we get

$$E_e = \int_{t_1}^{t_2} \left[v(t) - cx(t)^2 \right] dt \qquad (2.40)$$

Note that the R.H.S. of Eqn. (2.40) is definite integral with time t as a dummy variable. Hence, E_e is a function of the parameter c (not t) and E_e is minimum for some choice of c.

For minimization of E_e, a necessary condition is given as follows:

$$\frac{dE_e}{d_c} = 0 \qquad (2.41)$$

Putting Eqn. (2.40) in Eqn. (2.41), we get

$$\frac{d}{dc}\left\{ \int_{t_1}^{t_2} [v(t) - cx(t)]^2\, dt \right\} = 0$$

Expanding the squared term inside the integral, we get

$$\frac{d}{dc}\left\{ \int_{t_1}^{t_2} \left[v^2(t) + c^2x^2(t) - 2cv(t)x(t) \right] dt \right\} = 0$$

or $$\frac{d}{dc}\left[\int_{t_1}^{t_2} v^2(t)dt\right]+\frac{d}{dc}\left[\int_{t_1}^{t_2} c^2x^2(t)dt\right]-\frac{d}{dc}\left[2c\int_{t_1}^{t_2} v(t)x(t)dt\right]=0$$

or $$-2\int_{t_1}^{t_2} v(t)x(t)dt+2c\int_{t_1}^{t_2} x^2(t)dt=0$$

or $$c=\frac{\int_{t_1}^{t_2} v(t)x(t)dt}{\int_{t_1}^{t_2} x^2(t)dt}$$

$$=\frac{1}{E_x}\int_{t_1}^{t_2} v(t)x(t)dt \tag{2.42}$$

where $$E_x=\int_{t_1}^{t_2} x^2(t)\,dt$$

Up to now, we have seen a remarkable similarity between the behavior of vectors and signals as indicated by the following two equations:

$$c=\frac{\vec{v}\cdot\vec{x}}{\vec{x}\cdot\vec{x}}=\frac{1}{|\vec{x}|^2}\vec{v}\cdot\vec{x} \tag{i}$$

$$c=\frac{\int_{t_1}^{t_2} v(t)x(t)dt}{\int_{t_1}^{t_2} x^2(t)dt}$$

$$=\frac{1}{E_x}\int_{t_1}^{t_2} v(t)x(t)dt \tag{ii}$$

It is evident from these two parallel expressions that area under the product of two signals corresponds to the scalar or dot product of two vectors. In fact, the area under the product of $v(t)$ and $x(t)$ is called the inner product of $v(t)$ and $x(t)$.

The energy of a signal is the inner product of a signal with itself and corresponds to the vector length squared (which is the inner product of the vectors with itself).

If a signal $v(t)$ is approximated by another signal $x(t)$ as follows:

$v(t) \simeq c\,x(t)$ then the optimum value of c that minimizes the energy of the error signal in this approximation by Eqn. (2.42).

We have a signal $v(t)$ which contains a component $cx(t)$, where c is given by Eqn. (2.42). Note that in vector terminology, $cx(t)$ is the projection of $v(t)$ on $x(t)$. From the vector-signal analogy, we say that if the component of a signal $v(t)$ of the form $x(t)$ is zero (That is, $c = 0$), the signals $v(t)$ and $x(t)$ are orthogonal over the interval $[t_1, t_2]$. Therefore, we define the real signals $v(t)$ and $x(t)$ to be orthogonal over the interval $[t_1, t_2]$ if

$$\int_{t_1}^{t_2} v(t)x(t)dt = 0 \tag{2.43}$$

2.7.3 Orthogonality in Complex Signals

So far, our discussion was restricted to real functions of time "t." Now, we generalize our discussion to complex functions of time "t." We consider the same problem of approximating a function $v(t)$ by a function $x(t)$ over an interval $(t_1 \leq t \leq t_2)$:

$$v(t) \simeq cx(t) \tag{2.44}$$

In this case, both the coefficient c and the error

$$e(t) = v(t) - cx(t) \tag{2.45}$$

are complex (in general).

For the best approximation, we choose c such that we minimize E_e (Energy of the error signal) and given by

$$E_e = \int_{t_1}^{t_2} \left| v(t) - cx(t)^2 \right| dt \tag{2.46}$$

We also know that

$$\begin{aligned} |A + B|^2 &= (A + B)\,(A* + B*) \\ &= |A|^2 + |B|^2 + A*B + AB* \end{aligned} \tag{2.47}$$

After some manipulation in Eqn. (2.46) and using result of Eqn. (2.47), we can express the integral E_e in Eqn. (2.46) as follows:

$$E_e = \int_{t_1}^{t_2} \left| v(t)^2 \right| dt - \left| \frac{1}{\sqrt{E_x}} \int_{t_1}^{t_2} v(t)x*(t)dt \right|^2 + \left| c\sqrt{E_x} - \frac{1}{\sqrt{E_x}} \int_{t_1}^{t_2} v(t)x*(t)dt \right|^2$$

Since the first two terms on the R.H.S. are independent of c, it is clear that E_e is minimized by choosing c such that the third term is zero. This gives

$$c = \frac{1}{E_x} \int_{t_1}^{t_2} v(t)x*(t)dt \qquad \ldots(2.48)$$

This result needs to redefine orthogonality for the complex case as follows:

Two complex functions $x_1(t)$ and $x_2(t)$ are orthogonal over an interval $(t_1 \leq t \leq t_2)$ if

$$\int_{t_1}^{t_2} x_1(t)x_2*(t)dt = 0$$

or
$$\int_{t_1}^{t_2} x_1*(t)x_2(t)dt = 0 \qquad (2.49)$$

Either equality suffices. This is the general definition of orthogonality. This equation will reduce to Eqn. $\int_{t_1}^{t_2} v(t)x(t)dt = 0$, when the functions are real.

2.7.4 Energy of the Sum of Orthogonal Signals

We already know that the length of the sum of two orthogonal vectors is equal to the sum of the lengths squared of the two vectors. Thus, if vectors $\vec{x}$ and $\vec{y}$ are orthogonal, and if $\vec{z} = \vec{x} + \vec{y}$, then

$$|\vec{z}|^2 = |\vec{x}|^2 + |\vec{y}|^2$$

Similar results are also available for signals. The energy of the sum of the two orthogonal signals is equal to the sum of the energies of the two signals. Thus, if signals $x(t)$ and $y(t)$ are orthogonal over an interval $[t_1, t_2]$, and if

$$z(t) = x(t) + y(t), \text{ then}$$
$$E_z = E_x + E_y \qquad (2.50)$$

Real signals case is special case of complex signals

$$\begin{aligned}\int_{t_1}^{t_2} |z(t)|^2\, dt &= \int_{t_1}^{t_2} |x(t)+y(t)|^2\, dt \\ &= \int_{t_1}^{t_2} |x(t)|^2\, dt + \int_{t_1}^{t_2} |y(t)|^2\, dt + \int_{t_1}^{t_2} x(t)y*(t)dt \int_{t_2}^{t_2} x*(t)y(t)dt \\ &= \int_{t_1}^{t_2} |x(t)|^2\, dt + \int_{t_1}^{t_2} |y(t)|^2\, dt \qquad (2.51)\end{aligned}$$

This last result follows from the fact that because of orthogonality, the two integrals of the cross products $x(t)\, y*(t)$ and $x*(t)\, y(t)$ are zero. This result can be extended to the sum of any number of mutually orthogonal signals.

EXAMPLE 2.14

Determine the component in signal $v(t)$ of the form $\sin(t)$ for the square signal $v(t)$. It is shown in Figure 2.25. In other words, approximate $v(t)$ in terms of $\sin(t)$:

$$v(t) \simeq c \sin(t), 0 \le t \le 2\pi$$

So that the energy of the error signal is minimum.

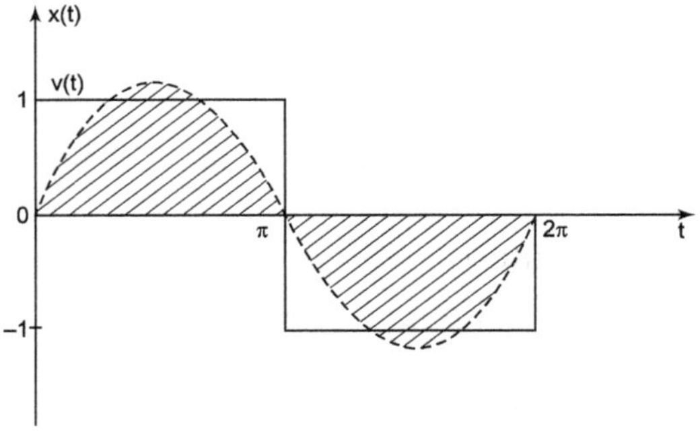

FIGURE 2.25 Approximation of a square signal in terms of a single sinusoid.

Solution: In this case,

$$x(t) = \sin(t) \tag{1}$$

Energy of signal $x(t)$ is determined as follows:

$$Ex = \int_0^{2\pi} \sin^2(t)dt$$

$$= \int_0^{2\pi} \left[\frac{1-\cos(2t)}{2}\right] dt$$

$$= \int_0^{2\pi} \frac{1}{2}dt - \frac{1}{2}\int_0^{2\pi} \cos(2t)dt$$

$$= \left[\frac{1}{2}t\right]_0^{2\pi} - \frac{1}{2}\left[\frac{\sin(2t)}{2}\right]_0^{2\pi}$$

$$= \frac{1}{2}[t]_0^{2\pi} - \frac{1}{4}\left[\sin(2t)\right._0^{2\pi}$$

$$= \pi - 0 = \pi$$

or $$E_x = \pi \tag{2}$$

Constant C is determined as follows:

$$c = \frac{\mathop{\equiv}\limits_{t_1}^{t_2} v(t)x(t)dt}{\mathop{\equiv}\limits_{t_1}^{t_2} x^2(t)dt}$$

$$= \frac{1}{E_x} \mathop{\equiv}\limits_{t_1}^{t_2} v(t)x(t)dt$$

$$= \frac{1}{\pi}\int_0^{2\pi} v(t)\sin(t)dt$$

or
$$c = \frac{1}{\pi}\int_0^{2\pi} v(t)\sin(t)dt$$

$$= \frac{1}{\pi}\left[\int_0^{\pi} \sin(t)dt + \int_{\pi}^{2\pi} -\sin(t)dt\right]$$

or
$$c = \frac{4}{\neq} \tag{3}$$

Therefore,
$$v(t) \simeq c\,\sin(t)$$

$$\simeq \frac{4}{\neq}\sin(t) \tag{4}$$

Eqn.(4) represents the best approximation of $v(t)$ by the function $\sin(t)$, which minimizes the error energy.

2.8 REPRESENTATION OF SIGNALS ON ORTHOGONAL BASIS

In this section, we discuss a method of representing a signal as a sum of orthogonal signals. Here again, we can benefit from the insight gained from a similar problem with vectors. We know that a vector can be represented as the sum of orthogonal vectors, which form the coordinate system of a vector space. The problem with signals is analogous. The results for signals are parallel to those for vectors. Here, we need review of vector representation.

2.8.1 Orthogonal Vector Space

Now, we consider a three-dimensional (3–D) Cartesian vector space described by three mutually orthogonal vectors x_1, x_2, and x_3. It is shown in Figure 2.26. First, we shall seek to approximate a three-dimensional vector (v) in terms of two mutually orthogonal vector x_1 and x_2:

$$v \simeq c_1x_1 + c_2x_2$$

The error in this approximation is given by

$$e = v - (c_1x_1 + c_2x_2)$$

or
$$v \simeq c_1x_1 + c_2x_2 + e$$

We can see from Figure 2.26 that the length of e is minimum when e is perpendicular to $x_1 - x_2$ plane, and c_1x_1 and c_2x_2 are the projections or components of v on x_1 and x_2, respectively. Therefore, the constants c_1 and c_2 are given by Eqn. (2.52)

$$c = \frac{v \cdot x}{x \cdot x} = \frac{1}{|x|^2} v \cdot x \tag{2.52}$$

Now let us determine the best approximation to v in terms of all three mutually orthogonal vectors x_1, x_2, and x_3:

$$v \simeq c_1x_1 + c_2x_2 + c_3x_3 \tag{2.53}$$

Figure 2.26 shows that a unique choice of c_1, c_2, and c_3 exists, for which Eqn. (2.52) is no longer an approximation but an equality:

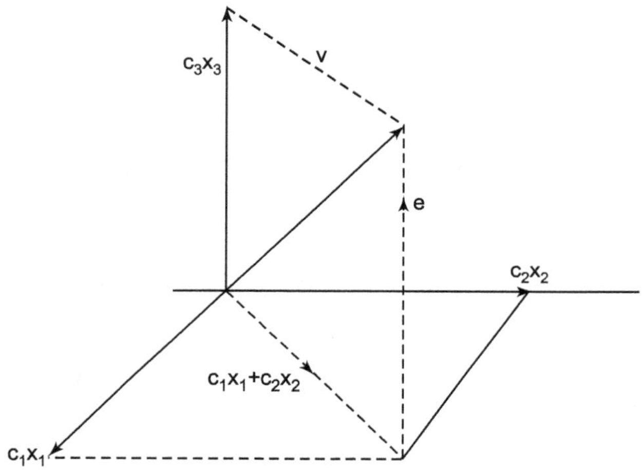

FIGURE 2.26 Representation of a vector in three-dimensional space.

$$v = c_1x_1 + c_2x_2 + c_3x_3$$

In this case, c_1x_1, c_2x_2, and c_3x_3 are the projections or components of v on x_1, x_2, and x_3, respectively. We can note here that the error in the approximation is zero when v is approximated in terms of three mutually orthogonal vectors x_1, x_2, and x_3. This is because v is a three-dimensional vector and the vectors x_1, x_2, and x_3 represent a complete set of orthogonal vectors in three-dimensional space. Here, the meaning of word "complete" is that it is impossible to find in this space another vector x_4 that is orthogonal to all three vectors x_1, x_2, and x_3. Any vector in this space can then be represented (with zero error) in terms of these three vectors. Such vectors are called basis vectors. If a set of vectors $\{x_i\}$ is not complete, the error in the approximation will generally not be zero.

Thus, in the three-dimensional case, it is generally not possible to represent a vector v in terms of only two basis vectors without an error.

The choice of basis vectors is not unique. In fact, a set of basis vectors corresponds to particular choice of coordinate system. Thus a three-dimensional vector v may be represented in many different ways which depends on the coordinate system used.

We can summarize, if a set of vectors $\{x_i\}$ is mutually orthogonal, that is, if

$$x_m \cdot x_n = \begin{cases} 0, & m \neq n \\ |x_m|^2, & m = n \end{cases}$$

and if this basis set is complete, a vector v in this space can be expressed as

$$v = c_1x_1 + c_2x_2 + c_3x_{:3}$$

where the constants c_i are given by

$$c_i = \frac{v \cdot x_i}{x_i \cdot x_i}$$

$$= \frac{1}{|x_i|^2} c \cdot x_i,\ i = 1, 2, 3$$

2.8.2 Orthogonal Signal Space

Here, we first define the orthogonality of a signal set $x_1(t)$, $x_2(t)$, ... $x_n(t)$ over the interval $[t_1, t_2]$ as follows:

$$\overset{t_2}{\underset{t_1}{\equiv}} x_m(t)x_n^*(t)dt = \begin{cases} 0, & \text{for } m \neq n \\ E_n, & \text{for } m = n \end{cases} \tag{2.54}$$

If the energies $E_n = 1$ for all values of n, then the set is normalized. Therefore, the set is called orthogonal set. An orthogonal set can be normalized by dividing $x_n(t)$ by $\sqrt{E_n}$ for all values of n.

Now, we are considering the problem of approximating a signal $v(t)$ over the interval $[t_1, t_2]$ by a set of N mutually orthogonal signals $x_1(t)$, ... $x_N(t)$:

$$v(t) \simeq c_1x_1(t) + c_2x_2(t) + \ldots\ldots + c_Nx_N(t) \tag{2.55}$$

or

$$v(t) \simeq \sum_{n=1}^{N} c_nx_n(t),\ t_1 \leq t \leq t_2 \tag{2.56}$$

In this approximation, energy of the error signal $e(t)$, *i.e.*, E_e is minimized by choosing C_n in the following manner:

$$C_n = \frac{\overset{t_2}{\underset{t_1}{\equiv}} v(t)x_n * (t)dt}{\overset{t_2}{\underset{t_1}{\equiv}} x_n^2(t)dt} \tag{2.57}$$

or $$C_n = \frac{1}{E_n}\int_{t_1}^{t_2} v(t)x_n^*(t)dt = n = 1,2,\ldots,N \tag{2.58}$$

Moreover, if the orthogonal set is complete, the error energy, $E_e \to 0$, and the representation in Eqns. (2.56) and (2.57) is no longer an approximation, but an equality

$$v(t) = c_1x_1(t) + c_2x_2(t) + \ldots\ldots + c_nc_n(t) + \ldots\ldots$$
$$= \sum_{n=1}^{\infty} c_nx_n(t), t_1 \le t \le t_2 \tag{2.59}$$

where the coefficients C_n are given by Eqns. (2.57) and (2.58). Because the error signal energy E_e approaches zero, it follows that the energy of $v(t)$ is now equal to the sum of the energies of its orthogonal components $c_1x_1(t)$, $c_2x_2(t)$, $c_3x_3(t)$, ….

The Series on the R.H.S. of Eqn. (2.57) is called the Generalized Fourier Series of signal $v(t)$ with respect to the set $\{x_n(t)\}$. When the set $\{x_n(t)\}$ is such that the error energy $E_e \to 0$ and $N \to \infty$ for every member of some particular class, we can say that the set $\{x_n(t)\}$ is complete on $[t_1, t_2]$ for that class of $v(t)$. Therefore, the set $\{x_n(t)\}$ is called a set of Basis functions or Basis Signals.

Thus, when the set $\{x_n(t)\}$ is complete, we have the equality of Eqn. (2.57). The equality here is not an equality in the ordinary sense, but in the sense that the error energy, that is, the energy of the difference between the two sides of Eqn. (2.57), approaches zero.

If the equality exists in the ordinary sense, the error energy is always zero, but the converse is not necessarily true. The error energy can approach zero even though $e(t)$, the difference between the two sides, is non-zero at some isolated instants.

This is because even if $e(t)$ *is* non-zero at such instants, the area under $e^2(t)$ is still zero. Thus, the Fourier Series on the R.H.S. of Eqn. (2.57) may differ from $v(t)$ at a finite number of points. In fact, when $v(t)$ has a jump discontinuity at $t = t_0$, the corresponding Fourier Series at t_0 converges to the mean of $v(t_0^+)$ and $v(t_0^-)$.

2.8.3 Parseval's Theorem

We already know that the energy of the sum of orthogonal signals is equal to the sum of their energies. Therefore, the energy of the R.H.S. of Eqn. (2.57) is the sum of the energies of the individual orthogonal components. The energy of a component $c_nx_n(t)$ is $c_n^2E_n$. Now, we equate the energies of the two sides of Eqn. (2.57) and we get

$$E_v = c_1^2 E_1 + c_2^2 E_2 + c_3^2 E_3 + \ldots.$$
$$= \bullet_n \quad c_n^2 E_n \tag{2.60}$$

This relation is called Parseval's Theorem.

Also, we know that the signal energy (the area under the squared value of a signal) is analogous to the square of the length of a vector in the vector-signal analogy. In vector space, we know that the square of the length of a vector is equal to the sum of the squares of the lengths of its orthogonal components. Parseval's Theorem [Eqn. 2.60] is the statement of this fact that is applied to signals.

2.9 SAMPLING OF CONTINUOUS-TIME SIGNALS

There are many ways to sample a continuous-time signal. Here, we will discuss only periodic sampling. It is also called *uniform sampling.*

If $s_\alpha(t)$ is a continuous-time signal, periodical measurement of continuous-time signal is called *periodic sampling or uniform sampling.*

By periodic sampling of continuous-time signal, we can get discrete-time signal.

Discrete-time signal, $$s_\alpha(nT_s) \equiv s_\alpha \big|_{t=nT_s}$$

where T is the sampling period and reciprocal of sampling period is termed as sampling frequency F_s.

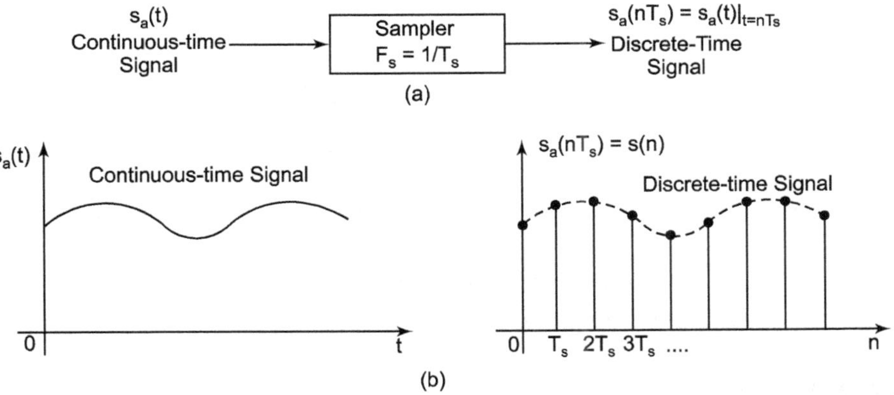

FIGURE 2.27 (*a*) Block diagram of a sampler, (*b*) Periodic sampling of continuous-time signal.

Periodic sampling is done according to sampling theorem given by Sannon, discussed in Sec. 2.9.2.

2.9.1 Nyquist Rate

Nyquist rate is defined as a *minimum sampling rate required for perfect reconstruction of sampled signal at the receiver.*

If any signal has the highest frequency component F_{max}, then

$$\text{Nyquist rate} = 2 \times F_{max}$$

2.9.2 Sampling Theorem

It is stated as follows: *For perfect reconstruction of sampled signal at receiver, sampling rate or sampling frequency should be greater than or equal to Nyquist rate of the message signal.*

According to the sampling theorem,

Sampling rate ≥ Nyquist rate, $2F_{max}$.

Periodic sampling establishes a relationship between the time variables t and n of continuous-time and discrete-time signals, respectively.

Consider a continuous-time signal, $s_\alpha(t) = A_s \cos(2\pi F_{max}\, t + \theta)$

Sampling periodically at a sampling rate $F_s = 1/T_s$ samples per second produces

$$\begin{aligned} s(n) \equiv s_\alpha(nT_s) &= A_s \cos(2\pi\, F_{max}\, nT_s + \theta) \\ &= A_s \cos\left(2\pi F_{max} n \frac{1}{F_s} + \theta\right) \\ &= A_s \cos\left(2\pi \frac{F_{max}}{F_s} n + \theta\right) \\ &= A_s \cos(2\pi f n + \theta), \ -\infty < n < \infty \end{aligned}$$

where $f = \dfrac{F_{max}}{F_s}$ is the frequency variable for discrete-time signals,

F_{max} is the frequency variable for continuous-time signals, and

F_s is the sampling rate.

2.9.3 Aliasing

When sampling frequency is less than Nyquist rate, then aliasing phenomenon occurs.

Nyquist rate = $2F_{max}$ = 2 × Highest frequency component of message signal

If sampling rate < Nyquist rate than it is called under sampling, and in this case, aliasing phenomenon occurs.

If sampling rate > Nyquist rate, then it is called over sampling, and in this case, no aliasing phenomenon occurs; in fact, this is a suitable and necessary condition for sampling process.

Aliasing phenomenon is defined as a phenomenon of high-frequency component in a spectrum of a signal seemingly taking on the identity of a lower frequency in the spectrum of its sampled version.

Figure 2.28 shows spectra of signals showing the sampling relations between analog and digital systems for a properly sampled input signal.

Figure 2.29 shows the effect of under sampling on the digital frequency response.

Aliasing problem occurs when sampling frequency $F_s < 2F_{max}$. In this case, sampling frequency F_s is not sufficiently high to prevent the shifting of high-frequency information into lower frequencies. Such transference of information from one band of frequencies to another is called Aliasing, and the resulting frequency response is called an aliased representation of the original signal.

There are two corrective measures which are used to eliminate aliasing.

1. a pre-alias low-pass filter is used before sampling for attenuating those high frequencies that are not essential for the transmission of information.
2. a pre-alias low-pass filtered signal is sampled at a rate slightly higher than the Nyquist rate ($F_s > 2F_{max}$).

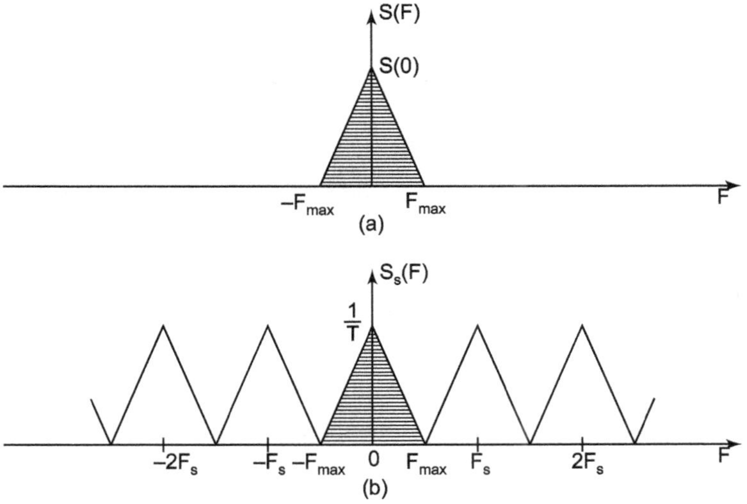

(*a*) Spectrum of a band-limited analog signal $s(t)$. (*b*) Spectrum of a sampled version of signal $s(t)$ for a sampling frequency $F_s = 2F_{max}$.

FIGURE 2.28 Spectrum of signals showing the sampling relations between analog and digital systems for a properly sampled input.

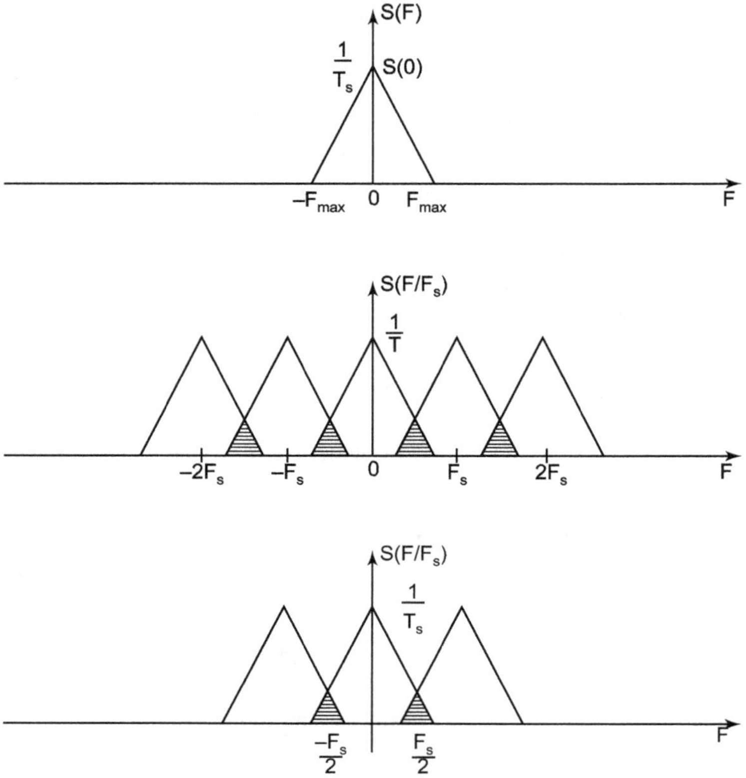

FIGURE 2.29 The effect of under sampling an analog signal on its digital frequency response showing aliasing around the folding frequency $F_s/2$.

EXAMPLE 2.15

Given the continuous-time signal $s_\alpha(t) = 5 \cos 200\, \pi t$. Determine

a. Minimum sampling rate (Nyquist rate) required to avoid aliasing.

b. If $F_s = 400$ Hz. What is the discrete-time signal obtained after sampling?

c. If $F_s = 150$ Hz. What is the discrete-time signal obtained after sampling?

d. What is frequency $0 < F < \frac{F_s}{2}$ of sinusoidal that yields samples identical to those obtained in part (*c*).

Solution:

a. If the signal is sampled at $F_s = 400$ Hz. Frequency of discrete-time signal,

$$f = \frac{\text{Frequency of analog signal, } F}{\text{Sampling frequency, } F_s} = \frac{100}{400} = \frac{1}{4}$$

The discrete-time signal,

$$s(n) = 5 \cos 2\pi f n = 5 \cos 2\pi \frac{1}{4} n = 5 \cos \frac{\pi n}{2}$$

b. The frequency of continuous-time signal,

$$F = 100 \text{ Hz.}$$

Hence, minimum sampling rate required to avoid aliasing is

$$2F = 2 \times 100 = 200 \text{ Hz}$$

c. If the signal is sampled at $F_s = 150$ Hz

Frequency of discrete-time signal, $f = \dfrac{F}{F_s} = \dfrac{100}{150} = \dfrac{2}{5}$

The discrete-time signal

$$s(n) = A_s \cos 2\pi f n = 5 \cos 2\pi \frac{2}{3} n = \cos \frac{4\pi}{3} n$$

$$= 5 \cos \left(2\pi - \frac{2\pi}{3}\right) n = 5 \cos \frac{2\pi n}{3}$$

d. For the sampling rate of $F_s = 150$ Hz

$$f = \frac{F}{F_s} \text{ or } F = f F_s = \times \frac{1}{3} \; 150 = 50 \text{ Hz}$$

The sinusoidal signal

$$y_\alpha(t) = 5 \cos 2nFt = 5 \cos 2\pi 50t = 5 \cos 100\pi t$$

Sampled at $F_s = 150$ Hz, yields identical samples

Hence, $F = 100$ Hz is an alias of $F = 50$ Hz for sampling rate $F_s = 150$ Hz.

EXAMPLE 2.16

Given the analog signal

$$s_\alpha(t) = 5 \cos 50\pi t + 20 \sin 300\pi t - 10 \cos 100\pi t$$

What is the Nyquist rate of this signal?

Solution: The frequencies present in the signal above

$$F_1 = \frac{\omega_1}{2\pi} = \frac{50\pi}{2\pi} = 25 \text{ Hz}$$

$$F_2 = \frac{\omega_2}{2\pi} = \frac{300\pi}{2\pi} = 150 \text{ Hz}$$

$$F_3 = \frac{\omega_3}{2\pi} = \frac{100\pi}{2\pi} = 50 \text{ Hz}$$

The highest frequency component of the given message signal

$$F_{max} = 150 \text{ Hz}$$

$$\therefore \qquad \text{Nyquist rate} = 2F_{max} = 2 \times 150 = 300 \text{ Hz.}$$

2.10 RECONSTRUCTION OF A SIGNAL FROM ITS SAMPLE VALUES

Interpolation is the commonly used procedure for reconstructing a signal from its sample values. Interpolation gives either approximate or exact reconstruction of the signal. One simple interpolation procedure is zero-order hold. Another useful form of interpolation is linear interpolation.

In linear interpolation, the adjacent samples (sample points) are connected by straight lines as shown in Figure 2.30. We can also use higher-order interpolation formulae for reconstructing the signal from its sample values.

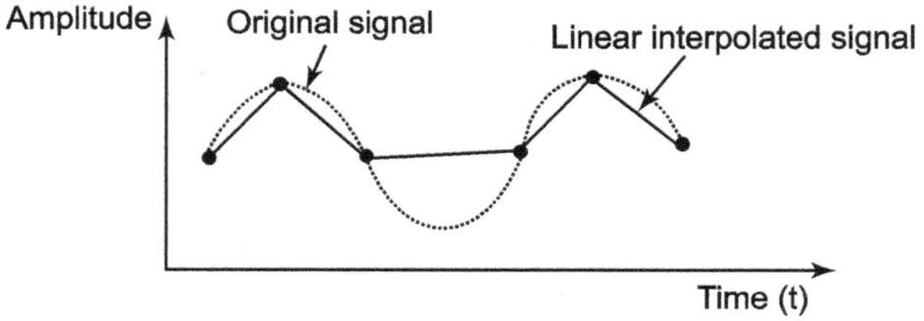

FIGURE 2.30 Linear interpolation between sample points.

For a band-limited signal, if the sampling instants are sufficiently close, then the signal can be reconstructed exactly by using a low-pass filter. Exact interpolation can be carried out between various sample points. The interpolation of the reconstruction of $s(t)$ as a process becomes evident when we consider the effect in the time domain of a low-pass filter. It is shown in Figure 2.31.

Output of low-pass filter, *i.e.*, reconstructed signal is given by

$$s_r(t) = s_s(t) * h(t) \tag{2.61}$$

where $s_s(t)$ is sampled signal, and it is given by

$$s_s(t) = \sum_{n=-\infty}^{\infty} s(nT_s)\delta(t - nT_s) \tag{2.62}$$

and $h(t)$ is impulse response of low-pass filter.

Reconstructed signal is given by

$$s_r(t) = \sum_{n=-\infty}^{\infty} s(nT_s)h(t - T_s) \tag{2.63}$$

Equation (2.63) describes a method to fit a continuous curve between the sample points $s(nT_s)$, and consequently, it is a interpolation formula.

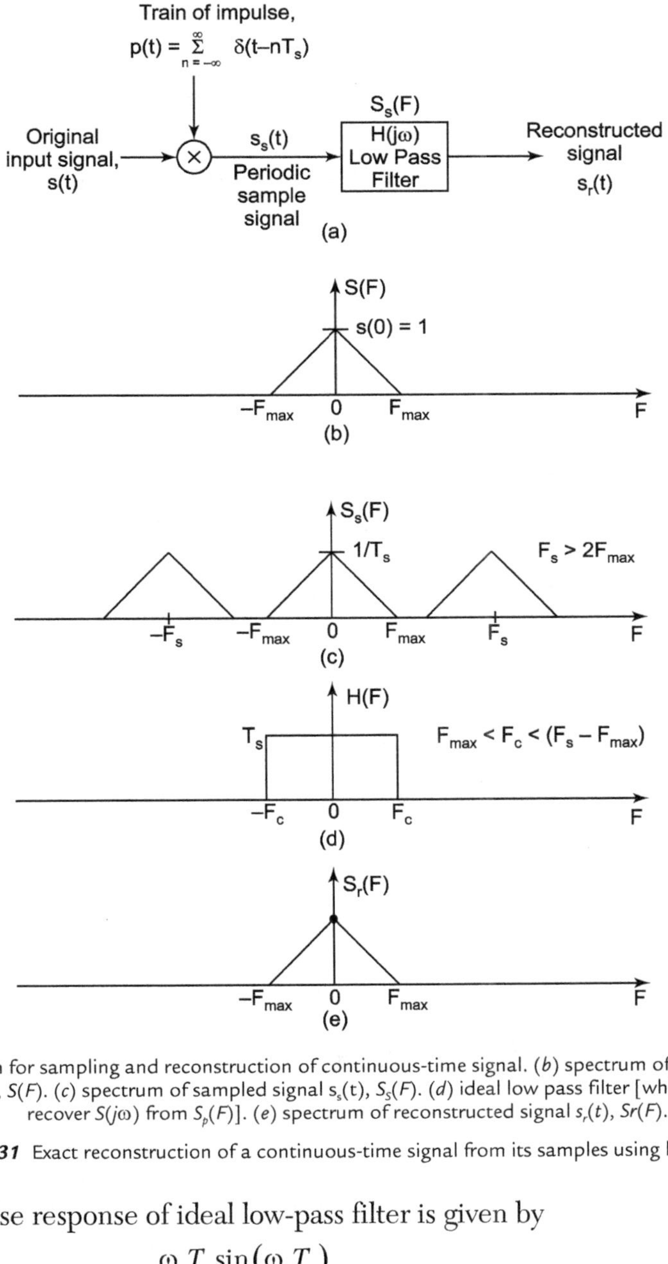

(a) System for sampling and reconstruction of continuous-time signal. (b) spectrum of original input signal $s(t)$, $S(F)$. (c) spectrum of sampled signal $s_s(t)$, $S_S(F)$. (d) ideal low pass filter [which is used to recover $S(j\omega)$ from $S_p(F)$]. (e) spectrum of reconstructed signal $s_r(t)$, $Sr(F)$.

FIGURE 2.31 Exact reconstruction of a continuous-time signal from its samples using low-pass filter.

Impulse response of ideal low-pass filter is given by

$$h(t) = \frac{\omega_c T_s \sin(\omega_c T_s)}{\pi \omega_c T_s} \tag{2.64}$$

where $\omega_c = 2\pi F_c$ is cut-off frequency of low-pass filter.

Substituting Eqn. (2.64) in Eqn. (2.63), we get

$$sr(t) = \sum_{n=-\infty}^{\infty} s(nT_s)h(t-nT_s)$$

$$= \sum_{n=-\infty}^{\infty} s(nT_s)\left\{\frac{\omega_c T_s}{\pi}\frac{\sin\left[\omega_c(t-nT_s)\right]}{\omega_c(t-nT_s)}\right\} \tag{2.65}$$

The reconstruction of a signal according interpolation formula of Eqn. (2.65) is shown in Figure 2.32, where $\omega_c = \frac{\omega_s}{2} = \frac{2\pi F_c}{2} = \pi F_s$.

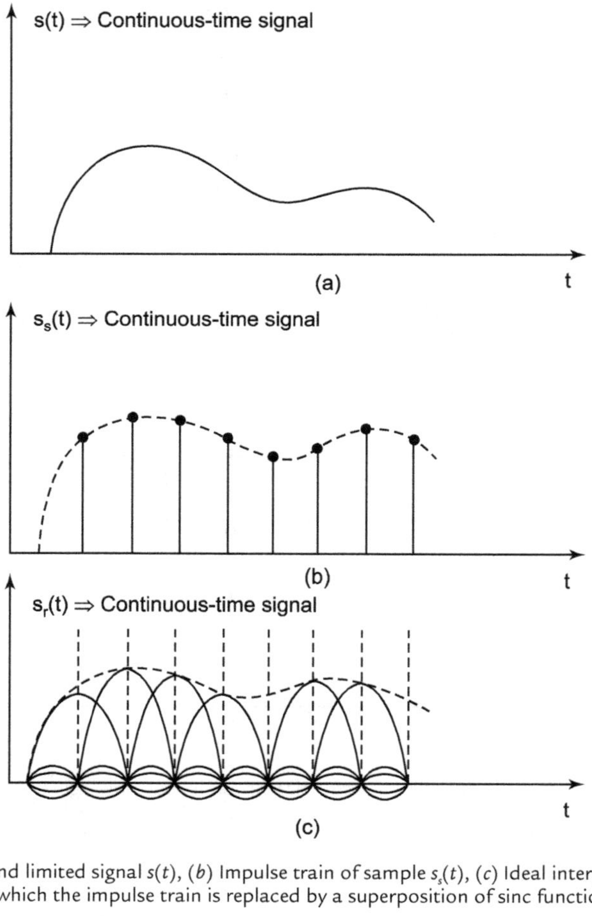

(*a*) Band limited signal $s(t)$, (*b*) Impulse train of sample $s_s(t)$, (*c*) Ideal interpolation in which the impulse train is replaced by a superposition of sinc functions.

FIGURE 2.32 Ideal interpolation for band-limited signal using sinc functions.

Interpolation using the $h(t)$, *i.e.*, impulse response of a low-pass filter as in Eqn. (2.64) is called band-limited interpolation. It implements exact reconstruction if $s(i)$ is band limited and it satisfies the condition of sampling theorem. According to sampling theorem for perfect reconstruction of a band-limited signal, sampling frequency should be greater than twice the higher frequency component of the signal. We prefer simpler interpolating function such as zero-order hold. It can be viewed as a form of interpolation between sample values in which the interpolating function $h(t)$ is the impulse response.

EXERCISES

1. Define discrete-time signals. How are these signals represented?
2. What are the elementary discrete-time signals?
3. What meant by manipulation of discrete-time signals? In how many ways, such types of manipulation of discrete-time sequences can be performed?
4. What are discrete-time systems? Give some examples of discrete-time systems.
5. Give classification of discrete-time systems.
6. Differentiate between the following systems:
   a. Static systems and dynamic systems.
   b. Time-invariant systems and time-varying systems.
   c. Linear systems and non-linear systems.
   d. Causal systems and non-causal systems.
   e. Stable systems and unstable systems.
7. Discuss convolution sum. Write properties of convolution sum for LTI systems. Summarize the computing of convolution sum.
8. What is the criterion of stability for LTI systems?
9. What are IIR and FIR LTI systems?
10. Differentiate between recursive and non-recursive discrete-time systems.
11. What is unit-sample (impulse) response?

12. What is correlation of two discrete-time signals? Give one example where correlation is used.
13. Differentiate auto-correlation and cross-correlation sequences.
14. What is sampling? Give the statement of sampling theorem.
15. Define Aliasing. How aliasing is eliminated?
16. Define inverse systems.
17. What is deconvolution? Discuss with the help on example.
18. Explain the reconstruction of a signal from its sample values.
19. Define interpolation which is used for reconstructing a signal from its sample values.

NUMERICAL EXERCISES

1. Determine the response (output) of the following systems to the input signal.

$$s(n) = \begin{cases} |n|, & -2 \le n \le 2 \\ 0, & \text{elsewhere} \end{cases}$$

   a. $y(n) = 5\,s(n)$ b. $y(n) = 2s(n-1)$

   c. $y(n) = 3s(n+1)$ d. $y(n) = \frac{1}{3}\,[s(n+1) + s(n) + s(n-1)]$

2. Check the systems for time invariance

   a. $y(n) = s(n) - s(n-1)$ b. $y(n) = ns(n)$

   c. $y(n) = 5s(-n)$ d. $y(n) = s(n)\cos\omega_0 n$

3. Check the systems for linearity

   a. $y(n) = ns(n)$ b. $y(n) = 2s(n^2)$

   c. $y(n) = As(n) + B$

4. Check the following systems for causality

   a. $y(n) = s(n) - s(n-1)$ b. $y(n) - As(n)$

   c. $y(n) = s(n) + 5s(n+3)$ d. $y(n) = s(n^2)$

   e. $y(n) = s(2n)$

5. Resolve the following sequences into a sum of weighted impulse sequences of

   a. $s(n) = \{2, 4, \underset{\uparrow}{1}, 3\}$ b. $s(n) = \{0, \underset{\uparrow}{1}, 3, 4\}$

   c. $s(n) = \{-1, 2, 3, \underset{\uparrow}{4}, 5\}$

6. Compute the convolution sum of the following sequences:

   a. $s(n) = \{1, 2, \underset{\uparrow}{4},\}$, $h(n)$ $\{1, \underset{\uparrow}{1}, 1, 1\}$

   b. $s(n) = \{1, \underset{\uparrow}{2}, -1,\}$, $h(n) = s(n)$

   c. $s(n) = \{1, \underset{\uparrow}{2}, 3, 4,\}$, $h(n)$ $\{\underset{\uparrow}{1}\}$

   d. $s(n) = \left(\frac{1}{3}\right)^n u\ (n), h(n) = \left(\frac{1}{4}\right)^n u(n)$

   e. $s(n) = A^n\ u(n), h(n) = B^n u(n)$ where $A = B$ and $A \neq B$

   f. $s(n) = \begin{cases} A^n, & -2 \leq n \leq 2 \\ 0, & \text{elsewhere} \end{cases}, h(n) = \begin{cases} 1, & 0 \leq n \leq 2 \\ 0, & \text{elsewhere} \end{cases}$

7. Check the system for stability with impulse response $h(n) - A^n u(n)$.

8. Compute the correlation of the sequences given in problem 6.

9. Given the analog signal, $s_\alpha(t) - 5 \cos 300\ \pi t$

   a. Determine the minimum sampling rate required to avoid aliasing.

   b. If signal is sampled at 600 Hz. What will be discrete time signal obtained after sampling?

   c. If signal is sampled at 225 Hz. What will be discrete-time signal obtained after sampling?

   d. What is the frequency $0 < F < \frac{F_s}{2}$ of a sinusoid that yields samples identical to those obtained in part (c).

5. Given a continuous-time signal

$$s_\alpha(t) = 5 \cos 80\pi t + 10 \sin 300\pi t - \cos 100\pi t$$

   What will be the Nyquist rate of this signal?

CHAPTER 3

THE Z-TRANSFORM

3.1 INTRODUCTION

Laplace transform is an extension of the continuous-time Fourier transform. This extension was motivated by the fact that the Laplace transform can be applied to a broader class of signals than the Fourier transform can. Since there are many signals for which the Fourier transform does not converge but it does for the Laplace transform.

z-transform for discrete-time signals is the discrete-time counterpart of the Laplace transform for continuous-time signals, and both of them have a similar relationship to the corresponding Fourier transform. The motivation for introducing generalization of Discrete-time Fourier transform (DTFT) are:

1. DTFT does not converge for all sequences but its generalization (z-transform) can be applied to a broader class signals.

2. A second advantage is that in analytical problems the z-transform notation is often conveniently used than the DTFT.

It is worth noting that transform techniques are used for the analysis of various signals and linear time-invariant (LTI) systems.

Comparison between Transforms and Logarithms: In various transforms, such as Laplace transform, Fourier transform, z-transform, etc., we use one transformation formula which converts signals or sequences from time-domain to another corresponding domain depending upon a particular transform, that domain is easy and suitable for computations and also we can extract more information in that domain which is necessary for signal processing.

We use an inverse transform formula to again convert back that signal into time-domain.

Purpose of Transformation: There are two purposes of transformation

a. For extracting more information from the transformed domain (e.g., frequency domain) for the purpose of signal processing.

b. For the purpose of simplification of computations in that domain. A logarithm is used for simplification and making easy computations and it is not used for the purpose of extracting any information from a sequence or signal that was done in transformation techniques.

As we know, transform techniques are used for the analysis of signals and LTI systems. The z-transform plays the same role in the analysis of discrete-time signals and LTI systems as the Laplace transform plays in the analysis of continuous-time signals and LTI systems.

3.2 DEFINITION OF THE *z*-TRANSFORM

z-transforms are of two types:

1. Two-sided z-transform
2. One-sided z-transform.

The two-sided z-transform of a discrete-time signal $s(n)$ is defined as the power series

$$S(z) = z\text{-transform of sequence} s(n)$$

$$= \mathcal{Z}[s(n)] = \sum_{n=-\infty}^{\infty} s(n)z^{-n} \tag{3.1}$$

where z is a complex variable.

$$z = re^{j\omega}$$

where $r = |z|$ = magnitude of z and w is the angle of z.

It is also called bilateral z-transform.

One-sided z-transform of a discrete-time signal $s(n)$ is defined as

$$S(Z) = \mathcal{Z}[s(n)] = \sum_{n=0}^{\infty} s(n)z^{-n} \tag{3.2}$$

It is also called unilateral z-transform.

The difference between two-sided z-transform and one-sided z-transform is that the lower limit of summation is zero for one-sided z-transform and $-\infty$ for two-sided z-transform. Generally, a one-sided z-transform is used for solving difference equations of discrete-time LTI systems.

For causal sequences the two-sided and one-sided z-transforms are equivalent. Two-sided z-transform can be defined only for a particular region of convergence (ROC). The ROC is not important for one-sided z-transform. One-sided z-transform is used for solving linear difference equations with non-zero initial conditions.

3.3 REGION OF CONVERGENCE (ROC)

For some value of z, the power series in Eq. (3.1) does not converge to a finite value.

The portion of the z-plane for which the series in Eq. (3.1) converges is called the ROC. The ROC depends upon the magnitude of z. The ROC cannot contain any poles, since the series becomes infinite at the poles. The ROC can be a circle, interior of a circle, exterior of a circle, an annulus, or the entire z-plane.

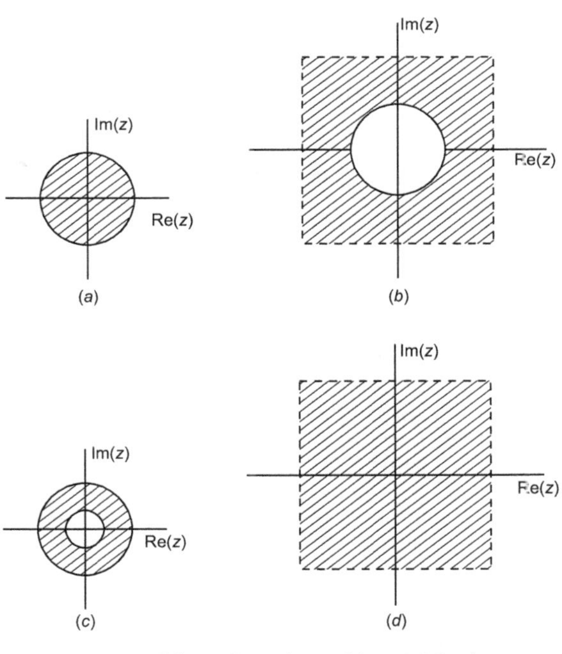

FIGURE 3.1 Possible configurations of the ROC for the z-transform. (a) Interior of a circle (b) Exterior of a circle (c) An annulus (d) The entire z-plane.

3.3.1 Possible Configurations of the ROC for the *z*-Transform

These configurations are:

1. interior of a circle [Figure 3.1(a)],
2. exterior of a circle [Figure 3.1(b)],

3. an annulus [Figure 3.1(*c*)], and
4. the entire z-plane [Figure 3.1(*d*)].

3.3.2 Properties of ROC for *z*-Transform

Properties of the ROC for the two-transform are given in Table 3.1.

TABLE 3.1.

	Discrete-time sequence, $s(n)$	ROC for its z-transform, $S(z)$
1.	Unit-sample sequence, $\delta(n)$	Entire z-plane
2.	Finite-duration causal sequence	Entire z-plane except $z = 0$
3.	Finite-duration sequence with $s(n) \neq 0$ for some $n < 0$ $s(n) = 0$ for all $n > 0$	Entire z-plane except for $z = \infty$
4.	Finite-duration sequence with $s(n) \neq 0$ for some $n < 0$ $s(n) = 0$ for some $n > 0$	Entire z-plane except for $z = 0$ and $z = \infty$
5.	Right-sided sequence $s(n) = 0$ for some $n < 0$	Outward from outermost pole
6.	Right-sided sequence $s(n) \neq 0$ for some $n < 0$	Outward from outermost pole, $z = \infty$ is excluded
7.	Left-sided sequence $s(n) = 0$ for all $n > 0$	Inward from innermost pole
8.	Left-sided sequence $s(n) \neq 0$ for some $n > 0$	Inward from innermost pole, $z = 0$ is excluded
9.	Two-sided sequence	Annulus

EXAMPLE 3.1

Determine the z-transforms of the following finite-duration sequences:

(*a*) $s(n) = \{\underset{\uparrow}{2}, 4, 5, 6, 1\}$ (*b*) $s(n) = \{1, 2, \underset{\uparrow}{3}, 4, 5\}$

(*c*) $s(n) = \{\underset{\uparrow}{0}, 0, 1, 2, 3\}$ (*d*) $s(n) = \delta(n)$

(*e*) $s(n) = \delta(n - k), k > 0$ (*f*) $s(n) = \delta(n + k), k > 0$

Solution:

(*a*) *z*-transform is defined as

$$S(z) = \mathcal{Z}[s(n)] = \sum_{n=-\infty}^{\infty} s(n)z^{-n}$$

Given $s(n) = \{2, 4, 5, 6, 1\}$ (↑ at 2)

or $$S(z) = \sum_{n=-\infty}^{\infty} s(n)z^{-n} = \sum_{n=0}^{4} s(n)z^{-n}$$

$$= s(0)z^{-0} + s(1)z^{-1} + s(2)z^{-2} + s(3)z^{-3} + s(4)z^{-4}$$

$$= 2.1 + 4z^{-1} + 5z^{-2} + 6z^{-3} + 1z^{-4}$$

$$= 2 + 4z^{-1} + 5z^{-2} + 6z^{-3} + z^{-4}$$

ROC: Entire *z*-plane except $z = 0$

(*b*) Given $s(n) = \{1, 2, 3, 4, 5\}$ (↑ at 3)

z-transform is defined as

$$S(z) = \mathcal{Z}[s(n)] = \sum_{n=-\infty}^{\infty} s(n)z^{-n} = \sum_{n=-2}^{2} s(n)z^{-n}$$

$$= s(-2)z^{2} + s(-1)z^{1} + s(0)z^{0} + s(1)z^{-1} + s(2)z^{-2}$$

$$= 1z^{2} + 2z^{1} + 3.1 + 4z^{-1} + 5z^{-2}$$

$$= z^{2} + 2z + 3 + 4z^{-1} + 5z^{-2}$$

ROC: Entire *z*-plane except $z = \infty$ and $z = 0$

(*c*) Given $s(n) = \{0, 0, 1, 2, 3\}$ (↑ at first 0)

z-transform is defined as

$$S(z) = \mathcal{Z}[s(n)] = \sum_{n=-\infty}^{\infty} s(n)z^{-n} = \sum_{n=0}^{4} s(n)z^{-n}$$

$$= s(0)z^{-0} + s(1)z^{-1} + s(2)z^{-2} + s(3)z^{-3} + s(4)z^{-4}$$

$$= 0.z^{0} + 0.z^{-1} + 1.z^{-2} + 2.z^{-3} + 3.z^{-4}$$

$$= z^{-2} + 2z^{-3} + 3z^{-4}$$

ROC: Entire z-plane except $z = 0$

(d) Given $s(n) = d(n) = \begin{cases} 1, & n = 0 \\ 0, & n \neq 0 \end{cases}$

z-transform is defined as

$$S(z) = \mathcal{Z}[s(n)] = \sum_{n=-\infty}^{\infty} s(n) z^{-n}$$
$$= \ldots + s(-2)z^{2} + s(-1)z^{-1} + s(0)z^{0} + s(1)z^{-1} + s(2)z^{-2} + \ldots$$
$$= \ldots + 0z^{2} + 0.z^{1} + 1.z^{0} + 0.z^{-1} + 0.z^{-2} + \ldots = 1$$

or $\quad S(z) = 1$

ROC: Entire z-plane

(e) Given $s(n) = \delta(n-k), k > 0$

z-transform is defined as

$$S(z) = \mathcal{Z}[s(n)] = \sum_{n=-\infty}^{\infty} s(n) z^{-n}$$
$$= \sum_{n=-\infty}^{\infty} \delta(n-k) z^{-n}$$

Substituting $m = n - k$
or $\quad n = m + k$, we get

$$S(z) = \sum_{m=-\infty}^{\infty} \delta(m) z^{-(m+k)} = \sum_{m=-\infty}^{\infty} \delta(m) z^{-n} z^{-k}$$
$$= z^{-k} \sum_{m=-\infty}^{\infty} \delta(m) z^{-m} = z^{-k}.1 = z^{-k}$$

where $\sum_{m=-\infty}^{\infty} \delta(m) z^{-m} = 1$

$$S(z) = \mathcal{Z}[\delta(n-k)] = z^{-k}$$

ROC: Entire z-plane except $z = 0$.

(f) Given $s(n) = \delta(n+k), k > 0$

z-transform is defined as

$$S(Z) = \sum_{m=-\infty}^{\infty} \delta(m) z^{-(m-k)} = \sum_{m=-\infty}^{\infty} \delta(m) z^{-m} z^{k}$$

Substituting $m = n + k$
or $n = m - k$, we get

$$S(z) = \sum_{m=-\infty}^{\infty} \delta(m) z^{-(m-k)} = \sum_{m=-\infty}^{\infty} \delta(m) z^{-m} z^{k}$$

$$= z^{k} \sum_{m=-\infty}^{\infty} \delta(m) z^{-m} = z^{k}.1 = z^{k}$$

where $\sum_{m=-\infty}^{\infty} \delta(m) z^{-m} = 11$

$$S(z) = \mathcal{Z}[\delta(n+k)] = z^{k}$$

ROC: Entire z-plane except $z = \infty$.

EXAMPLE 3.2

Determine the z-transform of the following discrete-time signals

(a) $s(n) = u(n)$	(b) $s(n) = A^{n} u(n)$
(c) $s(n) = -A^{n} u(-n-1)$	(d) $s(n) = \cos(\omega_0 n) u(n)$
(e) $s(n) = \sin(\omega_0 n) u(n)$	(f) $s(n) = A^{n} u(n) + B^{n} u(-n-1)$
(g) $s(n) = nu(n)$	(h) $s(n) = u(n-1)$.

Solution:

(a) Given $s(n) = u(n)$ and $u(n)$ is defined as

$$u(n) = \begin{cases} 1, & n \geq 0 \\ 0, & n < 0 \end{cases}$$

z-transform is defined as

$$S(n) = \mathcal{Z}[s(n)] = \sum_{n=-\infty}^{\infty} s(n) z^{-n}$$

$$= \sum_{n=-\infty}^{\infty} u(n) z^{-n} = \sum_{n=0}^{\infty} 1.z^{-n}$$

$$= z^{-0} + z^{-1} + z^{-2} + \ \ldots \ + z^{-\infty} = \frac{1}{1 - z^{-1}}$$

[Sum of geometric series for infinite number of points is given by

$$S_{n=\infty} = \frac{\text{First point}}{1 - \text{common ratio}} \Bigg]$$

This is a sum of geometric series for infinite number of points.
ROC: $|z^{-1}| < 1$
or $|z| > 1$

(b) Given $s(n) = A^n u(n)$

z-transform is defined as

$$S(z) = \mathcal{Z}[s(n) = \sum_{n=-\infty}^{\infty} s(n)z^{-n} = \sum_{n=-\infty}^{\infty} A^n u(n) z^{-n}$$

$$= \sum_{n=0}^{\infty} A^n z^{-n} = \sum_{n=0}^{\infty} \left(Az^{-1}\right)^n$$

where $u(n) = \begin{cases} 1, & n \geq 0 \\ 0, & n < 0 \end{cases}$

(This is a geometric series for infinite number of points)

or $\quad S(z) = (Az^{-1})^0 + (Az^{-1})^1 + (Az^{-2})^2 + \ldots + (Az^{-1})\infty$

$$= \frac{1}{1 - Az^{-1}}$$

ROC: $|Az^{-1}| < 1$
or $|z| > |A|$
where A is a scalar quantity. ROC of the z-transform of the sequence

$$s(n) = A^n u(n) \text{ is } |z| > |A|$$

This is a causal sequence.
The ROC is the exterior of a circle having radius $|A|$.

ROC of the z-transform of the sequence $s(n) = A^n u(n)$ can be shown graphically on the z-plane and is illustrated in Figure 3.2.

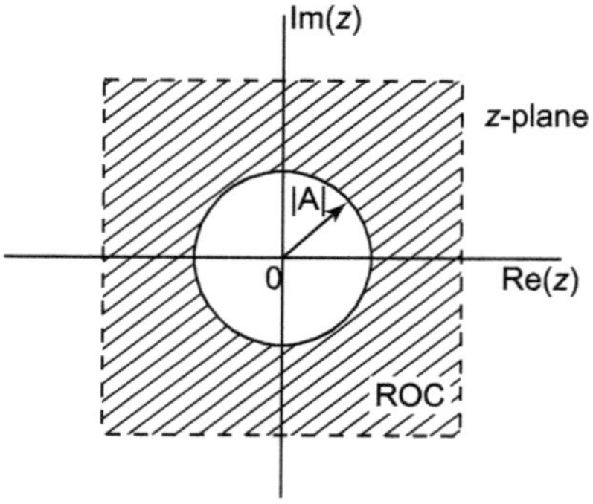

FIGURE 3.2 ROC of the z-transform of the sequence $s(n) = A^n u(n)$.

(c) Given $s(n) = -A_n u(-n-1)$

$$= \begin{cases} 0, & n \geq 1 \\ -A^n, & n < 0 \end{cases}$$

z-transform is defined as

$$s(z) = \mathcal{Z}[s(n)] = \sum_{n=-\infty}^{\infty} s(n)z^{-n}$$

$$= \sum_{n=-\infty}^{\infty} (-A^n u)z^{-n}$$

$$= \sum_{n=-\infty}^{-1} \left(A^{-1}z\right)z^{-n} = \sum_{n=-\infty}^{-1} \left(A^{-1}z^{-1}\right)^{-n}$$

Substituting $l = -n$

$$= -\sum_{l=\infty}^{1} \left(A^{-1}z\right)^{l} = \sum_{l=1}^{\infty} \left(A^{-1}z\right)^{l}$$

$$= -\left[A^{-1}z)^{1} + A^{-1}z)^{2} + A^{-1}z)^{3} + \ldots\right]$$

$$= \frac{-A^{-1}z}{-A^{-1}z}$$

(Sum of geometric series for infinite number of points).

ROC: $|A^{-1}z| < 1$
or $|z| < |A|$
The ROC is interior of a circle having radius $|A|$.
Above sequence,

$$s(n) = -A^n u(-n-1)$$

is a non-causal sequence.

ROC of the z-transform of the above sequence can be shown graphically on the z-plane. Its ROC is shown in Figure 3.3.

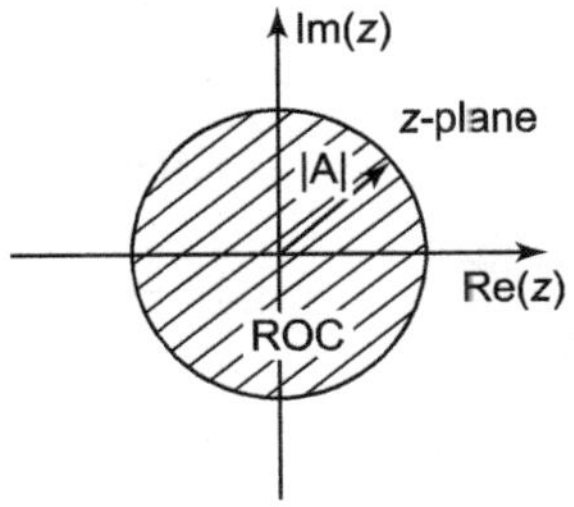

FIGURE 3.3 ROC of the z-transform of above sequence $s(n) = -A^n u(n-1)$

(d) Given $s(n) = \cos \omega_0 n u(n)$

$$= \left\{\frac{e^{j\omega_0 n} + e^{-j\omega_0 n}}{2}\right\} u(n)$$

z-transform is given as

$$S(z) = \mathcal{Z}[s(n)] = \sum_{n=-\infty}^{\infty} s(n)z^{-n}$$

$$= \sum_{n=-\infty}^{\infty} \cos\left(\omega_0 n\right) u(n)z^{-n} = \sum_{n=-\infty}^{\infty} \left\{\frac{e^{j\omega_0 n} + e^{-j\omega_0 n}}{2}\right\} u(n)z^{-n}$$

$u(n)$ is defined as

$$u(n) = \begin{cases} 1, & n \geq 0 \\ 0, & n < 0 \end{cases}$$

or

$$S(z) = \frac{1}{2}\sum_{n=0}^{\infty} e^{j\omega_0 z^n} + \frac{1}{2}\sum_{n=0}^{\infty} e^{-j\omega_0 z^{-n}}$$

$$= \frac{1}{2}\sum_{n=0}^{\infty}\left(e^{j\omega_0 n z^{-1}}\right)^n + \frac{1}{2}\sum_{n=0}^{\infty}\left(e^{j\omega_0 z^{-1}}\right)^n$$

$$= \frac{1}{2}\left[\frac{1}{1 - e^{j\omega_0 z^{-1}}}\right] + \frac{1}{2}\left[\frac{1}{1 - e^{-j\omega_0 z^{-1}}}\right]$$

$$= \frac{1 - z^{-1}\cos\omega_0}{1 - 2z^{-1}\cos\omega_0 + z^{-2}}, \text{ ROC}: |z| > 1$$

ROC: $\left|e^{j\omega_0 z^{-1}}\right| < 1 \Rightarrow |z| > \left|e^{j\omega_0}\right| > 1$

and $\left|e^{j\omega_0} z^{-1}\right| < 1 \Rightarrow |z| > \left|e^{j\omega_0}\right| > 1$

If we get $\left|e^{\pm j\omega_0}\right| = 1$.

ROC of the z-transform of the sequence $s(n) = \cos(\omega_0 n)\, u(n)$ is exterior to the unit circle. It can be shown graphically on z-plane. Its ROC is shown in Figure 3.4.

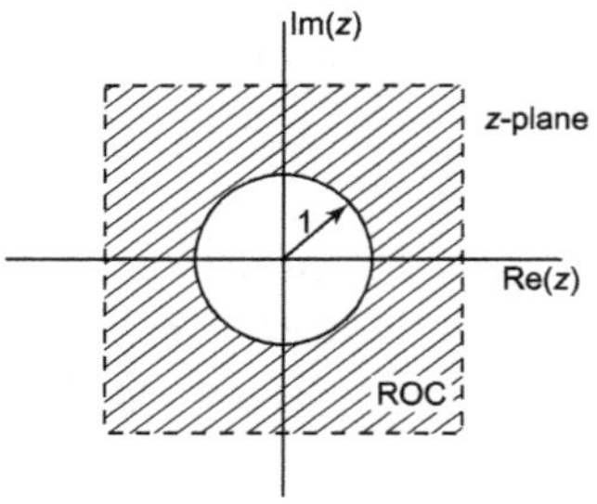

FIGURE 3.4 ROC of the transform of the sequence $s(n) = \cos(\omega_0 n)\, u(n)$.

(*e*) Given $s(n) = \sin(\omega_0 n)\, u(n) = \left[\frac{e^{j\omega_0 n} - e^{-j\omega_0 n}}{2j}\right] u(n)$

z-transform is given as

$$S(z) = \mathcal{Z}[s(n)] = \sum_{n=-\infty}^{\infty} s(n) z^{-n}$$

$$= \sum_{n=-\infty}^{\infty} \sin(\omega_0 n) u(n) z^{-n}$$

$$= \sum_{n=-\infty}^{\infty}\left[\frac{e^{j\omega_0 n} - e^{-j\omega_0 n}}{2j}\right] u(n) z^{-n}$$

$u(n)$ is defined as $u(n) = \begin{cases} 1, & n \geq 0 \\ 0, & n < 0 \end{cases}$

or
$$S(z) = \frac{1}{2j}\sum_{n=0}^{\infty}\left(e^{j\omega_0}z^{-1}\right) - \frac{1}{2j}\sum_{n=0}^{\infty}\left(e^{-j\omega_0}z^{-1}\right)^n$$
$$= \frac{1}{2j}\left[\frac{1}{1-e^{j\omega_0}z^{-1}}\right] - \frac{1}{2j}\left[\frac{1}{1-e^{-j\omega_0}z^{-1}}\right]$$
$$= \frac{z^{-1}\sin\omega_0}{1-2z^{-1}\cos\omega_0 + z^{-2}} \text{ROC}: |z| > 1$$

Its ROC is also shown in Figure 3.4.

(*f*) Given $s(n) = A^n u(n) + B^n u(-n-1)$

z-transform is defined as

$$S(z) = \mathcal{Z}[z(n)] = \sum_{n=-\infty}^{\infty} s(n)z^{-n}$$

or $$S(z) = \sum_{n=-\infty}^{\infty}\left[A^n u(n) + B^n u(-n-1)\right]z^{-n}$$

or $$S(z) = \sum_{n=-\infty}^{\infty} A^n z^{-n} u(n) + \sum_{n=-\infty}^{\infty} B^n z^{-n} u(-n-1) \quad (1)$$

where, $$u(n) = \begin{cases} 1, & n \geq 0 \\ 0, & n < 0 \end{cases}, \quad u(-n-1) = \begin{cases} 1, & n \leq -1 \\ 0, & n \geq 0 \end{cases}$$

Substituting these values in above Eq. (1), we get

$$S(z) = \sum_{n=0}^{\infty} A^n z^{-n} + \sum_{n=-\infty}^{-1} B^n z^{-n}$$

or $$S(z) = \underbrace{\sum_{n=0}^{\infty}\left(Az^{-1}\right)^n}_{\text{1st part}} + \underbrace{\sum_{n=-\infty}^{-1}\left(Bz^{-1}\right)^n}_{\text{2nd part}}$$

In second part, putting $m = -n$

$$S(z) = \sum_{n=0}^{\infty}\left(Az^{-1}\right)^n + \sum_{m=\infty}^{1}\left(Bz^{-1}\right)^{-m}$$

or $$S(z) = \sum_{n=0}^{\infty}\left(Az^{-1}\right)^n + \sum_{m=1}^{\infty}\left(Bz^{-1}\right)^{-m}$$

$$\text{or}\quad S(z) = \sum_{n=0}^{\infty}\left(Az^{-1}\right)^{n} + \sum_{m=1}^{\infty}\left(B^{-1}z\right)^{-m}$$

Ist series IInd series

The first power series converges if $|A^{z^{-1}}| < 1 \Rightarrow |z| > |A|$
The second power series converges if $|B^{-1}z| < 1 \Rightarrow |z| < |B|$
For convergence of $S(z)$, we consider two different cases:

Case I: If $|B| < |A|$. In this case, the two ROCs above do not overlap, as shown in Figure 3.5(a). Consequently, we cannot find values of z for which both power series In this case, we cannot converge simultaneously to determine $S(z)$.

Case II: If $|B| < |A|$. In this case, there is an annulus region in the z-plane where both power series converge simultaneously as shown in Figure 3.5(b).

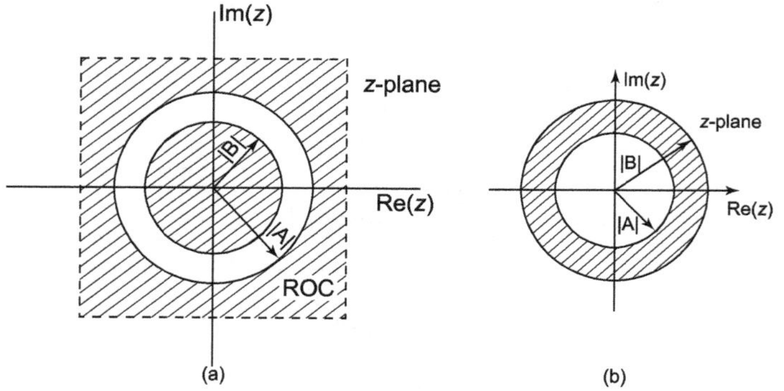

FIGURE 3.5 (a) if $|B| < |A|$ then $S(z)$ does not exist (b) ROC for a z-transform of the sequence $s(n) = A^{n}u(n) + B^{n}u(-n-1)$ if $|B| > |A|$.

In this case, we can obtain $S(z)$.

$$\begin{aligned} S(z) &= \frac{1}{1-Az^{-1}} + \frac{B^{-1}z}{1-B^{-1}z} \\ &= \frac{1-B^{-1}z+B^{-1}z-AB^{-1}}{1-Az^{-1}-B^{-1}z+AB^{-1}} \\ &= \frac{B-A}{B-ABz^{-1}-z+A} = \frac{B-A}{B+A-z-ABz^{-1}} \end{aligned}$$

The ROC of the $S(z)$ is $|A| < |z| < |B|$.

(g) Given $s(n) = r(n)$. It is a ramp sequence and given as $s(n) = nu(n)$ z-transform is defined as

$$S(z) = \mathcal{Z}[s(n)] = \sum_{n=-\infty}^{\infty} s(n)z^{-n} = \sum_{n=-\infty}^{\infty} r(n)z^{-n}$$

$$= \sum_{n=-\infty}^{\infty} nu(n)z^{-n}$$

From the definition of $u(n) = \begin{cases} 1, & n \geq 0 \\ 0, & n < 0 \end{cases}$, substituting this in Eq. (1), we get

$$S(z) = \sum_{n=0}^{\infty} nz^{-n}$$

Now expanding the above power series, we get

$$\begin{array}{rcl} S(z) & = & 0z^{-0} + 1z^{-1} + 2z^{-2} + 3z^{-3} + 4z^{-4} + \ldots \\ z^{-1}S(z) & = & 0z^{-1} + 1z^{-2} + 2z^{-3} + 3z^{-4} + \ldots \\ - & & - \quad - \quad - \\ \hline \end{array}$$

Subtracting,

$$S(z) - z^{-1}S(z) = z^{-1} + z^{-2} + z^{-3} + z^{-4} + \ldots$$

$$S(z)[1 - z^{-1}] = \frac{z^{-1}}{1 - z^{-1}}$$

[Sum of geometric series of infinite number of points]

$$S(z)[1 - z^{-1}] = \frac{z^{-1}}{1 - z^{-1}}$$

$$\text{or} \quad S(z) = \frac{z^{-1}}{\left(1 - z^{-1}\right)^2} = \frac{z}{(1 - z)^2}$$

ROC: $|z^{-1}| < 1$
or $|z| > 1$

(h) Given $s(n) = u(n - 1)$
z-transform is defined as

$$S(z) = \mathcal{Z}\left[s(n)\right] = \sum_{n=-\infty}^{\infty} s(n)z^{-n} = \sum_{m=-\infty}^{\infty} u(n - 1)z^{-1}$$

Substituting $m = n - 1$

or $\qquad n = m + 1$

or $$S(n) = \sum_{m=-\infty}^{\infty} u(m) z^{-(m+1)} = \sum_{n=-\infty}^{\infty} u(n-1) z^{-m} z^{-n}$$

$$= z^{-1} \sum_{m=-\infty}^{\infty} u(m) z^{-m}, \text{ from } u(m) \begin{cases} 1, & m \geq 0 \\ 0, & m < 0 \end{cases}$$

$$= z^{-1} \left[\sum_{m=-\infty}^{\infty} z^{-m} \right] = z^{-1} [z^{-0} + z^{-1} + z^{-2} + z^{-3} + \ldots]$$

$$= z^{-1} \frac{1}{1 - z^{-1}} = \frac{z^{-1}}{1 - z^{-1}} = \frac{1}{z - 1}$$

ROC: $|z^{-1}| < 1$ or $|z| > 1$.

3.4 PROPERTIES OF *z*-TRANSFORM

We already know that *z*-transform is a very powerful tool for the analysis of discrete-time signals and LTI systems. It possesses the following important properties.

1. Linearity.
2. Time shifting.
3. Scaling in the *z*-domain.
4. Time reversal.
5. Differentiation in the *z*-domain.
6. Convolution of two discrete-time sequences.
7. Correlation of two discrete-time sequences.
8. Multiplication of two discrete-time sequences.
9. Parseval's theorem for *z*-transform.
10. Conjugation of complex sequences.
11. The initial value theorem.

TABLE 3.2 Properties of z-transform

Property	Signal sequences $s(n)$, $s_1(n)$, $s_2(n)$	z-transform of sequences $S(z)$, $S_1(z)$, $S_2(z)$	ROC R, R_1, R_2
Linearly	$As_1(n) + Bs_2(n)$	$AS_1(z) + BS_2(z)$	At least the intersection of R_1 and R_2
Time shifting	$s(n - n_0)$	$z^{-n0}S(z)$	R, except for the possible addition or deletion of origin ($z = 0$)
Scaling of z-domain	$A^n s(n)$	$S(z/A)$	Scaled version of R, i.e., $\|A\|\ R$
Time reversal	$s(-n)$	$S(^{z-1})$	Inverted version of R (i.e., $\|A\|\ R^{-1}$ = the set of points z^{-1}, where z is in R)
Time expansion	$s(mn)$	$S(z^m)$	$R^{1/m}$ (i.e., the set of points $z^{1/m}$, where z in R)
Conjugation	$s^*(n)$	$S^*(z^*)$	R
Complex convolution	$s_1(n) * s_2(n)$	$S_1(z)S_2(z)$	At least the intersection of R_1 and R_2
Correlation	$(r)s_1s_2(l) = s_1(l) * s_2(-l)$	$Rs_1s_2(z) = S_1(z)\ S_2(1/z)$	At least the intersection of R_1 and R_2
First difference	$s(n) - s(n - 1)$	$(1 - ^{z-1})S(z)$	At least the intersection of R and $\|z\| > 0$
Accumulation	$\sum_{m=-1}^{\infty} s(m)$	$\frac{1}{1-z^{-1}}S(z)$	At least the intersection of R and $\|z\| > 0$
Differentiation in the z-domin	$ns(n)$	$-zds(z)/dz$	R

3.4.1 Linearity

A discrete-time system is said to be linear if it is homogeneous and satisfies the principle of superposition.

If the z-transforms for the two discrete-time sequences is given as

$$s_1(n) \xleftrightarrow{z} S_1(z)$$

and
$$s_2(n) \xleftrightarrow{z} S_2(z)$$

then according to property of linearity of z-transform.

$$s(n) = A_1 s_1(n) + A_2 s_2(n) \xleftrightarrow{z} S(z) = A_1 S_1(z) + A_2 S_2(z)$$

for any value of constants A_1 and A_2 and for any arbitrary discrete-time sequences $s_1(n)$ and $s_2(n)$.

EXAMPLE 3.3

Find the z-transform and ROC of the signal sequence

$$s(n) = [4(2^n) - 5(3^n)]u(n).$$

Solution:

Given: $s(n) = \left[4(2^n u(n)\right] - 3^n u(n)\left[3^n u(n)\right]$ (1)

Now the combined signal is given by,

$$s(n) = As_1(n) + Bs_2(n) \quad (2)$$

After comparing Eqs. (1) and (2), we get two sequences as

$$s_1(n) = (2^n)u(n)$$

and $s_2(n) = (3^n)u(n)$

Taking the z-transform of both sides of Eq. (2), we get

$$S(z) = AS_1(z) + BS_2(z)$$

$$= 4S_1(z) - 5S_2(z) \quad (3)$$

$$\mathcal{Z}\left[s_1(n)\right] = S_1(z) = \sum_{n=-\infty}^{\infty} s_1(n) z^{-n} = \sum_{n=-\infty}^{\infty} (2)^n u(n) z^{-n}$$

$$\left[\text{From } u(n) = \begin{cases} 1, & n \geq 0 \\ 0, & n < 0 \end{cases}\right]$$

$$= \sum_{n=0}^{\infty} (2)^n z^{-n} = \sum_{n=0}^{\infty} \left(2z^{-1}\right)^n = \frac{1}{1-2^{-1}}$$

ROC: $|z| > 2$

$$\mathcal{Z}[s_2(n)] = S_2 z = \sum_{n=-\infty}^{\infty} S_2(n) z^{-n} = \sum_{n=-\infty}^{\infty} 3^n u(n) z^{-n}$$

$$= \sum_{n=0}^{\infty} 3^n z^{-n} = \sum_{n=0}^{\infty} (3z^{-1})^n = \frac{1}{1-3z^{-1}}$$

ROC: $|z| > 3$

After substituting the value of $S_1(z)$ and $S_2(z)$ in Eq. (3), we get

$$S(z) = \frac{4}{1-2z^{-1}} - \frac{5}{1-3z^{-1}} \qquad \text{ROC: } |z| > 3$$

3.4.2 Time Shifting

The purpose of time shifting is that we want to see a particular signal in previous time or in future time by delaying the signal in time or advancing the signal in time, respectively. Consider a discrete-time sequence $s(n)$ and its z-transform $S(z)$.

If $s(n) \xleftrightarrow{z} S(z)$

Now if signal $s(n)$ is delayed by k units of time, then we get

$$s(n-k) \xleftrightarrow{z} z^k.S(z)$$

Now if signal $s(n)$ is advanced by k units of time, then we get

$$s(n+k) \xleftrightarrow{z} z^k.S(z)$$

The ROC of $z^{-k}S(z)$ is the same as that of $S(z)$ except for $z = 0$ and the ROC of $z^k S(z)$ is the same as that of $S(z)$ except for $z = \infty$.

EXAMPLE 3.4

Find z-transform of signal

$$s_2(n) = \{1,\ 2,\ \underset{\uparrow}{3},\ 4,\ 0,\ 1\}$$

from the z-transform of signal

$$s_1(n) = \{\underset{\uparrow}{1},\ 2,\ 3,\ 4,\ 0,\ 1\}$$

by using time shifting property of z-transform.

Solution:

Above two signal sequences $s_2(n)$ and $-s_1(n)$ can be related in time-domain by

$$s_2(n) = s_1(n+2) \qquad (1)$$

Taking the z-transform of Eq. (1), we get

$$S_2(z) = z^2 S_1(z) \qquad (2)$$

But $S_1(z) = \sum_{n=-\infty}^{\infty} s_1(n) z^{-n}$

or $S_1(z) = \sum_{n=0}^{\infty} s_1(n) z^{-n} = s_1(0) z^0 + s_1(1) z^{-1} + s_1(2) z^{-2} +$

$$+ s_1(3) z^{-3} + s_1(4) z^{-4} + s_1(5) z^{-5}$$

or $S_1(z) = 1 + 2z^{-1} + 3z^{-2} + 4z^{-3} + 0z^{-4} + 1z^{-5}$

or $S_1(z) = 1 + 2z^{-1} + 3z^{-2} + 4z^{-3} + z^{-5} \qquad (3)$

Substituting Eq. (3) in Eq. (2), we get

$$S_2(z) = z^2 S_1(z) = z^2\left[1 + 2z^{-1} + 3z^{-2} + 4z^{-3} + z^{-5}\right]$$

$$= z^2 + 2z + 3 + 4z^{-1} + z^{-3}$$

ROC: All values of z except $z = \infty$.

3.4.3 Scaling in *z*-Domain

Scaling in the z-domain means we want to multiply or divide z-domain parameter by any constant A. By scaling in z-domain, we can increase or decrease the value of z. *Scaling in z-domain (frequency domain) by a scalar quantity A is equivalent to scaling in time-domain by a scalar quantity 1/A.*

If $\quad s(n) \overset{z}{\longleftrightarrow} S(z) \qquad \text{ROC}: R_1 < |z| < R_2$

then $\quad A^n s(n) \overset{z}{\longleftrightarrow} S\left(\frac{z}{A}\right) \qquad \text{ROC}: |A| R_1 |z| < |A| R_2$

(This is the Division of z-domain parameter by A)

and $\quad A^{-n} s(n) \overset{z}{\longleftrightarrow} S(Az) \qquad \text{ROC}: \left|\frac{1}{A}\right| R_1 < |z| < \frac{1}{R_2}$

(This is the Multiplication of z-domain parameter by A)

For any constant A, real or complex.

3.4.4 Time Reversal

Time reversal of a discrete-time sequence means taking the mirror image of the signal or folding the signal.

If we have a discrete-time sequence $s(n)$, then its mirror image will be $s(-n)$.

If $$(sn) \xleftrightarrow{\mathbb{Z}} S(z) \qquad \text{ROC}: R_1 < |z| < R_2$$

$$\frac{1}{R_1} < |z| < \left|\frac{1}{A}\right| R_2 \text{ then} \qquad (s-n) \xleftrightarrow{\mathbb{Z}} S\left(\frac{1}{2}\right) \qquad \text{ROC}: R_1 < |z| < R_2$$

EXAMPLE 3.5

Find the z-transform of sequence $s(n) = u(-n)$.

Solution:

We know that, z-transform of unit-step sequence $u(n)$ is given by

$$S(z) = \mathcal{Z}[u(n)] = \frac{1}{1 - z^{-1}} \qquad \text{ROC}: |z| > 1$$

By using property of time reversal of z-transform, we get

$$\mathcal{Z}[s(-n)] = S\left(z^{-1}\right)$$

$$\mathcal{Z}[u(-n)] = S\left(z^{-1}\right) = \left.\frac{1}{1 - z^{-1}}\right|_{z = z^{-1}}$$

or $$= \frac{1}{1 - z} \qquad \text{ROC}: |z| > 1$$

3.4.5 Differentiation in the z-Domain

Here we want to differentiate z-transform of discrete-time sequence with respect to z. Differentiation of z-transform of a discrete-time sequence $s(n)$ with respect to variable z, that is, $dS(z)/dz$.

If $$s(n) \xleftrightarrow{\mathbb{Z}} S(z)$$

then $$n\, s(n) \xleftrightarrow{\mathbb{Z}} -z \frac{dS(s)}{dz}$$

In this case, both transforms will have the same ROC.

EXAMPLE 3.6

Find the sequence $s(n)$ where z-transform is given by

$$S(z) = \log(1 + Az^{-1}), \ |z| > |A.$$

Solution:

Taking the first derivative of $S(z)$, we get

$$\frac{dS(z)}{dz} = \frac{d}{dz}\left[\log\left(1 + Az^{-1}\right)\right]$$

or
$$\frac{dS(z)}{dz} = \frac{1}{\left(1 + Az^{-1}\right)}\left(-Az^{-2}\right)$$

or
$$\frac{dS(z)}{dz} = \frac{-Az^{-2}}{\left(1 + Az^{-1}\right)}$$

or
$$-z\frac{dS(z)}{dz} = \frac{+Az^{-1}}{\left(1 + Az^{-1}\right)} \qquad (1)$$

From the property of differentiation in z-domain of z-transform:

$$-z\frac{dS(z)}{dz} = \mathcal{Z}[ns(n)] \qquad (2)$$

Comparing Eqs. (1) and (2), we get

$$\mathcal{Z}[ns(n)] = \frac{Az^{-1}}{\left(1 + Az^{-1}\right)} \qquad (3)$$

From Eq. (3), we get

$$ns(n) = \mathcal{Z}^{-1}\left[\frac{Az^{-1}}{1 + Az^{-1}}\right] = AZ^{-1}\left[\frac{z^{-1}}{1 + Az^{-1}}\right]$$

$$= A(-A)^{n-1}u(n-1)$$

or
$$s(n) = (-1)^{n-1}\frac{A^n}{n}u(n-1).$$

3.4.6 Convolution of the Two Discrete-time Sequences

The convolution property is one of the most important property of the z-transform because it is used to convert the convolution of two discrete-time signals in the time-domain into multiplication of their z-transform.

We have two discrete-time sequences $s_1(n)$ and $s_2(n)$.

If
$$s_1(n) \xleftrightarrow{z} S_1(z)$$
$$s_2(n) \xleftrightarrow{z} S_2(z)$$

Then
$$s(n) = s_1(n) * s_2(n) = \sum_{k=-\infty}^{\infty} s_1(k)s_2(n-k) \xleftrightarrow{z} S(z)$$
$$= S_1(z)S_2(z)$$

$*$(asterix) shows the convolution operation between sequence $s_1(n)$ and $s_2(z)$.

The ROC of the z-transform of $s(ii) = s_1(n) * s_2(n)$ is at least the intersection of $S_1(z)$ and $S_2(z)$.

We now enumerate the steps involved in computation of the convolution of two signals using z-transform.

Step I: Compute the z-transform of individual signals which are to be convolved.

$$S_2(z) = \mathcal{Z}\left[s_1(n)\right]$$
$$S_2(z) = \mathcal{Z}\left[s_2(n)\right]$$

Step II: Multiply two z-transforms

$$S(z) = S_1(z)S_2(z)$$

Step III: Find the inverse z-transform of $S(z)$.

$$s(n) = \mathcal{Z}^{-1}[S(z)]$$

EXAMPLE 3.7

Find the convolution of sequences

$$S_1(z) = \{\underset{\uparrow}{1}, -2,\ 1\}$$

$$S_2(n) = \{\underset{\uparrow}{1},\ 1,\ 1\}.$$

Solution:

Step I: Determine z-transforms of individual signal sequences.

$$S_1(z) = \mathcal{Z}[s_1(n)] = \sum_{n=0}^{2} s_1(n)z^{-n}$$
$$= s_1(0)z^0 + s_1(1)z^{-1} + s_1(s)z^{-2}$$
$$= 1 - 2z^{-1} + 1z^{-2} = 1 - 2z^{-1} + z^{-2}$$

$$S_2(z) = \mathcal{Z}[s_2(n)] = \sum_{n=0}^{2} s_2(n)z^{-n}$$
$$= s_2(0)z^0 + s_2(1)z^{-1} + s_2(2)z^{-2}$$
$$= 1 \times 1 + 1 \times z - 1 + 1 \times z^{-2}$$
$$= 1 + z^{-1} + z^{-2}$$

Step II: Multiplication of $S_1(z)$ and $S_2(z)$

$$S(z) = S_1(z)S_2(z)$$

Step III: Taking inverse z-transform of $S(z)$

$$s(n) = \mathcal{Z}^{-1}[S(z)] = \mathcal{Z}^{-1}[1 - 3z^{-1} - 2z^{-3} + z^{-4}]$$
$$= \{\underset{\uparrow}{3}, 0, 0, 0, 2, 1\}$$

3.4.7 Correlation of Two Discrete-time Sequences

Correlation of two signals means a degree of similarity between two similar or dissimilar signals.

Correlation of two discrete-time sequences is equivalent to multiplication of z-transform of $s_1(n)$, $S_1(z)$ and z-transform of second sequence $s_2(n)$, $S_2(z)$ at $z = 1/z$.

If $\quad s_1(n) \xleftrightarrow{z} S_1(z)$

$\quad s_2(n) \xleftrightarrow{z} S_2(z)$

then $\quad r_1 s_1 s_1 = \sum_{n=-\infty}^{\infty} s_1(n)s_2(n-l) \xleftrightarrow{z} Rs_1s_2(z)$

$$= S_1(z)S_2\left(\frac{1}{z}\right)$$

We now enumerate the steps involved in computation of the correlation of two sequences using z-transform.

Step I: Compute the z-transform of individual sequences which are to be correlated

$$S_1(z) = \mathcal{Z}[s_1(n)]$$
$$S_2(z) = \mathcal{Z}[s_2(n)]$$

Step II: Determine $S_2\left(\frac{1}{z}\right)$

$$S_2\left(\frac{1}{z}\right) = S_2(Z)\big|_{z=1/z}$$

Step III: Multiply two-transform

$$S(z) = S_1(z)S_2\left(\frac{1}{z}\right)$$

Step IV: Find the inverse z-transform of $S(z)$

$$s(n) = \mathcal{Z}^{-1}[S(z)]$$

where $s(n)$ is the resultant of correlation between sequences $s_1(n)$ and $s_2(n)$.

EXAMPLE 3.8

Find the auto-correlation sequence of the signal.

$$s_1(n) = A^n u(n), \qquad -1 < A < 1.$$

Solution:

Auto-correlation sequence is given by

$$rs_1s_1(l) = \sum_{n=-\infty}^{\infty} s_1(n)s_1(n-l)$$

This sequence can be determined by z-transform and it can be represented as

$$Rrs_1s_1 = \mathcal{Z}\left[r_{s_1s_2}(l)\right] = \mathcal{Z}\left[\sum_{n=-\infty}^{\infty} s_1(n)s_1(n-l)\right]$$

$$= S_1(z)S_1\left(\frac{1}{z}\right) \qquad (1)$$

$$S_1(z) = \mathcal{Z}[s_1(n)] = \sum_{n=-\infty}^{\infty} s_2(n) z^{-n}$$

$$= \sum_{n=-\infty}^{\infty} A^n u(n) z^{-n} = \sum_{n=0}^{\infty} A^n z^{-n}, \qquad \text{since} \quad u(n) = \begin{cases} 1, & n \geq 0 \\ 0, & n < 0 \end{cases}$$

$$= \sum_{n=0}^{\infty} \left[Az^{-1} \right]^n$$

or

$$S_1(z) = \frac{1}{1 - Az^{-1}} \qquad \text{ROC}: |z| > |A \qquad (2)$$

It is a causal signal

Also we can determine

$$S_1\left(\frac{1}{2}\right) = S_1(z)\big|_{z=1/z} = \frac{1}{1 - Az} \qquad \text{ROC}: |z| > |A| \qquad (3)$$

It is a non-causal signal.

Substituting the Eqs. (3) and (2) in Eq. (1), we get

$$Rs_1s_1(z) = S_1(z) S_1\left(\frac{1}{z}\right) = \left(\frac{1}{1 - Az^{-1}}\right)\left(\frac{1}{1 - Az}\right)$$

$$= \frac{1}{1 - A\left(z + z^{-1}\right) + A^2}, \text{ROC:} |A| < |z| < \frac{1}{|A|}$$

Since the ROC of $Rs_1s_2\,(z)$ is a ring, $rs_1s_1(l)$ is a two-sided signal, even if $s_1(n)$ is causal.

Auto-correlation sequence,

$$rs_1s_1(l) = \mathcal{Z}^{-1}\left[R_{s_1s_1}(z)\right] = \mathcal{Z}^{-1}\left[\frac{1}{1 - A\left(z + z^{-1}\right) + A^2}\right]$$

$$= \mathcal{Z}^{-1}\left[\left(\frac{1}{1 - Az^{-1}}\right)\left(\frac{1}{1 - Az}\right)\right]$$

$$= \mathcal{Z}^{-1}\left[\frac{1}{\left(1 - A^2\right)\left(1 - Az^{-1}\right)} + \frac{1}{\left(1 - A^2\right)\left(1 - Az\right)}\right]$$

$$= \frac{1}{1 - A^2} \mathcal{Z}^{-1}\left[\frac{1}{1 - Az^{-1}} + \frac{1}{1 - Az}\right]$$

$$= \frac{1}{1 - A^2} A^{|l|}.$$

3.4.8 Multiplication of Two Discrete-time Sequences

The z-transform of a multiplication of two discrete-time sequences is the complex convolution of the z-transforms of the individual sequences.

If
$$s_1(n) \overset{z}{\longleftrightarrow} S_1(z)$$
$$s_2(n) \overset{z}{\longleftrightarrow} S_2(z)$$

then
$$s(n) = s_1(n)s_2(n) \overset{z}{\longleftrightarrow} S_2(z) = \frac{1}{2\pi j}\oint_c S_1(\upsilon)S_2\left(\frac{z}{\upsilon}\right)\upsilon^{-1}d\upsilon$$

where υ = other variable in frequency domain
z = standard variable in frequency domain for z-transform
j = $\sqrt{-1}$
π = standard parameter.

This is called complex convolution of $S_1(z)$ and $S_2(z)$.

Where c is the closed contour that encloses the origin and lies within the ROC common to both $S_1(\upsilon)$ and $S_2(_1/\upsilon)$.

This property of z-transform is not generally used, because the complexity of computations increases.

3.4.9 Parseval's Theorem

If we have two discrete-time sequences $s_1(n)$ and $s_2(n)$ and both are complex-valued sequences.

Then, according to Parseval's theorem

$$\sum_{n=-\infty}^{\infty} s_1(n)s_2^*(n) = \frac{1}{2\pi j}\oint_c S_1(\upsilon)S_2^*\left(\frac{1}{\upsilon *}\right)\upsilon^{-1}d\upsilon$$

ROC of $S_1(z)$: $R_{1l} < |z| < R_{1u}$
ROC of $S_2(z)$: $R_{2l} < |z| < R_{2u}$
For the above theorem following condition should be satisfied.

$$R_{1l}R_{2l} < 1 < R_{1u}R_{2u}$$

where $s_2^*(n)$ is the complex conjugate of $s_2(n)$, where $s_2(n)$ is a complex-valued sequence if

$$s_2(n) = s_{2R}(n) + js_{2I}(n)$$

then
$$s_2(n) = s_{2R}(n) - js_{2I}(n)$$

where $s_{2R}(n)$ and $s_{2I}(n)$ are the real part and imaginary part of signal $S_2(n)$, respectively.

$S_2^*(1/\upsilon^*)$ is the complex conjugate of z-transform of $s_2(n)$ at $z = 1/\upsilon^*$, where υ is another frequency variable and υ^* is the complex conjugate of υ.

$$j = \sqrt{-1}$$

l for lower limit
u for upper limit

$\oint_c$ represents a contour integration.

Parseval equations relate energy in a signal to energy in its frequency spectrum.

3.4.10 Conjugation of a Complex Sequence

Let us take a sequence which is complex valued.

If $$s(n) \overset{z}{\longleftrightarrow} (z)$$

then $$s*(n) \overset{z}{\longleftrightarrow} S*(z*)$$

ROC for both $S(z)$ and $S*(z*)$ will be the same.

where $s*(n)$ and $S*(z)$ are the complex conjugates of $s(n)$ and $S(z)$ respectively and $z*$ is complex conjugate of z.

3.4.11 Initial Value Theorem

If a discrete-time sequence $s(n) = 0$ for $n < 0$, that is, $s(n)$ is a causal sequence;

then $s(n)$ at $n=0$

$$= s(0) = \lim_{z \to 0}\{S(z)\} \tag{3.4}$$

This is called *initial value theorem.*

3.5 SOME COMMON z-TRANSFORM PAIRS

Some of the commonly used discrete-time sequences and their z-transforms are given in Table 3.3.

3.6 THE INVERSE z-TRANSFORM

The mechanism for transforming $S(z)$ back to a discrete-time sequence $s(n)$ is called the *inverse* z-transform.

Inverse z-transform is formally given as

$$s(n) = \frac{1}{2\pi j}\int_c S(z)z^{n-1}dz \tag{3.5}$$

Eq. (3.5) represents a contour integral over a closed path c that encloses the origin and lies within the ROC of $S(z)$.

Contour integration is a complex integration over a closed path in the ROC encompassing the origin $z = 0$ in the z-plane, exactly once in the counterclockwise direction, a circle of radius $c > R1$. Where $R1$ is the radius of convergence of z-transform.

From the Cauchy integral theorem, we know that

$$\frac{1}{2\pi j}\oint_c z^{n-1-m}dz = \begin{cases} 1, & m = n \\ 0, & m \neq n \end{cases} \tag{3.6}$$

where the integral is over any simple contour c encircling the origin.

Figure 3.6 shows the closed contour c.

TABLE 3.3 Some Common z-transform Pairs

Signal $s(n)$	z-transform of signal, $S(z)$	Region of convergence (ROC)
$\delta(n)$	1	All values of z
$u(n)$	$\frac{1}{1-z^{-1}}$	$\lvert z\rvert > 1$
$r(n) = nu\ (n)$	$\frac{z^{-1}}{\left(1-z^{-1}\right)^2}$	$\lvert z\rvert > 1$
$A^n u(n)$	$\frac{z^{-1}}{1-Az^{-1}}$	$\lvert z\rvert > \lvert A\rvert$
$nA^n u(n)$	$\frac{z^{-1}}{\left(1-Az^{-1}\right)^2}$	$\lvert z\rvert > \lvert A\rvert$

(Continued)

Signal $s(n)$	z-transform of signal, $S(z)$	Region of convergence (ROC)
$-A^n u(-n-1)$	$\frac{-1}{1-Az^{-1}}$	$\|z\| > \|A\|$
$-nA^n u(-n-1)$	$\frac{-1}{\left(1-Az^{-1}\right)^2}$	$\|z\| > \|A\|$
$(\cos \omega_0 n)\, u(n)$	$\frac{1-z^{-1}\cos\omega_0}{1-2z^{-1}\cos\omega_0+z^{-2}}$	$\|z\| > 1$
$(\sin \omega_0 n)\, u(n)$	$\frac{1-z^{-1}\sin\omega_0}{1-2z^{-1}\cos\omega_0+z^{-2}}$	$\|z\| > 1$
$(A^n \cos \omega_0 n)\, u(n)$	$\frac{1-Az^{-1}\cos\omega_0}{1-2Az^{-1}\cos\omega_0+A^2z^{-2}}$	$\|z\| > \|A\|$
$(A^n \sin \omega_0 n)\, u(n)$	$\frac{Az^{-1}\sin\omega_0}{1-2Az^{-1}\cos\omega_0+A^2z^{-2}}$	$\|z\| > \|A\|$
$u(n-1)$	$z^{-1}\left(\frac{1}{1-z^{-1}}\right)$	$\|z\| > 1$
$\delta(n-k)$	$z^{-k}.1 = z^{-k}$	All values of z except $z = 0$
$d(n+k)$	$z^{k}.1 = z^{k}$	All values of z except $z = \infty$

From the definition of z-transform

$$S(z) = \sum_{m=-\infty}^{\infty} s(m)z^{-m} \tag{3.7}$$

Multiplying Eq. (3.7) by z^{n-1} and integrating over a closed contour with the ROC and enclosing the origin, we get

$$\oint_c S(z)z^{n-1}dz$$

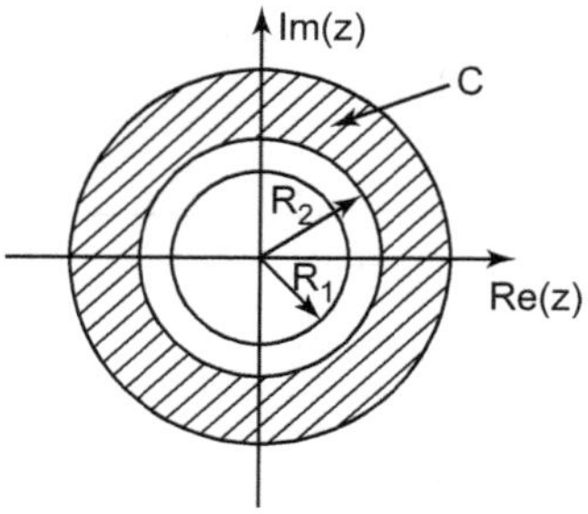

FIGURE 3.6 Closed contour c taken for the integral, in the ROC of $S(z)$ and encircling the origin $z - 0$ exactly in the origin-clockwise direction.

$$= \oint_c \sum_{m=-\infty}^{\infty} s(m) z^{n-1} dz \tag{3.8}$$

where contour c is taken exactly once in the counterclockwise direction.

Now the order of integration and summation can be interchanged as the series is convergent on this contour. It yields

$$\oint_c S(z) z^{n-1} dz = \sum_{m=-\infty}^{\infty} s(m) \oint_c z^{n-1-m} dz \tag{3.9}$$

$$= 2\pi\, j s(n) \quad \text{(on applying Cauchy integral theorem)}$$

Three methods are often used for the evaluation of the inverse z-transform in practice.

These methods are

1. The inverse z-transform by contour integration method.
2. The inverse z-transform by power series expansion method.
3. The inverse z-transform by partial fraction expansion method.

3.6.1 The Inverse z-Transform by Contour Integration Method

In this method, we use the Cauchy Residue Theorem for determining the inverse z-transform directly from the contour integral.

Cauchy Residue Theorem: If we have $S(z)$ which is a function of complex variable z and c is a closed path in the z-plane. Here we assume that $dS(z)/dz$ exists on and inside the contour c and $S(z)$ has no poles at $z = z_0$.

In general, the Cauchy Residue Theorem is stated as

$$\frac{1}{2\pi j} \oint_c \frac{S(z)}{(z - z_0)^k} dz = \begin{cases} \dfrac{1}{k-1} \left| \dfrac{d^k S(z)}{dz^{k-1}} \right|_{z=z_0} & \text{if } z_0 \text{ is inside } c \\ 0, & \text{if } z_0 \text{ is outside } c \end{cases} \tag{3.10}$$

For $k = 1$
$$\frac{1}{2\pi j} \oint_c \frac{S(z)}{(z - z_0)^k} dz = \begin{cases} S(z)|_{z=0} & \text{if } z_0 \text{ is inside } c \\ 0 & \text{if } z_0 \text{ is inside } c \end{cases} \tag{3.11}$$

The values on the RHS of Eqs. (3.10) and (3.11) are called *residues of pole at* $z - z_0$. Inverse z-transform of $S(z)$ is given by

$$
\begin{aligned}
s(n) &= \frac{1}{2\pi j}\oint_c S(z)z^{n-1}dz \\
&= \sum_{\text{All poles inside } c} [\text{Residue of } S(z)z^{n-1} \text{ at } z = z_0] \\
&= \sum_i (z - z_i) S(z) z^{n-1} |_{z=z_i} \qquad (3.12)
\end{aligned}
$$

where $\{z_i\}$ are simple poles.

Note: If $S(z)\, z^{n-1}$ has no poles inside c for one or more values of n. Then, $s(n) = 0$ for these values.

EXAMPLE 3.9

Evaluate the inverse z-transform of

$$S(z) = \frac{z}{z - A}, |z| > |A|$$

Using contour integration method.

Solution:

Given $$S(z) = \frac{z}{z - A}, |z| > |A|$$

The contour integration formula is given as

$$
\begin{aligned}
s(n) &= \frac{1}{2\pi j}\oint_c S(z)z^{n-1}dz \\
&= \sum_i (z - z_i) S(z) z^{n-1} |_{z=z_i}
\end{aligned}
$$

Putting the value of $S(z)$

$$
\begin{aligned}
s(n) &= \sum_i (z - z_i) \frac{z}{z - A} z^{n-1} |_{z=z_i} \\
&= \sum_i (z - z_i) \frac{z^n}{z - A} |_{z=z_i} \qquad (i)
\end{aligned}
$$

For $n \geq 0$: there is only one pole at $z = A$

$$s(n) = (z - A)\frac{z^n}{z - A}|_{z=A} = z^n\Big|_{z=A} = (A)^n \qquad (ii)$$

For $n < 0$

$n = -1$, Eq. (i) becomes

$$s(-1) = \sum_i (z - z_i)\frac{z^{-1}}{z - A}\bigg|_{z=z_i}$$

$$= \sum_i (z - z_i)\frac{1}{(z)(z - A)}\bigg|_{z=z_i}$$

$$= (z - 0)\frac{1}{(z)z - A}\bigg|_{z=0} + (z - A)\frac{1}{(z)z - A}\bigg|_{z=A}$$

$$= \frac{1}{z - A}\bigg|_{z=0} + \frac{1}{z}\bigg|_{z=A} = -\frac{1}{A} + \frac{1}{A} = 0$$

$n = -2$, Eq. (i) becomes

$$s(-2) = \sum_i (z - z_i)\frac{z^{-2}}{z - A}\bigg|_{z=z_i}$$

$$= \sum_i (z - z_i)\frac{1}{(z)^2(z - A)}\bigg|_{z=z_i}$$

$$= \frac{d}{dz}\left\{\frac{1}{z - A}\right\}_{z=0} + \frac{1}{z^2}\bigg|_{z=A}$$

$$= -\frac{1}{\left(z - A^2\right)}\bigg|_{z=0} + \frac{1}{z^2}\bigg|_{z=A} = -\frac{1}{A^2} + \frac{1}{A^2} = 0$$

Similarly, $n = -3$, $s(-3) = 0$
$n = -4$, $s(-4) = 0$ and so on.

Now we can show that $s(n) = 0$, $n < 0$

Thus $s(n) = A^n u(n)$

EXAMPLE 3.10

Determine the inverse z-transform of

$$S(z) = \frac{5z}{(z - 1)(z - 2)}$$

Using the contour integration method.

Solution:
The contour integral formula is given by

$$s(n) = \frac{1}{2\pi j}\oint_c S(z)z^{n-1}dz$$

$$= \sum_{\text{All poles inside } c} \text{Residue of } S(z)z^{n-1} \text{at poles of } S(z)$$

$$= \sum_i (z - z_i)S(z)z^{n-1}\Big|_{z=z_i}$$

$$= \sum_i (z - z_i)\frac{5z}{(z-1)(z-2)}z^{n-1}\Big|_{z=z_i}$$

$$= \sum_i (z - z_i)\frac{5z^n}{(z-1)(z-2)}\Bigg|_{z=z_i} \qquad (i)$$

$$= \frac{5z^n}{(z-2)}\Bigg|_{z=1} + \frac{5z^n}{(z-1)}\Bigg|_{z=2}$$

$$= -5(1)^n + 5(2)^n = 5(-1 + 2n), n^30$$

or $$s(n) = 5(-1 + 2^n), \text{ for } n \geq 0$$

For $n < 0$,
$n = -1$. The Eq. (i) becomes

$$s(n) = \sum_i (z - z_i)\frac{5z^n}{(z-1)(z-2)}\Bigg|_{z=z_i} \qquad (ii)$$

$$s(-1) = \sum_i (z - z_i)\frac{5z^{-1}}{(z-1)(z-2)}\Bigg|_{z=z_i}$$

$$= 5\sum_i (z - z_i)\frac{1}{z(z-1)(z-2)}\Bigg|_{z=z_i}$$

$$= 5\left[\frac{1}{(z-1)(z-2)}\Bigg|_{z=0} + \frac{1}{z(z-2)}\Bigg|_{z=1} + \frac{1}{z(z-1)}\Bigg|_{z=2}\right]$$

$$= 5\left[\frac{1}{(0-1)(0-2)} + \frac{1}{1(1-2)} + \frac{1}{2(2-1)}\right]$$

$$= 5\left[\frac{1}{2} - 1 + \frac{1}{2}\right] = 5[1-1] = 0$$

$$n = -2 s(-2) = \sum_i (z - z_i) \frac{5z^{-2}}{(z-1)(z-2)}\bigg|_{z=z_i}$$

$$= 5\sum_i (z - z_i) \frac{1}{z^2(z-1)(z-2)}\bigg|_{z=z_i}$$

$$= 5\left[\frac{d}{dz}\left\{\frac{1}{(z-1)(z-2)}\right\}\bigg|_{z=0} + \frac{1}{z^2(z-2)}\bigg|_{z=1} + \frac{1}{z^2(z-1)}\bigg|_{z=2}\right]$$

$$= \frac{3}{4} - 1 + \frac{1}{4} = 1 - 1 = 0$$

$$n = -3, s(-3) = 0$$

$$n = -4, s(-4) = 0 \text{ and so on.}$$

Hence, $s(n) = 0, n < 0$

But $\quad s(n) = 5(2^n - 1)$, for $n \geq 0$

$\therefore \quad s(n) = 5(2^n - 1)u(n)$.

3.6.2 The Inverse z-Transform by Power Series Expansion Method

For rational z-transforms a power series expansion can be obtained by long division.

For this method, we have a z-transform $S(z)$ with its corresponding ROC. Now we can expand $S(z)$ into a power series of the form

$$S(z) = \sum_{n=-\infty}^{\infty} C_n z^{-n} \tag{3.13}$$

which converges in the given ROC.

Inverse z-transforms of $S(z)$ is given by

$$s(n) = \mathcal{Z}^{-1}[S(z)] = C_n \text{ for all } n \tag{3.14}$$

EXAMPLE 3.11

Determine the inverse z-transform of

$$S(z) = \frac{2}{2 - 3z^{-1} + z^{-2}} \text{ by long division method.}$$

When (a) ROC: $|z| > 1$

(b) ROC: $|z| < \frac{1}{2}$.

Solution:

(a) In this part ROC is $|z| > 1$. Since the ROC is the exterior of a circle whose radius is 1. Now $s(n)$ will be causal sequence and it requires a power series expansion in negative power of z.

Long division is performed as:

$$2 - 3z^{-1} + z^{-2} \overline{\big)\, 2 \qquad\qquad } \Big(1 + \frac{3}{2}z^{-1} + \frac{7}{4}z^{-2} + \ldots.$$

$$\underline{\begin{array}{ccc} 2 & -3z^{-1} & +z^{-2} \\ - & + & - \end{array}}$$

$$\begin{array}{ccc} 3z^{-1} & -z^{-2} & \\ 3z^{-1} & -\frac{9}{2}z^{-2} & +\frac{3}{2}z^{-3} \\ - & + & - \end{array}$$

$$\begin{array}{ccc} \frac{7}{2}z^{-2} & -\frac{3}{2}z^{-3} & \\ \frac{7}{2}z^{-2} & -\frac{21}{4}z^{-3} & +\frac{7}{4}z^{-4} \\ - & + & - \end{array}$$

$$\frac{15}{4}z^{-3} - \frac{7}{4}z^{-4}$$

$$S(z) = \frac{2}{2 - 3z^{-1} + z^{-2}} = 1 + \frac{3}{2}z^{-1} + \frac{7}{4}z^{-2} + \ldots$$

But $\quad S(z)=\sum_{n=0}^{\infty} s(n)z^{-n} = s(0)z^{-0} + s(1)z^{-1} + 2(2)z^{-2} + \ldots$

$$s(n)=\{s(0),s(1),s(2),\ \ldots.\}=\left\{\underset{\uparrow}{1},\frac{3}{2},\frac{7}{4},\ldots\right\}$$

(*b*) In this case, the ROC is the interior of a circle of radius 1/2.

Consequently, then signal $s(n)$ is non-causal and it requires a power series expansion in positive powers of z.

Now long division is performed as

$$z^{-2}-3z^{-1}+2\ \overline{\Big)\ \cancel{2}-6z+4x^{2}\ \Big(}\ 2z^{2}+6z^{3}+14z^{4}+\ldots$$

$$\begin{array}{llll}
 & \cancel{2} & & \\
 - & + & - & \\
\hline
 & \cancel{6z}-4z^{2} & & \\
 & \cancel{6z} & -18z^{2} & +12z^{3} \\
 - & + & - & \\
\hline
 & 14\cancel{z^{2}} & -12z^{3} & \\
 & \cancel{14}z^{-1} & -42z^{3} & +28z^{4} \\
 - & + & - & \\
\hline
 & & 30z^{3} & -28z^{4}
\end{array}$$

$$S(z)=2z^{2}+\ 6z^{3}+\ 14z^{4}+\ \ldots \tag{1}$$

z-transform for $n \le 0$

$$S(z)=s(0)z^{-0}+s(-1)z^{-1}+s(-2)z^{2}+s(-3)z^{3}+s(-4)z^{4}+\ldots$$

Comparing Eq. (1) with expansion of z-transform for $n \le 0$, we get the sequence

$$s(n)=\{\ldots 14,\ 6,\ 2,0,\ \underset{\uparrow}{0}\}$$

This is a non-causal sequence.

EXAMPLE 3.12

Find $s(0)$, $s(1)$, $s(2)$, $s(3)$, $s(4)$ for

$$S(z) = \frac{z^2 - 1}{z^3 + 2z + 4} \text{ by long division method.}$$

Solution:

Long division is performed as follows:

$$
\begin{array}{r|l|l}
z^3 + 2z + 4 & z^2 - 1 & z^{-1} - 3z^{-3} - 4z^{-4} \\
 & z^2 + 2 + 4z^{-1} & \\
 & - \quad - \quad - & \\
\hline
 & -3 - 4z^{-1} & \\
 & -3 \qquad - 6z^{-2} - 12z^{-3} & \\
 & + \qquad + \quad + & \\
\hline
 & -4z^{-1} + 6z^{-2} + 12z^{-3} & \\
 & -4z^{-1} - 6z^{-2} + 12z^{-3} & \\
 & + \quad + \quad + & \\
\hline
 & -4z^{-1} + 6z^{-2} + 12z^{-3} & \\
 & -4z^{-1} \qquad - 8z^{-3} - 16z^{-4} & \\
 & + \qquad + \quad + & \\
\hline
 & 6z^{-2} + 20z^{-3} + 16z^{-4} &
\end{array}
$$

Thus

$$S(z) = z^{-1} - 3z^{-3} - 4z^{-4} + \ldots \tag{1}$$

By the definition of z-transform

$$S(z) = \sum_{n=-\infty}^{\infty} s(n)z^{-n}$$

$$= \ldots s(-4)z^4 + s(-3)z^3 + s(-2)z^2 + s(-1)z^1$$

$$+ s(0)z^0 + s(1)z^{-1} + s(2)z^{-2} + s(3)z^{-3} + s(4)z^{-4} + \ldots \tag{2}$$

Comparing Eqs. (1) and (2), we get

$$s(0)=0$$
$$s(1)=1$$
$$s(2)=0$$
$$s(3)=-3$$
$$s(4)=-4$$

3.6.3 The Inverse z-Transform by Partial Fraction Expansion Method

In this method, we attempt to express the function $S(z)$ as a linear combination of $S_1(z)$, $S_2(z)$, $S_3(z)$ where $S_1(z)$, $S_2(z)$... are additive parts of function $S(z)$ recovered by partial fraction, it is shown in Eq. (3.15)

$$S(z) = A_1S_1(z) \; + A_2S_2(z) + ... + A_kS_k(z) \tag{3.15}$$

where $S_1(z)$, $S_2(z)$, ... $S_k(z)$ are the expressions with inverse z-transforms $s_1(n)$, $s_2(n)$, ... $s_k(n)$. Then inverse z-transform of $S(z)$ will be

$$\begin{aligned} s(n) &= \mathcal{Z}^{-1}[s(z)] \\ &= \mathcal{Z}^{-1}\left[A_1S_1(z)A_2S_2(z)...A_kS_k(z)\right]^{-1} \\ &= \mathcal{Z}^{-1}\left[A_1S_1(z)\right]\mathcal{Z}^{-1}\left[A_2S_2(z)\right] + ...\mathcal{Z}^{-1}\left[A_kS_k(z)\right] \\ &= A_1s_1(n) + A_2s_2(n) + ... + A_ks_k(n) \end{aligned} \tag{3.16}$$

This method is particularly useful if $S(z)$ is a rational function given as follows.

$$S(z) = \frac{N(z)}{D(z)} = \frac{B_0 + B_1z^{-1} + ... + B_mz^{-m}}{A_0 + A_1z^{-1} + ... + B_nz^{-N}} \tag{3.17}$$

Here A_0 should be equal to 1.

The partial fraction method is applicable only for proper rational transfer function. A rational transfer function is called proper if

$$AN \neq 0 \quad \text{and} \quad M < N.$$

EXAMPLE 3.13

Determine the inverse z-transform by using partial fraction expansion method. Given

$$S(z) = \frac{z}{(z-1/2)(z-1/4)}.$$

Solution:

Partial fraction expansion of $S(z)$ is given as

$$S(z) = \frac{z}{(z-1/2)(z-1/4)} = \frac{\alpha_1}{(z-1/2)} + \frac{\alpha_2}{(z-1/4)} \qquad (i)$$

$$z = \frac{1}{2}, \qquad \alpha_1 = \frac{z}{z-1/4} = \frac{1/2}{1/2-1/4} = 2$$

$$z = \frac{1}{4} \qquad \alpha_2 = \frac{z}{z-1/2} = \frac{1/4}{1/4-1/2} = 1$$

Substituting the value of α_1 and α_2 in Eq. (*i*), we get

$$S(z) = \frac{2}{(z-1/2)} - \frac{1}{(z-1/4)} \qquad (ii)$$

Taking the inverse z-transform of both sides of Eq. (*ii*), we get

$$s(n) = \mathcal{Z}^{-1}\left[\frac{2}{z-1/2}\right] - \mathcal{Z}^{-1}\left[\frac{1}{z-1/4}\right]$$

$$= \mathcal{Z}^{-1}\left[\frac{2z^{-1}}{1-1/2z^{-1}}\right] - \mathcal{Z}^{-1}\left[\frac{z^{-1}}{1-1/4z^{-1}}\right]$$

$$= 2\left(\frac{1}{2}\right)^{n-1} u(n-1) - \left(\frac{1}{4}\right)^{n-1} u(n-1)$$

$$= 4\left[\left(\frac{1}{2}\right)^{n} - \left(\frac{1}{4}\right)^{n}\right]u(n-1).$$

EXAMPLE 3.14

Obtain inverse z-transform using partial fraction expansion method. Given

$$S(z) = \frac{4z^2 - 2z}{z^3 - 5z^2 + 8z - 4}.$$

Solution:

Here $S(z)$ is a proper rational function. So we can apply partial fraction expansion method for inverse z-transform.

$$S(z) = \frac{4z^2 - 2z}{z^3 - 5z^2 + 8z - 4} = \frac{4z^2 - 2z}{(z-1)(z-2)^2}$$

$$= \frac{\alpha_1}{(z-1)} + \frac{\alpha_2}{(z-2)} + \frac{\alpha_3}{(z-2)^2} \qquad (i)$$

$z - 1 = 0$, or $z = 1$

$$\alpha_1 = \frac{4z^2 - 2z}{(z-2)^2} = \frac{4(1)^2 - 2\times 1}{(1-2)^2} = 2$$

$z - 2 = 0$,

$$\alpha_2 = \frac{d}{dz}\left[\frac{4z^2 - 2z}{z-1}\right] = \frac{(z-1)(8z-2) - \left(4z^2 - 2z\right)(1)}{(z-1)^2}$$

or $z = 2$

$$= \frac{(2-1)(8\times 2 - 2) - \left[4(2)^2 - 2\times 2\right]}{(2-1)^2} = \frac{14-12}{1} = 2$$

$z - 2 = 0$, or $z = 2$

$$\alpha_3 = \frac{4z^2 - 2z}{z-1} = \frac{4(2)^2 - 2\times 2}{2-1} = \frac{16-4}{1} = 12$$

Putting the values of α_1, α_2 and α_3 in Eq. (*i*), we get

$$S(z) = \frac{2}{(z-1)} + \frac{2}{(z-2)} + \frac{12}{(z-2)^2} \qquad (ii)$$

$$\Rightarrow \quad S(z) = \frac{2z^{-1}}{1-z^{-1}} + \frac{2z^{-1}}{1-2z^{-1}} + \frac{12z^{-2}}{\left(1-2z^{-1}\right)^2} \qquad (iii)$$

Taking the inverse z-transform of both sides of Eq. (*iii*), we get

$$s(n) = \mathcal{Z}^{-1}\left[\frac{2z^{-1}}{1-z^{-1}}\right] + \mathcal{Z}^{-1}\left[\frac{2z^{-1}}{1-2z^{-1}}\right] + \mathcal{Z}^{-1}\left[\frac{12z^{-2}}{\left(1-2z^{-1}\right)^2}\right]$$

$$= 2(1)^{n-1} + 2(2)^{n-1} + 6(n-1)(2)^{n-1}, n \geq 0$$

TABLE 3.4 Inverse *z*-transform

z-transform *s*(*n*) *S*(*z*)	Inverse *z*-transform *s*(*n*)	ROC
1	$\delta(n)$	All z
$\frac{z}{z-1}$	$u(n),\ n \geq 0$	$\lvert z \rvert > 1$
$\frac{z}{z-1}$	$-u(n),\ n < 0$	$\lvert z \rvert > 1$
$\frac{z}{z-A}$	$A^n,\ n \geq 0$	$\lvert z \rvert > \lvert A \rvert$
$\frac{z}{z-A}$	$-A^n,\ n < 0$	$\lvert z \rvert > \lvert A \rvert$
$\frac{Az}{(z-A)^2}$	$nA^n,\ u(n)$	$\lvert z \rvert > \lvert A \rvert$
$\frac{z}{z-1}$	$-A^n u,\ (-n-1)$	$\lvert z \rvert > \lvert A \rvert$
$\frac{Az}{(z-A)^2}$	$-nA^n u,\ (-n-1)$	$\lvert z \rvert > \lvert A \rvert$
$\frac{z^2 - z\cos\omega_0}{z^2 - 2z\cos\omega_0 + 1}$	$(\cos\omega_0 n)\ u(n)$	$\lvert z \rvert > 1$
$\frac{z\sin\omega_0}{z^2 - 2z\cos\omega_0 + 1}$	$(\sin\omega_0 n)\ u(n)$	$\lvert z \rvert > 1$
$\frac{z^2 - Az\cos\omega_0}{z^2 - 2z\ Az\cos\omega_0 + A^2}$	$(A^2\cos\omega_0 n)\ u(n)$	$\lvert z \rvert > \lvert A \rvert$
$\frac{Az\cos\omega_0}{z^2 - 2Az\cos\omega_0 + A^2}$	$(A^2\sin\omega_0 n)\ u(n)$	$\lvert z \rvert > \lvert A \rvert$

3.7 SYSTEM FUNCTION

The z-transform of a discrete-time system's unit-sample response $h(n)$ is called the system function or transfer function of the system. It is denoted by $H(z)$.

$$H(z) = \mathcal{Z}[h(n)] = \sum_{n=-\infty}^{\infty} h(n)z^{-n}$$

In other words, system function can be defined as the ratio of two-transform of output and z-transform of input keeping all initial conditions zero, that is, initially the system is relaxed.

$$H(z) = \frac{Y(z)}{S(z)} = \frac{z\text{-transform of output}}{z\text{-transform of input}} \tag{3.18}$$

The output of a relaxed LTI system to an input sequence $s(n)$ can be obtained by computing the convolution of $s(n)$ with $h(n)$ of the system, where $h(n)$ is the unit-sample response of the system.

$$y(n) = \sum_{k=-\infty}^{\infty} s(k)h(n-k) = s(n) * h(n)$$

Here * denotes convolution operation.
Above relationship can be represented in the z-domain as

$$Y(z) = S(z)H(z)$$

or
$$H(z) = \frac{Y(z)}{S(z)}$$

where $S(z) = z$-transform of the input sequence, $s(n)$
$Y(z) = z$-transform of the output sequence, $y(n)$.
$H(z) = z$-transform of unit-sample response, $h(n)$.

$H(z)$ represents the z-domain characterization of a system, whereas $h(n)$ is the corresponding time-domain characterization of the system.

Transfer function for a system which is described by a linear constant-coefficient difference equation given as follows:

$$y(n) = \sum_{k=1}^{N} A_k y(n-k) + \sum_{k=0}^{M} B_k s(n-k) \tag{3.19}$$

Taking the z-transform of both sides of above Eq. (3.19), we get

$$Y(z) = \sum_{k=1}^{N} A_k z^{-k} Y(z) + \sum_{k=0}^{M} B_k z^{-k} S(n)$$

or $$\frac{Y(z)}{S(z)} = \frac{\sum_{k=0}^{M} B^k z^{-k}}{1 + \sum_{k=1}^{N} A_k z^{-k}} \tag{3.20}$$

Note that LTI system will always have rational z-transforms.

EXAMPLE 3.15

Determine the system function $H(z)$ and unit-sample response $h(n)$ of the system whose difference equation is

$$y(n) = \frac{1}{2} y(n-1) + 2s(n)$$

where $y(n)$ and $s(n)$ are the output and input of the system, respectively.

Solution:

Taking the z-transform of the above difference equation

$$Y(z) = \frac{1}{2} z^{-1} Y(z) + 2S(z)$$

or $$Y(z)\left[1 - \frac{1}{2} z^{-1}\right] = 2S(z)$$

or $$H(z) = \frac{Y(z)}{S(z)} = \frac{2}{1 - \frac{1}{2} z^{-1}}$$

This system function has a pole at $z = \frac{1}{2}$ and zero at $z = 0$.

$$H(z) = \frac{2}{1 - \frac{1}{2}} z^{-1}$$

$$h(n) = \text{Inverse } z\text{-transform of } H(z) = \mathcal{Z}^{-1}\left[\frac{2}{1 - 1/2z^{-1}}\right]$$

$$= 2\left(\frac{1}{2}\right)^n u(n)$$

This is the unit-sample response of the system.

3.8 POLES AND ZEROS OF RATIONAL z-TRANSFORMS

An important family of z-transforms is those for which $S(z)$ is a rational function.

Rational transfer function is a ratio of two polynomials of z, that is, $S(z) = P(z)/Q(z)$.

The zeros of a z-transform $S(z)$ are the values of z for which $S(z) = 0$.

The poles of a z-transform $S(z)$ are the values of z for which $S(z) = \infty$.

We can represent $S(z)$ graphically by pole-zero plot in the complete z-plane. Pole is located by X and zero by o. From the definition of ROC, the ROC of a z-transform should not contain any pole.

EXAMPLE 3.16

Determine the pole-zero plot for the signal $s(n) = (2)^n u(n)$.

Solution:

$$S(z) = \mathcal{Z}[s(n)]$$

$$= \sum_{n=-\infty}^{\infty} s(n) z^{-n}$$

$$= \sum_{n=-\infty}^{\infty} (2)^n u(n) z^{-n}$$

$$= \frac{1}{1 - 2z^{-1}} \qquad \text{ROC}: |z| > |2|$$

Thus $S(z)$ has one zero at $z = 0$ and one pole at $z = 2$. The pole-zero plot is shown in Figure 3.7.

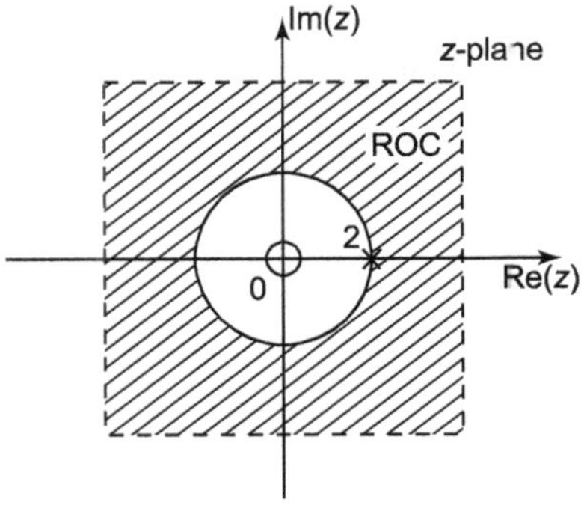

FIGURE 3.7 Pole-zero plot for the sequence $s(n) = (2)^n u(n)$. $s(n) = (2)^n u(n)$

3.9 SOLUTION OF DIFFERENCE EQUATIONS USING z-TRANSFORM

One-sided z-transform is used to solve difference equations with initial conditions. The difference equations relating the two time-domain signals can be converted into an equivalent algebraic equation with the help of using one-sided z-transform. This algebraic equation can be easily solved to obtain the z-transform of the desired signal. The signal in the time-domain is obtained by taking inverse z-transform of the resulting z-transform.

EXAMPLE 3.17

Determine the step response of a system given by

$$y(n) = Ay\,(n-1) + s(n), -1 < A < 1$$

When the initial condition is $y(-1) = 1$.

Solution:

By taking the one-sided z-transform of both sides of the above difference equation

$$y(n) = Ay\,(n-1) + s(n). \quad \text{We obtain}$$

$$Y(z) = A\left[z^{-1}Y(z) \ + y(-1)\right] + S(z)$$

or
$$Y(z) = A\left[z^{-1}Y(z) \ + \ 1\right] + S(z)$$

or
$$Y(z)\left[1 - Az^{-1}\right] = A + S(z) \qquad (1)$$

But for step response, $s(n) = u(n)$

Then,
$$S(z) = \mathcal{Z}[u(n)] = \frac{1}{1 - z^{-1}} \qquad (2)$$

Substituting Eq. (2) in Eq. (1), we get

$$Y(z)\left[1 - Az^{-1}\right] = A + \frac{1}{1 - z^{-1}}$$

or
$$Y(z) = \frac{A}{1 - Az^{-1}} + \frac{1}{\left(1 - z^{-1}\right)\left(1 - Az^{-1}\right)} \qquad (3)$$

or
$$Y(z) = \frac{A}{1 - Az^{-1}} + \frac{\alpha_1}{1 - z^{-1}} + \frac{\alpha_2}{1 - Az^{-1}} \qquad (4)$$

[By partial fraction expansion]

$1 - z^{-1} = 0, z - 1 \qquad \alpha_1 = \dfrac{A}{1 - Az^{-1}} = \dfrac{1}{1-A}$

$1 - Az^{-1} = 0, z^{-1} = 1/A \qquad \alpha_2 = \dfrac{1}{1 - z^{-1}} = \dfrac{1}{1 - \dfrac{1}{A}} = \dfrac{A}{A-1} = \dfrac{A}{1-A}$

Substituting the values of α_1 and α_2 in Eq. (4), we get

$$Y(n) = \frac{A}{1 - Az^{-1}} + \frac{\left(\dfrac{1}{1-A}\right)}{1 - z^{-1}} + \frac{\left(\dfrac{-A}{1-A}\right)}{1 - Az^{-1}} \tag{5}$$

Taking the inverse z-transform of Eq. (5), we get

$$Y(n) = A^{n+1}u(n) = \frac{1}{(1-A)}u(n) + \frac{-1}{(1-A)}A^{n+1}u(n).$$

$$= A^{n+1}u(n) + \left(\frac{1 - A^{n+1}}{(1-A)}\right)u(n)$$

$$= \frac{1}{1-A}\left[1 - A^{n+2}\right]u(n)$$

EXAMPLE 3.18

Solve the following difference equation by using z-transform method

$$s(n + 2) + 3s(n + 1) + 2s(n) = 0$$

Initial conditions are $s(0) = 0$ and $s(1) = 1$.

Solution:

Given the difference equation

$$s(n+2) + 3s(n+1) + 2s(n) = 0 \tag{1}$$

Taking the z-transform of both sides of the above equation, we get

$$\left[z_1 S(z) \ - z_1 s(0) - zs(1)\right] + 3\left[zS(z) \ - zs(0)\right] + 2[S(z)] = 0$$

or $$\left[z^2 S(z) \ - z^2.0 \ - z.1\right] + 3\left[zS(z) - z.0\right] + 2\left[S(z)\right] = 0$$

or $$\left[Z3S(z) - 3\right] + 3zS(z) + 2S(z) = 0$$

or $$S(z)\left[z^2 + 3z + 2\right] = z$$

$$S(z) = \frac{z}{z^2 + 3z + 2} \tag{2}$$

Taking the inverse z-transform of above z-transform by partial fraction expansion method

$$S(z) = \frac{z}{z^2+3z+2} = \left\{\frac{\alpha_1}{(z+1)} + \frac{\alpha_2}{(z+2)}\right\} z \quad (3)$$

$$= \frac{z}{z+1} - \frac{z}{z+2} = \frac{1}{1+z^{-1}} - \frac{1}{1+2z^{-1}}$$

$$S(z) = \frac{1}{1+z^{-1}} - \frac{1}{1+2z^{-1}} \quad (4)$$

Taking the inverse z-transform of above Eq. (4), we get

$$s(n) = \mathcal{Z}^{-1}\left[\frac{1}{1+z^{-1}}\right] - \mathcal{Z}^{-1}\left[\frac{1}{1+2z^{-1}}\right]$$

$$= (-1)^n u(n) - (-2)^n (n).$$

EXAMPLE 3.19

Find the response of the system

$$s(n+2) - 3s(n+1) + 2s(n) = \delta(n) \quad (1)$$

When all the initial conditions are zero.

Solution:

Taking z-transform of both sides of above equation, we get

$$z^2 S(z) - 3z^1 S(z) + 2S(z) = 1$$

or $\quad S(z)\left[z^2 - 3z + 2\right) = 1$

or $\quad S(z) = \frac{1}{z^2-3z+2} = \frac{1}{(z-2)(z-1)} = \frac{\alpha_1}{z-2} + \frac{\alpha_2}{z-1}$

or $\quad S(z) = \frac{1}{z-2} - \frac{1}{z-1} \quad (2)$

(By partial fraction expansion)

Taking inverse z-transform of both sides of Eq. (2), we get

$$s(n) = \mathcal{Z}^{-1}\left[\frac{1}{z-2}\right] - \mathcal{Z}^{-1}\left[\frac{1}{z-1}\right]$$

$$= \mathcal{Z}\left[\frac{z^{-1}}{1-2z^{-1}}\right] - \mathcal{Z}^{-1}\left[\frac{z^{-1}}{1-z^{-1}}\right]$$

$$= (2)^{n-1} - (1)^{n-1} = 1 + (2)^{n-1}$$

3.10 ANALYSIS OF LINEAR TIME-INVARIANT (LTI) SYSTEMS IN THE z-DOMAIN

We have already studied the system function of LTI systems or the transfer function of the LTI system.

Also, we know that system function is directly related to unit-sample response (i.e., Impulse response) of the LTI system. Discrete-time LTI systems are described by their difference equations. System function can be found only for initially relaxed systems. In this section, we describe the use of the system function in the determination of the response of the system to some excitation signal. Furthermore, we extend this method of analysis to non-relaxed systems. Here, we shall focus our attention on the important class of pole-zero systems represented by linear constant-coefficient difference equations with arbitrary initial conditions.

We also consider the topic of stability of LTI systems. Finally, we provide a detailed analysis of second-order discrete-time systems. These second-order systems form the basic building blocks in the realization of higher-order systems.

3.10.1 Response of LTI Systems with Rational System Functions

General form of a linear constant-coefficient difference equation of discrete-time LTI system is given by

$$y(n) = \sum_{k=1}^{N} A_k y(n-k) + \sum_{k=0}^{M} B_k s(n-k) \tag{3.21}$$

System function corresponding to Eq. (3.21) can be obtained by using time shifting property of z-transform.

$$\mathcal{Z}\, y(n) = \mathcal{Z}\left\{-\sum_{k=1}^{N} A_k y(n-k) + \sum_{k=0}^{M} B_k s(n-k)\right\}.$$

or
$$Y(z) = -\sum_{k=1}^{N} A_k Y(z) z^{-k} + \sum_{k=0}^{M} B_k S(z) z^{-k}$$

or
$$Y(z) + \sum_{k=1}^{N} B_k S(z) z^{-k} = Y(z) + \sum_{k=1}^{N} B_k S(z) z^{-k}$$

or
$$Y(z) + \sum_{k=1}^{N} A_k z^{-k} = \sum_{k=0}^{M} B_k S(z) z^{-1}$$

or
$$\frac{Y(z)}{S(z)}=\frac{\sum_{k=0}^{M}B_k z^{-k}}{1+\sum_{k=1}^{N}A_k z^{-k}}$$

or
$$H(z)=\frac{Y(z)}{S(z)}=\frac{\sum_{k=0}^{M}B_k z^{-k}}{1+\sum_{k=1}^{N}A_k z^{-k}} \tag{3.22}$$

Therefore, a linear time-invariant system described by a constant-coefficient difference equation has a rational system function. This is the general form for the system function of a system described by a linear constant-coefficient difference equation.

We represent system function $H(z)$ as a ratio of two polynomials $D(z)/C(z)$, where $D(z)$ is the numerator polynomial that contains the zeros of $H(z)$, and $C(z)$ is the denominator polynomial that determines the poles of $H(z)$.

Furthermore, let us assume that the input signal $s(n)$ has a rational z-transform $S(z)$ of the form.

$$S(z)=\frac{N(z)}{Q(z)} \tag{3.23}$$

Most of the signals of practical interest have rational z-transforms.

If the system is initially relaxed, that is, the initial conditions for the difference equation are zero, $y(-1) = y(-2) = y(-3) = ... = y(-N) - 0$, the z-transform of the output of the system has the form.

$$\begin{aligned} Y(z) &= H(z)S(z) \\ &= \frac{D(z)}{C(z)}\cdot\frac{N(z)}{Q(z)} \end{aligned} \tag{3.24}$$

Now suppose that the system contains simple poles $p_1, p_2, ..., p_N$, and the z-transform of the input signal contains poles $q_1, q_2, ... q_3$, where $p_k \neq q_m$ for all $k = 1, 2, ...\ N$ and $m - 1, 2, ..., L$. In addition, we assume that the zeros of the numerator polynomials $D(z)$ and $N(z)$ do not coincide with the poles $\{p_k\}$ and $\{q_k\}$, so that there is no pole-zero cancelation.

Then a partial fraction expansion of $Y(z)$ gives

$$Y(z)=\underbrace{\sum_{k=1}^{N}\frac{C_k}{1-p_k z^{-1}}}_{\text{Ist part}}+\underbrace{\sum_{k=1}^{L}\frac{Q_k}{1-q_k z^{-1}}}_{\text{IInd part}} \tag{3.25}$$

On taking inverse z-transform of Eq. (3.25), we obtain

$$Y(n) = \underbrace{\sum_{k=1}^{N} C_k (p_k)^n u(n)}_{\text{Ist part}} + \underbrace{\sum_{k=1}^{L} Q_k (q_k)^n u(n)}_{\text{IInd part}} \tag{3.26}$$

We observe from Eq. (3.26) that the output sequence $y(n)$ can be subdivided into two parts. The first part is a function of the poles $\{p_k\}$ of the system and is called the natural response of the system. The influence of the input signal on this part of the response is through the scale factors $\{C_k\}$.

The second part of the response is a function of the poles $\{q_k\}$ of the input signal and is called the forced response of the system. The influence of the system on this response is exerted through the scale factors $\{Q_k\}$.

It is to be emphasized that the scale factors $\{C_k\}$ and $\{Q_k\}$ are functions of both sets of poles $\{q_k\}$ and $\{q_k\}$. For example, if $S(z) = 0$. so that the input is zero, then $Y(z) = 0$, and consequently, the output is zero. Clearly, then the natural response of the system is zero. This implies that the natural response of the system is different from the zero-input response.

When $S(z)$ and $H(z)$ have one or more poles in common or when $S(z)$ and/or $H(z)$ contain multiple-order poles, then $Y(z)$ will have multiple-order poles. Consequently, the partial fraction expansion of $Y(z)$ will contain factors of form $1/(1 - p_l^{z-1})k$, $k = 1, 2, \ldots, m$, where m *is* the pole order. The inversion of these factors will produce terms of the form $n^{k-1} p_l^n$ in the output $y(n)$ of the system.

3.10.2 Response of Pole-Zero Systems with Non-Zero Initial Conditions

We suppose that the signal $s(n)$ is applied to the pole-zero system at $n = 0$. Thus the signal $s(n)$ is assumed to be causal. The effects of all previous input signals to the system are reflected in the initial conditions $y(-1)$, $y(-2)$, ..., $y(-N)$. Since the input $s(n)$ is causal and we are interested in determining the output $y(n)$ for $n \geq 0$, we can use the unilateral or one-sided z-transform. One-sided z-transform allows us to deal with the initial conditions.

Now taking the one-sided z-transform of both sides of Eq. (3.21), we get

$$\mathcal{Z}[y(n)] = \mathcal{Z}_u\left[-\sum_{k=1}^{N} A_k y(n-k) + \sum_{k=0}^{M} B_k s(n-k)\right]$$

or

$$Y^{+}(z) = -\sum_{k=1}^{N} A_k z^{-k}\left[Y^{+}(z) + \sum_{n=1}^{k} y(-n)z^{n}\right]$$

$$+\sum_{k=0}^{M} B_k z^{-k} S^{+}(z) \tag{3.27}$$

Since $s(n)$ is a causal, we can set $S+(z) = S(z)$.

In any case Eq. (3.27) may be expressed as

$$Y^{+}(z) = \underbrace{\frac{\sum_{k=0}^{M} B_k z^{-k}}{1+\sum_{k=1}^{N} A_k z^{-k}}}_{\text{I}} S(z) - \underbrace{\frac{\sum_{k=1}^{N} A_k z^{-k}\sum_{n=1}^{k} y(-n)z^{n}}{1+\sum_{k=1}^{N} A_k z^{-k}}}_{\text{II}} \tag{3.28}$$

$$Y^{+}(z) = H(z)S(z) + \frac{N_0(z)}{C(z)}$$

where

$$N_0(z) = -\sum_{k=1}^{N} A_k z^{-k}\sum_{n=1}^{k} y(-n)z^{n} \tag{3.29}$$

From Eq. (3.28), it is apparent that the output of the system with non-zero initial conditions can be subdivided into two parts.

The first part is the zero-state response of the system. It is defined in z-domain as

$$Y_{zs} = H(z)S(z) \tag{3.30}$$

and the second part is the zero-input response. It is given in z-domain as

$$Y_{zi}^{+}(z) = \frac{N_0(z)}{C(z)}$$

Hence, the total response is the sum of their two output components. These output components can be expressed in the time-domain by determining the inverse z-transforms of $Y_{zs}(z)$ and $Y_{zi}(z)$, separately, and then adding the results. Thus

$$y(n) = y_{zs}(n) + y_{zi}(n) \tag{3.31}$$

Since the denominator of $Y_{zi}^{+}(z)$, is $C(z)$, its poles are $p_1, p_2, p_3, \ldots, p_n$. Consequently, the zero-input response has the form

$$Y_{zi}(n) = \sum_{k=1}^{N} E_k (p_k)^n u(n) \tag{3.32}$$

This can be added to Eq. (3.26) and the terms involving the poles can be combined to yield the total response in the form

$$y(n) = \sum_{k=1}^{N} C_k (p_k)^n u(n) + \sum_{k=1}^{L} Q_k (q_k)^n u(n) + \sum_{k=1}^{N} E_k (p_k)^n u(n)$$

$$= \sum_{k=1}^{N} [C_k + E_k](p_k)^n u(n) + \sum_{k=1}^{L} Q_k (q_k)^n u(n)$$

$$= \sum_{k=1}^{N} C_k^{'} (p_k)^n u(n) + \sum_{k=1}^{L} Q_k (q_k)^n u(n) \quad (3.33)$$

where, $C_k = C_k + E_k$ (3.34)

This discussion indicates clearly that the effect of initial conditions is to alter the natural response of the system through modification of the scale factor $\{C_k\}$.

There are no new poles introduced by the non-zero initial conditions. Furthermore, there is no effect on the forced response of the system. These important points can be understood by following example.

EXAMPLE 3.20

Find the unit-step response of the discrete-time system described by the following difference equation

$$y(n) = 0.9y(n-1) - 0.81y(n-2) + s(n)$$

under the following initial conditions:

$(a)\ y(-1) = y(-2) = 0$ $(b)\ y(-1) = y(-2) = 1.$

Solution:

The system function of above system described difference equation

$$y(n) = 0.9y(n-1) - 0.81y(n-2) + s(n)$$

is determined by taking z-transform as

$$\mathcal{Z}[Y(n)] = \mathcal{Z}[0.9y(n-1) - 0.81y(n-2) + s(n)]$$

$$Y(z) = 0.9z^{-1}Y(z) - 0.81z^{-2}Y(z) + S(z)$$

or $$\frac{Y(z)}{S(z)} = \frac{1}{1 - 0.9z^{-1} + 0.81z^{-2}}$$

or $$H(z) = \frac{Y(z)}{S(z)} = \frac{1}{1 - 0.9z^{-1} + 0.81z^{-2}} \quad (1)$$

This system has two complex-conjugate poles at

$$p_1 = 0.9e^{j\pi/3} \text{ and } p_2 = 0.9e^{-j\pi/3}$$

The z-transform of unit-step sequence $u(n)$ is

$$S(n) = \mathcal{Z}[u(n)]$$

$$= \frac{1}{1 - z^{-1}} \tag{2}$$

Substituting Eq. (2) in Eq. (1), we get

$$Y_{zs}(z) = \frac{1}{\left[1 - 0.9e^{\frac{j\pi}{3}} z^{-1}\right]\left[1 - 0.9e^{\frac{j\pi}{3}} z^{-1}\right]\left[1 - z^{-1}\right]}$$

$$= \frac{(0.542 - j0.049)}{\left(1 - 0.9e^{\frac{j\pi}{3}} z^{-1}\right)} + \frac{(0.542 + j0.049)}{\left(1 - 0.9e^{\frac{j\pi}{3}} z^{-1}\right)} + \frac{1.099}{\left(1 - z^{-1}\right)} \tag{3}$$

We can determine the zero-state response by taking inverse z-transform of Eq.(3)

$$y_{zs}(n) = \mathcal{Z}^{-1}\left[Y_{zs}(z)\right]$$

$$= \left[1.099 + 1.088(0.9)^n \cos\left(\frac{\pi}{3} n = 5.2°\right)\right] u(n) \tag{4}$$

(a) Since the initial conditions are zero in this case, we conclude that

$$y(n) = y_{zs}(n)$$

(b) For the initial conditions

$$y(-1) = y(-2) = 1$$

the additional component in the z-domain is given

$$Y_{zi}(z) = \frac{N_0(z)}{C(z)}$$

$$= \frac{0.09 - 0.81z^{-1}}{1 - 0.9z^{-1} + 0.81z^{-2}}$$

$$= \frac{(0.026 + j0.4936)}{\left(1 - 0.9e^{\frac{j\pi}{3}} z^{-1}\right)} + \frac{(0.026 - j0.4936)}{\left(1 - 0.9e^{\frac{j\pi}{3}} z^{-1}\right)} \tag{5}$$

Consequently, the zero-input response is determined as

$$y_{zi}(n) = Z^{-1}[Y_{zi}(z)]$$

$$= Z^{-1}\left[\frac{(0.026 + j0.4936)}{\left(1 - 0.9e^{\frac{j\pi}{3}} z^{-1}\right)} + \frac{(0.026 - j0.4936)}{\left(1 - 0.9e^{\frac{j\pi}{3}} z^{-1}\right)}\right]$$

$$= 0\,988(0\,9)^n \cos\left(\frac{\pi}{3}n + 87°\right)u(n)u(n) \quad (6)$$

In this case, the total response has the z-transform

$$Y(z) = Y_{zs}(z) + Y_{zi}(z) \quad (7)$$

Substituting Eqs. (3) and (5) in Eq. (7), we get

$$y(z) = \frac{1.099}{\left(1 - z^{-1}\right)} + \frac{(0.568 + j0.445)}{\left(1 - 0.9e^{\frac{\pi}{3}} z^{-1}\right)} + \frac{(0.568 - j0.445)}{\left(1 - 0.9e^{-\frac{j\pi}{3}}\right)} \quad (8)$$

Total response of the system in this case is obtained by taking inverse z-transform of Eq. (8)

$$y(n) = 1.099u(\text{n}) + 1.44(0.9)^n \cos\left(\frac{\pi}{3}n + 38°\right)u(n) \quad (9)$$

3.10.3 Transient and Steady-State Responses

We have already studied that the response of a system to a given input can be separated into two components. These components are

Natural Response and Forced Response.

The natural response of a causal system has the form

$$y_{nr}(n) = \sum_{k=1}^{N} C_k \left(p_k\right)^n u(n) \quad (3.35)$$

where $\{p_k\}$ $k = 1, 2, 3, \ldots, N$ are the poles of the system and $\{C_k\}$ are the scale factors. These scale factors depend upon the initial conditions and on the characteristics of the input sequence.

If $|p_k| < 1$ for all values of k, then, natural response $y_{nr}(n)$ decays to zero as n approaches ∞. In such a case, we refer to the natural response of the system as the transient response.

The rate at which natural response $y_{nr}(n)$ decays toward zero depends on the magnitude of the pole positions. If all the poles have small magnitudes, the decay is very rapid.

On the other hand, if one or more poles are located near the unit circle, the corresponding terms in natural response $y_{nr}(n)$ will decay slowly toward zero and the transient will persist for a relatively longer time.

The forced response of the system has the form

$$fr(n) = \sum_{k=1}^{L} Q_k \left(q_k\right)^n u(n) \tag{3.36}$$

where $\{q_k\}$, $k = 1, 2, 3, ..., L$ are the poles in the forcing function and are the scale factors. These scale factors depend on the input sequence and on the characteristics of the system. If all the poles of the input signal fall inside the unit circle, forced response $y_{fr}(n)$ will decay toward zero as n approaches ∞, just as in the case of natural response.

It is not surprising, since the input signal is also a transient signal. On the other hand, when the causal input signal is a sinusoid, the poles fall on the unit circle and consequently, the forced response is also a sinusoid that persists for all $n > 0$. In this case, the forced response is called the steady-state response of the system. Thus, for the system to sustain a steady-state output for $x \geq 0$ the input signal must persist for all $n \geq 0$.

Transient and steady-state response can be understood by the following example.

EXAMPLE 3.21

Find the transient and steady-state responses of the discrete-time system characterized by the difference equation.

$$y(n) = 0.5y(n-1) + s(n)$$

when the input signal is

$$s(n) = 10\cos\left(\frac{\pi n}{4}\right)u(n)$$

The system is initially at rest, that is, it is initially relaxed.

Solution:
The system function can be found by taking z-transform of the above difference equation.

$$\mathcal{Z}[y(n)] = \mathcal{Z}\left[0.5y(n-1) + s(n)\right]$$

or
$$Y(z) = 0.5z^{-1}Y(z) + S(z)$$

or
$$Y(z)[1-0.5z^{-1}] = S(z)$$

or
$$\frac{Y(z)}{S(z)} = \frac{1}{1-0.5z^{-1}}$$

or
$$H(z) = \frac{Y(z)}{S(z)} = \frac{1}{1-0.5z^{-1}} \tag{1}$$

This system has a pole at $z = 0.5$.

The z-transform of the input signal can be determined by using the table of z-transform as

$$S(z) = \mathcal{Z}[s(n)] = \mathcal{Z}\left[10\cos\left(\frac{\pi n}{4}\right)u(n)\right]$$

$$= \frac{10\left[1-\left(\frac{1}{\sqrt{2}}\right)z^{-1}\right]}{1-\sqrt{2}z^{-1}+z^{-2}} \tag{2}$$

Eq. (1) can be written as

$$Y(z) = H(z)S(z) = \frac{1}{1-0.5z^{-1}}S(z) \tag{3}$$

Substituting Eq. (2) in Eq. (3), we get,

$$Y(z) = \frac{1}{\left(1-0.5z^{-1}\right)}\frac{10\left[1-\left(\frac{1}{\sqrt{2}}\right)z^{-1}\right]}{1-\sqrt{2}z^{-1}+z^{-2}}$$

$$= \frac{1}{\left(1-0.5z^{-1}\right)}\frac{10\left[1-\left(\frac{1}{\sqrt{2}}\right)z^{-1}\right]}{\left(1-e^{\frac{j\pi}{4}}z^{-1}\right)\left(1-e^{-\frac{j\pi}{4}}z^{-1}\right)}$$

or
$$Y(z)=\frac{6.3}{1-0.5z^{-1}}+\frac{6.78e^{-j28.7^\circ}}{\left(1-e^{\frac{j\pi}{4}}z^{-1}\right)}+\frac{6.78e^{j28.7^\circ}}{\left(1-e^{-\frac{j\pi}{4}}z^{-1}\right)} \tag{4}$$

(Using partial fraction expansion)

The natural or transient response is determined as

$$yfr(n)=\text{Inverse } z\text{-transform}\left[\frac{6.3}{1-0.5z^{-1}}\right]$$

$$=6.3(0.5)^n u(n) \tag{5}$$

The forced or steady-state response is determined as

$$yfr(n)=\text{Inverse } z\text{-transform}\frac{6.78e^{-j28.7^\circ}}{\left(1-e^{\frac{j\pi}{4}z^{-1}}\right)}+\frac{6.78e^{j28.7^\circ}}{\left(1-e^{-\frac{j\pi}{4}z^{-1}}\right)}$$

$$=\left[6.78e^{-28.7^\circ}\left(e^{\frac{j\pi n}{4}}\right)+6.78e^{28.7^\circ}\left(e^{-\frac{j\pi n}{4}}\right)\right]u(n)$$

$$=13.56\cos\left(\frac{\pi}{4}n-28.7^\circ\right)u(n)$$

Thus, we see that the steady-state response persists for all $n \geq 0$, just as the input signal persists for all $n \geq 0$.

3.10.4 Causality and Stability

A causal LTI system is one whose unit-sample (impulse) response $h(n)$ satisfies the condition.

$$h(n)=0, n<0$$

We have also shown that the ROC of the z-transform of a causal sequence is the exterior of a circle. Consequently, an LTI system is causal if and only if the ROC of the system function is the exterior of a circle of radius $r < \infty$, including the point $z = \infty$.

The stability of an LTI system can also be expressed in terms of the characteristics of the system function $H(z)$. A necessary and sufficient condition for LTI discrete-time system to be bounded input bounded output (BIBO) stable is

$$\sum_{n=-\infty}^{\infty} |h(n)| < \infty$$

In turn, this condition implies that $H(z)$ must contain the unit circle within its ROC.

Indeed, since $\quad H(z) = \sum_{n=-\infty}^{\infty} h(n)z^{-n}$

it follows that

$$|H(z)| \leq \sum_{n=-\infty}^{\infty} |h(n)z^{-n}| = \sum_{n=-\infty}^{\infty} |h(n)||z^{-n}|$$

when evaluate on the unit circle (i.e., $|z| = 1$),

$$|H(z)| \leq \sum_{n=-\infty}^{\infty} |h(n)|$$

Hence, if the system is BIBO stable, the unit circle is contained in the ROC of $H(z)$. The converse is also true. Therefore, an LTI system is BIBO stable if and only if the ROC of the system function $H(z)$ includes the unit circle.

We know that the conditions for causality and stability are different and that one does not imply the other.

For example, a causal system may be stable or unstable, just as a non-causal system may be stable or unstable. Similarly, an unstable system may be either causal or non-causal, just as a stable system may be causal or non-causal.

For a causal system, however, the condition on stability can be narrowed to some extent. Indeed, a causal system is characterized by a system function $H(z)$ having a ROC as the exterior of some circle of radius r.

For a stable system, the ROC must include the unit circle. Consequently, casual and stable system must have system functions that converges for $|z > r$ < 1. Since the ROC cannot contain any poles of $H(z)$, it follows that a causal LTI system is BIBO stable if and only if all the poles of $H(z)$ are inside the unit circle, $|z| = 1$.

EXAMPLE 3.22

An LTI (discrete-time) system is characterized by the system function

$$H(z) = \frac{3-4z^{-1}}{1-3.5z^{-1}+1.5z^{-2}}$$

Specify the ROC of $H(z)$.
Also, find unit-sample response $h(n)$ for the following conditions:

a. The system is stable

b. The system is causal

c. The system is anti-causal.

Solution:

System function

$$H(z) = \frac{3-4z^{-1}}{1-3.5z^{-1}+1.5z^{-2}}$$

$$= \frac{3-4z^{-1}}{\left(1-\frac{1}{2}z^{-1}\right)\left(1-3z^{-1}\right)}$$

$$= \frac{1}{\left(1-\frac{1}{2}z^{-1}\right)} + \frac{2}{\left(1-3z^{-1}\right)} \qquad (i)$$

The system has poles at $z = \frac{1}{2}$ and $z = 3$

a. Since the system is stable. Therefore its ROC must include the 1 unit circle and hence it is $\frac{1}{2} < |z| < 3$. Consequently, $h(h)$ is non-causal.

In this case, unit-sample response $h(n)$ is given by

$$h(n) = \left(\frac{1}{2}\right)^n u(n) - 2(3)^n u(-n-1)$$

b. Since the system is causal, its ROC is $|z| > 3$. In this case

$$h(n) = \left(\frac{1}{2}\right)^n u(n) + 2(3)^n u(n)$$

This system is unstable.

c. If the system is anti-causal, its ROC is $|z| < 0.5$. Hence

$$h(n) = -\left[\left(\frac{1}{2}\right)^n + 2(3)^n\right] u(-n-1)$$

In this case, the system is unstable.

3.10.5 Pole-zero Cancelations

When a z-transform of signal has a pole that is at the same location as zero, the pole canceled by zero. Consequently, the term containing that pole in the inverse z-transform vanishes. Such pole-zero cancelations are very important in the analysis of pole-zero systems.

Pole-zero cancelations can occur either in the system function itself or in the product of the system function with the z-transform of the input signal. Pole-zero cancelations in system function itself reduce the order of the system by one. Pole-zero cancelations in the case of product of $H(z)$ and $S(z)$, where $S(z) = \mathcal{Z}\{s(n)\}$, suppress the pole of the system by the zero in the input signal or vice versa. Thus by properly selecting the position of the zeros of the input signal, it is possible to suppress one or more system modes (poles factors) in the response of the system.

Similarly, by proper selection of the zeros of the system function $H(z)$, it is possible to suppress one or more modes of the input signal from the response of the system.

When the zero is located very near the pole but not exactly at the same location the term in the response has a very small amplitude. For example, non-exact pole-zero cancelations can occur in practice as a result of insufficient numerical precision used in representing the coefficient of the system. Consequently, one should not attempt to stabilize an inherently unstable system by placing a zero in the input signal at the location of the pole.

EXAMPLE 3.23

Find the unit-sample (impulse) response $h(n)$ of the system characterized by the difference equation.

$$y(n) = 2.5y(n-1) - y(n-2) + s(n) - 5s(n-1) + 6s(n-2).$$

Solution:
System function $H(z)$ can be determined by taking z-transform of both sides of above difference equation.

$$\mathcal{Z}[y(n)] = \mathcal{Z}[2.5y(n-1) - y(n-2) + s(n) - 5s(n-1) + 6s(n-2)]$$

or $$Y(z) = 2.5z^{-1}Y(z) - z^{-2}Y(z) + S(z) - 5z^{-1}S(z) + 6z^{-2}S(z)$$

or $$H(z) = \frac{Y(z)}{S(z)} = \frac{1-5z^{-1}+6z^{-2}}{1-2.5z^{-1}+z^{-2}} \qquad (1)$$

Eq. (1) can also be written as

$$H(z) = \frac{1-5z^{-1}+6z^{-2}}{\left(1-\frac{1}{2}z^{-1}\right)\left(1-2z^{-1}\right)} \qquad (2)$$

This system has poles at $p_1 = 2$ and $p_2 = \frac{1}{2}$. Consequently,

$$Y(z) = H(z)S(z)$$

or $$Y(z) = \frac{1-5z^{-1}+6z^{-2}}{\left(1-\frac{1}{2}z^{-1}\right)\left(1-2z^{-1}\right)}S(z) \qquad (3)$$

Since $s(n) = \delta(n)$

$$\mathcal{Z}[s(n)] = \mathcal{Z}[\delta(n)]$$

or $$S(z) = 1 \qquad (4)$$

Substituting Eq. (4) in Eq. (3), we get

$$y(z) = \frac{1-5z^{-1}+6z^{-2}}{\left(1-\frac{1}{2}z^{-1}\right)\left(1-2z^{-1}\right)}\cdot 1$$

$$= z\left(\frac{A}{z-1/2} + \frac{B}{z-2}\right) = z\left(\frac{5/2}{z-1/2} + \frac{0}{z-2}\right)$$

$$= \frac{\frac{5}{2}z}{z-\frac{1}{2}} = \frac{2.5z}{z-\frac{1}{2}}$$

The fact that $B = 0$ indicates that there exists a zero at $z = 2$ which cancels the pole at $z = 2$. In fact, the zeros at occur $z = 2$ and $z = 3$.

Consequence, $H(z)$ reduces to

$$H(z) = \frac{1 - 3z^{-1}}{1 - \frac{1}{2}z^{-1}}$$

$$= \frac{z - 3}{z - \frac{1}{2}}$$

$$= 1 - \frac{2.5z^{-1}}{1 - \frac{1z^{-1}}{2}} \tag{5}$$

Taking the inverse z-transform of both sides of Eq. (5), we get

$$h(n) = \mathcal{Z}^{-1}[H(z)]$$

$$= \mathcal{Z}^{-1}\left[1 - \frac{2.5z^{-1}}{1 - 1/2z^{-1}}\right]$$

$$= \delta(n) - 2.5\left(\frac{1}{2}\right)^{n-1} u(n-1) \tag{6}$$

The reduced-order system was obtained by canceling the common pole and zero. Then it is characterized by the difference equation.

$$y(n) = \frac{1}{2}y(n-1) + s(n) - 3n(n-1) \tag{7}$$

Although the original system is also BIBO stable due to pole-zero cancelations, in a practical implementation of the second-order system, we may encounter instability due to imperfect cancelation of the pole and the zero.

EXAMPLE 3.24

Find the response of the discrete-time system described by difference equation given as

$$y(n) = \frac{5}{6}y(n-1) - \frac{1}{6}y(n-2) + s(n)$$

to the input signal

$$s(n) = \delta(n) - \frac{1}{3}\delta(n-1)$$

Solution:

The system function of this system can be found by taking z-transform of both sides of difference equation.

$$y(n) = \frac{5}{6}y(n-1) - \frac{1}{6}y(n-2) + s(n)$$

as

$$Y(z) = \frac{5}{6}z^{-1}Y(z) - \frac{1}{6}y^{-2}Y(z) + S(z)$$

or

$$H(z) = \frac{Y(z)}{S(z)} = \frac{1}{1 - \frac{5}{6}z^{-1} + \frac{1}{6}z^{-2}} \quad (1)$$

Eq. (1) can be written as

$$H(z) = \frac{1}{\left(1 - \frac{1}{2}z^{-1}\right)\left(1 - \frac{1}{3}z^{-1}\right)} \quad (2)$$

This system has two poles at $z = 1/2$ and $z = 1/3$.

The z-transform of input signal can be determined as

$$S(z) = \mathcal{Z}[s(n)]$$

$$= \mathcal{Z}\left[\delta(n) - \frac{1}{3}\delta(n-1)\right]$$

$$= 1 - \frac{1}{3}z^{-1} \quad (3)$$

In this case, the input signal contains a zero at $z = 1/3$ which cancels the pole at $z = 1/3$. Consequently,

$$Y(z) = H(z)S(z)$$

$$Y(z) = \frac{1}{1 - \frac{1}{2}z^{-1}} \quad (4)$$

Now, taking the inverse z-transform of Eq. (4), we get the response of the system

$$y(n) = \left(\frac{1}{2}\right)^n u(n) \quad (5)$$

Clearly, the mode $(1/3)^n$ is suppressed from the output as a result of the pole-zero cancelation.

3.10.6 Multiple-order Pole and Stability

A necessary and sufficient condition for a causal LTI system to be BIBO stable is that all its poles be inside the unit circle. The input signal is bounded if its z-transform contains poles $\{q_k\}$, $k = 1, 2, 3, ..., L$, which satisfy the condition $|q_k| \leq 1$ for all values of k. We note that the forced response of the system, given here:.

$$yfr(n) = \sum_{k=1}^{L} Q_k (q_k)^n u(n)$$

is also bounded, even when the input signal contains one or more distinct poles on the unit circle.

In view of the fact that a bounded input signal may have poles on the unit circle, it might appear that a stable system may also have poles on the unit circle. This is not the case, however, since such a system produces an unbounded response when executed at the same position on the unit circle.

EXAMPLE 3.25

Find the step response of the causal system described by the difference equation

$$y(n) = y(n-1) + s(n) \tag{1}$$

Solution:

System function of above system can be determined by taking z-transform of both side of Eq. (1) as

$$\mathcal{Z}[y(n)] = \mathcal{Z}\left[y(n-1) + s(n)\right]$$

or $$Y(z) = z^{-1}Y(z) + S(z)$$

or $$\frac{Y(z)}{S(z)} = \frac{1}{1-z^{-1}}$$

or $$H(z) = \frac{Y(z)}{S(z)} = \frac{1}{1-z^{-1}}$$

or $$Y(z) = \frac{1}{1-z^{-1}} S(z) \tag{2}$$

For step response, we put

$$s(n) = u(n)$$

$$\mathcal{Z}[s(n)] = \mathcal{Z}[u(n)]$$

$$S(n) = \frac{1}{1 - z^{-1}} \tag{3}$$

Substituting Eq. (3) in Eq. (2), we get

$$Y(z) = \frac{1}{1 - z^{-1}} \cdot \frac{1}{1 - z^{-1}} = \frac{1}{\left(1 - z^{-1}\right)^2} \tag{4}$$

$Y(z)$ contains a double pole at $z = 1$. Taking the inverse z-transform of Eq.(4)

$$y(n) = \mathcal{Z}^{-1}[Y(z)]$$

$$= \mathcal{Z}^{-1}\left[\frac{1}{\left(1 - z^{-1}\right)^2}\right]$$

or $$y(n) = (n + 1)u(n)$$

This is ramp sequence.

Thus $y(n)$ is unbounded, even when the input is bounded. Consequently, the system is unstable.

Example 3.25 demonstrates clearly that BIBO stability requires that the system poles be strictly inside the unit circle, that is, $|z| = 1$. If the system poles are all inside the unit circle and the excitation sequence $s(n)$ contains one or more poles that coincide with the poles of the system, the output $Y(z)$ will contain multiple-order poles. Such multiple-order poles result in a output sequence that contains terms of the form

$$C_k n^b (p_k)^n u(n)$$

where $0 \le b \le m - 1$ and m is the order of the pole. If $|p_k| < 1$, these terms decay to zero as $n \to \infty$ because the exponential factor $(p_k)^n$ dominates the terms n^b.

Consequently, no bounded input signal can produce an unbounded output signal if the system poles are all inside the unit circle.

Finally, the only useful systems which contain poles on the unit circle are the digital oscillators. We call such systems marginally stable.

3.10.7 The Schur-Cohn Stability Test

We know that the stability of a system is determined by the position of the poles. The poles of the system are the roots of the denominator polynomial of $H(z)$, namely,

$$C(z) = 1A_1z^{-1} + A_2z^{-2} + \ldots + A_Nz^{-N} \tag{3.37}$$

When the system is causal all the roots of $C(z)$ must lie inside the unit circle for the system to be stable.

There are several computational procedures that help in determining if any of the roots of $C(z)$ be outside the unit circle. These procedures are called stability criteria.

Now, we describe the Schur-Cohn test procedure for the stability of a system characterized by the system function $H(z) = D(z)/C(z)$. Before we describe the Schur-Cohn test, we need to establish some useful notation.

We denote a polynomial of degree m by

$$Cm(z) = \sum_{k=0}^{m} A_m(k)z^{-k}, A_m(0) = 1 \tag{3.38}$$

The reciprocal or reverse polynomial $D_m(z)$ of degree *in* is defined as

$$D_m(z) = z^{-m}Cm\left(z^{-1}\right)$$

$$= \sum_{k=0}^{m} A_m(m-k)z^{-k} \tag{3.39}$$

We observe that the coefficients of $D_m(z)$ are the same as those of $C_m(z)$, but in reverse order.

In the Schur-Cohn Stability test, to determine if the polynomial $C(z)$ has all its roots inside the unit circle, we compute a set of coefficients.

These coefficients are called reflection coefficients, α_1, α_2, ... ,α_N from the polynomials $C_m(z)$. First, we set

$$C_N(z) = C(z)$$

and

$$\alpha_N = A_N(N) \tag{3.40}$$

Then we determine the lower-order polynomial $C_m(z)$, $m = N, N-1, N-2, \ldots, 1$, according to the recursive equation

$$C_{m-1}(z) = \frac{C_m(z) - \alpha_m D_m(z)}{1 - \alpha_m^2} \tag{3.41}$$

Where the coefficients are defined as

$$\alpha_n = A_m(m) \tag{3.42}$$

The Schur-Cohn Stability test states that the polynomial $C(z)$ given by Eq. (3.37) has all its roots inside the unit circle if and only if the reflection coefficients α_m satisfy the condition $|\alpha_m| < 1$ for all $m = 1, 2, \ldots, N$.

EXAMPLE 3.26

Find if the system having the system function

$$H(z) = \frac{1}{1 - \frac{7}{4}z^{-1} - \frac{1}{2}z^{-2}} \text{ is stable.}$$

Solution:

We begin with $C_2(z).C_2(z)$ is defined as

$$C_s(z) = 1 - \frac{7}{4}z^{-1} - \frac{1}{2}z^{-2}$$

Hence, $\alpha_2 = \frac{1}{2}$

Now $D_2(z) = -\frac{1}{2} - \frac{1}{4}z^{-1} + z^{-2}$

and $C_1(z) = \frac{C_2(z) - \alpha_2 D_2(z)}{1 - \alpha_2^2}$

$$= 1 - \frac{7}{2}z^{-1}$$

Therefore, $\alpha_1 = -\frac{7}{2}$

Since $|\alpha_1| > 1$, it follows that the system is unstable. The fact is easily established in this example, since the denominator is easily factored to yield the two poles at $p_1 = -2$ and $p_2 = 1/4$. However, for higher degree polynomials, the Schur-Cohn test provides a simpler test for stability than direct factoring of $H(z)$.

The Schur-Cohn Stability test can be easily programmed in a digital computer. It is very efficient in terms of arithmetic operations. Specially, it requires only N^2 multiplications to determine the coefficients $\{\alpha_m\}$, $m = 1, 2, \ldots, N$. The recursive equation $C_{m-1}(z) = \dfrac{C_m(z) - \alpha_m D_m(z)}{1-\alpha_m^2}$ can be expressed in terms of the polynomial coefficients corresponding to equal powers. Indeed, it is easily established that equation:

$$C_{m-1(z)} = \frac{C_m(z) - \alpha_m D_m(z)}{1-\alpha_m^2}$$

is equivalent to the following algorithm

$$A_N(k) = A_k, k = 1,2,...N \tag{7}$$

$$\alpha_N = A_N(N) \tag{8}$$

Then, for $m = N, N-1, ..., N$, compute

$$\alpha_m = A_m(m)$$

$$A_{m-1}(0) = 1$$

and
$$A_{m-1(k)} = \frac{A_m(k) - \alpha_m D_m(k)}{1-\alpha_m^2}, k = 1, 2, ..., m-1 \tag{9}$$

where
$$D_m(k) = A_m(m-k),\ k = 0,1,2 \tag{10}$$

This recursive algorithm for the computation of coefficients $\{\alpha_m\}$ finds application in various signal processing problems, especially in speech signal processing.

3.10.8 Stability of Second-order Systems

Here, we discuss a detailed analysis of a system having two poles. As we know that two-pole systems form the basic building blocks for the realization of higher-order systems.

Let us consider a causal two-pole system described by the second-order difference equation.

$$y(n) = -A_1 y(n-1) - A_2 y(n-2) + B_0 s(n) \tag{3.43}$$

The system function can determined by taking z-transform of both sides of above equation as

$$\mathcal{Z}[y(n)] = \mathcal{Z}\left[-A_1 y(n-1) - A_2 y(n-2) + B_0 s(n)\right]$$

or
$$Y(z) = -A_1 z^{-1} Y(z) - A_2 z^{-2} Y(z) + B_0 S(z)$$

or $$Y(z)[1 + A_1z^{-1} + A_2z^{-2}] = B_0S(z)$$

or $$\frac{Y(z)}{S(z)} = \frac{B_0}{1 + A_1z^{-1} + A_2z^{-2}} \tag{3.44}$$

Eq. (3.44) can be written as

$$H(z) = \frac{Y(z)}{S(z)} = \frac{B_0z^2}{z^2 + A_1z + A_2} \tag{3.45}$$

This system has two zeros at the origin ($z = 0$) and two poles at

$$p_1 = z = -\frac{A_1}{2} + \sqrt{\frac{A_1^2 - 4A_2}{4}}$$

$$p_2 = -\frac{A_1}{2} - \sqrt{\frac{A_1^2 - 4A_2}{4}}$$

This system is BIBO stable if the poles lie inside the unit circle, that is, $|z| = 1$, that is if $|p_1| < 1$ and $|p_2| < 1$. These conditions can be related to the values of the coefficients A_1 and A_2. In particular, the roots of a quadratic equation satisfy the relations

$$A_1 = -(p_1 + p_2) \tag{3.46}$$

$$A_2 = p_1p_2 \tag{3.47}$$

From Eqs. (3.46) and (3.47), we easily obtain the conditions that A_1 and A_2 must satisfy for stability. First, A_2 must satisfy the condition.

$$|A_2| = |p_1p_2| = |p_1||p_2| < 1 \tag{3.48}$$

The condition for A_1 can be expressed as

$$|A_1| < 1 + A_2 \tag{3.49}$$

The conditions in Eqs. (3.48) and (3.49) can also be derived from the Schur-Cohn Stability test. From the following recursive equation:

$$A_N(k) = A_k, k = 1, 2, \ldots, N$$

$$\alpha_N = A_N(N)$$

$$A_{m-1}(k) = \frac{A_m(k) - \alpha_m B_m(k)}{1-\alpha_m^2}$$

where $B_m(k) = A_m(m-k), k = 0,\ 1,\ ....,m,$

We obtain that

$$\alpha_1 = \frac{A_1}{1+A_2} \tag{3.50}$$

and $\alpha_2 = A_2$ (3.51)

The system is stable if and only if $|\alpha_1| < 1$ and $|\alpha_2| < 1$. Consequently,

$$-1 < A_2 < 1$$

or, equivalently,

$$A_1 < 1 + A_2$$

$$A_1 > -1 - A_2$$

which are in agreement with Eq. (3.50). Therefore, a two-pole system is stable if and only if the coefficients A_1 and A_2 satisfy conditions given Eqs. (3.49) and (3.50).

The stability conditions are given in Eqs. (3.49) and (3.50), define a region in the coefficient plane (A_1, A_2), which is in the form of a triangle as shown in Figure 3.8. The system is stable if and only if the point (A_1, A_2) lies inside the triangle. This triangle is called the stability triangle.

Characteristics of Second-order (two-pole) Discrete-time Systems: The characteristics of the two-pole system depend on the location of the poles or, equivalently, on the location of the point (A_1, A_2) in the stability triangle. The poles of the system may be real or complex conjugate.

These poles depend on the value of the discriminant $\Delta = A_1^2 - 4A_2$. The parabola $A_2 = A_1^2/4$ splits the stability triangle into two regions (as shown in Figure 3.8. The region below the parabola ($A_1^2 > 4A_2$)

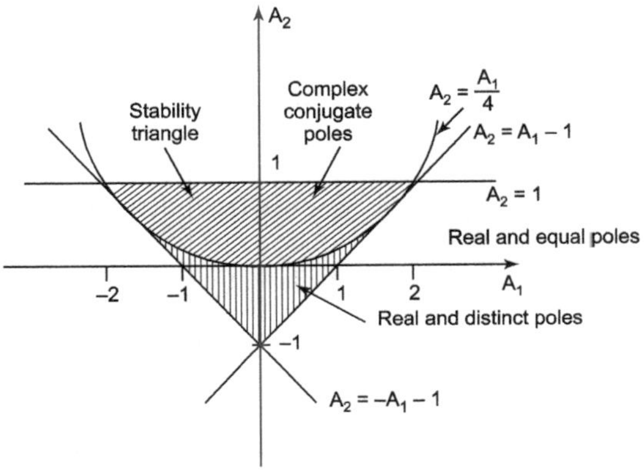

FIGURE 3.8 Region of stability (stability triangle) in the (A_1, A_2) coefficient plane for a second-order system.

corresponds to real and distinct poles. The points on the parabola $(A_1^2 > 4A_2)$ result in real and equal (double) poles. Finally, the points above the parabola correspond to complex-conjugate poles.

Now, we discuss the behavior of the system by using unit-sample responses for the following three cases:

1. Real and distinct poles $\left(A_1^2 > 4A_2\right)$
2. Real and equal poles $\left(A_1^2 = 4A_2\right)$
3. Complex-conjugate poles $\left(A_1^2 < 4A_2\right)$.

Real and distinct poles $\left(A_1^2 > 4A_2\right)$**:** Since poles p_1 and p_2 are real and $p_1 \neq p_2$, the system function can be expressed in the form

$$H(z) = \frac{A_1}{1 = p_1 z^{-1}} + \frac{A_2}{1 - p_s z^{-1}} \tag{3.52}$$

where

$$\left.\begin{aligned} A_1 &= \frac{B_0 p_1}{p_1 - p_2} \\ A_2 &= \frac{B_0 p_2}{p_1 - p_2} \end{aligned}\right\} \tag{3.53}$$

Consequently, the unit-sample response $h(n)$ can be determined by taking inverse z-transform of Eq. (5.52)

$$\begin{aligned} h(n) &= \mathcal{Z}^{-1}\left[H(z)\right] \\ &= \mathcal{Z}^{-1}\left[\frac{A_1}{1 - p_1 z^{-1}} + \frac{A_2}{1 - p_2 z^{-1}}\right] \\ &= A_1 (p_1)^n u(n) + A_2 (p_2)^n u(n) \\ &= [A_1 (p_1)^n + A_2 (p_2)^n] u(n) \end{aligned} \tag{3.54}$$

Putting the values of Eq. (3.53) in Eq. (3.54), we get

$$h(n) = \frac{B_0}{p_1 - p_2}\left(p_1^{n+1} - p_2^{n+1}\right) u(n) \tag{3.55}$$

Therefore, the unit-sample (impulse) response $h(n)$ is the difference of two decaying exponential sequences. Figure 3.9 shows a typical graph when the poles are distinct.

$$h(n) = \frac{1}{(p_1 - p_2)}\left[p_1^{n+1} - p_2^{n+1}\right]u(n)$$

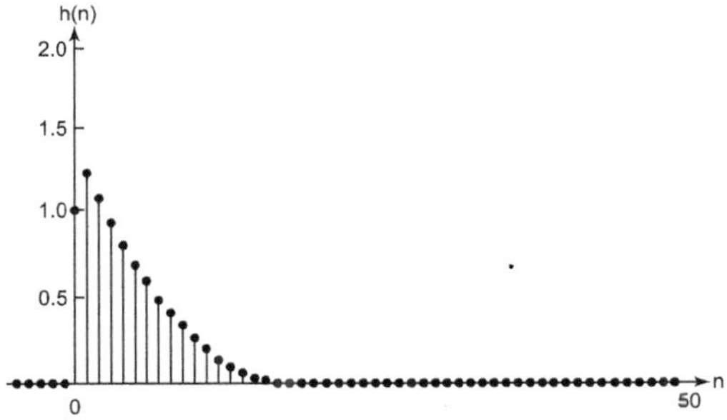

FIGURE 3.9 Plot of impulse response $h(n)$ given by Eq. (3.55) with poles $p_1 = 0.5$ and $p_2 = 0.75$.

Real and Equal poles $(A_1^2 = 4A_2)$. In this case $p_1 = p_2 = p = -A_1/2$. The system function is given by

$$H(z) = \frac{B_0}{\left(1 - pz^{-1}\right)^2} \tag{3.56}$$

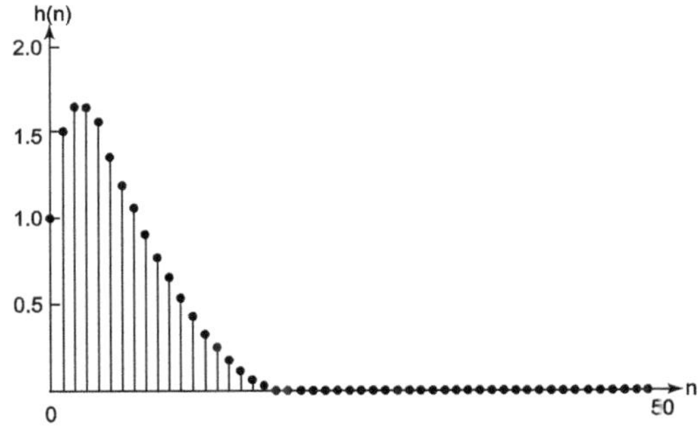

FIGURE 3.10 Plot of $h(n)$ given by Eq. (3.57) with $p = 3/4$, $h(n) = (n + 1)\, p^n\, u(n)$

Now, unit-sample response $h(n)$ of the system is determined as

$$h(n) = \mathcal{Z}^{-1}[h(n)]$$

$$= \mathcal{Z}^{-1}\left[\frac{B_0}{\left(1 - pz^{-1}\right)^2}\right]$$

$$= B_0(n+1)p^n u(n) \tag{3.57}$$

Here, we observe that $h(n)$ is the product of a ramp sequence and a real decaying exponential sequence. The graph of impulse response given by Eq. (3.57) is shown in Figure 3.10.

Complex-conjugate poles $(A_1^2 = 4A_2)$. Since the poles are complex conjugate, the system function $H(z)$ can be expressed as

$$H(z) = \frac{D}{1 - pz^{-1}} + \frac{D*}{1 - p*z^{-1}} \tag{3.58}$$

or

$$H(z) = \frac{D}{1 - re^{j\omega_0}z^{-1}} + \frac{D*}{1 - re^{-j\omega_0}z^{-1}} \tag{3.59}$$

Since, we know that $p = re^{'j\omega_0}$ and $0 < \omega_0 < \pi$. Note that when the poles are complex conjugates, the parameters A_1 and A_2 are related to r and ω_0 according to

$$\left.\begin{aligned} A_1 &= -2r\cos\omega_0 \\ A_2 &= r^2 \end{aligned}\right\} \tag{3.60}$$

The constant D in the partial fraction expansion of $H(z)$ is easily shown to be

$$\begin{aligned} D &= \frac{B_0 p}{p - p*} = \frac{B_0 r e^{j\omega_0}}{r\left(e^{j\omega_0} - e^{-j\omega_0}\right)} \\ &= \frac{B_0 e^{j\omega_0}}{j2\sin\omega_0} \end{aligned} \tag{3.61}$$

Consequently, the unit-sample (impulse) response of the system with complex-conjugate poles is

$$\begin{aligned} h(n) &= \frac{B_0 r^n}{\sin\omega_0}\left[\frac{e^{j(n+1)\omega_0} - e^{-j(n+1)\omega_0}}{2j}\right]u(n) \\ &= \frac{B_0 r^n}{\sin\omega_0}\sin(n+1)\omega_0 u(n) \end{aligned} \tag{3.62}$$

In this case unit-sample (impulse) response $h(n)$ has an oscillatory behavior with an exponentially decaying envelope when $r < 1$. The angle ω_0 of the poles determines the frequency of oscillation and the distance r of the poles from the origin determines the rate of decay. When r is close to unity, the decay is slow. When r is close to the origin, the decay is fast. A typical graph of impulse response $h(n)$ is shown in Figure 3.11.

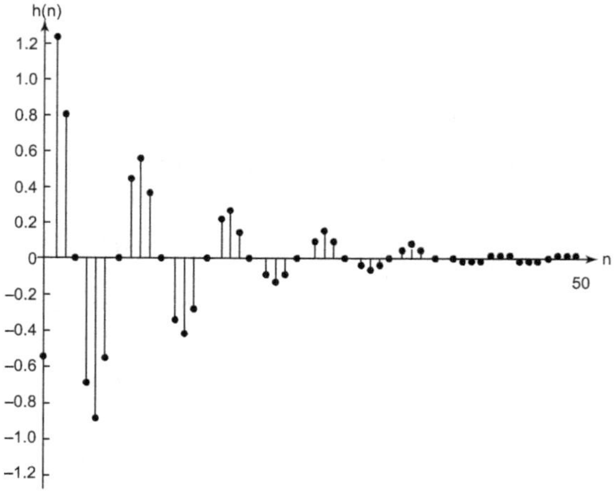

FIGURE 3.11 Plot of impulse response $h(n)$ given by Eq. (3.62) with $B_0 = 1$, $\omega_0 = \pi/4$, $r = 0.9$,

$$h(n) = \left[\frac{B_0 r^n}{\sin\omega_0}\right]\sin(n+1)\omega u(n).$$

EXERCISES

1. Define z-transform with ROC.
2. Why ROC is required in z-transform determination?
3. Give some possible configurations of the ROC for the z-transform.
4. Give the properties of the ROC for the z-transform.
5. Discuss various properties of the z-transform.
6. What is inverse z-transformation? Discuss all the methods of the inverse z-transform.
7. Define a system function. What is the relationship between system function and unit-sample response, $h(n)$ of a system?
8. What is meant by poles and zeros of a rational z-transform?
9. How z-transform and inverse z-transform are used for solving difference equations of a discrete-time system?

NUMERICAL EXERCISES

1. Find the z-transform with ROC for the following discrete-time sequences;

   (*a*) $s(n)=\left(\frac{1}{3}\right)^n u(n)$ (*b*) $s(n)=\left(\frac{1}{3}\right)^n u(-n-1)$

   (c) $s(n)=\left(\frac{1}{3}\right)^n u(-n)$ (*d*) $\delta(n)$

   (*e*) $\delta(n-1)$ (*f*) $\delta(n+1)$

2. Determine the z-transform and their ROC for the following sequences:

   (*a*) $s(n)=\{\underset{\uparrow}{2}, 4, 5, 6, 8\}$ (*b*) $s(n)=\{1, 2, \underset{\uparrow}{3}, 4, 5\}$

   (c) $s(n)=\{\underset{\uparrow}{0}, 0, 3, 4, 5, 6\}$

3. Determine the z-transform and its ROC of following sequences:

$s(n) = A^n u(n) + B^n u(-n-1)$

where A and B are constant scalar quantities.

4. Determine the z-transform and the ROC of the following signals:

(*a*) $s(n) = [4(3)^n - 5(4)^n]\, u(n)$
(*b*) $s(n) = A(\cos \omega_0 n)\, u(n)$
(*c*) $s(n) = A(\cos \omega_0 n)\, u(n)$
(*d*) $s(n) = A^n(\cos \omega_0 n)\, u(n)$
(*e*) $s(n) = A^n(\sin \omega_0 n)\, u(n)$
(*f*) $s(n) = nA^n u(n)$
(*h*) $s(n) = u(n)$

5. Find the inverse z-transform $s(n)$ for the following one-sided z-transforms is given by:

(*a*) $S(z) = \dfrac{z(z+3)}{z^2 + 3z + 2}$

(*b*) $S(z) = \dfrac{2z^2}{2z^2 - 2z + 1}$

(*c*) $S(z) = \dfrac{z^2 + 2}{z^2 + 1}$

(*d*) $S(z) = \dfrac{z(4z-3)}{2z^2 - 3z + 1}$

6. Solve the following difference equations using one-sided z-transform:

(*a*) $y(n) = \dfrac{1}{3}\, y(n-1)\, \dfrac{1}{4} - y(n-2) = 0,\ y(-1) = y(-2) = 1$

(*b*) $y(n) = 0.5y(n-1) + s(n)$, where input $s(n) = \left(\dfrac{1}{3}\right)^n u(n),\ y(-1) = 1$

(*c*) $y(n) = 0.25y(n-2) + s(n)$

Input $s(n) = u(n)$, $y(-1) = 0$ and $y(-2) = 1$.

7. Determine the unit-sample response and system function of the following causal systems:

(*a*) $y(n) = 0.7y(n-1) - 0.1\, y(n-2) + 2s(n) - s(n-2)$
(*b*) $y(n) = y(n-1) - 0.5y(n-2) + s(n) + s(n-1)$
(*c*) $y(n+2) - 3y(n+1) + 2y(n) + s(n)$

where $y(n)$ = output of a system, and $s(n)$ = input of a system.

CHAPTER 4

Frequency Analysis Using DTFT

4.1 INTRODUCTION TO DISCRETE-TIME FOURIER TRANSFORM (DTFT)

In this chapter, we will discuss about the Fourier transform of discrete-time signals, that is, discrete-time Fourier transform (DTFT). There are many similarities in the analysis of continuous-time and discrete-time signals using the Fourier series and there are also important differences between continuous-time Fourier series (CTFS) and discrete-time Fourier series (DTFS). For example, the Fourier series representation of a discrete-time periodic signal is finite series but the Fourier series representation of the continuous-time periodic signal is infinite series. Also, we will see in this chapter, there are corresponding differences between continuous-time Fourier transform (CTFT) and DTFT.

Here we will extend the Fourier series description of discrete-time periodic signals in order to develop a Fourier transform representation for discrete-time non-periodic signals.

4.2 DEVELOPMENT OF THE DISCRETE-TIME FOURIER TRANSFORM (DTFT)

DTFT is a transformation tool that transforms a discrete-time signal from time-domain to frequency domain. In this section, we will develop an expression for DTFT for discrete-time signals. We have already learned that the

Fourier series coefficient for a continuous-time periodic square wave can be viewed as samples of an envelope function. As the fundamental period of the continuous-time periodic square wave increases, its samples become more and more finely spaced. Here a non-periodic signal s(t) is used to construct a periodic signal $\tilde{s}(t)$ that equals s(t) over one fundamental period. As the period approaches infinity, $\tilde{s}(t)$ will be equal to s(t) over larger and larger intervals of time and the Fourier series representation for $\tilde{s}(t)$ converges to the Fourier transform representation for $\tilde{s}(t)$. In this section, we will develop an analogous procedure for deriving an expression for DTFT for discrete-time non-periodic signals.

Consider a general discrete-time sequence $s(n)$ of finite duration. That is,

$$\begin{aligned} s(n) &\neq 0, \text{ for } -N_1 \leq n \leq N_2 \\ &= 0, \text{ otherwise} \end{aligned} \tag{4.1}$$

From this non-periodic signal, we can construct a periodic signal or sequence $\tilde{s}(n)$ for which $s(n)$ is of one period. A discrete-time non-periodic signal $s(n)$ is shown in Figure 4.1(a) and a periodic signal $\tilde{s}(n)$ which is constructed from $s(n)$ for which $s(n)$ is of one fundamental period is shown in Figure 4.1(b).

If fundamental period N_0 approaches infinity then $\tilde{s}(n) = s(n)$ for any finite value of n.

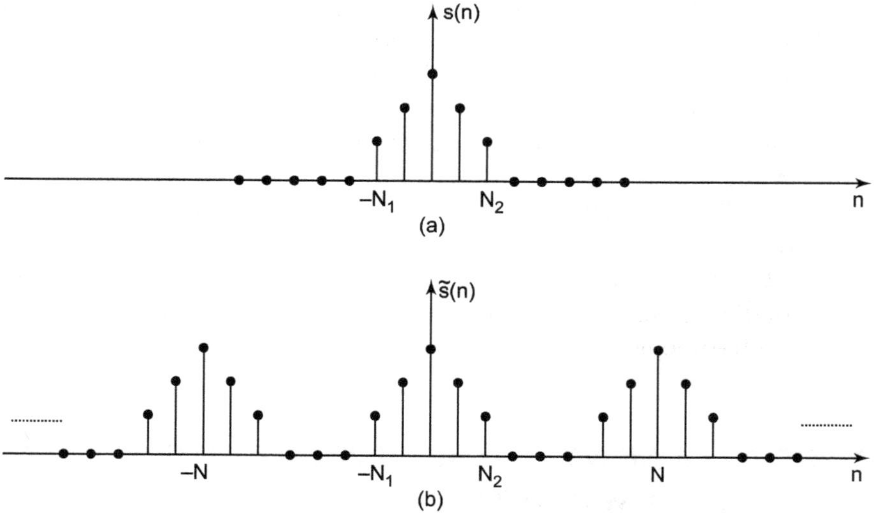

FIGURE 4.1 (a) Finite duration discrete-time signal $s(n)$
(b) Periodic signal $s(n)$ constructed to be equal to $s(n)$ over one period.

Fourier series representation of $\tilde{s}(n)$ can be expressed as

$$\tilde{s}(n) = \sum_{k=<N_0>} A_k e^{jk\omega_0 n} \tag{4.2}$$

where A_k is the Fourier series coefficient and $\omega_0 = 2\pi/N_0$ is the fundamental frequency. N_0 is the fundamental period.

Fourier series coefficients A_k are given by

$$A_k = \frac{1}{N_0} \sum_{n=<N_0>} \tilde{s}(n) e^{-jk\omega_0 n} \tag{4.3}$$

Since $s(n) = \tilde{s}(n)$ over one period that includes the interval $N_1 \le n \le N_2$. Therefore replacing $\tilde{s}(n)$ by $s(n)$ in Eq. (4.3), we get

$$A_k = \frac{1}{N_0} \sum_{n=N_1}^{N_2} s(n) e^{-jks\omega_0 n} \tag{4.4}$$

But $s(n)$ is zero outside the interval $-N_1 \le n \le N_2$.

Defining the function

$$S(e^{j\omega}) = \sum_{n=-\infty}^{\infty} s(n) e^{-j\omega n} \tag{4.5}$$

We see that the Fourier series coefficients A_k are directly proportional to samples of $S(e^{j\omega})$

$$A_k = \frac{1}{N_0} S\left(e^{jk\omega_0}\right) \tag{4.6}$$

Now combining Eqs. (4.2) and (4.6), we get

$$\tilde{s} = \sum_{k=<N_0>} A_k e^{jk\omega_0 n} = \sum_{k=<N_0>} \frac{1}{N_0} S\left(e^{jk\omega_0}\right) e^{jk\omega_0 n} \tag{4.7}$$

Since $\omega_0 = \dfrac{2\pi}{N_0}$ or $\dfrac{1}{N_0} = \dfrac{\omega_0}{2\pi}$, Eq. (4.7) can be rewritten as

$$\tilde{s}(n) = \sum_{k=<N_0>} \frac{\omega_0}{2\pi} S(e^{jk\omega_0}) e^{jk\omega_0 n}$$

$$= \frac{1}{2\pi} \sum_{k=<N_0>} S\left(e^{jk\omega_0}\right) e^{jk\omega_0 n_\omega} \tag{4.8}$$

As fundamental period N_0 increases, fundamental frequency ω_0 decreases. As N_0 approaches infinity, Eq. (4.8) passes to an integral. Eq. (4.8) can be graphically interpreted in Figure 4.2.

From Eq. (4.5), $S(e^{j\omega})$ is periodic in w with period 2π and therefore $\omega > n$ is also periodic in ω with the same period. Thus, the product $S(e^{j\omega})e^{j\omega_0 n}$ will also be periodic in w with period 2π. Each term in the summation in Eq. (4.8) represents the area of a rectangle of height $S(e^{jkw\omega_0})e^{j\omega_0 n}$ and width ω_0. As fundamental frequency ω_0 approaches zero, the summation in Eq. (4.8) becomes an integral. The summation is carried out over N_0 at consecutive intervals of width ω_0 but the total interval of integration will always be of width 2π.

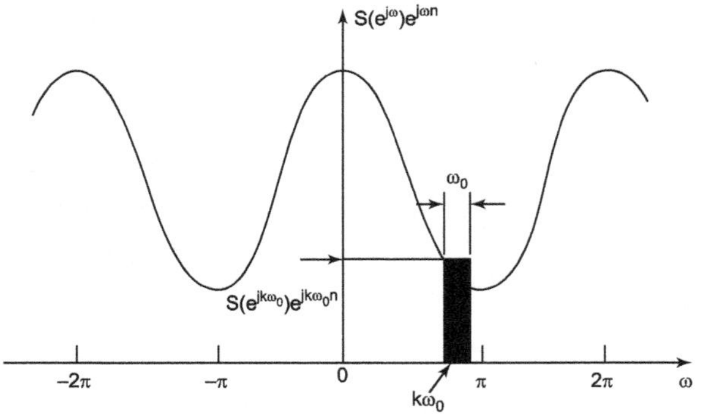

FIGURE. 4.2 Graphical representation of equation $\tilde{s}(n) = \frac{1}{2\pi}\sum_{k=<N_0>} S\left(e^{jk\omega 0}\right)e^{jk\omega_0 n}\omega_0$

Therefore, as $N_0 \to \infty$ or $\omega_0 \to 0$, $\tilde{s}(n) = s(n)$, the Eq. (4.8) becomes

$$s(n) = \frac{1}{2\pi}\int_{2\pi} S\left(e^{j\omega}\right)e^{j\omega n}d\omega \tag{4.9}$$

where

$$S\left(e^{j\omega}\right) = \sum_{n=-\infty}^{\infty} s(n)e^{-j\omega n} \tag{4.10}$$

The function $S(e^{j\omega})$ is called the DTFT. Eq. (4.9) is called the DTFT synthesis equation. Eq. (4.10) is referred to as the DTFT analysis equation. These equations show how a non-periodic sequence can be thought of as a linear combination of complex exponential functions. The DTFT has many similarities with the CTFT such as linearity, convolution property, etc. The major differences between the DTFT and CTFT are:

i. DTFT is periodic in w with period 2π but CTFT is not periodic.

ii. DTFT has a finite interval of integration in the synthesis equation but CTFT has an infinite interval of integration in the synthesis equation.

4.3 CONVERGENCE OF THE DTFT

Here we will discuss about the convergence of the infinite summation in the DTFT analysis equation given as

$$S\left(e^{j\omega}\right)=\sum_{n=-\infty}^{\infty} s(n)e^{-j\omega n} \tag{4.11}$$

The conditions on $s(n)$ that guarantee the convergence of this sum are direct counterparts of the convergence conditions for the CTFT. Eq. (4.11) will converge either if signal $s(n)$ is absolutely summable, that is,

$$\sum_{n=-\infty}^{\infty} |s(n)| < \infty \tag{4.12}$$

or, if the sequence has finite energy, that is,

$$\sum_{n=-\infty}^{\infty} |s(n)|^2 < \infty \tag{4.13}$$

But, there are no issues associated with DTFT synthesis equation given by

$$s(n)=\frac{1}{2\pi}\int_{2\pi} S(e^{j\omega})e^{j\omega n}d\omega \tag{4.14}$$

as the integral in this synthesis equation is over a finite interval of integration. This is a similar situation as for the DTFS synthesis equation which involves a finite sum. Consequently, there is no issue of convergence associated with DTFT synthesis equation.

EXAMPLE 4.1

Determine DTFT of a discrete-time signal $s(t)=A^n u(n)$, $|A|<1$.

Solution:

DTFT of $s(n)$ is given by

$$S\left(e^{j\omega}\right)=\text{DTFT}\{s(n)\}=\sum_{n=-\infty}^{\infty} A^n u(n)e^{-j\omega n}$$

$$=\sum_{n=-\infty}^{\infty} A^n u(n)e^{-j\omega n} \tag{1}$$

But $u(n)$ is the unit step sequence and it is defined as

$$u(n)=\begin{cases}1, & n\geq 0\\ 0, & n<0\end{cases} \tag{2}$$

Substituting Eq. (2) in Eq. (1), we obtain

$$S\left(e^{j\omega}\right)=\sum_{n=0}^{\infty} A^n 1e^{-j\omega n}=\sum_{n=-\infty}^{\infty}\left(Ae^{-j\omega}\right)^n$$

(This is a geometric progression of infinite number of terms.)

$$=\frac{1}{1-Ae^{-j\omega}}$$

or $$S\left(e^{j\omega}\right)=\frac{1}{1-Ae^{-j\omega}}$$

Note that DTFT of any discrete-time sequence $s(n)$ is periodic function in ω with period 2π.

EXAMPLE 4.2

Determine the DTFT of the discrete-time signal

$$s(n)=A^{|n|}, |A|<1.$$

Solution:

DTFT of the discrete-time signal is determined;

$$S\left(e^{j\omega}\right)=\text{DTFT}\{s(n)\}=\sum_{n=-\infty}^{\infty} s(n)e^{-j\omega n}$$

$$=\sum_{n=-\infty}^{\infty} A^{|n|}e^{-j\omega n}=\underbrace{\sum_{n=0}^{\infty} A^{n}e^{-j\omega n}}_{\text{I}}+\underbrace{\sum_{n=-\infty}^{-1} A^{-n}e^{-j\omega n}}_{\text{II}}$$

Substituting $n = -m$ in the second summation, we obtain

$$S\left(e^{j\omega}\right)=\sum_{n=-\infty}^{\infty} A^{|n|}e^{-j\omega n}+\sum_{m=1}^{\infty} A^{m}e^{j\omega m}$$

Both of the summations given in Eq. (1) are infinite geometric progressions that we can evaluate in closed form, producing.

$$S\left(e^{j\omega}\right)=\sum_{n=0}^{\infty}\left(Ae^{-j\omega}\right)^{n}+\sum_{m=1}^{\infty}\left(Ae^{j\omega}\right)^{m}$$

$$=\frac{1}{1-Ae^{-j\omega}}+\frac{Ae^{j\omega}}{1-Ae^{j\omega}}$$

$$=\frac{1-Ae^{j\omega}+Ae^{j\omega}\left(1-Ae^{-j\omega}\right)}{\left(1-Ae^{-j\omega}\right)\left(1-Ae^{j\omega}\right)}$$

$$=\frac{1-Ae^{j\omega}+Ae^{j\omega}-A^{2}}{1-Ae^{-j\omega}-Ae^{j\omega}+A^{2}}$$

$$=\frac{1-A^{2}}{1-2A\left(\dfrac{e^{j\omega}+e^{-j\omega}}{2}\right)+A^{2}}$$

or $$S\left(e^{j\omega}\right)=\frac{1-A^{2}}{1-2A\cos\omega+A^{2}} \tag{2}$$

This discrete-time signal $s(n)$ is shown in Figure 4.3 and its DTFT is shown in Figure 4.4.

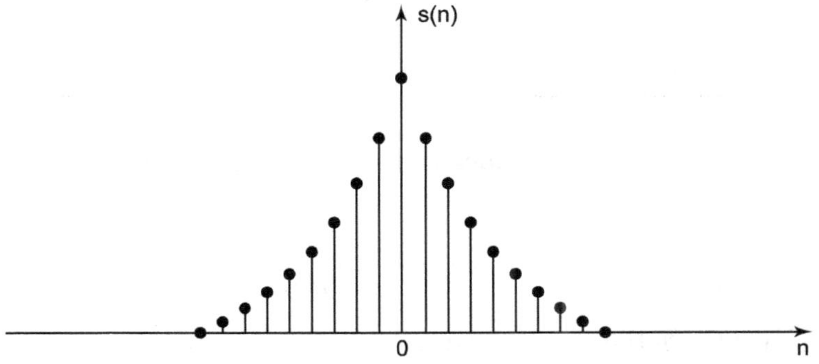

FIGURE 4.3 Discrete-time signal $s(n) = A^{|n|}$, $|A| < 1$.

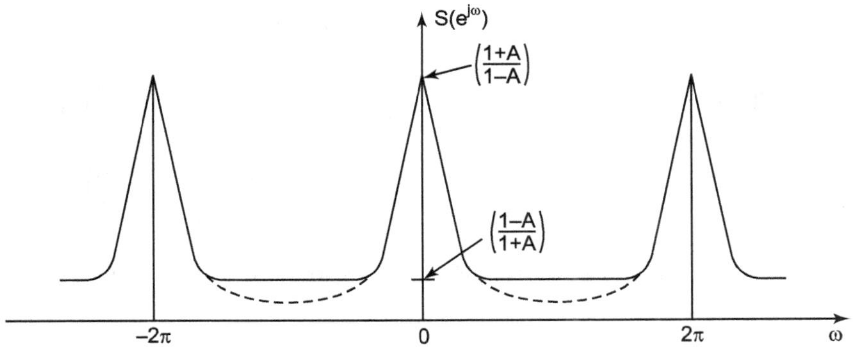

FIGURE 4.4 DTFT of the discrete-time signal $= A^{|n|}$, $|A| < 1$.

Here $S\left(e^{j\omega}\right) = \dfrac{1 - A^2}{1 - 2A\cos\omega + A^2}$ is a real-valued function of ω.

EXAMPLE 4.3

Determine the DTFT of the discrete-time rectangular pulse given as

$$s(n) = \begin{cases} 1, & |n| \le N_1 \\ 0, & |n| > N_1 \end{cases}$$

Solution:

This rectangular pulse is shown in Figure 4.5.

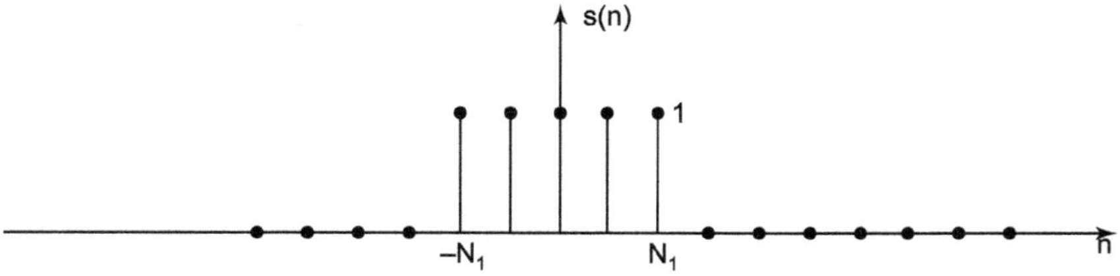

FIGURE 4.5 Discrete-time rectangular pulse.

$$s(n) = \begin{cases} 1, & |n| \le N_1 \\ 0, & |n| > N_1 \end{cases}$$

DTFT of this rectangular pulse is determined as

$$S\left(e^{j\omega}\right) = \text{DTFT}\,[s(n)] = \sum_{n=-\infty}^{\infty} s(n)e^{-j\omega n}$$

$$= \sum_{n=-N_1}^{N_1} 1e^{-j\omega n} \sum_{n=-N_1}^{N_1} e^{-j\omega n} \tag{1}$$

This is a geometric series or progression of a finite number of terms. Eq. (1) can be rewritten as

$$S\left(e^{j\omega}\right) = \sum_{n=N_1}^{N_1} e^{-j\omega n} = \frac{\sin\omega\left[N_1 + (1/2)\right]}{\sin(\omega/2)}$$

The DTFT of rectangular pulse is shown in Figure 4.6.

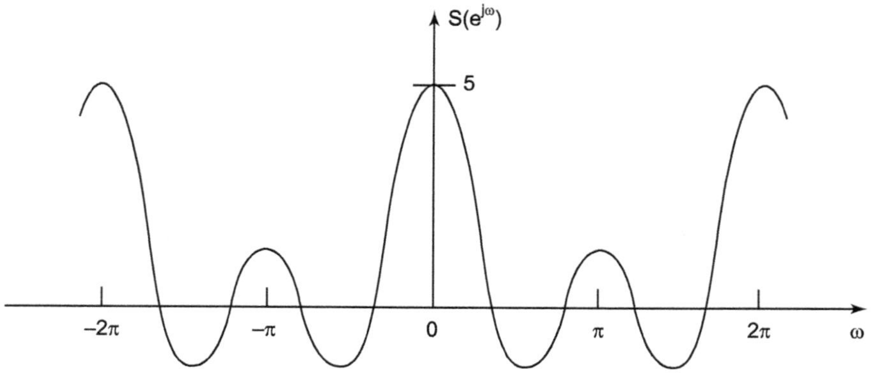

FIGURE 4.6 DTFT of rectangular pulse shown in Figure 4.5 for $N_1 = 2$.

DTFT of a discrete-time rectangular pulse is periodic with period 2π but the CTFT of the continuous-time rectangular pulse is not periodic.

4.4 FOURIER TRANSFORM OF DISCRETE-TIME PERIODIC SIGNALS

In this section, we will study about the Fourier transform for discrete-time periodic signals. Fourier transform for discrete-time periodic signals is interpreted as the Fourier transform of a periodic signal as an impulse train in the frequency domain.

To derive an expression for Fourier transform for discrete-time periodic signals, we consider the signal

$$s(n) = e^{j\omega n} \tag{4.15}$$

We know that the CTFT of $e^{j\omega_0 n}$ can be interpreted as an impulse at $\omega = \omega_0$. Therefore, we expect the same type of Fourier transform that results for the discrete-time signal of Eq. (4.15). However, the DTFT must be periodic in w with period 2π Fourier transform of $s(n) = e^{j\omega n}$ should have impulses at ω_0, $\omega_0 \pm 2\pi$, $\omega_0 \pm 4\pi$ and so on. Fourier transform of $s(n)$ is the impulse train. It is given by Eq. (4.16) and illustrated in Figure 4.7.

$$S\left(e^{j\omega}\right) = \sum_{m=-\infty}^{\infty} 2\pi\delta\left(\omega - \omega_0 - 2\pi m\right) \tag{4.16}$$

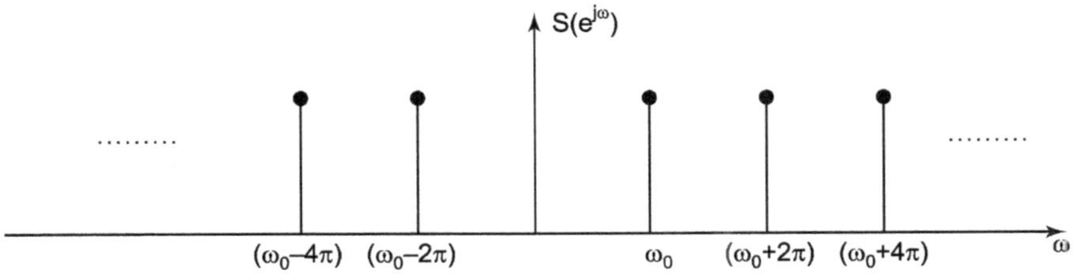

FIGURE 4.7 Illustration of Fourier transform of signal $s(n) = e^{j\omega n}$.

For checking the validity of expression given in Eq. (4.16), we will determine the inverse Fourier transform.

Substituting the Eq. (4.16) into the DTFT synthesis equation given as

$$\frac{1}{2\pi}\int_{2\pi} S\left(e^{j\omega}\right)e^{j\omega n}d\omega$$

$$= \frac{1}{2\pi}\int_{2\pi}\left[\sum_{m=-\infty}^{\infty} 2\pi\delta\left(\omega - \omega_0 - 2\pi m\right)\right]e^{j\omega n}d\omega \tag{4.17}$$

We know that any interval of length 2π includes exactly one impulse in the summation given by Eq. (4.16). Therefore, if the interval of integration is so chosen that it includes the impulse located at $\omega_0 + 2n(l)$, then Eq. (4.17) becomes

$$\frac{1}{2\pi}\int_{2\pi} S(e^{j\omega})e^{j\omega n}d\omega = e^{j(\omega_0\delta\omega l)=e^{j\omega n}} \tag{4.18}$$

Fourier series representation for a discrete-time periodic signal $s(n)$ is given by

$$s(n) = \sum_{k=<N_0>} A_k e^{jk\omega n} \tag{4.19}$$

and the DTFT is given by

$$S\left(e^{j\omega}\right) = \sum_{k=-\infty}^{\infty} 2\pi A_k\delta\left(\omega - \frac{2\pi k}{N_0}\right) \tag{4.20}$$

From the above analysis, we conclude that the Fourier transform of a discrete-time periodic signal can be directly constructed from its Fourier series coefficients A_k.

EXAMPLE 4.4

Determine the DTFT of the discrete-time periodic signal

$$s(n) = \cos\omega_0 n$$

with fundamental frequency $\omega_0 = 2\pi/5$.

Solution:

Using Euler's relation, signal $s(n) - \cos\omega_0 n$ can be written as

$$s(n) = \cos\omega_0 n \frac{e^{j\omega_0 n} + e^{-j\omega_0 n}}{2} \tag{1}$$

$$= \frac{1}{2}e^{j\omega_0 n} + \frac{1}{2}e^{-j\omega_0 n}$$

But we know that DTFT of a period signal $s(n) = e^{j\omega_0 n}$ is given by

$$S(e^{j\omega}) = \text{DTFT }\{s(n)\} = \text{DTFT}\{e^{j\omega_0 n}\}$$

$$= \sum_{m=-\infty}^{\infty} 2\pi\delta\left(\omega - \omega_0 - 2\pi m\right) \tag{2}$$

Similarly, DTFT of Eq. (1) can be determined as

$$S\left(e^{j\omega}\right) = \text{DTFT}\{s(n)\} = \text{DTFT}\left\{\frac{1}{2}e^{j\omega_0 n} + \frac{1}{2}e^{-j\omega_0 n}\right\}$$

$$= \frac{1}{2}\text{DTFT}\left\{e^{js\omega_0 n}\right\} + \frac{1}{2}\text{DTFT}\left\{e^{-js\omega_0 n}\right\}$$

$$=\frac{1}{2}\sum_{m=-\infty}^{\infty} 2\pi\delta\,(\omega-\omega_0-2\pi m)+\frac{1}{2}\sum_{m=-\infty}^{\infty} 2\pi\delta\,(\omega+\omega_0-2\pi m)$$

$$\left(\text{Substituting } \omega_0=\frac{2\pi}{5}\right)$$

or $$S\left(e^{j\omega}\right)=\sum_{m=-\infty}^{\infty}\pi\delta\left(\omega-\frac{2\pi}{5}-2\pi m\right)+\sum_{m=-\infty}^{\infty}\left(\omega+\frac{2\pi}{3}-2\pi 3\right) \tag{3}$$

$$S\left(e^{j\omega}\right)=\sum_{m=-\infty}^{\infty}\pi\delta\left(\omega-\frac{2\pi}{5}-2\pi m\right)+\sum_{m=-\infty}^{\infty}\left(\omega+\frac{2\pi}{5}-2\pi m\right)$$

$$=\left\{\ldots+\pi\delta\left(\omega-\frac{2\pi}{5}+2\pi\right)\right\}+\pi\delta\left(\omega-\frac{2\pi}{5}\right)$$

$$+\pi\delta\left(\omega-\frac{2\pi}{5}-2\pi\right)+\ldots\}$$

$$+\left\{\ldots+\pi\delta\left(\omega+\frac{2\pi}{5}+2\pi\right)\right\}+\pi\delta\left(\omega-\frac{2\pi}{5}\right)$$

$$+\pi\delta\left(\omega+\frac{2\pi}{5}-2\pi\right)+\ldots\}$$

or $$S\left(e^{j\omega}\right)=\pi\delta\left(\omega-\frac{2\pi}{5}\right)+\pi\delta\left(\omega+\frac{2\pi}{3}\right), -\pi\le\omega\le\pi \tag{4}$$

Here $S(e^{j\omega})$ repeats periodically with a period of 2π. DTFT of $s(n)=\cos\omega_0 n$ is shown in Figure 4.8.

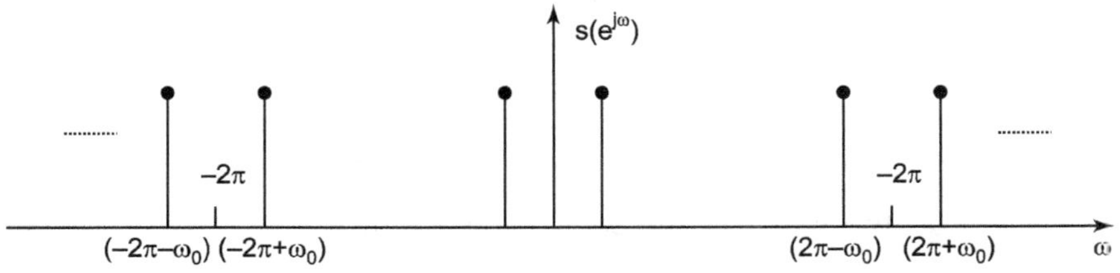

FIGURE 4.8 DTFT of $s(n)=\cos\omega_0 n$.

EXAMPLE 4.5

Determine the DTFT of the discrete-time periodic impulse train given by

$$s(n)=\sum_{k=-\infty}^{\infty}\delta(n-kN_0).$$

Solution:

Discrete-times periodic impulse train $s(n)=\sum_{k=-\infty}^{\infty}\delta(n-kN_0)$ is sketched in Figure 4.9.

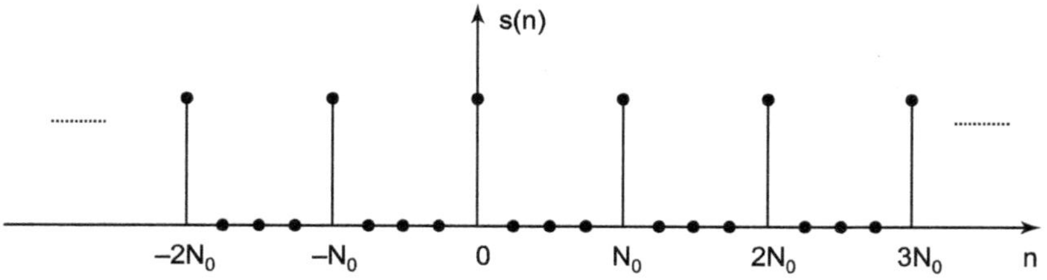

FIGURE 4.9 Discrete-time periodic impulse train $s(n)=\sum_{k=-\infty}^{\infty}\delta(n-kN_0)$.

Now, we can calculate the Fourier series coefficients for this discrete-time periodic impulse train $s(n)=\sum_{k=-\infty}^{\infty}\delta(n-kN_0)$ directly from equation given below

$$A_k = \sum_{n=<N_0>} s(n)e^{-jk\omega_0 n} \tag{1}$$

Substituting $s(n)=\sum_{k=-\infty}^{\infty}\delta\,(n-kN_0)$ and choosing the interval of summation as $0 \le n \le N_0 - 1$ Eq. (1), we obtain

$$A_k = \frac{1}{N_0}\sum_{n=0}^{N_0-1}\sum_{k=-\infty}^{\infty}\delta(n-kN_0)e^{-jk\omega_0 n} = \frac{1}{N_0} \tag{2}$$

DTFT of $s(n)$ is determined as

$$S(e^{j\omega}) = \text{DTFT}\{s(n)\} = \text{DTFT}\left\{\sum_{k=-\infty}^{\infty}\delta(n-kN_0)\right\}$$

$$= \sum_{k=-\infty}^{\infty} 2\pi A_k\delta\left(\omega-\frac{2\pi k}{N_0}\right) \tag{3}$$

Substituting Eq. (2) in Eq. (3), we obtain

$$S(e^{j\omega}) = \frac{2\pi}{N_0}\sum_{k=-\infty}^{\infty}\delta\left(\omega-\frac{2\pi k}{N_0}\right) \tag{4}$$

DTFT of discrete-time periodic impulse train $s(n)=\sum_{k=-\infty}^{\infty}\delta(\omega-kN_0)$ is shown in Figure 4.10.

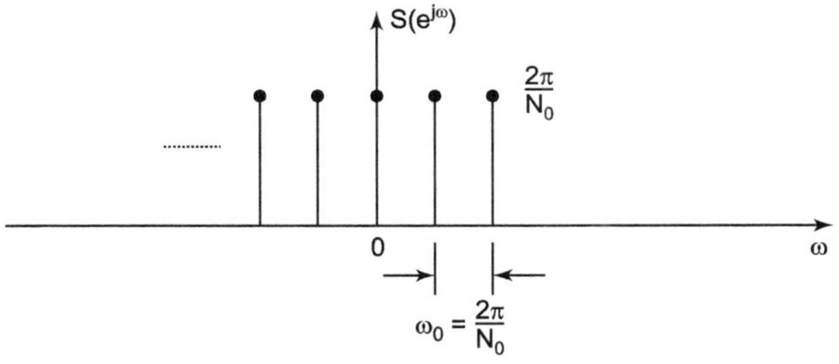

FIGURE 4.10 DTFT of discrete-time periodic impulse train.

$$s(n) = \sum_{k=-\infty}^{\infty} \delta(n - kN_0)$$

4.5 PROPERTIES OF THE DTFT

In this section, we will study about the various properties of the Fourier transform of discrete-time signals. Fourier transform of discrete-time signals is also referred to as DTFT. These properties are often useful in reducing the complexity in the evaluation of the Fourier transform and inverse Fourier transform. Also, we will see some of the similarities and differences between CTFT and DTFT. The derivation of DTFT properties is essentially identical to its continuous-time counterpart, that is, CTFT. Also, because of the close relationship between the Fourier series and Fourier transform, many of the DTFT properties translate directly into corresponding DTFS properties.

Here we will use one notation similar to that used for CTFT to indicate the pairing of a signal and its Fourier transform. That is

$$S(e^{j\omega}) = \text{DTFT }\{s(n)\}$$

$$s(n) = \text{Inverse DTFT } S(e^{j\omega})$$

or $$s(n) \xleftrightarrow{\text{DTFT}} S(e^{j\omega})$$

In this section, we will discuss the following properties of the DTFT:

1. Periodicity of the DTFT;
2. Linearity of the DTFT;
3. Time-shifting and frequency shifting;

4. Complex conjugation and conjugate symmetry;
5. Differencing and accumulation in time-domain;
6. Time reversal of a discrete-time sequence;
7. Time expansion;
8. Differentiation in frequency domain;
9. Parseval's relation for DTFT;
10. The convolution property for DTFT;
11. The multiplication property for DTFT.

4.5.1 Periodicity of the DTFT

We have already discussed that DTFT of a discrete-time signal is always periodic in w with period 2π. A periodic function is a function that repeats its value after some fixed value of the independent variable. This fixed value of the independent variable is called the period of the periodic function.

The DTFT of $s(n)$ *is* always periodic in π with period 2π, that is,

$$s(n) \xleftrightarrow{\text{DTFT}} S\left(e^{j\omega}\right)$$

$$SS\left(e^{j(\omega+2\pi)}\right) = S(e^{j\omega}) \tag{4.21}$$

This is in contrast to CTFT, which in general is not a periodic function.

4.5.2 Linearity of DTFT

The DTFT is also a linear transformation tool as CTFT.

$$s_1 n \xleftrightarrow{\text{DTFT}} \left\{S_1(e^{j\omega})\right\}$$

$$s_2 n \xleftrightarrow{\text{DTFT}} \left\{S_2(e^{j\omega})\right\}$$

where $s_1(n)$ and $s_2(n)$ are discrete-time sequences whose DTFTs are $S_1(e^{j\omega})$ and $S_2(e^{j\omega})$, respectively.

From the property of linearity, it is true for the above two discrete-time sequences $s_1(n)$ and $s_2(n)$ given as

$$A_1 s_1(n) + A_2 s_2(n) \xleftrightarrow{\text{DTFT}} A_1 S_1(n) + A_2 S_2(n)\{S_1(e^{j\omega})\} \tag{4.22}$$

where A_1 and A_2 are constants.

4.5.3 Time-Shifting and Frequency Shifting

Here we will first study the effect of time-shifting of a discrete-time sequence $s(n)$ on its DTFT $S(e^{j\omega})$ and then the effect of frequency shifting.

If $$s(n) \xleftrightarrow{\text{DTFT}} \{S(e^{j\omega})\}$$

then $$s(n-n_0) \xleftrightarrow{\text{DTFT}} e^{-j\omega_0\{S(e^{j\omega})\}} \tag{4.23}$$

where $s(n - n_0)$ is the time-shifted version of $s(n)$. Now will see the effect of frequency shifting.

If $$s(n) \xleftrightarrow{\text{DTFT}} \{S(e^{j\omega})\}$$

then $$s^{j\omega_0} \xleftrightarrow{\text{DTFT}} S(e^{j\omega-\omega_0}) \tag{4.24}$$

where $S(e^{(j\omega-\omega_0)})$ is the frequency-shifted version of $S(e^{j\omega})$.

4.5.4 Complex Conjugation and Conjugate Symmetry

We can obtain complex conjugation of a complex discrete-time signal $s(n)$ by reversing the sign of the imaginary part of the complex signal $s(n)$.

Let $s(n)$ be the complex discrete-time signal, $s(n) = s_R(n) + js_I(n)$ and its DTFT is also complex, that is,

$$S(e^{j\omega}) = S_R(j\omega) + jS_I(e^{j\omega})$$

where $s_R(n)$ and $s(n)$ are real and imaginary parts of $s(n)$, and $S_R(e^{j\omega})$ and $S_I(e^{j\omega})$ are also real and imaginary parts of $S(e^{jw})$, respectively.

Now complex conjugation of $s(n)$ is given by

$$s*(n) = s_R(n) - js(n)$$

and complex conjugation of $S(e^{jw})$ is given by

$$S*(e^{j\omega}) = S_R(e^{j\omega}) - jS_I(e^{j\omega})$$

If $$s(n) \xleftrightarrow{\text{DTFT}} S(e^{j\omega})$$

then $$s*(n)\ s*(n) \xleftrightarrow{\text{DTFT}} S*(e^{-j\omega}) \tag{4.25}$$

Also, if signal $s(n)$ is a real-valued function, then its DTFT $S(e^{jw})$ will be conjugate symmetric. That is

$$S(e^{j\omega}) = S*(e^{-j\omega}) \text{ for } s(n) \text{ real} \tag{4.26}$$

From Eq. (4.26), we can say that $\mathrm{Re}\{S(e^{j\omega})\}$ is an even function of w and $\mathrm{Im}\{S(e^{j\omega})\}$ is an odd function of w. Similarly, the magnitude of $\{S(e^{j\omega})\}$ is an even function and the phase angle $\angle\{S(e^{j\omega})\}$ is an odd function.

Also, we can decompose sequence $s(n)$ into even and odd parts. Furthermore,

$$\mathrm{E_V}\{s(n)\} = s_E(n) \xleftrightarrow{\text{DTFT}} \mathrm{Re}\, S(e^{j\omega}) = S_R(e^{j\omega})$$

and
$$Od\{s(n)\} = s_0(n) \xleftrightarrow{\text{DTFT}} j\,\mathrm{Im}\{S(e^{j\omega})\} = jS_I(e^{j\omega})$$

where $\mathrm{E}_v\{s(n)\}$ and $Od\{s(n)\}$ are the even and odd parts, respectively, of $s(n)$. Specifically, if $s(n)$ is real and even, its DTFT is also real and even.

4.5.5 Differencing and Accumulation in Time

Accumulation is the discrete-time counterpart of integration. Integration is used for continuous-time signals. The inverse of accumulation is referred to as first differencing.

Let $s(n)$ be a discrete-time signal with DTFT $S(e^{j\omega})$. The first differencing of signal $s(n)$ is given by $s(n) - s(n-1)$. DTFT of first differencing of signal $s(n)$ can be determined by using the properties of linearity and time-shifting as

$$s(n) - s(n-1) \xleftrightarrow{\text{DTFT}} S(e^{j\omega}) - e^{-j\omega}S(e^{j\omega})$$

$$= (1 - e^{-j\omega})S(e^{j\omega}) \tag{4.27}$$

The accumulation of signal $s(n)$ is given by

$$y(n) = \sum_{m=-\infty}^{\infty} s(m) \tag{4.28}$$

Eq. (4.28) can also be expressed as

$$y(n) - y(n-1) = s(n) \tag{4.29}$$

Thus, we can conclude that DTFT of $y(n)$ should be related to the DTFT of $s(n)$ division by $(1 - e^{-j\omega})$. This is not perfectly correct but the precise relationship is given by

$$\sum_{m=-\infty}^{\infty} s(m) \xleftrightarrow{\text{DTFT}} \frac{1}{1-e^{-j\omega}} + \pi S\left(e^{j0}\right) \sum_{k=-\infty}^{\infty} \delta(\omega - 2\pi k) \tag{4.30}$$

The impulse train on the RHS of Eq. (4.30) shows the dc or average value that can result from summation.

EXAMPLE 4.6

Determine the DTFT of the unit-step function $s(n) = u(n)$ using the accumulation property of DTFT.

Solution:

We know that

$$g(n) = s(n) \xleftrightarrow{\text{DTFT}} S\left(e^{j\omega}\right) = 1$$

Also, we have studied that the unit-step function $u(n)$ is the running sum of the unit-impulse function &(n). This relation is given as

$$S\left(e^{j\omega}\right) = \text{DTFT} \sum_{m=-\infty}^{\infty} s(m)$$

Now, taking the DTFT of both sides of Eq. (2) and using the accumulation property of DTFT, we obtain

$$S\left(e^{j\omega}\right) = \text{DTFT}\left\{\sum_{m=-\infty}^{\infty} s(m)\right\}$$

or

$$S\left(e^{j\omega}\right) = \frac{1}{\left(1-e^{-j\omega}\right)} G\left(e^{j\omega}\right) + \pi G\left(e^{j0}\right) \sum_{k=-\infty}^{\infty} \delta(\omega - 2\pi k) \qquad (3)$$

But from Eq. (1), we have

$$G(e^{j\omega}) = 1 \qquad \text{and} \qquad G(e^{j0}) = 1$$

Substituting these values in Eq. (3), we obtain

$$S\left(e^{j\omega}\right) = \frac{1}{\left(1-e^{-j\omega}\right)} + \pi \sum_{k=-\infty}^{\infty} \delta(\omega - \omega\pi k) \qquad (4)$$

4.5.6 Time Reversal of Discrete-Time Signals

Taking a mirror image or folding of a discrete-time sequence is called time reversal. Consider a discrete-time signal or sequence $s(n)$ whose mirror image is given by $s(n)$.

If $$s(n) \xleftrightarrow{\text{DTFT}} S\left(e^{j\omega}\right)$$

then $$s(-n) \xleftrightarrow{\text{DTFT}} s(e^{-j\omega}) \qquad (4.31)$$

$s(n)$

Now we can prove this property as

$$y(n) = s(-n) \qquad (4.32)$$

Taking DTFT of both sides of Eq. (4.32), we obtain

$$y(e^{j\omega}) = \text{DTFT}\{y(n)] = \sum_{n=-\infty}^{\infty} s(-n) e^{-j\omega n} \qquad (4.33)$$

Now substituting $m = -n$ in Eq. (4.33), we obtain

$$y\left(e^{j\omega}\right) = \sum_{m=-\infty}^{\infty} s(m) e^{-j\omega(-m)} = S(e^{-j\omega}) \qquad (4.34)$$

That is,

$$s(-n) \xleftrightarrow{\text{DTFT}} S\left(e^{-j\omega}\right) \tag{4.35}$$

Hence, it is to be proved.

4.5.7 Time Expansion of Discrete-Time Signals

Time expansion of a discrete-time signal is equivalent to multiplication of independent variable, that is, time of a signal by an integer scalar quantity A, which is greater than unity.

We have already derived time-expansion property of CTFT which is given by

$$s(At) \xleftrightarrow{\text{CTFT}} \frac{1}{|A|} S\left(\frac{j\omega}{A}\right) \tag{4.36}$$

We cannot slow down the signal by choosing $A < 1$. On the other hand, if we let A be an integer other than ± 1. For example $A = 2$ then $s(2n)$, we cannot merely speed up the original signal. That is since n can take on only integer values the signal $s(2n)$ consists of the even samples of original signal $s(n)$ alone.

Time-expanded signal is given by

$$s_{(k)}(n) = \begin{cases} s(n/k) & \text{if } n \text{ is a multiple of } k \\ 0 & \text{if } n \text{ is not multiple of } k \end{cases} \tag{4.37}$$

where k is a positive integer.

For $k = 2$, the sequence $s_{(k)}(n)$ is obtained from the original sequence $s(n)$ by placing $k - 1$ zeroes between successive values of the original sequence. Now, we can say that $s_{(k)}(n)$ is a slowed-down version of $s(n)$.

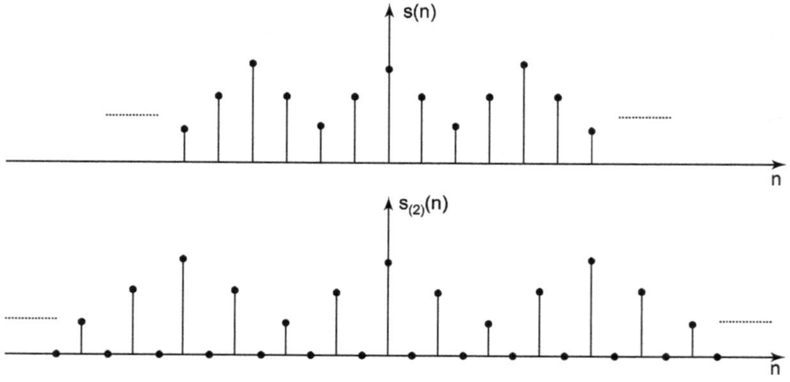

FIGURE 4.11 The signal $s_{(2)}(n)$ obtained from $s(n)$ by inserting one zero between successive values of the original signal $s(n)$.

One sequence $s(n)$ and its slowed-down version $s_{(k)}(n)$ for $k = 2$ are shown in Figure 4.11.

Since $s_{(k)}(n)$ equals zero unless n is multiple of k.

For $n = mk$, DTFT of $s_{(k)}(n)$ is given by

$$S_{(k)}(e^{j\omega}) = \text{DTFT}\{s_{(k)}(n)\} = \sum_{n=-\infty}^{\infty} s_{(k)}(n)e^{-j\omega n}$$

$$= \sum_{n=-\infty}^{\infty} s_{(k)}(mk)e^{-j\omega mk}$$

Furthermore, since $S_{(k)}(mk)$, we find that

$$S_{(k)}(ej\omega) = \sum_{m=-\infty}^{\infty} s(m)e^{-(j\omega)m} = S\left(e^{jk\omega}\right)$$

That is

$$s(k)(n) \xleftrightarrow{\text{DTFT}} S\left(e^{kj\omega}\right) \tag{4.38}$$

As the discrete-time signal is spread out and slowed-down in time by taking $k > 1$, its DTFT is compressed. For example, since $S(e^{j\omega})$ is periodic in w with period $2n$ and $S(e^{jk\omega})$ is also periodic in w with period $2\pi/k$.

Now we can conclude that there is an inverse relationship between time and frequency domains. In other words, we can say that as the value of k increases, $s_{(k)}(n)$ spreads out while its DTFT is compressed.

EXAMPLE 4.7

Determine the DTFT of the discrete-time sequence $s(n)$ shown in Figure 4.12 using a time-expansion property of DTFT.

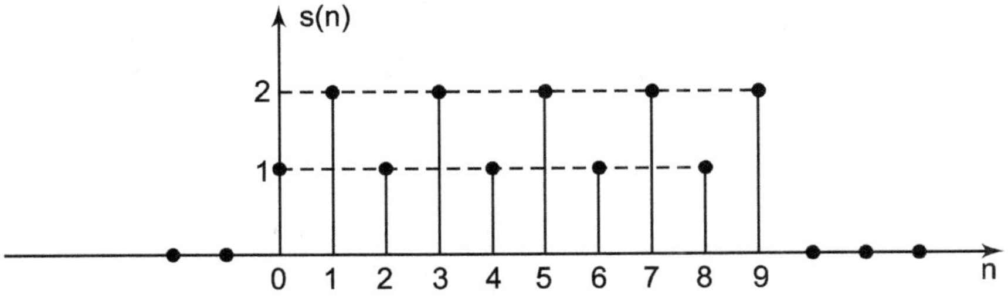

FIGURE 4.12 Discrete-time sequence $s(n)$.

Solution:

This sequence $s(n)$ can be related to the simpler sequence $g(n)$ shown in Figure 4.13(a).

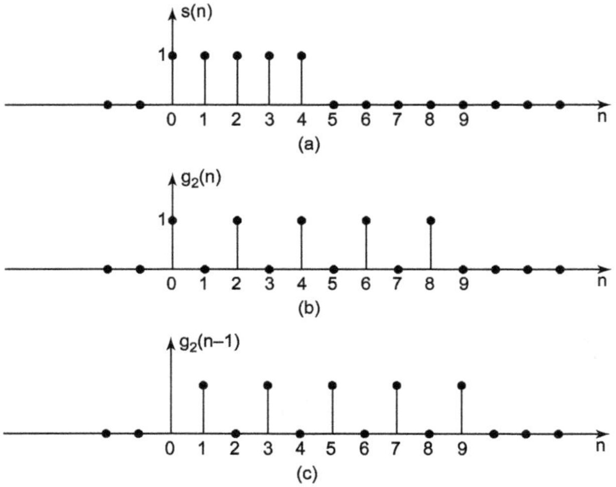

FIGURE 4.13 (a) Simpler sequence $g(n)$. (b) Time-expanded version of sequence $g(n)$. (c) Time-shifted version of $g_{(2)}(n)$ by one unit to the right.

The discrete-time sequence $s(n)$ is related to the discrete-time sequence $g(n)$ as

$$s(n) = g_{(2)}(n) + 2g_{(2)}(n-1) \tag{1}$$

where
$$g_2(n) = \begin{cases} g(n/2), & \text{if } n \text{ is even} \\ 0, & \text{if } n \text{ is odd} \end{cases} \tag{2}$$

and $g_2(n-1)$ is the shifted version *of* $g_{(2)}(n)$ by one unit to the right. The signals $g_{(2)}(n)$ and $2g_{(2)}(n-1)$ are depicted in Figure 4.13(*b*) and 4.13(*c*), respectively.

Here, we relate given sequence $s(n)$ in terms of a simpler sequence $g(n)$ which is a discrete-time rectangular pulse:

$$\text{DTFT of } g(n) = G(e^{j\omega}) = \sum_{n=-\infty}^{\infty} g(n)e^{-j\omega n}$$

$$= \sum_{n=0}^{4} 1e^{-j\omega n} = 1\frac{\left[1-\left(e^{-j\omega}\right)^5\right]}{\left[1-e^{-j\omega}\right]}$$

$$= \frac{1-e^{-j5\omega}}{1-e^{-j\omega}} = \frac{e^{-j5\omega/2}\left[\dfrac{e^{j5\omega/2}-e^{-5j\omega/2}}{2j}\right]}{e^{-j\omega/2}\left[\dfrac{e^{j\omega/2}-e^{-j\omega/2}}{2j}\right]}$$

or

$$G(e^{j\omega}) = e^{j2\omega}\left\{\frac{\sin\left(\frac{5\omega}{2}\right)}{\sin\left(\frac{\omega}{2}\right)}\right\} \tag{3}$$

$$s(n)G(e^{j\omega}) \xleftrightarrow{\text{DTFT}} G(e^{j\omega}) = e^{j2\omega}\left\{\frac{\sin\left(\frac{5k\omega}{2}\right)}{\sin\left(\frac{k\omega}{2}\right)}\right\}$$

then from time-expansion property

$$g_{(k)}(n) \xleftrightarrow{\text{DTFT}} G(e^{jk\omega}) = e^{j2k\omega}\left\{\frac{\sin\left(\frac{5k\omega}{2}\right)}{\sin\left(\frac{k\omega}{2}\right)}\right\}$$

For $k = 2$,

$$g_{(2)}(n) \xleftrightarrow{\text{DTFT}} G(e^{j2\omega}) = e^{-j2\times 2\omega}\left\{\frac{\sin\left(\frac{5\times 2\omega}{2}\right)}{\sin\left(\frac{2\omega}{2}\right)}\right\}$$

$$= e^{j4\omega}\left\{\frac{\sin(5\omega)}{\sin(\omega)}\right\} \tag{4}$$

Now using the linearity and time-shifting properties, we get

$$2g_{(2)}(n) \xleftrightarrow{\text{DTFT}} e^{-j\omega}\left\{2G\left(e^{j2\omega}\right)\right\}$$

$$= 2e^{-j\omega}\left\{e^{-j4\omega}\left[\frac{\sin(5\omega)}{\sin(\omega)}\right]\right\}$$

$$= 2e^{-j5\omega}\left[\frac{\sin(5\omega)}{\sin(\omega)}\right] \tag{5}$$

Combining Eqs. (4) and (5), we have

$$s(n) = g_{(2)}(n) + 2g_{(2)}(n)(n-1) \xleftrightarrow{\text{DTFT}} e^{-j4\omega}\left[\frac{\sin(5\omega)}{\sin(\omega)}\right] \xleftrightarrow{\text{DTFT}}$$

$$+ 2e^{-j5\omega}\left[\frac{\sin(5\omega)}{\sin(\omega)}\right]$$

$$= e^{-j4\omega}(1-23^{-j\omega})\left[\frac{\sin(5\omega)}{\sin(\omega)}\right]$$

4.5.8 Differentiation in Frequency Domain

Here we will study the differentiation of DTFT of. $s(n)$, $S(e^{j\omega})$ with respect to ω. Let

$$s(n) \xleftrightarrow{\text{DTFT}} S(e^{j\omega})$$

Using the DTFT analysis equation $S\left(e^{j\omega}\right) = \sum_{n=-\infty}^{\infty} s(n)e^{-j\omega n}$ and differentiating both sides with respect to ω, we obtain

$$\frac{dS\left(e^{j\omega}\right)}{d\omega} = \frac{d}{d\omega}\left[\sum_{n=-\infty}^{\infty} s(n)e^{-j\omega n}\right]$$

$$= \sum_{n=-\infty}^{\infty} s(n)\left[-jne^{-j\omega n}\right]$$

$$= -j\sum_{n=-\infty}^{\infty} ns(n)e^{-j\omega n}$$

or $$\frac{jdS\left(e^{j\omega}\right)}{d\omega} = -j^2\sum_{n=-\infty}^{\infty} ns(n)e^{-j\omega n}$$

or $$\frac{jdS\left(e^{j\omega}\right)}{d\omega} = \sum_{n=-\infty}^{\infty} (ns(n))e^{-j\omega n} = \text{DTFT}\{ns(n)\}$$

or $$ns(n) \xleftrightarrow{\text{DTFT}} \frac{jdS\left(e^{j\omega}\right)}{d\omega} \tag{4.39}$$

4.5.9 Parseval's Relation for DTFT

Parseval's relation for DTFT states that the total energy in a discrete-time signal s(n) may be determined either by computing the energy per unit time $|s(n)|^2$ and summing overall time or by computing the energy per unit frequency $|S(e^{j\omega})|2/2\pi$ and integrating over a full 2ω interval of distinct discrete-time frequencies. In analogous with the continuous-time signal, $|S(e^{j\omega})|$ is called the energy-density spectrum of the signal $s(n)$. Parseval's relation for DTFT is given as

$$\sum_{n=-\infty}^{\infty} | s(n) |^2 = \frac{1}{2\pi} \int_{2\pi} \left|S(e^{j\omega})\right|^2 d\omega \tag{4.40}$$

4.5.10 The Convolution Property

Here we will discuss the importance of the DTFT with regard to its effect on the convolution operation and analysis of discrete-time LTI systems. By using DTFT, convolution of two discrete-time signals $s_1(n)$ and $s_2(n)$ is converted into multiplication of DTFT of individual discrete-time signals.

If $$s_1(n) \xleftrightarrow{\text{DTFT}} S_1(e^{j\omega})$$

and $$s_2(n) \xleftrightarrow{\text{DTFT}} S_2(e^{j\omega})$$

then from convolution property of DTFT

$$s(n) = s_1(n) * s_2(n) \xleftrightarrow{\text{DTFT}} S(e^{j\omega}) = S_1(e^{j\omega}) S_2(e^{j\omega}) \tag{4.41}$$

Now we are going to apply DTFT for representing and analyzing discrete-time LTI systems. Specifically, if $s(n)$, $h(n)$ and $y(n)$ are the input, impulse response and output, respectively, of a discrete-time LTI system.

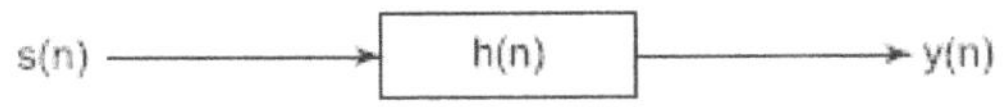

FIGURE 4.14 Discrete-time LTI system.

Output $y(n)$ is determined by convolving $s(n)$ and $h(n)$ given as

$$y(n) = s(n) * h(n) \tag{4.42}$$

By using DTFT, Eq. (4.42) can be expressed as

$$Y(e^{j\omega}) = S(e^{j\omega}) * H(e^{j\omega}) \tag{4.43}$$

where $S(e^{j\omega})$, $H(e^{j\omega})$ and $Y(e^{j\omega})$ are the DTFTs of $s(n)$, $h(n)$ and $y(n)$, respectively.

Now, combining Eqs. (4.42) and (4.43), we get

$$y(n) = s(n) * h(n) \xleftrightarrow{\text{DTFT}} Y(e^{j\omega}) = S(e^{j\omega}) H(ej^{j\omega}) \tag{4.44}$$

where $H(e^{j\omega})$ is the DTFT of the impulse response $h(n)$ of the discrete-time LTI system. It is also called frequency response of discrete-time LTI system.

EXAMPLE 4.8

Determine the frequency response of a discrete-time LTI system with impulse response $h(n) = 5(n - n_0)$. Also determine output for this system.

Solution:
The frequency response of a discrete-time LTI system is equal to the DTFT of the impulse response $\delta(n)$ of the system. The frequency response is determined as

$$H(e^{j\omega}) = \text{DTFT}\{h(n)\} = \sum_{n=-\infty}^{\infty} h(n)e^{-j\omega n}$$

$$= \sum_{n=-\infty}^{\infty} \delta\left(n - n_0\right)e^{-j\omega n} = e^{-j\omega n_0} \quad (1)$$

We know from convolution property of DTFT

$$y(n) = h(n) * s(n) \xleftrightarrow{\text{DTFT}} = H(e^{j\omega})S(e^{j\omega})$$

or

$$Y(^{j\omega}) = H(e^{j\omega})S(e^{j\omega}) \quad (2)$$

Substituting Eq. (1) in Eq. (2), we obtain

$$Y(e^{j\omega}) = e^{-j\omega n_0} S(e^{j\omega}) \quad (3)$$

Output $y(n)$ of above discrete-time LTI system is determined by taking the inverse DTFT of Eq. (3). Taking inverse DTFT, we get

$$y(n) = \text{Inverse DTFT } \{Y(e^{j\omega})\}$$

$$= \text{Inverse DTFT}\{e^{-j\omega n_0} S(e^{j\omega})\} = s(n - n_0) \quad (4)$$

Note in this example output $y(n)$ is equal to the shifted version of input $s(n)$ by a constant time n_0. The frequency response $H(e^{j\omega}) = e^{-j\omega n_0}$ is purely time-shifted and has unity magnitude at all frequencies. Its phase characteristics are equal to $-\omega n_0$, that is, it is linear with frequency.

EXAMPLE 4.9
Determine the impulse response $h(n)$ of a discrete-time ideal low pass filter whose frequency response $H(e^{j\omega})$ is shown in Figure 4.15.

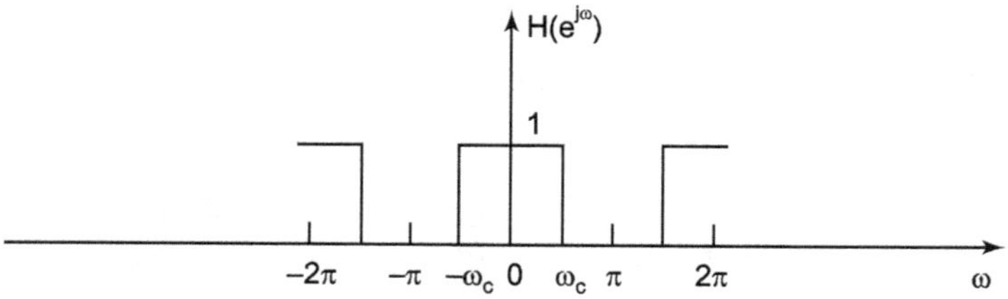

FIGURE 4.15 Frequency response of a discrete-time ideal low pass filter.

Solution:
Impulse response $h(n)$ of discrete-time ideal low pass filter is equal to inverse DTFT of the frequency response $He(e^{j\omega})$.

$$h(n) = \text{Inverse DTFT}\{H(e^{j\omega})\}$$

$$= \frac{1}{2\pi}\int_{2\pi} H\left(e^{j\omega}\right)e^{j\omega n}\,d\omega \tag{1}$$

In particular, using $-\pi \le \omega \le \pi$ as the interval of integration in Eq. (1), we obtain

$$h(n) = \frac{1}{2\pi}\int_{-\pi}^{\pi} H\left(e^{j\omega}\right)e^{j\omega n}\,d\omega$$

$$= \frac{1}{2\pi}\int_{-\omega_c}^{\omega_c} 1e^{j\omega n}\,d\omega = \frac{1}{2\pi}\left[\frac{1}{jn}e^{j\omega n}\right]_{-\omega_c}^{\omega_c}$$

$$= \frac{1}{n\pi}\left[\frac{e^{j\omega n} - e^{-j\omega n}}{2j}\right] = \frac{1}{n\pi}\sin\omega_c n$$

or $$h(n) = \frac{\sin\omega_c n}{\pi n} \tag{2}$$

The impulse response $h(n)$ is shown in Figure 4.16.

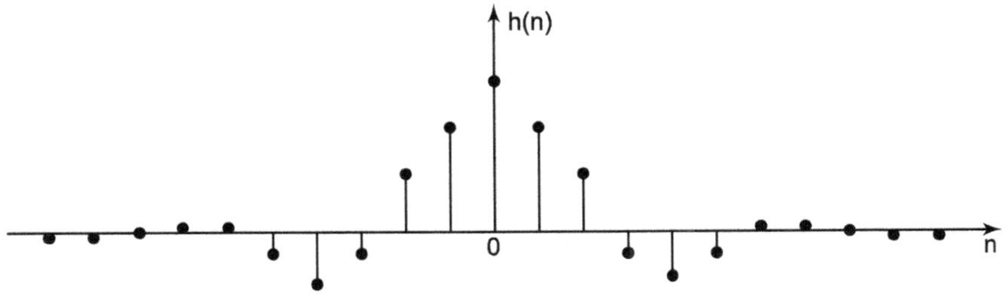

FIGURE 4.16 Impulse response of a discrete-time ideal low pass filter.

EXAMPLE 4.10

Determine the output $y(n)$ of a discrete-time LTI system with impulse response $h(n) = A^n u(n)$ with $|A| < 1$ to an input $s(n) = B^n u(n)$ with $|B| < 1$.

Solution:

Output $y(n)$ is determined by using convolution property of DTFT as

$$y(n) = h(n) * s(n) \xleftrightarrow{\text{DTFT}} Y(e^{j\omega}) = H(e^{j\omega})S(e^{j\omega}) \xleftrightarrow{\text{DTFT}}$$

or $$Y(e^{j\omega}) = H(e^{j\omega})S(e^{j\omega}) \tag{1}$$

Now we will determine $H(e^{j\omega})$ and $S(e^{j\omega})$ as

$$H(e^{j\omega}) = \text{DTFT}\{h(n)\} = \sum_{n=-\infty}^{\infty} h(n)e^{-j\omega n} = \sum_{n=-\infty}^{\infty} A^n u(n)e^{-j\omega n}$$

$$\left[\text{But } u(n) \text{ is defined as } u(n) = \begin{cases} 1, & n \geq 0 \\ 0, & n < 0 \end{cases}\right]$$

$$= \sum_{n=0}^{\infty} A^n e^{-j\omega n} = \sum_{n=0}^{\infty} \left(Ae^{-j\omega}\right)^n$$

or $$H(e^{j\omega}) = \frac{1}{1 - Ae^{-j\omega}} \tag{2}$$

Similarly, we can determine $S(e^{j\omega})$ as

$$S(e^{j\omega}) = \text{DTFT}\{s(n)\} = \sum_{n=-\infty}^{\infty} s(n)e^{-j\omega n}$$

$$= \sum_{n=-\infty}^{\infty} B^n u(n)e^{-j\omega n} = \sum_{n=0}^{\infty} B^n e^{-j\omega n}$$

$$= \sum_{n=0}^{\infty} \left(Be^{-j\omega}\right)^n = \frac{1}{1 - Be^{-j\omega}}$$

or $$S\left(e^{j\omega}\right) = \frac{1}{1 - Be^{-j\omega}} \tag{3}$$

Substituting Eqs. (2) and (3) in Eq. (1), we get

$$Y(e^{j\omega}) = H(e^{j\omega})S(e^{j\omega})$$

$$= \left(\frac{1}{1 - Ae^{-j\omega}}\right)\left(\frac{1}{1 - Be^{-j\omega}}\right) = \frac{1}{\left(1 - Ae^{-j\omega}\right)\left(1 - Be^{-j\omega}\right)}$$

(using partial fraction expansion method)

or $$Y(e^{j\omega}) = \frac{\alpha_1}{1 - Ae^{-j\omega}} + \frac{\alpha_2}{1 - Be^{-j\omega}} \tag{4}$$

Determination of values of α_1 and α_2:

$$1 - Ae^{-j\omega} = 0 \qquad \text{or} \qquad e^{-j\omega} = \frac{1}{A}$$

$$\alpha_1 = \frac{1}{\left(1 - Be^{-j\omega}\right)} = \frac{1}{\left(1 - B\frac{1}{A}\right)} = \frac{A}{(A - B)}$$

$$1 - Be^{-j\omega} = 0 \qquad \text{or} \qquad e^{-j\omega} = \frac{1}{B},$$

$$\alpha_2 = \frac{1}{\left(1 - Ae^{-j\omega}\right)} = \frac{1}{\left(1 - A\frac{1}{B}\right)} = \frac{B}{(B-A)} = \frac{-B}{(A-B)}$$

Substituting the values of α_1 and α_2 in Eq. (1) we get

$$Y\left(e^{j\omega}\right) = \frac{\left(\frac{A}{A-B}\right)}{1 - Ae^{-j\omega}} + \frac{\left(-\frac{B}{A-B}\right)}{1 - Be^{-j\omega}} \tag{5}$$

For determining output $y(n)$, we take inverse DTFT of both sides of Eq. (1), we get

$$y(n) = \text{Inverse DTFT } [Y(e^{j\omega})\}$$

$$= \text{INVERSE DTFT } \frac{\left(\frac{A}{A-B}\right)}{1 - Ae^{-j\omega}} + \frac{\left(-\frac{B}{A-B}\right)}{1 - Be^{-j\omega}}$$

$$= \left(\frac{A}{A-B}\right) \text{INVERSE DTFT } \left\{\frac{1}{1 - Ae^{-j\omega}}\right\}$$

$$-\left(\frac{B}{B-A}\right) \text{INVERSE DTFT } \left\{\frac{1}{1 - Be^{-j\omega}}\right\}$$

$$= \left(\frac{A}{A-B}\right) A^n u(n) - \left(\frac{B}{B-A}\right) B^n u(n)$$

or

$$y(n) = (A - B)\{A^{n+1} u(n) - B^{n+1} u(n)\} \tag{6}$$

Note that for $\alpha_1 = \alpha_2$, the partial fraction expansion of Eq. (4) is not valid.

4.5.11 The Multiplication Property

The multiplication property of DTFT is used in the sampling and communication theory.

Consider $s(n)$ is a sequence that is a product of two sequences $s_1(n)$ and $s_2(n)$.

$$s(n) = s_1(n) s_2(n)$$

Let

$$s_1(n) \xleftrightarrow{\text{DTFT}} S_1(e\omega)$$

$$s_2(n) \xleftrightarrow{\text{DTFT}} S_2(e\omega)$$

and $$s(n) \xleftrightarrow{\text{DTFT}} S(e\omega)$$

Taking the DTFT of both sides of Eq. (4.45) we get

$$S(e^{j\omega}) = \text{DTFT}\{s(n)\}$$

$$= \sum_{n=-\infty}^{\infty} s(n)e^{-j\omega n} = \sum_{n=-\infty}^{\infty} s_1(n)s_2(n)e^{-j\omega n} \tag{4.46}$$

Since from inverse DTFT

$$s_1(n) = \frac{1}{2\pi}\int_{2\pi} S_1\left(e^{j\theta}\right)e^{j\theta n}d\theta \tag{4.47}$$

Substituting Eq. (4.47) in Eq. (4.46), we obtain

$$S\left(e^{j\omega}\right) = \sum_{n=-\infty}^{\infty} s_1(n)s_2(n)e^{-j\omega n}$$

$$= \sum_{n=-\infty}^{\infty} s_1(n)\left\{\frac{1}{2\pi}\int_{2\pi} S_1\left(e^{j\theta}\right)e^{j\theta n}d\theta\right\}e^{-j\omega n} \tag{4.48}$$

Interchanging the order of summation and integration in Eq. (4.48), we get

$$S\left(e^{j\omega}\right) = \frac{1}{2\pi}\int_{2\pi} S_1\left(e^{j\theta}\right)\left[\sum_{n=-\infty}^{\infty} s_2(n)e^{-j(\omega-\theta)n}\right]d\theta \tag{4.49}$$

But $$S_2\left(e^{j\omega-\theta}\right) = \sum_{n=-\infty}^{\infty} s_2(n)e^{-j(\omega-\theta)n} \tag{4.50}$$

Substituting the Eq. (4.50) in Eq. (4.49), we get

$$S\left(e^{j\omega}\right) = \frac{1}{2\pi}\int_{2\pi} S_1\left(e^{j\theta}\right)S_2^{\left(e^{j(\omega-\theta)}\right)d\theta} \tag{4.51}$$

Eq. (4.51) corresponds to a periodic convolution of $S_1(e^{j\omega})$ and $S_2(e^{j\omega})$. Integral in Eq. (4.51) can be evaluated over an interval of length 2π. The usual form of convolution is often referred to as non-periodic convolution as the integral ranges from $-\infty$ to ∞.

EXAMPLE 4.11

Determine the DTFT of the multiplication of two discrete-time signals given as $s(n) = S_1(n)\, s_2(n)$.

where $$s_1(n) = \frac{\sin(3\pi n/4)}{\pi n} \text{ and } s_2(n) = \frac{\sin(\pi n/2)}{\pi n}$$

Solution:

From the multiplication property of DTFT, we know that DTFT of $s(n)$, that is, $S(e^{j\omega})$ is the periodic convolution of $S_1(e^{j\omega})$ and $S_2(e^{j\omega})$. Here integration is done over any interval of length $2n$. In this periodic convolution, we have chosen the interval $-\pi < \theta < \pi$. The multiplication property of DTFT is given as

$$s(n) = s_1(n)s_2(n) \xleftarrow{\text{DTFT}} S\left(e^{j\omega}\right) = \frac{1}{2\pi}\int_{2\pi} S_1\left(e^{j\theta}\right)S_2^{\left(e^{j(\omega-\theta)}\right)d\theta}$$

or
$$S\left(e^{j\omega}\right) = \frac{1}{2\pi}\int_{2\pi} S_1\left(e^{j\theta}\right)S_2\left(e^{j(\omega-\theta)}\right)d\theta \tag{1}$$

Eq. (1) represents a periodic convolution, it resembles a non-periodic convolution except that the integration is limited to the interval $-\pi < \theta \leq \pi$.

Eq. (1) can be converted into an ordinary convolution by defining

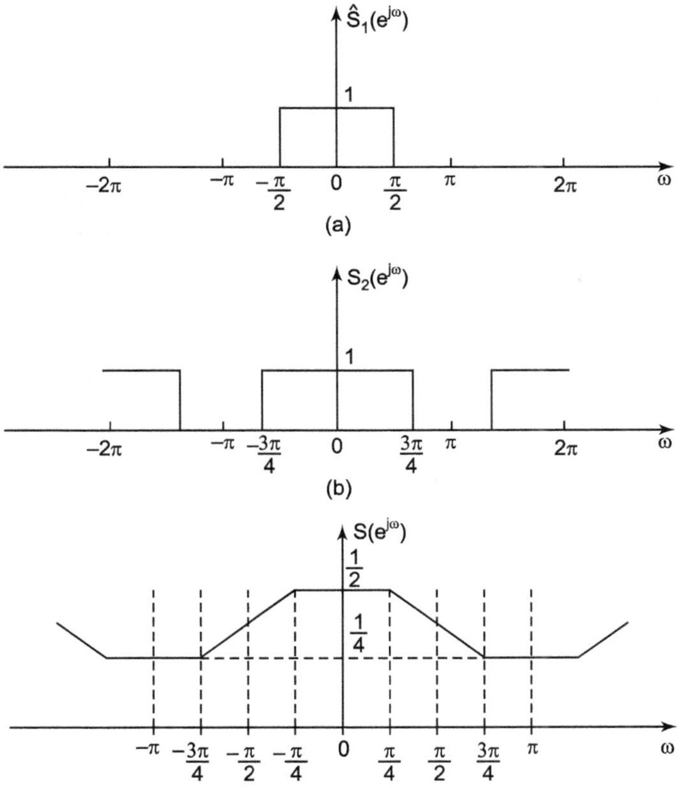

FIGURE 4.17 (a) $\hat{S}_1(e^{j\omega})$ shows one period of $S_1(e^{j\omega})$. (b) DTFT of $S_2(n)$, that is, $S_2(e^{j\omega})$. (c) Resultant of the periodic convolution of $\hat{S}_1(e^{j\omega})$ and $S_2(e^{j\omega})$.

$$\hat{S}_1\left(e^{j\omega}\right)=\begin{cases}S_1(e^{j\omega}), & \text{for } -\pi<\omega\leq\pi\\ 0, & \text{otherwise}\end{cases} \tag{2}$$

Substituting Eq. (2) in Eq. (1), we get

$$S\left(e^{j\omega}\right)=\frac{1}{2\pi}\int_{-\pi}^{\pi}S_1\left(e^{j\theta}\right)S_2\left(e^{j(\omega-\theta)}\right)d\theta$$

$$=\frac{1}{2\pi}\int_{-\infty}^{\infty}\hat{S}_1(e^{j\theta})S_2\left(e^{j(\omega-\theta)}\right)d\theta \tag{3}$$

Thus, $S(e^{j\omega})$ is $1/2\pi$ the non-periodic convolution of the rectangular pulse $\hat{S}_1(e^{j\omega})$ and periodic square wave $S_2(e^{j\omega})$. Both $\hat{S}_1(e^{j\omega})$ and $S_2(e^{j\omega})$ are shown in Figure 4.17. The result of this convolution is the DTFT of $s(n)$, that is, $S(e^{j\omega})$ shown in Figure 4.17(c).

4.6 TABULATION OF PROPERTIES OF DTFT

A number of important properties of the DTFT are summarized in Table 4.1.

TABLE 4.1 Summary of Important Properties of DTFT

DTFT property	Non-periodic signal	DTFT
	$s(n)$ $s_1(n)$ $s_2(n)$	$S(e^{j\omega})$, $S_1(e^{j\omega})$ and $S_2(e^{j\omega})$ These are periodic with period 2π
Linearity	$A_1 s_1(n) + A_2 s_2(n)$	$A_1 s_1(e^{j\omega}) + A_2 s_2(e^{j\omega})$ where A_1 are A_2 scalar constants $e^{-j\omega n_0}S(e^{j\omega})$
Time-shifting	$s(n-n_0)$	where n_0 is constant time
Frequency shifting	$e^{j\omega_0 n}s(n)$	$S(e^{j(\omega-\omega_0)})$
Complex conjugation	$s^*(n)$	$S^*(e^{-j\omega})$
Time reversal	$s(-n)$	$S(e^{-j\omega})$
Time expansion	$s_{(k)}(n)$ $\begin{cases}s(n/k), & \text{if } n \text{ is multiple of } k\\ 0, & \text{if } n \text{ is a not multiple of } k\end{cases}$	$S(e^{-jk\omega})$

Convolution of two discrete-time signal	$s_1(n) * s_2(n)$	$s_1(e^{j\omega})\, s_2(e^{j\omega})$
Multiplication of two discrete-time signal	$s_1(n)\, s_2(n)$	$\frac{1}{2\pi}\int_{2\pi} S_1\left(e^{j\theta}\right)S_2\left(e^{j(\omega-\theta)}\right)d\theta$
First difference of a discrete-time signal in time	$s(n) - s(n-1)$	$(1 - e^{-j\omega})\, S(e^{j\omega})$
Accumulation of a discrete-time signal	$\sum_{k=-\infty}^{n} s(k)$	$\frac{1}{1-e^{-j\omega}}S\left(e^{j\omega}\right)+\pi S\left(e^{j0}\right) \times \sum_{k=-\infty}^{\infty}\delta(\omega-2\pi k)$
Differentiation in frequency	$ns(n)$	$\frac{jdS\left(e^{j\omega}\right)}{d\omega}$
Conjugate symmetry for real signal	$s(n)$ real	$\begin{cases} S\left(e^{j\omega}\right)=S*\left(e^{-j\omega}\right) \\ \mathrm{Re}\left\{S\left(e^{j\omega}\right)\right\}=\mathrm{Re}\left\{S\left(e^{-j\omega}\right)\right\} \\ \mathrm{Im}\left\{S\left(e^{j\omega}\right)\right\}=-\left\{S\left(e^{-j\omega}\right)\right\} \\ \left\|s\left(e^{j\omega}\right)\right\|=\left\|S\left(e^{-j\omega}\right)\right\| \\ \angle S\left(e^{j\omega}\right)=-\angle S\left(e^{-j\omega}\right) \end{cases}$
Symmetry for real and even signals	$s(n)$ real and even	$s(e^{j\omega})$ real and even
Symmetry for real and odd signals	$s(n)$ real and even	$s(e^{j\omega})$ purely imaginary and odd
Decomposition of a real signal into even and odd pats.	$s_E(n) = \mathrm{Ev}\{s(n)\}$ $s_0(n) = \mathrm{Od}\{s(n)\}$ Here, $s(n)$ is real signal	$\mathrm{Re}\ \{S(e^{j\omega})\}$ $j\mathrm{Im}\ \{S(e^{j\omega})\}$

Parseval's relation for non-periodic signals

$$\sum_{n=-\infty}^{\infty}\left|s(n)\right|^2=\frac{1}{2\pi}\int_{2\pi}\left|S(e^{j\omega})\right|^2 d\omega$$

4.7 TABULATION OF DTFT PAIRS

A summary of the basic DTFT pairs is shown in Table 4.2.

TABLE 4.2 Summary of Basic DTFT Pairs

Discrete-time signal $s(n)$	**DTFT $S(e^{j\omega})$**
Exponential function $e^{j\omega_0 n}$	$2\pi \sum_{m=-\infty}^{\infty} \delta(\omega-\omega_0-2\pi m)$ where $\omega_0 = \frac{2\pi m}{N}$
Cosine function	$\frac{2\pi}{2}\left[\sum_{m=-\infty}^{\infty} \delta(\omega-\omega_0-2\pi m)+\sum_{m=-\infty}^{\infty} \delta(\omega-\omega_0-2\pi m)\right]$
$\cos\omega_0 n = \frac{e^{j\omega_0 n}-e^{-j\omega_0 n}}{2}$	$\frac{2\pi}{2j}\left[\sum_{m=-\infty}^{\infty} \delta(\omega-\omega_0-2\pi m)+\sum_{m=-\infty}^{\infty} \delta(\omega+\omega_0-2\pi m)\right]$
$s(n) = 1\ e^{j\omega_0 n}\vert\omega 0=0$	$2\pi\left[\sum_{m=-\infty}^{\infty} \delta(\omega-\omega_0-2\pi m)+\sum_{m=-\infty}^{\infty} \delta(\omega-2\pi m)\right]$
$s(n) = 2\pi \sum_{k=<N_0>} A_k e^{j\omega_0 n}$	$S(e^{j\omega}) = 2\pi \sum_{k=-\infty}^{\infty} A_k \delta\left(\omega-\frac{2\pi km}{N_0}\right)$
Periodic square wave	$S(e^{j\omega}) = 2\pi \sum_{k=-\infty}^{\infty} A_k \delta\left(\omega-\frac{2\pi km}{N_0}\right)$
$s(n) = \begin{cases} 1, & \lvert n\rvert \le N_1 \\ 1, & N_1 < \lvert n\rvert \le N_0/2 \end{cases}$ and for periodic signal $s(n + N_0)\ s(n)$	Where, $\begin{cases} \frac{\sin[(2\pi k/N_0)(N_1+1/2)]}{N_0 \sin[2\pi k/2N_0]} \\ k \ne 0, \pm N_0, \pm 2N_0 .... \\ (2N_1+1)/N_0, \\ k = 0, \pm N_0, 2N_0 .... \end{cases}$
$\sum_{k=-\infty}^{\infty} \delta(n-kN_0)$	$\frac{2\pi}{N_0}\sum_{k=-\infty}^{\infty} \delta\left(\omega-\frac{2\pi km}{N_0}\right)$
$s(n) = A^n u(n),\ \lvert A\rvert < 1$	$S(e^{j\omega}) = \frac{1}{1-Ae^{-j\omega}}$

$s(n)=\begin{cases}1, & \lvert n\rvert \le N_1 \\ 1, & \lvert n\rvert > N_1\end{cases}$	$S\left(e^{j\omega}\right)=\dfrac{\sin\left[\omega\left(N_1+\frac{1}{2}\right)\right]}{\sin\left(\frac{\omega}{2}\right)}$
$s(n)=\dfrac{\sin W_n}{\pi n}=\dfrac{W}{\pi}\left(\dfrac{\sin W_n}{W_n}\right)$ $=\dfrac{W}{\pi}\sin c\left(\dfrac{Wn}{\pi}\right)$ for $0 < W < \pi$	$S\left(e^{j\omega}\right)=\begin{cases}1, & 0\le\lvert\omega\rvert\le W \\ 0, & W<\lvert\omega\rvert\le\pi\end{cases}$ where is $S(e^{j\omega})$ is periodic with period 2π
Impulse function $\delta(n)$	1
Unit-step function $u(n)$	$\dfrac{1}{1-e^{j\omega}}+\sum_{k=-\infty}^{\infty}\pi\delta(\omega-2\pi k)$
Shifted version of $\delta(n)$, $\delta(n-n_0)$, where n_0 is constant time	$e^{j\omega n_0}$
$(n+1)A^n u(n)$, $\lvert A\rvert < 1$	$\dfrac{1}{\left(1-Ae^{-j\omega}\right)^2}$
for $\lvert A\rvert < 1$	$\dfrac{1}{\left(1-Ae^{-j\omega}\right)^m}$

4.8 DUALITY

There is a duality between the CTFT analysis equation and the CTFT synthesis equation. Both equations are given as

CTFT analysis equation,

$$S\left(e^{j\omega}\right)=\int_{-\infty}^{\infty} S(t)e^{-j\omega t}\,dt \tag{4.52}$$

and CTFT synthesis equation

$$s(t)=\frac{1}{2\pi}\int_{-\infty}^{\infty} S(j\omega)d^{j\omega t}\,d\omega \tag{4.53}$$

But there is no corresponding duality between the DTFT analysis equation and DTFT synthesis equation. These equations are given as

DTFT analysis equation,

$$S\left(e^{j\omega}\right) = \sum_{n=-\infty}^{\infty} s(n) e^{-j\omega n} \tag{4.54}$$

and DTFT synthesis equation

$$s(n) = \frac{1}{2\pi} \int_{2\pi} S\left(e^{j\omega}\right) e^{j\omega n} d\omega \tag{4.55}$$

However, there is a duality between the DTFS equation. These equations are given as DTFS analysis equation,

$$A_k = \frac{1}{N_0} \sum_{n=<N_0>} s(n) e^{-jk\omega_0 n} \tag{4.56}$$

and DTFS synthesis equation,

$$s(n) = \sum_{k=<N_0>} A_k e^{-jk\omega_0 n} \tag{4.57}$$

where A_k is the Fourier series coefficients of a discrete-time periodic signal $s(n)$.

Also, there is a duality relationship between the DTFT and the CTFS.

4.8.1 Duality in the DTFS

Since the Fourier series coefficients A_k of the periodic signal $s(n)$ are themselves a periodic sequence, we can expand the sequence A_k in a Fourier series. The duality property for DTFS implies that the Fourier series coefficients for the periodic sequence A_k are the values of $(1/N_0)s(-n)$.

In other words, the duality property for DTFS implies that the Fourier series coefficients for the periodic sequence A_k is proportional to the values of the original sequence which is reversed or folded in time.

Proof:

To prove the duality property for DTFS, we consider two periodic sequences with period N_0. These two sequences are related through the summation given as,

$$s_1(m) = \frac{1}{N_0} \sum_{l=<N_0>} s_2(l) e^{-jl\omega_0 n} \tag{4.58}$$

where $$\omega_0 = \frac{2\pi}{N_0}$$

If we let $m = k$ and $l = n$, Eq. (4.58) becomes

$$s_1(k) = \sum_{n=N_0>} s_2(n)e^{-jn\omega_0 k}$$

$$= \sum_{n=<N_0>} s_2(n)e^{-jn\omega_0 k} \tag{4.59}$$

Comparing Eq. (4.59) with DTFS analysis equation given as

$$A_k = \frac{1}{N_0} \sum_{n=<N_0>} s(n)e^{-jk\omega_0 n} \tag{4.60}$$

We see that the sequence $s_1(k)$ corresponds to the Fourier series coefficients of the signal $s_2(n)$.

Eq. (4.60) can be represented as

$$s(n) \xleftrightarrow{\text{DTFT}} A_k$$

Similarly, Eq. (4.59) can be represented as

$$s_2(n) \xleftrightarrow{\text{DTFT}} S_1(k) \tag{4.61}$$

Alternatively, if we let $m = n$ and $l = -k$, Eq. (4.58) becomes

$$s_1(n) = \frac{1}{N_0} \sum_{k=<N_0>} s_2(-k)e^{jk\omega_0 n}$$

$$= \sum_{k=<N_0>} \frac{1}{N_0} s_2(-k)e^{jk\omega_0 n} \tag{4.62}$$

Now, comparing Eq. (4.62) with DTFS synthesis equation given as

$$s(n) = \sum_{k=<N_0>} A_k e^{jk\omega_0 n} \tag{4.63}$$

We obtain, $$s_1(n) = s_1(n) \xleftrightarrow{\text{DTFT}} \frac{1}{N} s_2(-k) \tag{4.64}$$

Now, we can conclude that $\frac{1}{N_0} S_2(-k)$ corresponds to the sequence of the Fourier series coefficients of discrete-time signal $s_1(n)$.

Thus it is to be proved.

As in continuous-time, this duality implies that every property of the DTFS has a dual character.

For example, we have two DTFS pairs given as

$$s(n - n_0) \xleftrightarrow{\text{DTFS}} A_k e^{jk\omega_0 n_0} \tag{4.65}$$

$$e^{jk\omega_0 n} \xleftrightarrow{\text{DTFS}} A_{k-m} \tag{4.66}$$

Both DTFS pairs are dual.

Duality is often useful in reducing the complexity of the calculations involved in determining Fourier series representations.

EXAMPLE 4.12

Verify the duality of the following periodic signal with a period of $N_0 = 9$

$$s(n) = \begin{cases} 1/9\dfrac{\sin(5\pi n/9)}{\sin(\pi n/9)}, & n \neq \text{multiple of } 9 \\ 5/9, & n = \text{multiple of } 9 \end{cases} \tag{1}$$

Solution:

We have already studied that the Fourier series coefficients of a discrete-time square (rectangular) wave is given as

$$A_k = \begin{cases} 1/N_0 \dfrac{\sin\left[2\pi k\left(N_1 + \frac{1}{2}\right)\Big/ N_0\right]}{\sin\left(\dfrac{\pi k}{N_0}\right)}, & k \neq 0, \pm N_0, \pm 2N_0, \ldots \\ \dfrac{2N_1 + 1}{N_0}, & k \neq 0, \pm N_0, \pm 2N_0, \ldots \end{cases} \tag{2}$$

Duality, then suggest that the Fourier series coefficients for $s(n)$ must be in the form of a rectangular square wave. The rectangular (square) wave with period $N_0 = 9$ is such that

$$g(n) = \begin{cases} 1, & |n| \leq 2 \\ 0, & 2 < |n| \leq 4 \end{cases} \tag{3}$$

The Fourier series coefficients B_k for discrete-time square wave $g(n)$ is determined as

$$B_k = \begin{cases} \dfrac{1}{9}\dfrac{\sin\left(\dfrac{5\pi n}{9}\right)}{\sin\left(\dfrac{\pi k}{N_0}\right)}, & \text{for } k \neq \text{multiple of } 9 \\ \dfrac{5}{0} & \text{for } k = \text{multiple of } 9 \end{cases} \tag{4}$$

The DTFS analysis equation for sequence $g(n)$ is given as

$$B_k = \frac{1}{N_0} \sum_{n=<N_0>} g(n) e^{-jk\omega_0 n}$$

$$= \frac{1}{9} \sum_{n=-2}^{2} (1) e^{-jk(2\pi/9)n}$$

$$= \frac{1}{9} \sum_{k=-2}^{2} e^{-jk(2\pi n)/9} \tag{5}$$

Interchanging the names of the variables k and n and putting $B_n = s(n)$, we obtain

$$B_n = \frac{1}{9} \sum_{k=-2}^{2} e^{-(jn2\pi k)/9}$$

or

$$s(n) = \frac{1}{9} \sum_{k=-2}^{2} e^{-(j2\pi nk)/9} \tag{6}$$

Also letting $k = -k'$ in the sum of RHS of Eq. (1), we get

$$s(n) = \frac{1}{9} \sum_{k'=-2}^{2} e^{(j2\pi nk')/9} \tag{7}$$

or

$$s(n) = \sum_{k'=-2}^{2} \frac{1}{9} e^{(j2\pi k')/9} \tag{8}$$

Finally, we can say that RHS of Eq. (8) has the form of a DTPS synthesis equation for $s(n)$.

$$s(n) = \sum_{k=<N_0>} A_k e^{-jk\omega_0 n} \tag{9}$$

We thus conclude that the Fourier series coefficients of sequence $s(n)$ are given by

$$A_k = \begin{cases} \frac{1}{9}, & |k| \leq 2 \\ 0, & 2 < |k| \leq 4 \end{cases} \tag{10}$$

These Fourier series coefficients A_k, of course, are periodic with period $N_0 = 9$.

4.8.2 Duality Between the DTFT and CTFTS

We have already studied duality in DTFS. In this section, we will study the property of duality between the DTFT and CTFS. Here, we will compare

CTFS analysis and synthesis equations with DTFT analysis and synthesis equations. These equations are given as follows:

CTFS synthesis equation,

$$s(t) = \sum_{m=-\infty}^{\infty} A_k e^{jk\omega_0 t} \tag{4.67}$$

and CTFS analysis equation,

$$A_k = \frac{1}{T_0} \int_{T_0} S(t) e^{jk\omega_0 dt} \tag{4.68}$$

DTFT synthesis equation,

$$s(n) = \frac{1}{2\pi} \int_{2\pi} S\left(e^{j\omega}\right) e^{j\omega n} d\omega \tag{4.69}$$

DTFT analysis equation,

$$S\left(e^{j\omega}\right) = \sum_{n=-\infty}^{\infty} s(n) e^{-j\omega n} \tag{4.70}$$

Eqs. (4.69) and (4.68) are very similar. Similarly, Eqs. (4.70) and (4.67) are also very similar. Now we can interpret Eqs. (4.69) and (4.70) as a Fourier series representation of the periodic frequency response $S(e^{j\omega})$. In particular, since $S(e^{j\omega})$ is a periodic function of co with period $2n$ and it has a Fourier series representation as a weighted sum of harmonically related periodic exponential functions of w, all of which have the common period of $2n$.

DTFT of $s(n)$, $S(e^{j\omega})$ can be represented in a Fourier series as a weighted sum of the exponential functions $e^{j\omega}n$, $n = 0, \pm 1, \pm 2, \ldots$. From equation $S(e^{j\omega}) = \sum_{n=-\infty}^{\infty} s(n) e^{-j\omega n}$, we see that the nth Fourier series coefficients in this expansion (that is, the Fourier series coefficient multiplying $e^{j\omega}n$ is $s(-n)$. Since the fundamental period of $S(e^{j\omega})$ is 2π, equation $s(n) = \frac{1}{2} \int_{2\pi} S(e^{j\omega}) e^{j\omega n} d\omega$ can be interpreted as the Fourier series analysis equation for the Fourier series coefficient $s(n)$ (That is, for the Fourier series coefficient multiplying $e^{-j\omega n}$ in the expression for $S(e^{j\omega})$ in equation $S(e^{j\omega}) = \sum_{n=-\infty}^{\infty} s(n) e^{-j\omega n}$.

EXAMPLE 4.13

Determine the DTFT of the sequence $s(n) = \dfrac{\sin(\pi n/2)}{\pi n}$.

Solution:

To determine the DTFT of the sequence $s(n) = \frac{\sin(\pi n/2)}{\pi n}$, we will exploit the duality between the DTFT synthesis equation and the CTFS analysis equation.

To use duality, we must first identify a continuous-time signal g(t) with period' $T_0 = 2\pi$ and Fourier series coefficients $A_k = s(k)$. If g(t) is a periodic square wave with period 2π, that is,

$$g(t) = \begin{cases} 1, & |t| \le T_1 \\ 0, & T_1 < |t| \le \pi \end{cases} \tag{1}$$

then the Fourier series coefficients of g(t) are

$$A_k = \frac{\sin(kT_1)}{k\pi} \tag{2}$$

If we take $T_1 = \frac{\pi}{2}$, then

$$A_k = \frac{\sin\left(k\frac{\pi}{2}\right)}{k\pi} = s(k) \tag{3}$$

In this case, CTFS analysis equation for g(t) is

$$\frac{\sin\left(\frac{\pi k}{2}\right)}{\pi n} = \frac{1}{2\pi}\int_{-\pi}^{\pi} g(t)e^{-jkt}$$

$$= \frac{1}{2\pi}\int_{-\pi}^{\pi} (1)e^{-jkt} \tag{4}$$

Renaming k to n and t as to in Eq. (4), we get

$$\frac{\sin\left(\frac{\pi k}{2}\right)}{\pi n} = \frac{1}{2\pi}\int_{-\pi/2}^{\pi/2} (1)e^{-jn\omega}d\omega \tag{5}$$

Replacing n by –n on both sides of Eq. (1), we get

$$\frac{\sin\left(\frac{\pi k}{2}\right)}{\pi n} = \frac{1}{2\pi}\int_{-\pi/2}^{\pi/2} (1)e^{-jn\omega}d\omega \tag{6}$$

The RHS of Eq. (6) has the form of the DTFT synthesis equation for sequence $s(n)$, where

$$S\left(e^{j\omega}\right)=\begin{cases}1, & |\omega|\leq\pi/2\\ 0, & \pi/2<|\omega|\leq\pi\end{cases}$$

4.9 DISCRETE-TIME LTI SYSTEMS CHARACTERIZED BY LINEAR CONSTANT-COEFFICIENT DIFFERENCE EQUATIONS

The general form of a linear constant-coefficient difference equation for a discrete-time LTI system is given as

$$\sum_{k=0}^{N}A_k y(n-k)=\sum_{k=0}^{M}B_k s(n-k) \tag{4.71}$$

This difference equation is generally called nth order difference equation.

Here, A_k and B_k are the constant coefficients, $s(n)$ and $y(n)$ are the inputs and outputs of the discrete-time LTI system, respectively.

Here, we will determine the frequency response $s(e^{j\omega})$ for an LTI system described by Eq. (4.71). In determining the frequency response $H(e^{j\omega})$ of a discrete-time LTI system, we will take advantage of several important properties of the DTFT.

There are two related methods for determining the frequency response $H(e^{j\omega})$ of the discrete-time LTI system.

In the first method, we explicitly use the fact that complex exponential functions are eigenfunctions of LTI systems. Specifically, if $s(n) = e^{j\omega n}$ is the input to a discrete-time LTI system then the output of this system is given by the multiplication of $(e^{j\omega})$ and $He^{j\omega n}$.

Now, substituting $s(n) = e^{j\omega n}$ and $y(n) = H(e^{j\omega})\, e^{j\omega n}$ in Eq. (4.71), we get

$$\sum_{k=0}^{N}A_k y(n-k)=\sum_{k=0}^{M}B_k s(n-k)$$

or

$$\sum_{k=0}^{N}A_k H(e^{j\omega})e^{j\omega(n-k)}=\sum_{k=0}^{M}A_k{}^{j\omega(n-k)}$$

or

$$H(e^{j\omega})\sum_{k=0}^{N}A_k e^{j\omega n}e^{-j\omega k}=\sum_{k=0}^{M}B_k e^{j\omega n}e^{-j\omega k}$$

or

$$H(e^{j\omega}) = \frac{\sum_{k=0}^{M} B_k e^{-j\omega k}}{\sum_{k=0}^{N} A_k e^{-j\omega k}} \tag{4.72}$$

In the second method, we make use of the convolution, linearity, and time-shifting properties of DTFT. Let $S(e^{j\omega})$, $Y(e^{j\omega})$ and $H(e^{j\omega})$ be the DTFT of the input $s(n)$, output $y(n)$ and impulse response $h(n)$, respectively. Figure 4.18 shows a discrete-time LTI system.

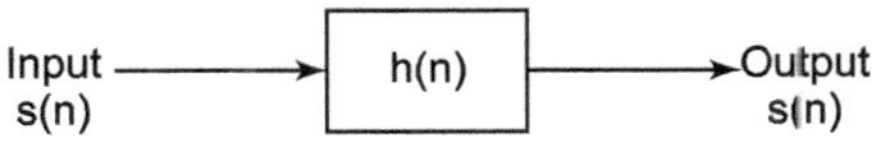

FIGURE 4.18 Discrete-time LTI system

From the convolution property of the DTFT,

$$y(n) = s(n) * h(n) \xleftrightarrow{\text{DTFT}} Y(e^{j\omega}) = S(e^{j\omega})H(e^{j\omega})$$

$$Y(e^{j\omega}) = S(e^{j\omega})S(e^{j\omega}) \tag{7.73}$$

or

$$H(e^{j\omega}) = \frac{Y(e^{j\omega})}{S(e^{j\omega})} \tag{7.74}$$

Taking the DTFT of both sides of Eq. (4.71) and using the properties of linearity and time-shifting of DTFT, we get

$$\sum_{k=0}^{N} A_k y(n-k) = \sum_{k=0}^{M} B_k s(n-k) \tag{4.75}$$

$$\text{DTFT}\left\{\sum_{k=0}^{M} A_k y(n-k)\right\} = \text{DTFT}\left\{\sum_{k=0}^{M} B_k y(n-k)\right\}$$

or

$$\sum_{k=0}^{N} A_k \text{DTFT}\{y(n-k)\} = \sum_{k=0}^{M} B_k \text{DTFT}\{s(n-k)\}$$

or

$$\sum_{k=0}^{N} A_k e^{-jk\omega_0} Y(e^{j\omega}) = \sum_{k=0}^{M} B_k e^{-jk\omega_0 S(e^{j\omega})}$$

or

$$S(e^{j\omega})\sum_{k=0}^{N} A_k e^{-jk\omega_0} = S(e^{j\omega})\sum_{k=0}^{M} A_k e^{-jk\omega_0}$$

or

$$\frac{Y(e^{j\omega})}{S(e^{j\omega})} = \frac{\sum_{k=0}^{M} B_k e^{-jk\omega}}{\sum_{k=0}^{N} B_k e^{-jk\omega}}$$

From Eqs. (4.74) and (4.76), we obtain

$$H\left(e^{j\omega}\right)=\frac{Y\left(e^{j\omega}\right)}{S\left(e^{j\omega}\right)}=\frac{\sum_{k=0}^{M}B_k e^{-jk\omega}}{\sum_{k=0}^{N}B_k e^{-jk\omega}}$$

We see from Eq. (4.77) that frequency response $H(e^{j\omega})$ is a ratio of two polynomials in the variable $e^{-j\omega}$. The coefficient of the numerator polynomial is the same as the ones that appear on the RHS of Eq. (4.71) and the coefficients of the denominator polynomial are the same as the ones that appear on the LHS of Eq. (4.71). Therefore, the frequency response of an LTI system described by Eq. (4.71) can be written by inspection.

EXAMPLE 4.14

Determine the frequency response and impulse response of a causal discrete-time LTI system that is characterized by the difference equation given as

$$y(n)-Ay(n-1)=s(n) \tag{1}$$

Solution:

To determine frequency response, we take DTFT of both sides of Eq. (1), we obtain

$$\text{DTFT}\,\{y(n)-Ay(n-\ 1)\}=\text{DTFT}\{s(n)\}$$

or

$$Y(e^{j\omega})-Ae^{-j\omega}Y(e^{j\omega})\ =S(e^{j\omega})$$

or

$$Y(e^{j\omega})[1-Ae^{-j\omega}]=S(e^{j\omega})$$

or

$$\frac{Y\left(e^{j\omega}\right)}{S\left(e^{j\omega}\right)}=\frac{1}{1-Ae^{-j\omega}} \tag{2}$$

But we know that frequency response $H(e^{j\omega})$ is given by

$$H(e^{j\omega})=\frac{Y\left(e^{j\omega}\right)}{S\left(e^{j\omega}\right)} \tag{3}$$

From Eqs. (2) and (3), we obtain

$$H\left(e^{j\omega}\right)=\frac{1}{1-Ae^{-j\omega}} \tag{4}$$

Now we will determine impulse response $h(n)$ as

$$h(n)=\text{Inverse DTFT}\ [H(e^{j\omega})]$$

$$= \text{INVERSE DTFT}\left[\frac{1}{1-Ae^{-j\omega}}\right] = A^n u(n)$$

Thus, the impulse response of the LTI system is

$$h(n) = A^n u(n) \tag{5}$$

EXAMPLE 4.15

Determine the frequency response $H(e^{j\omega})$ and impulse response $h(n)$ of a causal discrete-time LTI system that is characterized by the difference equation given as

$$y(n) - \frac{3}{4}y(n-1) + \frac{1}{8}y(n-2) = 2s(n) \tag{1}$$

Solution:

To determine frequency response, taking DTFT of both sides of Eq. (1), we get

$$\text{DTFT}\left[y(n) - \frac{3}{4}y(n-1) + \frac{1}{8}y(n-2)\right] = \text{DTFT}\{2(s)\}$$

or $$y(e^{j\omega}) = 2S(e^{j\omega})$$

or $$y\left(e^{j\omega}\right)\left[1 - \frac{3}{4}e^{-j\omega} + \frac{1}{8}e^{-j2\omega}\right] = 2S\left(e^{j\omega}\right)$$

or $$\frac{\gamma\left(e^{j\omega}\right)}{S\left(e^{j\omega}\right)} = \frac{2}{1 - \frac{3}{4}e^{-j\omega} + \frac{1}{8}e^{-j2\omega}} \tag{2}$$

But frequency response is given by

$$H\left(e^{j\omega}\right) = \frac{\gamma\left(e^{j\omega}\right)}{S\left(e^{j\omega}\right)}$$

From Eqs. (2) and (3), we get frequency response,

$$H\left(e^{j\omega}\right) = \frac{2}{1 - \frac{3}{4}e^{-j\omega} + \frac{1}{8}e^{-j2\omega}} \tag{4}$$

Now, we will determine impulse response $h(n)$ by taking inverse DTFT of Eq. (1) as

$$h(n) = \text{Inverse DTFT}\{H(e^{j\omega})\}$$

$$= \text{INVERSE DTFT}\left\{\frac{2}{1-\frac{3}{4}e^{-j\omega}+\frac{1}{8}e^{-j2\omega}}\right\} \quad (5)$$

By using partial fraction expansion method:

$$\frac{2}{1-\frac{3}{4}e^{-j\omega}+\frac{1}{8}e^{-j2\omega}} = \frac{2}{\left(1-\frac{3}{4}e^{-j\omega}\right)+\left(1-\frac{1}{8}e^{-j\omega}\right)}$$

$$= \frac{a_1}{\left(1-\frac{3}{4}e^{-j\omega}\right)} + \frac{a_2}{\left(1-\frac{1}{4}e^{-j\omega}\right)} \quad (6)$$

Determination of α_1 and α_2:

$$1-\frac{1}{2}e^{j\omega} = 0 \qquad \text{or} \qquad e^{-j\omega} = 2$$

$$a_1 = \frac{2}{\left(1-\frac{1}{4}e^{-j\omega}\right)} = \frac{2}{\left(1-\frac{1}{4}2\right)} = 4$$

$$1-\frac{1}{2}e^{-j\omega} = 0 \qquad \text{or} \qquad e^{-j\omega} = 4$$

$$a_2 = \frac{2}{\left(1-\frac{1}{4}e^{-j\omega}\right)} = \frac{2}{\left(1-\frac{1}{4}4\right)} = -2$$

Putting the value of α_1 and α_2 in Eq. (1) we get

$$\frac{2}{1-\frac{3}{4}e^{-j\omega}+\frac{1}{8}e^{-j2\omega}} = \frac{4}{\left(1-\frac{1}{2}e^{-j\omega}\right)} - \frac{2}{\left(1-\frac{1}{4}e^{-j\omega}\right)} \quad (7)$$

Substituting Eq. (7) in Eq. (5), we obtain

$$h(n) = \text{Inverse DTFT}\left\{\frac{4}{\left(1-\frac{1}{2}e^{-j\omega}\right)} - \frac{2}{\left(1-\frac{1}{4}e^{-j\omega}\right)}\right\}$$

$$= \text{Inverse DTFT}\left\{\frac{4}{\left[\left(1-\frac{1}{2}e^{-j\omega}\right)\right]}\right\}$$

$$-\text{Inverse DTFT}\left\{\frac{2}{\left[\left(1-\frac{1}{2}e^{-j\omega}\right)\right]}\right\}$$

or
$$h(n) = 4\left(\frac{1}{2}\right)^n u(n) - 2\left(\frac{1}{2}\right)^n u(n) \qquad (8)$$

This is the impulse response of a given system.

EXAMPLE 4.16

Determine the output $y(n)$ of a causal discrete-time LTI system that is characterized by the difference equation

$$y(n) - \frac{3}{4}y(n-1) + \frac{1}{8}y(n-2) = -2s(n) \qquad (1)$$

for input
$$s(n) = \left(\frac{1}{4}\right)^n u(n) \qquad (2)$$

Solution:

From Eqs. (1) and (2), we get

$$y(n) - \frac{3}{4}y(n-1) + \frac{1}{8}y(n-2) = -2s(n)$$

or
$$y(n) - \frac{3}{4}y(n-1) + \frac{1}{8}y(n-2) = 2\left(\frac{1}{4}\right)^n u(n) \qquad (3)$$

Now taking the DTFT of both sides of Eq. (3) we obtain

$$\text{DTFT}\left\{y(n) - \frac{3}{4}y(n-1) + \frac{1}{8}y(n-2)\right\} = \text{DTFT}\left\{2\left(\frac{1}{4}\right)^n u(n)\right\}$$

Now taking the DTFT of both sides of Eq. (3) we obtain

$$\text{DTFT}\left\{y(n) - \frac{3}{4}y(n-1) + \frac{1}{8}y(n-2)\right\} = \text{DTFT}2\left\{\left(\frac{1}{4}\right)^n u(n)\right\}$$

or $$\text{DTFT}\left\{y(n)-\frac{3}{4}\text{DTFT}\{y(n-1)\}+\frac{1}{8}\text{DTFT}\{y(n-2)\}\right\}$$

$$\text{DTFT}\left\{y\left(\frac{1}{4}\right)^n u(n)\right\}$$

or $$y\left(e^{j\omega}\right)\left(-\frac{3}{4}e^{-j\omega}+\frac{1}{8}e^{-j2\omega}\right)=\frac{2}{\left(1-\frac{1}{4}e^{-j\omega}\right)}$$

$$Y\left(e^{j\omega}\right)=\frac{2}{\left(1-\frac{3}{4}e^{-j\omega}+\frac{1}{8}e^{-j2\omega}\right)\left(1-\frac{1}{4}e^{-j\omega}\right)}$$

$$=\frac{2}{\left(1-\frac{1}{2}e^{-j\omega}\right)\left(1-\frac{1}{4}e^{-j\omega}\right)\left(1-\frac{1}{4}e^{-j\omega}\right)}$$

$$=\frac{2}{\left(1-\frac{1}{2}e^{-j\omega}\right)\left(1-\frac{1}{4}e^{-j\omega}\right)^2} \tag{4}$$

By using partial fraction expansion method

$$Y\left(e^{j\omega}\right)=\frac{2}{\left(1-\frac{1}{2}e^{-j\omega}\right)\left(1-\frac{1}{4}e^{-j\omega}\right)^2}$$

$$=\frac{a_1}{\left(1-\frac{1}{2}e^{-j\omega}\right)}+\frac{a_2}{\left(1-\frac{1}{4}e^{-j\omega}\right)}+\frac{a_3}{\left(1-\frac{1}{4}e^{-j\omega}\right)}$$

$$=\frac{8}{\left(1-\frac{1}{2}e^{-j\omega}\right)}+\frac{(-4)}{\left(1-\frac{1}{4}e^{-j\omega}\right)}+\frac{(-2)}{\left(1-\frac{1}{4}e^{-j\omega}\right)} \tag{5}$$

To determine, we take inverse DTFT of Eq. (5), we get

$$y(n)=\text{Inverse DTFT }\{Y(e^{j\omega})\}$$

$$=\text{Inverse DTFT}\left\{\frac{2}{\left(1-\frac{1}{2}e^{-j\omega}\right)\left(1-\frac{1}{4}e^{-j\omega}\right)^2}\right\}$$

$$= \text{Inverse DTFT}\left\{\frac{8}{\left(1-\frac{1}{2}e^{-j\omega}\right)} - \frac{4}{\left(1-\frac{1}{4}e^{-j\omega}\right)} - \frac{2}{\left(1-\frac{1}{4}e^{-j\omega}\right)}\right\}$$

$$= 8 \times \text{inverse DTFT}\left\{\frac{1}{\left(1-\frac{1}{2}e^{-j\omega}\right)}\right\}$$

$$-4 \times \text{inverse DTFT}\left\{\frac{1}{\left(1-\frac{1}{4}e^{-j\omega}\right)}\right\} - 2 \times \text{inverse DTFT}\left\{\frac{1}{\left(1-\frac{1}{4}e^{-j\omega}\right)^2}\right\}$$

or $$y(n) = 8\left(\frac{1}{2}\right)^n u(n) - 4\left(\frac{1}{4}\right)^n u(n) - 2(n+1)\left(\frac{1}{4}\right)^n u(n) \tag{6}$$

This is the output of the system for a given input.

EXERCISES

1. What is discrete-time Fourier Transform (DTFT)?
2. Distinguish between CTFT and DTFT.
3. Distinguish between discrete-time Fourier transform (DTFT) and discrete-time Fourier sequence (DTFS).
4. Write a short note on the development of discrete-time Fourier transform (DTFT).
5. Are there any issues associated with DTFT? Discuss.
6. Give DTFT analysis and synthesis equations.
7. Discuss Fourier transform for discrete-time signals.
8. Discuss various properties of DTFT in brief.
9. Tabulate various properties of DTFT.
10. Tabulate various basic DTFT pairs.
11. What is periodicity? Discuss property of periodicity for DTFT.

12. What do you mean by the linearity property of DTFT?
13. Discuss properties of time-shifting and frequency shifting of DTFT.
14. State and explain differencing and accumulation properties of DTFT.
15. Distinguish time reversal from time expansion.
16. Give the expression of Parseval's relation to DTFT.
17. Discuss the following properties of DTFT :
    - **a.** The convolution property of DTFT.
    - **b.** The multiplication property of DTFT.
18. Discuss the following:
    - **a.** Duality in the DTFS analysis and synthesis equations.
    - **b.** Duality between the DTFT and the CTFT.

NUMERICAL EXERCISES

1. Determine the DTFT of following signals:

a. $s(n) = u(n-2) - u(n-6)$

b. $s(n) \left(\frac{1}{3}\right)^{|n|} u(n-2)$

c. $s(n) = \left(\frac{1}{3}\right)^{|n|} \cos\left[\frac{\pi}{8}(n-1)\right]$

d. $s(n) = \sin\left(\frac{\pi}{2}n\right) + \cos(n)$

e. $s(n) = \left(\frac{\sin \pi n/5}{\pi n}\right)\cos\left(\frac{7\pi}{2}n\right)$

Ans.

a. $S(e^{j\omega}) = e^{-2j\omega} + e^{-3j\omega} + e^{-4j\omega} + e^{-5j\omega}$

b. $S(e^{j\omega}) = \dfrac{e^{2j\omega}}{9\left(1 - \frac{1}{3}e^{-j\omega}\right)}$

c. $S(e^{j\omega}) = \dfrac{1}{2j}\left[\dfrac{e^{j\pi/4}e^{j\omega}}{1 - \left(\frac{1}{2}\right)e^{j\pi/8}e^{j\omega}} + \dfrac{e^{j\pi/4}e^{j\omega}}{1 - \left(\frac{1}{2}\right)e^{j\pi/8}e^{j\omega}}\right]$

d. $S(e^{j\omega}) = \frac{\pi}{j}\left[\delta\left(\omega-\frac{\pi}{2}\right)-\left(\omega+\frac{\pi}{2}\right)\right]+\pi[\delta(\omega-1)+\delta(\omega+1)]$

e. in $0 \le |\omega| \le (e)$ $S(e^{j\omega}) = \begin{cases} 1, & 3\pi/10 < |\omega| \le 7\pi/10 \\ 0, & \text{elsewhere} \end{cases}$

2. Find the inverse DTFT of the following DTFTs:

a. $S(e^{j\omega}) = 1 + 3e^{-j\omega} + 2e^{-j2\omega} - 4e^{-j3\omega} + e^{-j10\omega}$

b. $S(e^{j\omega}) = \cos^2\omega + \sin^2 3\omega$

c. $S(e^{j\omega}) = \begin{cases} 1, & \pi/4 \le |\omega| \le 3\pi/4 \\ 0, & 3\pi/4 \le |\omega| \le \pi, 0 \le |\omega| < \pi/4 \end{cases}$

d. $S(e^{j\omega}) = \sum_{k=-\infty}^{\infty} (-1)^k \delta\left(\omega - \frac{\pi k}{2}\right)$

Ans.

a. $s(n) = \delta(n) + 3\delta(n-1) + 2\delta(n-2) - 4\delta(n-3) + \delta(n-10)$

b. $s(n) = \delta(n) = \frac{1}{4}\delta(n-2) + \frac{1}{4}\delta(n+2) - \frac{1}{4}\delta(n-3) - \frac{1}{4}\delta(n+3)$

c. $s(n) = 1 - e^{j\pi n/2} + e^{j\pi n} - e^{j3\pi n/2}$

d. $s(n) = \frac{1}{\pi n}\left[\sin\left(\frac{3\pi n}{4}\right) - \sin\left(\frac{\pi n}{4}\right)\right]$

3. Find the response of a discrete-time LTI system with impulse response $h(n) = (1/2)^n u(n)$ for input $s(n) = (3/4)^n u(n)$

Use DTFT analysis and synthesis equations.

$$\left[\textbf{Ans. } y(n) = 3\left(\frac{3}{4}\right)^n u(n) - 2\left(\frac{1}{2}\right)^n u(n)\right]$$

4. The difference equation of a causal discrete-time LTI system is given as

$$y(n) = \frac{1}{2}y(n-1) + s(n)$$

a. Find frequency response $H(e^{j\omega})$ for the system

b. Find output response to the input given as

$$s(n) = \left(\frac{1}{2}\right)^n u(n)$$

$$\left[\textbf{Ans. a. } H(e^{j\omega}) = \frac{1}{1 + \frac{1}{2}e^{-j\omega}}\right.$$

$$\textbf{b. } y(n) = \frac{1}{2}\left(\frac{1}{2}\right)^n u(n) + \frac{1}{2}\left(-\frac{1}{2}\right)^n u(n)\Bigg]$$

5. Two discrete-time LTI systems are cascaded. These two systems have frequency responses

$$H_1\left(e^{j\omega}\right) = \frac{2 - e^{-j\omega}}{1 + \frac{1}{2}e^{-j\omega}}$$

$$H_1\left(e^{j\omega}\right) = \frac{1}{1 + \frac{1}{2}e^{-j\omega} + \frac{1}{4}e^{-j2\omega}}$$

a. Find the difference equation describing the overall system

b. Find the impulse response of the overall system

$$\Bigg[\textbf{Ans. a. } y(n) + \frac{1}{8}y(n-3) = 2s(n) - s(n-1)$$

$$\textbf{b. } h(n) = \frac{4}{3}\left(-\frac{1}{2}\right)^n u(n) + \frac{1 + j\sqrt{3}}{3}\left(\frac{1}{2}e^{-j/120}\right)^n u(n)$$

$$+\frac{1 + j\sqrt{3}}{3}\left(\frac{1}{2}e^{-j/120}\right)^n u(n)\Bigg]$$

6. Determine the impulse response of the inverse system and the difference equation that characterize the inverse system. Following difference equations for causal discrete-time LTI systems are given:

$$\textbf{a. } y(n) + s(n) - \frac{1}{2}s(n-1) \qquad \textbf{b. } y(n) = -\frac{1}{2}y(n-1) + s(n)$$

$$\Bigg[\textbf{Ans. a. } y(n) = -\frac{1}{4}y(n-1) = s(n) \qquad \textbf{b. } y(n) + \frac{1}{2}s(n-1)\Bigg]$$

7. Find a linear constant-coefficient difference equation relating the input and output. We have given impulse response for the system

$$h(n) = \left(\frac{1}{2}\right)^n u(n) + \frac{1}{2}\left(\frac{1}{4}\right)^n u(n)$$

Ans. $y(n)=\frac{3}{4}y(n-1)-\frac{1}{8}y(n-2)+\frac{3}{2}s(n)-\frac{1}{2}s(n-1)$

8. A block diagram implementation of a causal discrete-time LTI system is depicted in Figure 4.19.

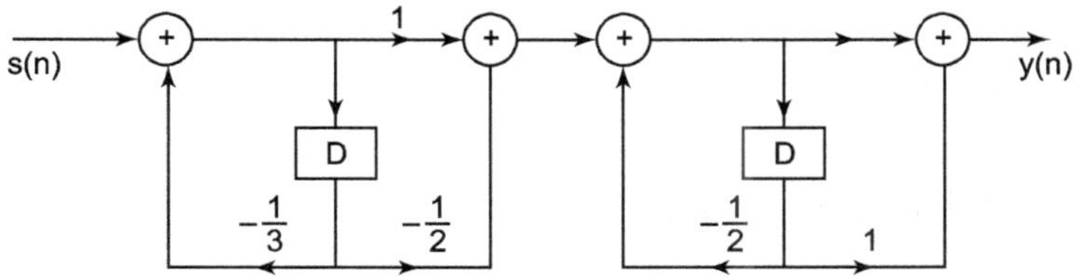

FIGURE 4.19.

a. Find a difference equation relating input $s(n)$ and output $y(n)$

b. Determine frequency response of the system

c. Determine the system's impulse response.

[**Ans. a.** $y(n)=\frac{1}{4}y(n-2)+\frac{1}{4}s(n)+\frac{7}{8}s(n-1)-\frac{1}{2}s(n-2)$

b. $H\left(e^{j\omega}\right)=\dfrac{\frac{1}{4}+\frac{7}{8}e^{-j\omega}-\frac{1}{2}e^{2j\omega}}{1+\frac{1}{2}e^{-2j\omega}}$

c. $h(n)=2\delta(n)-\frac{21}{16}\left(-\frac{1}{2}\right)^n u(n)+\frac{7}{16}\left(\frac{1}{2}\right)^n u(n)$]

9. We have given the fact that

$$A^n \xleftrightarrow{\text{DTFT}} \frac{1-A}{1-2A\cos\omega+A^2}, |A<1|$$

Use the property of duality to determine the Fourier series coefficients of the following continuous-time signal with period $T=1$.

$$s(t)=\frac{1}{5-4\cos(2\pi t)}$$

$$\left[\textbf{Ans.} \text{ Fourier series coefficients of the signal are } \frac{1}{5-4\cos(2\pi t)}\right]$$

10. A discrete-time LTI system with impulse response $h_1(n) = (1/3)^n\, u(n)$ is connected in parallel with another causal discrete-time LTI system with impulse response $h_2(n)$. The resulting parallel interconnection has the frequency response

$$H\left(e^{j\omega}\right) = \frac{-12+5e^{j\omega}}{12-7e^{-j\omega}+e^{-j2\omega}}$$

Find impulse response $h_2(n)$.

$$\left[\textbf{Ans.}\ h_2(n) = -2\left(\frac{1}{4}\right)^n u(n)\right]$$

CHAPTER 5

DISCRETE FOURIER TRANSFORMS (DFTS)

5.1 INTRODUCTION

Frequency-domain analysis of discrete-time signals is usually and conveniently performed on a digital signal processor. This digital signal processor may be a general-purpose digital computer or specially designed digital hardware. Fourier transform of discrete-time signal $s(n)$ is called discrete-time Fourier transform (DTFT) and denoted by $S(e^{j\omega})$ or $S(\omega)$. DTFT $S(\omega)$ is a continuous function of frequency (ω). Therefore, this representation is not a computationally convenient representation of $\{s(n)\}$. We represent a sequence by samples of its continuous spectrum. This frequency-domain representation of a signal is called discrete Fourier transform. It is a very powerful tool for frequency-domain analysis of discrete-time signals.

The discrete Fourier transform (DFT) is itself a sequence rather than a function of a continuous variable and it corresponds to equally spaced frequency samples of the DTFT of a signal. DFT plays a central role in the implementation of various DSP algorithms. Fourier series representation of the periodic, sequence corresponds to the DFT of the finite length sequence. DFT is used for transforming discrete-time sequence $s(n)$ of finite length into discrete-frequency sequence $S(K)$ of finite length. By DFT, discrete-time sequence $s(n)$ is transformed into corresponding discrete-frequency sequence $S(K)$.

Continuous-time Fourier transform (CTFT) of a continuous-time signal is given by

$$S(j\omega) = \int_{-\infty}^{\infty} S(t)e^{-j\omega t}\,d\omega \tag{5.1}$$

This is an analysis equation of CTFT.

It is also called the Fourier integral. Inverse CTFT is given by the following synthesis equation

$$s(t) = \frac{1}{2\pi}\int_{-\infty}^{\infty} S(t)e^{-j\omega t}\,d\omega \tag{5.2}$$

CTFT is used for non-periodic continuous-time signals. It produces a continuous spectrum of $s(t)$.

Similarly, DTFT of a discrete-time signal $s(n)$ is given by

$$s(\omega) \text{ or } S\left(e^{j\omega}\right) = \sum_{n=-\infty}^{\infty} s(n)e^{-j\omega n} \tag{5.3}$$

This is called the analysis equation of DTFT of a discrete-time sequence. Inverse DTFT is given by the following synthesis equation

$$s(n) = \frac{1}{2\pi}\int_{2\pi} S(e^{j\omega})e^{j\omega n}\,d\omega \tag{5.4}$$

It also produces a continuous spectrum of $s(n)$.

DFT of a discrete-time sequence $s(n)$ is obtained by performing the sampling operation in both the time and frequency domains. But DTFT of a discrete-time sequence $s(n)$ is obtained by performing the sampling operation in the time-domain only.

DFT of a discrete-time sequence $s(n)$ is related to DTFT of the same sequence by

$$S(K) = S(\omega)\,|_{\omega=2\pi\,\mathrm{K/N}}$$

where $S(K)$ is the DFT of $s(n)$ and is given by

$$S(K) = \sum_{n=-\infty}^{\infty} s(n)e^{-k2\pi\,Kn/N}$$

and $S(\omega)$ is the DTFT of the sequence $s(n)$ and is given by

$$S(\omega) = \sum_{n=-\infty}^{\infty} s(n)e^{-j\omega n}$$

DTFT gives a continuous spectrum but DFT gives a discrete spectrum.

5.2 DEFINITION OF DFT

DFT is used for transforming a discrete-time sequence $s(n)$ of a finite length of N-points into a discrete-frequency sequence of the same finite length as $s(n)$.

DFT is defined as

$$S(K) = \text{DFT} = \sum_{N=0}^{N-1} s(n)e^{-(j2\pi/N)n} \tag{5.5}$$

where $K = 0,\ 1,\ \ldots N-1.$

and *inverse* DFT (IDFT) is defined as

$$s(n) = \text{IDFT}\ [S(K)] = \sum_{N=0}^{N-1} S(K)e^{-(j2\pi/N)n} \tag{5.6}$$

where $n = 0, 1, \ldots. N - 1$.

$S(K)$ is the DFT of the signal $s(n)$ and is given by

$$S(K) = S_R(K) + jS_I(K)$$

Magnitude or amplitude of $S(k)$ is given by

$$|S(K)| = \sqrt{S_R^2(K) + S_I^2(K)}$$

Phase of $S(K)$ *is* given by

$$\angle S(K) = \tan^{-1}\left[\frac{S_I(K)}{S_R(K)}\right]$$

where $S_R(K)$ is the real part of the $S(K)$ and $S_I(K)$ *is* the imaginary part of $S(K)$.

Spectrum drawn between the amplitude of $S(K)$ and discrete-frequency variable (K) is called amplitude or magnitude spectrum of DFT of $s(n)$, that is, $S(K)$. Spectrum drawn between phase of $S(K)$ and discrete-frequency variable (K) is called phase spectrum of $S(K)$. Both spectrums are shown by the example given below.

EXAMPLE 5.1

Compute the eight-point DFT of the signal

$$s(n) = \{\underset{\uparrow}{1},1,1,\ 1,\ 1,\ 1,\ 0,\ 0\}$$

Also, sketch its magnitude and phase spectrums.

Solution:
DFT of a signal $s(n)$ is given by

$$S(K) = \sum_{n=0}^{N-1} s(n) W_N^{nK} = \sum_{n=0}^{7} s(n) W_N^{nK}$$

where $$W_N = e^{-j2\pi/N} = e^{-j2\pi/8} = e^{-j\omega/4}$$

$$= s(0)W_N^{0K} + s(1)W_N^{1K} + s(2)W_N^{2K} + s(3)W_N^{3K}$$

$$+s(4)W_N^{4K} + s(5)W_N^{5K} + s(6)W_N^{6K} + s(7)W_N^{7K}$$

$$= 1 + W_N^{K} + W_N^{2K} + W_N^{3K} + W_N^{4K} + W_N^{5K}$$

$$= 1 + e^{-j\pi K/4} + e^{-j2\pi K/4} + e^{-j3\pi K/4} + e^{-j4\pi K/4} + e^{-j5\pi K/4}$$

$K = 0$, $S(0) = 6$
$K = 1$, $S(1) = -0.7071 - j1.7071$
$K = 2$, $S(2) = 1 - j$
$K = 3$, $S(3) = 0.7071 + j0.2929$
$K = 4$, $S(4) = 0$
$K = 5$, $S(5) = 0.7071 - j0.2929$
$K = 6$, $S(6) = 1 + j$
$K = 7$, $S(7) = -0.7071 + j1.7071$
DFT of $s(n)$ is given by

$$S(K) = \{S(0), S(1), S(2), S(3), S(4), S(5), S(6), S(7)\}$$

$$= \{6, -0.7071 - j1.7071,\ 1 - j, -0.7071 + j0.2929, 0,$$

$$0.7071 - j0.2029, 1 + j, -0.7071 + j1.7071\}$$

$$\text{Amplitude} = \sqrt{(\text{Real part})^2 + (\text{Imaginary part})^2}$$

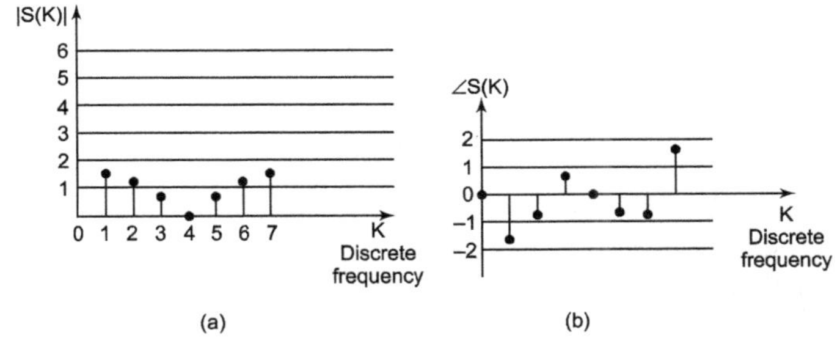

FIGURE 5.1 *(a)* Amplitude spectrum and *(b)* phase spectrum.

$$\text{Phase} = \tan^{-1}\left(\frac{\text{Imaginary part}}{\text{Real part}}\right)$$

Amplitude of $S(K)$ is given by

$$S(K) = \{6,\ 1.8478, 1.4142, 0.7654, 0, 0.7654, 1.4142, 1.8478\}$$

Phase of $S(K)$ is given by

$$\angle S(K) = \{0, -1.9635, -0.785, 0.3927, 0, -0.3927 > 0.785, 1.9635\}$$

EXAMPLE 5.2

Determine the N-point DFT of a finite-duration sequence given by

$$s(n) = \begin{cases} 1, & 0 \le n \le L-1 \\ 0, & \text{otherwise} \end{cases}$$

Solution:

DFT of sequence $s(n)$ is given by

$$S(K) = \text{DTFT } [s(n)] \sum_{n=0}^{N} s(n) e^{-(j2\pi k/N)n}$$

Now,

$$S(K) = \begin{cases} 1, & 0 \le n \le L-1 \\ 0, & \text{otherwise} \end{cases}$$

$$S(K) = \sum_{n=0}^{L-1} 1.e^{(-j2\pi K/N)n}$$

$$= e^{-(j2\pi K/N)^0} + e^{-(j2\pi K/N)1} + \ldots + e^{-(j2\pi K/N)(L-1)}$$

$$= 1 + e^{-j2pK/N} + \ldots + \ + e^{(-j2pK/N)(L-1)}$$

This is a geometric progression of the form

$$= \frac{A\left[1-R^{M+1}\right]}{[1-R]} \tag{1}$$

where A is the first term of GP.

R is the common ratio of GP.

Putting the values of A, R, and M in Eq. (1), we get

$$S(K) = \frac{1\left[1-\left(e^{-j2\pi K/N}\right)^{L-1+1}\right]}{\left[1-\left(e^{-j2\pi K/N}\right)\right]} = \frac{1-e^{-j2\pi KL/N}}{1-e^{-j2\pi K/N}}$$

where $K = 0, 1, \ldots, N-1$

or $$S(K) = \frac{1-e^{-j2\pi KL/N}}{1-e^{-j2\pi K/N}}$$ where $K = 0, 1, \ldots, N-1$

This is the N-point DFT of the sequence $s(n)$.

EXAMPLE 5.3

Find the DFT of the sequence

$$s(n) = \begin{cases} 1, & 2 \le n \le 6 \\ 0, & \text{for } n = 0,1,7,8,9 \end{cases}$$

Given that $N = 10$.

Solution:

DFT of $s(n)$ is given by

$$S(K) = \sum_{n=0}^{N-1} s(n) e^{-(j2\pi K/N)n}$$

Let $e^{-j(2\pi/N)} = W$, then

$$S(K) = \sum_{n=0}^{N-1} s(n) W^{-nK} = \sum_{n=2}^{6} s(n) W^{-nK}$$

where $$W = e^{-j(2\pi/N)}$$

$$= \sum_{n=0}^{6} W^{-nK} - \sum_{n=0}^{1} W^{-nK}$$

$$= \frac{1-\left(W^{-K}\right)^{6+1}}{1-W^{-K}} - \frac{1-(W^{-K})^{1+1}}{1-W^{-K}}$$

$$= \frac{1-W^{-7K}}{1-W^{-K}} - \frac{1-W^{-K}}{1-W^{-K}}$$

or
$$S(K) = \frac{1-W^{-7K}-1+W^{-2K}}{1-W^{-K}} - \frac{W^{-2K}-W^{-7K}}{1-W^{-K}} \tag{1}$$

Substituting $W = e^{j2\pi/N} + e^{j2\pi/10} = e^{j\pi/5}$ in Eqn. (1), we get

$$S(K) = \frac{e^{-j(\pi K/5)^2} - e^{-j(\pi K/5)^7}}{1-e^{-j(\pi K/2)(1)}} = e^{-j(4\pi K/5)} \frac{\sin\left(\frac{\pi K}{2}\right)}{\sin\left(\frac{\pi K}{10}\right)}$$

5.3 THE DFT AS A LINEAR TRANSFORMATION TOOL

DFT and IDFT both are linear transformations on sequences $s(n)$ and $S(K)$, respectively, where $s(n)$ and $S(K)$ are discrete-time sequence and discrete-frequency sequence, respectively.

DFT and Inverse DFT (IDFT) satisfy the principle of superposition. DFT is a linear transformation tool just like z-transform. The same property also applies to both DFT and its inverse DFT.

The DFT as a linear transformation tool is explained and studied by using matrices. DFT and IDFT can be computed by using matrices. Computation of N-point DFT requires $N \times N$ complex multiplications and $N(N-1)$ complex additions.

N-point vector S_N of frequency samples is given by

$$S_N = \begin{bmatrix} S(0) \\ S(1) \\ \vdots \\ S(N-1) \end{bmatrix} \tag{5.7}$$

N-point vector s_N of signal sequence $s(n)$ is given by

$$s_N = \begin{bmatrix} s(0) \\ s(1) \\ \vdots \\ s(n-1) \end{bmatrix} \tag{5.8}$$

DFT $$s_N = \sum_{n=0}^{N-1} s(n) e^{-j(2\pi K/N)n} = \sum_{n=0}^{N-1} s(n) W_N^{Kn}$$

where $K = 0, 1, \ldots, N-1$

IDFT, $$S(n) = \frac{1}{N}\sum_{K=0}^{N-1} s(K) e^{-(j2\pi K/N)n} = \frac{1}{N}\sum_{n=0}^{N-1} S(K) W_N^{-Kn}$$

where $K = 0, 1, \ldots, N-1$

where $W_N = e^{-j2\pi/N}$ is an Nth root of unity.

N-point DFT in matrix form is defined as

$$S_N = WN.\, s_N \tag{5.9}$$

where W_N is the matrix of the linear transformation and is a $N \times N$ symmetric matrix. W_N is given by

$$W_N = \begin{bmatrix} 1 & 1 & 1 & \cdots & 1 \\ 1 & W_N^1 & W_N^1 & \cdots & W_N^{N-1} \\ : & W_N^2 & W_N^4 & \cdots & W_N^{2(N-1)} \\ : & : & : & & : \\ 1 & W_N^{N-1} & W_N^{2(N-1)} & \cdots & W_N^{(N-1)(N-1)} \end{bmatrix} \tag{5.10}$$

N-point IDFT in matrix form is defined as

$$s_N = W_N^{-1} S_N = \frac{1}{N} W_N^* S_N \tag{5.11}$$

where W_N^* denotes the complex conjugate of matrix W_H and W_N^{-1} is the inverse of W_N.

The relationship between W_N and its complex conjugate will be

$$W_N \cdot W_N^* = N \cdot I_N \tag{5.12}$$

where I_N is an $N \times N$ identity matrix.

EXAMPLE 5.4

Determine the DFT of the four-point discrete-time sequence $s(n) = \{1, 2, 3, 4\}$ using DFT transformation matrix. Also determine IDFT from DFT.

Solution:

Here we require a 4 × 4 DFT transformation matrix because $s(n)$ has four points.

$$W_4 = \begin{bmatrix} W_4^0 & W_4^0 & W_4^0 & W_4^0 \\ W_4^0 & W_4^1 & W_4^2 & W_4^3 \\ W_4^0 & W_4^2 & W_4^4 & W_4^6 \\ W_4^0 & W_4^3 & W_4^6 & W_4^9 \end{bmatrix} \tag{1}$$

where $\qquad W_4^0 = [e^{-j2\pi/4}]^0 = 1.$

Now we will use the property of periodicity and symmetry of W_4.

The property of periodicity for W_N is given by

$$W_N^{K+N} = W_N^K$$

From this property, WN is repeated after a period N.

The property of symmetry for W_N is given by

$$W_N^{K+N/2} = -W_N^K$$

From this property, W_N is repeated with inverse sign after a half period $\frac{N}{2}$.

where $W_N = e^{-j2\pi/N}$ is called *phase rotation factor* or *twiddle factor* for DFT.

From the property of periodicity

$$W_4^4 = W_4^{(0+4)} = W_4^0$$

$$W_4^6 = W_4^{(2+4)} = W_4^2 \qquad \text{because } N = 4$$

$$W_4^9 = W_4^{(1+2\times4)} = W_4^1$$

Substituting these values in Eq. (1), we get

$$W_4 = \begin{bmatrix} 1 & 1 & 1 & 1 \\ 1 & W_4^1 & W_4^2 & W_4^3 \\ 1 & W_4^2 & W_4^0 & W_4^2 \\ 1 & W_4^3 & W_4^2 & W_4^1 \end{bmatrix} \tag{2}$$

$$W_4^1 = \left(e^{-j2\pi/4}\right) = -j$$

$$W_4^2 = \left(e^{-j2\pi/4}\right)^2 = -1$$

$$W_4^3 = \left(e^{-j2\pi/4}\right)^3 = j$$

Substituting the above values in Eq. (2), we get

$$W_4 = \begin{bmatrix} 1 & 1 & 1 & 1 \\ 1 & -j & -1 & j \\ 1 & -1 & 1 & -1 \\ 1 & j & -1 & -j \end{bmatrix}$$

Then $$S_4 = W_{4S_4} = \begin{bmatrix} 1 & 1 & 1 & 1 \\ 1 & -j & -1 & j \\ 1 & -1 & 1 & -1 \\ 1 & j & -1 & -j \end{bmatrix} \begin{bmatrix} 1 \\ 2 \\ 3 \\ 4 \end{bmatrix}$$

$$= \begin{bmatrix} 1+2+3+4 \\ 1-2j-3+4j \\ 1-2+3-4 \\ 1+2j-3-4j \end{bmatrix} \begin{bmatrix} 10 \\ -2+2j \\ -2 \\ -2-2j \end{bmatrix}$$

Inverse DFT of S_4 is determined by

$s_N = \dfrac{1}{N} W_N^* S_N$ where W_N^* is the complex conjugate W_N and $N = 4$:

$$s_4 = \frac{1}{4} W_4 * S_4 = \frac{1}{4} \begin{bmatrix} 1 & 1 & 1 & 1 \\ 1 & -j & -1 & -j \\ 1 & -1 & 1 & -1 \\ 1 & -j & -1 & j \end{bmatrix} \begin{bmatrix} 10 \\ -2-zj \\ -2 \\ -2-2j \end{bmatrix}$$

$$= \begin{bmatrix} 10-2+2j-2-2-2j \\ 10-2j+2j^2+2+2j+2j^2 \\ 10+2-2j-2-2+2j \\ 10+2j=2j^2+2-2j-2j^2 \end{bmatrix}$$

$$= \frac{1}{4}\begin{bmatrix} 4 \\ 8 \\ 12 \\ 16 \end{bmatrix} = \begin{bmatrix} 1 \\ 2 \\ 3 \\ 4 \end{bmatrix}$$

Note: W_N^* is determined from W_N by changing the sign of in W_N.

5.4 PROPERTIES OF DFT

Most of the properties of the DFT and z-transform have some similarities because they have some relationship with each other. DFT also differs in some properties such as the circular convolution property. This property is explained in detail in Figure 5.6.

Circular convolution is the special property of DFT of a discrete-time sequence. There are several properties of the DFT that play an important role in practical techniques for processing signals. These properties are given as follows:

1. Periodicity.
2. Linearity.
3. Symmetry properties.
4. Circular convolution of two sequences.
5. Time reversal of a sequence.
6. Circular time-shift of a sequence.
7. Circular frequency-shift of a sequence.
8. Circular correlation of two sequences.
9. Multiplication of two sequences.
10. Parseral's theorem.

These properties are discussed one by one in subsequent sections.

5.4.1 Periodicity

If a discrete-time signal is periodic then its DFT will also be periodic. If a signal or sequence repeats its waveform after N number of samples then it is called a periodic signal or sequence and N is called the period of signal.

If $s(n)$ is a discrete-time signal and $S(K)$ is the N-point DFT of $s(n)$. Then

$$s(n+N) = s(n) \text{ for all values of } n \tag{5.13}$$

This is true for periodic signals only.

$$S(K+N) = S(K) \text{ for all values of } K \tag{5.14}$$

DFT of $s(n)$ is also periodic.

This property is used in some applications of DSP such as in Fast Fourier Transform (FFT) algorithms used for linear filtering, power spectrum estimation, etc.

5.4.2 Linearity

DFT is also a linear transform just like transform. It satisfies the principle of superposition.

If $$s_1(n) \xleftrightarrow{\text{N}_{.\text{pt}}-DRT} S_1(K)$$

$$s_1(n) \xleftrightarrow{\text{N}_{.\text{pt}}-DRT} S_2(K)$$

Then for any real-valued or complex-valued constants A_1 and A_2

$$A_1 s_1(n) + A_2 s_2(n) \xleftrightarrow{\text{N}_{.\text{pt}}-DRT} A_1 S_1(K) + A_2 S_2(K)$$

5.4.3 Symmetry Properties of the DFT

If a signal or sequence repeats its waveform in a negative direction after $N/2$ number of samples, then it is called symmetric sequence or signal. Symmetry properties of the DFT are studied in two cases:

Case I. Signal $s(n)$ and its DFT $S(K)$ both are complex and periodic sequences.

These properties are given in Table 5.1.

We know that for periodic signals

$$s(n) = s(N - n) \text{ and } S(K) = S(N - K)$$

It concludes that if a discrete-time sequence $s(n)$ is periodic then its DFT $S(K)$ is also periodic in the same period.

TABLE 5.1 Symmetry properties of the DFT for $s(n)$ and $S(K)$ both are periodic and complex sequences

N-point discrete-time sequence $s(n)$, $0 < n < N - 1$	N-point DFT $S(K)$, $0 \le K \le N - 1$
Complex conjugate of $s(n) \Rightarrow s(n)$	$S^*(N - K)$
$s^*(N - n)$	$S^*(K)$
Real part of complex sequence $\Rightarrow s_R(n)$	Conjugate even part of $S(K)$ $\Rightarrow S_{ce}(K) = \frac{1}{2}[S(K) + S^*(N - K)]$
Imaginary part of complex sequence $\Rightarrow js_I(n)$	Conjugate odd part of $S(K)$ $\Rightarrow S_{CO}(K) = \frac{1}{2}[S(K) + S^*(N - J)]$
Conjugate even part of $s(n)$ $\Rightarrow S_{ce}(n) = \frac{1}{2}[s(n) + s^*(N - n)]$	$S_R(K)$
Conjugate odd part of $s(n)$ $\Rightarrow s_{co}(n) = \frac{1}{2}[s(n) + s^*(N - n)]$	$jS_1(K)$
where $s^*(n)$ is the complex conjugate of $s(n)$ and $S^*(K)$ is the complex conjugate of $S(K)$	

Case II. Now in this case $s(n)$ is a real and periodic sequence but its DFT is a complex and periodic sequence of N samples.

Now $S(K)$ satisfies the following symmetry conditions.

$$S(K) = S^*(N - K) \tag{5.15}$$

Real part of $S(K)$, $S_R(K) = S_R(N - K)$ (5.16)

Imaginary part of $S(K)$, $S_I(K) = S_I(N - K)$ (5.17)

Both $S_R(K)$ are $S_R(N - K)$ even functions and $S_I(K)$ and $S_I(N - K)$ are odd functions.

Amplitude of $S(K), |S(K)| = |S(N - K)|$ (5.18)

Phase of $S(K), \angle S(K) = -\angle S(N - K)$ (5.19)

Both $|S(K)|$ and $|S(N-K)|$ are even.
and $S(K)$ and $S(N-K)$ are odd.
Conjugate even part of

$$s(n) = \frac{1}{2}[s(n) + s(N-n)] \xleftrightarrow{N_{.pt}-DRT} S_R(K)$$

Conjugate odd part of

$$s(n) = \frac{1}{2}[s(n) - s(N-n)] \xleftrightarrow{N_{.pt}-DRT} jS_I(K)$$

where $S_R(K)$ is the real part of $S(K)$
and $S_I(K)$ is the imaginary part of $S(K)$.
Note. Complex discrete-time sequence is represented as

$$s(n) = s_R(n) + js_I(n)$$

and complex DFT of $s(n)$ is represented as

$$S(K) = S_R(K) + jS_I(K)$$

$S(K)$ is also represented in polar form as

$$S(K) = |S(K)| \; .ej\angle S(K)$$

where $|S(K)|$ is the amplitude of $S(K)$
$\angle S(K)$ is the phase of $S(K)$.

5.4.4 Circular Convolution

This is the special property of the DFT. This property is not satisfied by z-transform.

Circular convolution is defined as

$$s_3(m) = \sum_{n=0}^{N-1} s_1(n) s_2[(m-n)]_N \tag{5.20}$$

where $m = 0, 1, ..., N-1$

where $s_2[(m-n)]_N$ is a circular shift of sequence $s_2(n)$ and $s_1(n)$ and $s_2(n)$ are two discrete-time sequences of the finite length of N-point.

Circular convolution of two-discrete-time sequences $s_1(n)$ and $s_2(n)$ is denoted by

$$s_2(n) = s_1(n) \circledN s_2(n)$$

$$= s_2(n) \circledN s_2(n) \tag{5.21}$$

Here Ⓝ denotes N-pt. circular convolution of two sequences. Circular convolution like linear convolution is a commutative operation.

Multiplication of the DFTs of the two sequences is equivalent to the circular convolution of the two sequences in the time domain.

The basic difference between circular convolution and linear convolution is that, in circular convolution, the folding and shifting operations are performed in a circular fashion by computing the index of one of the sequences *modulo N*. Modulo N operation is not used in linear convolution. This operation is used for the circular shifting of sequence $s(n)$. It can be explained with the help of an example given below:

If $$s_1(n) \xleftrightarrow{\text{N.pt}-DRT} S_1(K)$$

and $$s_2(n) \xleftrightarrow{\text{N.pt}-DRT} S_2(K)$$

then from the special property of the DFT

$$s_1(n) \circledN s_2(n) = \sum_{n=-}^{N-1} s_1(n) s_2\left[(m-n)\right]_N \xleftrightarrow{\text{N.pt}-DRT} S_1(K) \cdot S_2(K)$$

Example for Modulo N-operation

In general, the circular shift of the sequence $s(n)$ can be represented as the index modulo N.

$$s'(n) = s(n-K, \text{modulo } N) = s[(n-K)]_N$$

if If = 2 and $N = 4$, then

$$s'(n) = s[(n-2)]_4$$

$n = 0,$	$s'(0) = s[(0-2)]_4 = s(2)$
$n = 1,$	$s'(1) = s[(1-2)]_4 = s[(-1)]_4 = s(3)$
$n = 2,$	$s'(2) = s[(2-2)]_4 = s[(0)]_4 = s(0)$
$n = 3,$	$s'(3) = s[(3-2)]_4 = s[(1)]_4 = s(1)$

Step in computation of circular convolution.

There are four steps in the computation of circular convolution just as those involved in the computation of ordinary linear convolution. These four steps are:

1. Folding one of two sequences.
2. Shifting the folded sequence.

3. Multiplying the two sequences for obtaining the product sequence.
4. Summing the values of the product sequences.

This is illustrated by Example 5.5.

EXAMPLE 5.5

Compute the circular convolution of the two discrete-time sequences $s_1(n) = \{1, 2, 1, 2\}$ and $s_2(n) = \{3, 2, 1, 4\}$.

Solution:

Circular convolution is defined as

$$s_3(m) = \sum_{n=0}^{N-1} s_1(n)s_2[(m-n)]_N$$

where $$s_1(n) = \{s_1(0), s_1(1), s_1(2), s_1(3)\} = \{1,2,1,2\}$$

and $$s_2(n) = \{s_2(0), s_1(1), s_2(2), s_2(3)\} = \{3,2,1,4\}$$

Computation of circular convolution using a graph as shown in Figure 5.2. Circular convolution is defined as

$$s_3(m) = \sum_{n=0}^{N-1} s_1(n)s_2[(m-n)]_N$$

From Figure 5.2, we can compute following:

$m = 0,$ $$s_3(0) = \sum_{n=0}^{3} s_1(n)s_2[(-n)]_4 = 3+8+1+4 = 16$$

$m = 1,$ $$s_3(1) = \sum_{n=0}^{3} s_1(n)s_2[(1-n)]_4 = 2+6+4+2 = 14$$

$m = 2,$ $$s_3(2) = \sum_{n=0}^{3} s_1(n)s_2[(2-n)]_4 = 2+4+3+8 = 17$$

$m = 3,$ $$s_3(3) = \sum_{n=0}^{3} s_1(n)s_2[(3-n)]_4 = 4+2+2+6 = 14$$

Now the circularly convoluted sequence

$$s_3(n) = \{s_3(0), s_3(1), s_3(2), s_3(3)\} = \{16,14,17,14\}$$

EXAMPLE 5.6

Compute the $s_3(n)$ corresponding to circular convolution of the sequences $s_1(n)$ and $s_2(n)$ by using DFT and IDFT approach.

Given $s_1(n) = \{2, 1, 2, 1\}$ and $s_2(n) = \{1, 2, 3, 4\}$

Solution:

DFT of $s_1(n)$ is given by

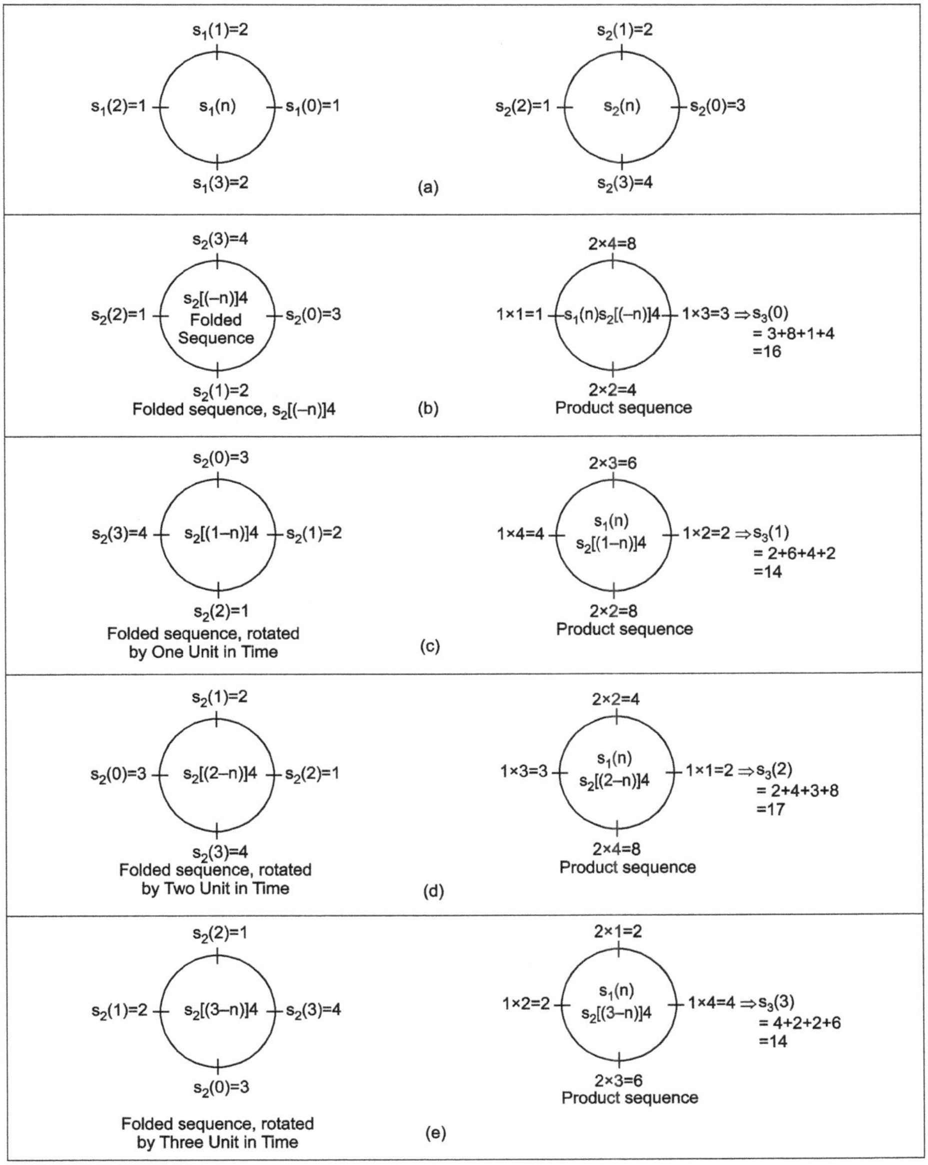

FIGURE 5.2 Circular convolution of the sequences $s_1(n)$ and $s_2(n)$ given in problem 5.5.

$$S_1(K) = \sum_{n=0}^{3} s_1(n)e^{-j2\pi Kn/4}, \quad K = 0,1,2,3$$

$$= s_1(0)e^{-j2\pi K.0/4} + s_1(1)e^{-j2\pi K.1/4} + s_1(2)e^{-j2\pi K.2/4}$$

$$+s_1(3)e^{-j2\pi K.3/4}$$

$$= 2(1) + 1.e^{-j2\pi K/4} + 2.e^{-j4\pi K/4} + 1.e^{-j6\pi K/4}$$

$$= 2 + 1.e^{-j\pi K/2} + 2e^{-j\pi K} + e^{-3\pi K/2}$$

where $K = 0, 1, 2, 3$

Thus, $K = 0$, $\quad S_0(1) = 2 + e^{-j\pi .0/2} + 2e^{-j\pi .0} + e^{-j3\pi .0/2}$

$$= 2 + 1 + 2 + 1 = 6$$

$K = 1$, $\quad S_1(1) = 2 + e^{-j\pi .1/2} + 2e^{-j\pi .1} + e^{-j3\pi .1/2} = 0$

$K = 2$, $\quad S_1(1) = 2 + e^{-j\pi .2/2} + 2e^{-j\pi .2} + e^{-j3\pi .2/2} = 2$

$K = 3$, $\quad S_1(3) = 2 + e^{-j\pi .3/2} + 2e^{-j\pi .3} + e^{-j3\pi .3/2} = 0$

$$S_1(K) = \{S_1(0), S_1(1), S_1(2), S_1(3)\} = \{6, 0, 2, 0\}$$

DFT of $s_2(n)$ is given by

$$s_2(K) = \sum_{n=0}^{3} s_1(n)e^{-j2\pi Kn/4}, \quad K = 0,1,2,3$$

$$= s_2(0)e^{-j2\pi K.0/4} + s_2(1)e^{-j2\pi K.1/4} + s_2(2)e^{-j2\pi K.2/4}$$

$$+s_2(3)e^{-j2\pi K.3/4}$$

$$= 1e^0 + 2e^{-j2\pi K/4} + 3e^{-j4\pi K/4} + 4e^{-j6\pi K/4}$$

$$= 1 + 2e^{-j\pi K/2} + 3e^{-j\pi J} + 4e^{-j3\pi K/2}$$

where $K = 0, 1, 2, 3$

$K = 0$, $\quad S_2(1) = 1 + 2e^{-j\pi .0/2} + 3e^{-j\pi .0} + 4e^{-j3\pi .0/2}$

$$= 1 + 2 + 3 + 4 = 10$$

$K = 1$, $\quad S_2(1) = 1 + 2e^{-j\pi .1/2} + 3e^{-j\pi .1} + 4e^{-j3\pi .1/2} = -2 + j2$

$K = 2$, $\quad S_2(2) = 1 + 2e^{-j\pi .2/2} + 3e^{-j\pi .2} + 4e^{-j3\pi .2/2} = -2$

$K = 3$, $\quad S_2(3) = 1 + 2e^{-j\pi .3/2} + 3e^{-j\pi .3} + 4e^{-j3\pi .3/2} = -2 - j2$

$$S_2(K) = \{S_2(0), S_2(1), S_2(2), S_2(3)\}$$

$$= \{10, -2 + j2, -2 - 2, j2\}$$

But from the important property of the DFT that circular convolution of two discrete-time sequences is equivalent to multiplication of DFTs of individual sequences:

$$S_3(K) = S_1(K).S_2(K)$$
$$= \{6,0,2,0\}\{10,-2+j2,-2,-2-j2\}$$
$$= \{60,0,-4,0\}$$
$$S_3(K) = \{60,0,-4,0\}$$

Taking inverse DFT of $S_3(K)$ for computing $s_3(n)$:

$$s_3(n) = \text{IDFT }[S_3(K)]$$
$$= \frac{1}{4}\sum_{K=0}^{3} S_3(K)e^{-j2\pi Kn/4}, \quad n = 0,1,2,3$$
$$= \frac{1}{4}\Big[S_3(0)e^{(j2\pi .n/4)} + s_3(1)e^{(j4\pi\, 1.n/4)} + s_3(2)e^{(j2\pi\, 2.n/4)}$$
$$+ s_3(3)e^{(j2\pi\, 3.n/4)}\Big]$$
$$= \frac{1}{4}\Big[60e^0 + 0e^{j2\pi n/4} + (-4)e^{j4\pi n/4} + 0.3^{j6\pi n/4}\Big]$$
$$= \frac{1}{4}\Big[60 + 4e^{j\pi n}\Big]15 - e^{j\pi n}$$

or $\qquad s_3(n) = 15 - e^{j\pi n}$

$n = 0, \qquad s_3(0) = 15 - e^{j\pi n.0} = 15 - 1 = 14$

$n = 1, \qquad s_3(1) = 15 - e^{j\pi\, 1} = 16$

$n = 2, \qquad s_3(2) = 15 - e^{j\pi\, 2} = 14$

$n = 3, \qquad s_3(3) = 15 - e^{j\pi\, 3} = 16$

$$s_3(n) = \{s_3(0), s_3(1), s_3(2), s_3(3)\} = \{14,16,14,16\}$$

This is the result of the output sequence from the circular convolution of two sequences by the DFT method.

5.4.5 Time Reversal of a Sequence

Reversing the N-point discrete-time sequence is equivalent to reversing the DFT values.

If $$s(n) \xleftrightarrow{\text{N.pt}-\text{DFT}} S(K)$$

Now we are reversing the sequence $s(n)$ in time

$$s[(-n)]_N = s(N-m) \xleftrightarrow{\text{N.pt}-\text{DFT}} S[(-K)]_N = S(N-K)$$

5.4.6 Circular Time-Shift of a Sequence

If $$s(n) \xleftrightarrow{\text{N.pt}-\text{DFT}} S(K)$$

then, $$s[(n-m)]_N \xleftrightarrow{\text{N.pt}-\text{DFT}} e^{-j2\pi Km/N} S(K)$$

where $s[(n-m)]N$ is a circularly time-shifted version of $s(n)$.

5.4.7 Circular Frequency-Shift of a Sequence

If $$s(n) \xleftrightarrow{\text{N.pt}-\text{DFT}} S(K)$$

then, $$e^{-j2\pi nm/N} s(n) \xleftrightarrow{\text{N.pt}-\text{DFT}} S[(K-m)]_N$$

where $S[(K-m)]_N$ is a circularly frequency shifted version of $S(K)$.

5.4.8 Circular Correlation of Two Sequences

Let us consider two complex-valued discrete-time sequences $s_1(n)$ and $s_2(n)$.

If $$s_1(n) \xleftrightarrow{\text{N.pt}-\text{DFT}} S_1(K)$$

and $$s_2(n) \xleftrightarrow{\text{N.pt}-\text{DFT}} S_2(K)$$

Then,

$$r_{s_1 s_2(m)} = \sum_{n=0}^{N-1} s_1(n) s_2^*[(n-m)]_N \xleftrightarrow{\text{N.pt}-\text{DFT}} S_1(K) S_2^*(K)$$

where $s_2^*(n-m)]_N$ is the complex conjugate of $s_2[(n-m)]_N$ and $S_2^*(K)$ is the complex conjugate of $S_2(K)$

$r_{s_1 s_2 (m)}$ is called circular correlation sequence.

5.4.9 Multiplication of Two Sequences

If $$s_1(n) \xleftrightarrow{\text{N}_{.\text{pt}}\text{-DFT}} S_1(K)$$

and $$s_2(n) \xleftrightarrow{\text{N}_{.\text{pt}}\text{-DFT}} S_2(K)$$

then, $$s_1(n)\cdot s_2(n) \xleftrightarrow{\text{N}_{.\text{pt}}\text{-DFT}} \frac{1}{N} S_1(K) \circledN S_2(K)$$

Ⓝ denotes circular convolution.

This property is dual of circular convolution and defined as

$$s_1(n) \circledN s_2(n) \xleftrightarrow{\text{N}_{.\text{pt}}\text{-DFT}} S_1(K)S_2(K)$$

5.4.10 Parseval's Theorem

For complex-valued sequences $s_1(n)$ and $s_2(n)$

If $$s_1(n) \xleftrightarrow{\text{N}_{.\text{pt}}\text{-DFT}} S_1(K)$$

and $$s_2(n) \xleftrightarrow{\text{N}_{.\text{pt}}\text{-DFT}} S_2(K)$$

Then, $$\sum_{n=0}^{N-1} s_1(n)s_2^*(n) = \frac{1}{N}\sum_{n=0}^{N-1} S_1(n)S_2^*(n)$$ [This expression is a general form of Parseval's theorem.]

If $$s_1(n) = s_2(n), \text{then } S_1(K) = S_2(K) = S(K)$$

Then the above theorem will be

$$\sum_{n=0}^{N-1} s(n)s^*(n) \xleftrightarrow{\text{N}_{.\text{pt}}\text{-DFT}} \sum_{K=0}^{N-1} S(K)S^*(K)$$

But we know that $s(u)\, s^*(n) = |s(n)|^2$

and $$SS(K)\cdot S^*(K) = |S(K)|^2$$

Putting these values in above expression, we get

$$\sum_{n=0}^{N-1} |s(n)|^2 = \sum_{K=0}^{N-1} |S(K)|^2 \tag{5.22}$$

When $s_1(n) = s_2(n)$ and $S_1(K) = S_2(K)$ it is called special form of Parseval's theorem.

5.5 TABULATION OF PROPERTIES OF DFT

All the properties of the DFT are given in Table 5.2.

TABLE 5.2

Properties of DFT	Time-domain	frequency-domain
Notation	$s_1(n)$, $s_2(n)$ Two discrete-time sequences	$S_1(K)$ and $S_2(K)$ Two discrete-frequency sequences
Periodicity	$s_1(n) = s_1(N+n)$ and $s_2(n) = s_2(N+n)$	$S_1(K) = S_1(N+K)$ and $S_2(K) = S_2(N+K)$
Linearity	$A_1s_1(n) + A_2s_2(n)$	$A_1S_1(K) + A_2S_2(K)$
Time–Reversal (Folding)	$s_1(N-n)$ and $s_2(N-n)$	$S_2(N-K)$ and $S_2(N-K)$
Circular time shift	$s_1[(n-m)N$ and $s_2[(n-m)]_N$	$e^{j2\pi Km/N}\, S_1(K)$ and $e^{j2\pi Km/N}\, S_2(K)$
Circular Frequency Shift	$e^{j2\pi nm/N}\, s_1(n)$ and $e^{j2\pi nm/N}\, s_2(n)$	$S_1[(K-m)]_N$ and $S_2[(K-m)]_N$
Complex conjugate	$s_1^*(n)$ and $s_2^*(n)$	$S_1^*(N-K)$ and $S_2^*(N-K)$
Circular convolution	$s_1^*(n)$ Ⓝ $s_2^*(n)$	$S_1(K)\, S_2(K)$
Circular correlation	$s_1^*(n)$ Ⓝ $s_2^*(n)$	$S_1(K)\, S_2^*(K)$
Multiplication of two sequences	$s_1(n)\, s_2(n)$	$S_1(K)$ Ⓝ $S_2(K)$
Parseval's theorem	$\sum_{n=0}^{N-1} s(n)s_2^*(n)$	$\frac{1}{N}\sum_{K=0}^{N-1} S(K)S^*(K)$

5.6 RELATIONSHIP BETWEEN DFT AND z-TRANSFORM

Now we will develop the relationship between DFT and z-transform, z-transform of a discrete-time sequence of finite duration is defined as

$$S(z) = \sum_{n=-\infty}^{\infty} s(n)z^{-n} \tag{5.23}$$

Consider a finite-duration sequence $s(n)$, $0 \leq n \leq N-1$.
Now Eq. (5.23) reduces to

$$S(z) = \sum_{n=0}^{N-1} s(n)z^{-n} \tag{5.24}$$

DFT is given by

$$S(K) = \sum_{n=0}^{N-1} s(n)e^{-(j2\pi K/N)_n} \tag{5.25}$$

Substituting $z = e^{j2\pi K/N}$ in Eq. (5.24), we get

$$S(z)\Big|_{z=e^{+j2\pi K/N}} = \sum_{n=0}^{N-1} s(n)e^{-(j2\pi K/N)_n} \tag{5.26}$$

Comparing Eqs. (5.26) and (5.25), we get

$$S(K) = S(z)\Big|_{z=e^{+j2\pi K/N}} \tag{5.27}$$

Now z-transform is uniquely determined by its Appoint DFT. $S(z)$ can be expressed as a function of the DFT, $S(K)$ as follows:

$$S(z) = \sum_{n=0}^{N-1} s(n)z^{-n} \qquad \text{[From Eqn. (5.24)]}$$

But inverse DFT is given by

$$s(n) = \frac{1}{N}\sum_{K=0}^{N-1} S(K)e^{j2\pi Kn/N} \tag{5.28}$$

Substituting the value of $s(n)$ in above Eq. (5.24), we get

$$S(z) = \sum_{n=0}^{N-1}\left[\frac{1}{N}\sum_{K=0}^{N-1} S(K)e^{j2\pi Kn/N}\right]z^{-n}$$

Rearranging the summations, we get

$$S(z) = \frac{1}{N}\sum_{K=0}^{N-1} S(K) \sum_{n=0}^{N-1} \left(e^{j2\pi K/N} z^{-1}\right)^n$$

$$= \frac{1}{N}\sum_{K=0}^{N-1} S(K) \left[\frac{1-\left(e^{-j2\pi K/N} z^{-1}\right)^{N-1+1}}{1-e^{j2\pi K/N} z^{-1}}\right]$$

$$= \frac{1}{N}\sum_{K=0}^{N-1} S(K) \left[\frac{1-e^{-j2\pi Kn/N} z^{-N}}{1-e^{j2\pi K/N} z^{-1}}\right]$$

But $e^{j2\pi K} = 1$

or
$$S(z) = \frac{1-z^{-N}}{N}\sum_{K=0}^{N-1} \frac{S(K)}{1-e^{j2\pi K/N} z^{-1}} \tag{5.29}$$

Eq. (5.29) is a very important result and gives the relationship between z-transform and DFT of a finite-duration discrete-time sequence.

5.7 LINEAR CONVOLUTION USING DFT

The product of two DFTs is equivalent to the circular convolution of the corresponding time-domain sequences. Here we have no use of circular convolution because our objective is to determine the output of a linear filter to a given input sequence. The purpose of linear convolution is linear filtering the input of a linear system.

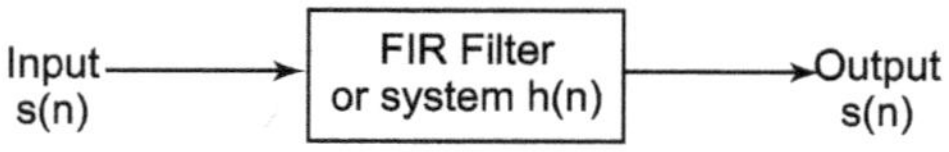

FIGURE 5.3 FIR filter.

The output of an FIR filter is given by the linear convolution of $s(n)$ and $h(n)$ as

$$y(n) = \sum_{k=-\infty}^{\infty} h(k)s(n-k)$$

Here, k is a discrete-time index.

Here, we are considering a finite-duration input sequence $s(n)$ of length L and the unit-sample response of FIR filter $h(n)$ of length M. Output of this FIR filter will be given by

$$y(n) = \sum_{k=0}^{M-1} h(k)s(n-k) \tag{5.30}$$

The duration of $y(n)$ will be $N \geq L + M - 1$.

The frequency-domain equivalent of Eq. (5.30) will be

$$Y(\omega) = S(\omega)H(\omega)$$

DFT,

$$Y(K) = Y(\omega)\big|_{\omega=2\pi K/N} = S(\omega)H(\omega)\big|_{\omega=2\pi K/N} \tag{5.31}$$

$$K = 0, 1, \ldots, N-1$$

DFT of size $N \geq L + M - 1$ required to represent $y(n)$ in the frequency domain

$$Y(K) = S(K)H(K) \tag{5.32}$$

$$K = 0,1,2,\ldots,N-1$$

Here K is a discrete-frequency index, where $S(K)$ is the DFT of $s(n)$ and represented by

$$S(K) = \sum_{n=0}^{N-1} s(n)e^{-(j2\pi K/N)_n}, \quad K=0,1,2,\ldots,N-1$$

and $H(K)$ is the DFT of $h(n)$ and represented by

$$H(K) = \sum_{n=0}^{N-1} h(n)e^{-(j2\pi K/N)_n}, \quad K = 0,1,2,..,N-1$$

Since $s(n)$ and $h(n)$ have a duration less than N. If DFT is to be used for performing linear convolution then we are required to increase the length of $s(n)$ arid $h(n)$ to N by padding these sequences with zero.

After padding these sequences with zero, the N-point circular convolution of $s(n)$ with $h(n)$ must be equivalent to the linear convolution of $s(n)$ with $h(n)$.

Thus we see that with zero as padding, the DFT can be used to perform linear filtering.

EXAMPLE 5.7

Compute linear convolution of $s(n)$ and $h(n)$ using DFT method

Given $$s(n)=\begin{cases}1, & n=0\\ \frac{1}{2}, & n=1\\ 0, & \text{otherwise}\end{cases}$$ and $$h(n)=\begin{cases}\frac{1}{2}, & n=0\\ 1, & n=1\\ 0, & \text{otherwise}\end{cases}$$

Solution:

Sample points of $s(n)$, $N_s = 2$

Sample points of $h(n)$, $N_h = 2$

Sample points of $y(n) = s(n)^*h(n)$, $N_y = N_s + N_h - 1 = 2 + 2 - 1 = 3$

To avoid time-aliasing, we convert two-sample sequences into three-sample sequences by padding with zero. Figure 5.4 illustrates the graphical representation of sequences $s(n)$ and $h(n)$.

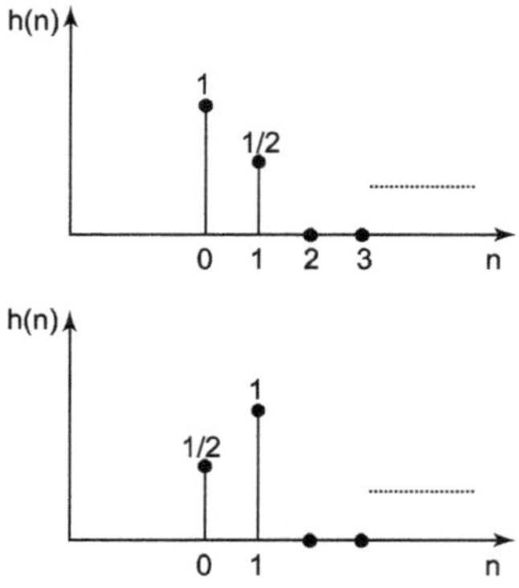

FIGURE 5.4 Graphical representation of $s(n)$ and $h(n)$.

Three-point DFT sequences are given by

$$S(K)=\sum_{n=0}^{2} s(n)e^{-(j2\pi K/3)},\quad K=0,1,2$$

$$=s(0)e^{-j2\pi K.0/3}+s(1)e^{-j2\pi K1/3}+s(2)e^{-j2\pi K2/3}$$

$$=1.1+\frac{1}{2}e^{-j2\pi K/3}+0.e^{-j4\pi K/3}=1+0.5e^{-j2\pi K/3}$$

$K=0,\qquad S(0)=1+0.5e^{-(j2\pi 3/0)}=0.5+1=1.5$

$K=1,\qquad S(1)=1+0.5e^{-(j2\pi/3)1}=1+0.5e^{-j2\pi/3}$

$K=2,\qquad S(2)=1+0.5e^{-(j2\pi/3)2}=1+0.5e^{-j4\pi/3}$

$$H(K)=\sum_{n=0}^{2} s(n)e^{-(j2\pi K/3)_n},\quad K=0,1,2$$

$$=h(0)e^{-(j2\pi K/3)^0}+h(1)e^{-(j2\pi K/3)^1}+h(2)e^{-(j2\pi K/3)^2}$$

$$=\frac{1}{2}e^0+1.e^{-j2\pi K/3}+0.e^{-j4\pi K/3}+1+0.5e^{-j2\pi K/3}$$

$K = 0, \qquad H(0) = 0.5 + e^{-(j2\pi/3)^0} = 1 + 0.5 = 1.5$

$K = 1, \qquad H(1) = 0.5e^{-(j2\pi/3)1} = 0.5e^{-j2\pi/3}$

$K = 2, \qquad H(2) = 0.5e^{-(j2\pi/3)2} = 0.5 + e^{-j4\pi/3}$

Since a time-domain convolution is equivalent to a frequency-domain multiplication

$$Y(K) = S(K) \cdot H(K), \quad K = 0, 1, 2$$

$K = 0, \qquad Y(0) = S(0) \cdot H(0) = 1.5 \times 1.5 = 2.25$

$K = 1, \qquad Y(1) = S(1) \cdot H(1)$

$$= (1 + 0.5e^{-j2\pi/3}) \,.\, (0.5 + e^{-j2\pi/3})$$

$$= 0.5 + 0.25e^{-j2\pi/3} + e^{-j2\pi/3} + 0.5e^{-j4\pi/3}$$

$$= 0.5 + 1.25e^{-j2\pi/3} + 0.5e^{-j4\pi/3}$$

$K = 2, \qquad Y(2) = S(2) \cdot H(2)$

$$= \left(1 + 0.5e^{-j4\pi/3}\right) \left(0.5 + e^{-j4\pi/3}\right)$$

$$= 0.5 + 0.25e^{-j4\pi/3} + e^{-j4\pi/3} + 0.5^{e-j8\pi/3}$$

$$= 0.5 + 1.25^{e-j4\pi}/3 + 0.5^{e-j8\pi/3}$$

The inverse DFT is given by

$$y(n) = \frac{1}{N}\sum_{K=0}^{N-1} Y(K)e^{(j2\pi K/N)_n}, \quad 0 \le n \le N-1 \quad (N = 3)$$

$$y(n) = \frac{1}{3}\sum_{K=0}^{2} Y(K)e^{j2\pi Kn/3}, \quad n = 0,1,2$$

$$= \frac{1}{3}\left[Y(0)e^{j2\pi\,0.n/3}\right] + Y(1)e^{j2\pi\,1.n/3} + Y(2)e^{j2\pi\,2.n/3}$$

$$= \frac{1}{3}\left[2.25 + \left(0.5 + 1.25e^{j2\pi/3} + 0.5e^{j4\pi/3}\right)e^{j2\pi n/3}\right.$$

$$\left. + \left(0.5 + 1.25e^{j4\pi/3} + 0.5e^{j8\pi/3}\right)e^{j4\pi n/3}\right]$$

$$n = 0, 2 \qquad y(0) = \frac{1}{3}\Big[2.25 + \left(0.5 + 1.25e^{j2\pi/3} + 0.5e^{j4\pi/3}\right)e^{j2\pi.0/3}$$
$$+\left(0.5 + 1.25e^{-j4\pi/3} + 0.5e^{-j8\pi/3}\right)e^{j4\pi.0/3}\Big]$$
$$= \frac{1}{3}\Big[2.25 + 0.5 + 1.25e^{-j2\pi/3} + 0.5e^{-j4\pi/3}$$
$$+0.5\,1.25e^{-j4\pi/3} + 0.5e^{-j8\pi/3}\Big]$$
$$= \frac{1}{3}\Big[3.25 + 1.25e^{-j2\pi/3} + 1.75e^{-j4\pi/3} + 0.5e^{-j8\pi/3}\Big]$$
$$= 0.5$$

In similar manner, we can compute

$$y(1) = 1.25 \text{ at } n = 1 \text{ and } y(3) = 0.50 \text{ at } n = 2$$

Hence $$y(n) = \begin{cases} 0.50, & n = 0 \\ 1.25, & n = 1 \\ 0.50, & n = 2 \\ 0, & \text{elsewhere} \end{cases}$$

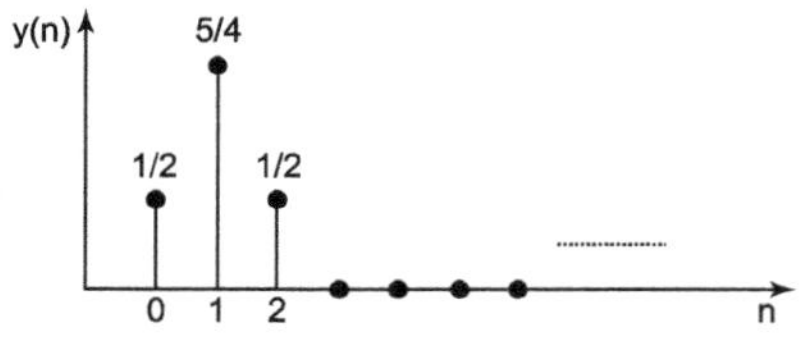

FIGURE 5.5 Graphical representation of resultant of convolution of $s(n)$ and $h(n)$.

Figure 5.5 illustrates the graphical representation of the resultant convolution of $s(n)$ and $h(n)$.

5.8 PITFALLS IN USING DFT

The DFT is primarily concerned with the analysis and processing of discrete-time periodic signals and it is an approximation of the CTFT. FFT is a computationally efficient and speedier method of performing DFT. FFT can be used directly for numerical spectral analysis of sampled versions of continuous-time signals. Most of the problems of using DFT arise due to a misunderstanding of what this approximation involves.

Important characteristics of DFT are:

1. CTFT of a periodic function is also periodic.
2. Both DFT and IDFT operations are done over a finite number of samples of the continuous-time signal.

3. Both DFT and IDFT produce discrete samples which are equally periodic with period N.
4. DFT of convolution between two discrete-time sequences is equal to the product of their individual DFTs.

Figure 5.6(a) illustrates the sinusoidal continuous-time signal $s(t) = \cos 2\pi f_0 t$ and its Fourier transform $S(f)$. $S(f)$ consists of two impulses at $f = \pm f_0$ because $s(t)$ has a single frequency.

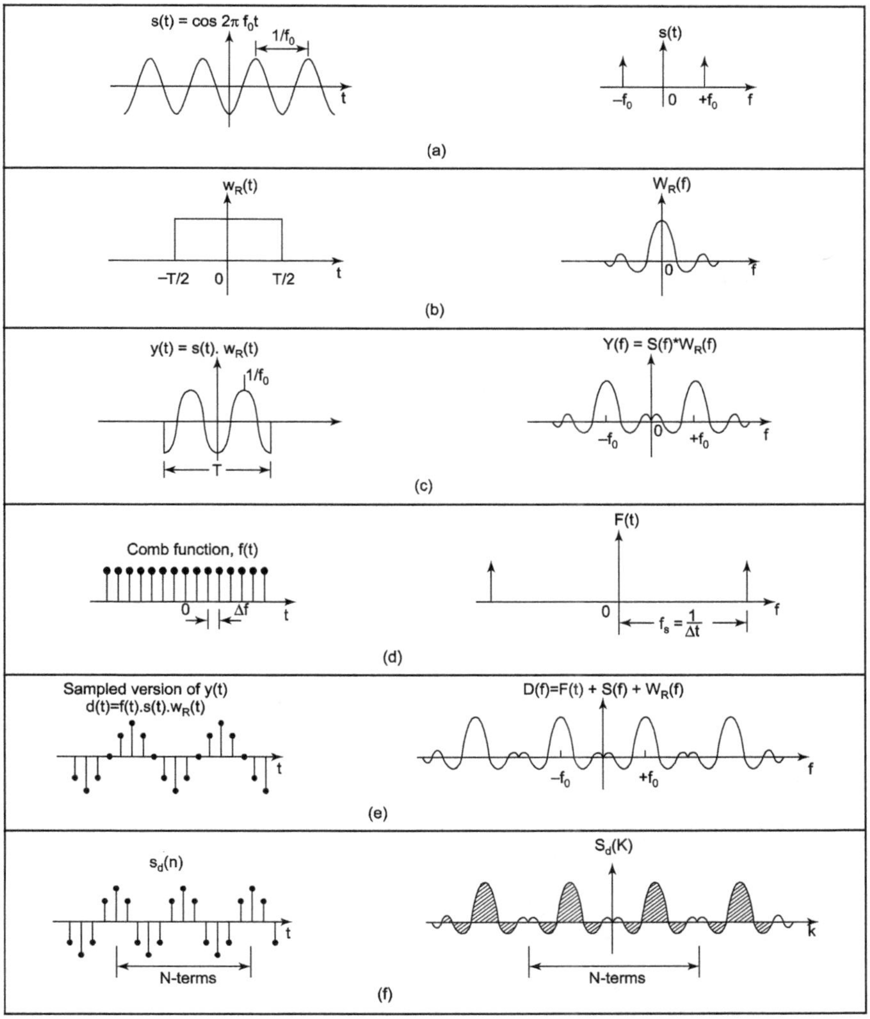

FIGURE 5.6 Discrete Fourier transform (DFT) of discrete-time sequence $s(n)$, $S(K)$ viewed as a corruptive estimate of continuous Fourier transform of continuous-time signal $s(t)$.

Figure 5.6(b) shows the rectangular window function $w_R(t)$ and its Fourier transform $W_R(f)$. The spectrum of window function extends from $-\infty$ to ∞. Here the window function $w_R(t)$ is a time-limited function.

Figure 5.6(c) represents the multiplication of $s(t)$, and $w_R(t)$ that is, $y(t)$ and its Fourier transform $Y(f) = S(f) * W_R(f)$. This product results in the "truncation" of continuous-time signal $s(t)$. The product in time-domain {; corresponds to convolution in the frequency domain. Star (*) denotes the convolution operation.

Figure 5.6(d) illustrates a comb function $f(t)$ and its Fourier transform $F(f)$. Comb function $f(t)$ is a periodic train of impulses with a time interval of Δt between the impulses. Fourier transform of comb function, that is, $F(f)$ is the corresponding Fourier spectra which are again a set of periodic impulses separated by $f_s = \frac{1}{\Delta t}$. It is the sampling frequency due to Nyquist.

Figure 5.6(e) illustrates the sampled version of product $y(f) = s(t)w_R(t)$ and it is denoted $d(t)$. Sampling is done by multiplying $y(t)$ with comb function $y(t)$ then $d(t) = y(t)\,f(t) = s(t)\,w_R(t)\,f(t)$ and corresponding Fourier transform

$$D(f) = Y(f)^* F(f) = S(f)^* W_R(f)^* F(f)$$

This figure shows that the sampled truncated cosine wave $s(t)$ produces continuous spectra from $-\infty$ to ∞.

Figure 5.6(f) shows the discrete-time sequence $s_d(n)$ and its DFT. DFT of sequence $s(n)$ is given by

$$S_d(K) = \sum_{n=0}^{N-1} s_d(n) e^{-j2\pi nK/N}$$

Note. The dotted pattern represents the DFT of $s_d(n)$, that is, $s_d(K)$.

$S_d(K)$ are the sample values of $S(f)$, where $S(f)$ is the Fourier transform of continuous-time signal $s(t)$.

5.8.1 Problems of Pitfalls in Using DFT

For convenience, we have taken cosine sinusoidal signal $s(t)$ of single frequency f_0. We expect an arbitrary waveform.

In this sub-section we are going to discuss the problems which arise in using DFT and their remedies.

1. Truncation error.
2. Aliasing error.

3. Spectral leakage.
4. Improper use of DFT.
5. Picket-Fence effect.

Now these problems will be discussed one by one in detail.

1. **Truncation Error:** This error is caused by using the rectangular window function $w_R(t)$. But a time-limited signal has an infinite frequency spectra and spectrum of $w_R(t)$ is spread into frequencies other than $\pm f_0$.

   The resultant $y(f)$ which is product of $s(t)$ and $w_R(t)$ gives a truncated version in time-domain. Fourier transform of $y(t)$, that is, $Y(f)$ is the spectral function which spread into frequencies other than only those due to $W_R(f)$. These are not useful and it is called truncation error.
2. **Aliasing Error:** We have already studied aliasing error in Chapter 2. To avoid aliasing effect, sampling rate should be greater than or equal to Nyquist rate. Nyquist rate is equal to twice the highest-frequency component of the signal. However, due to truncation of signal $s(t)$, the frequencies spread into the frequency-domain for larger than the allowable range. Even after sampling a signal higher than the Nyquist rate, the aliasing error cannot be entirely eliminated.
3. **Spectral Leakage:** We have one more effect other than truncation and aliasing error. It is called spectral leakage. Leakage of energy take place when the original signal $s(t)$ is truncated and sampled. It means energy in the original signal frequency f_0 will now leak into other frequencies. This leakage is due to truncating the original signal $s(t)$. Due to this leakage, we get an undesirable modification of the total spectrum. The spectral leakage effect cannot always be isolated from the aliasing effect because leakage leads to aliasing if the higher frequency component of the composite spectrum shifts beyond the folding frequency $f_s/2$.
4. **Improper Use of DFT:** This problem is shown in Figure 5.7(a). In this figure, the waveforms are continuous-time in shape. For correct use of DFT separate the waveform of $s(t)$ with duration T_1 which is equal to impulse duration T. Here, $y(t) - s(t) * h(t)$ where $*$ denotes convolution operation between $s(t)$ and $h(t)$ and results in $y(t)$. We have no signal between T and $T + T_1$. If this gap is not provided during computation by FFT, the aliasing effect will arise and its other associated problems will also occur.

5. **Picket-Fence Effect:** This problem is caused by the inability of the DFT to observe the spectrum as a continuous function $y(t)$ because DFT computation is limited to integer multiples of the fundamental frequency f_0, where f_0 is given by $f/N = f_s/N$.

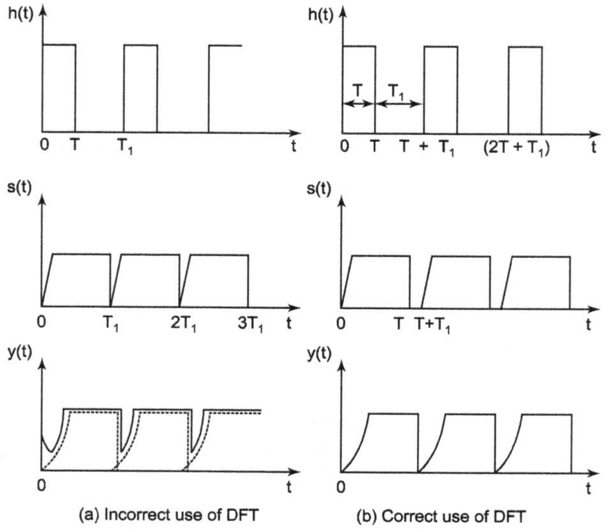

FIGURE 5.7 Correct and incorrect use of DFT.

The observation of spectrum with the DFT is equivalent to looking at it through a sort of "Fence." It means we observe the exact waveform of $y(t)$ only at discrete points of signals.

5.8.2 Remedies to Problems of Pitfalls in Using DFT

Now, we shall discuss very briefly the remedies to problems to pitfalls in using a discrete Fourier transform.

1. Remedy to truncation error.
2. Remedy to aliasing error.
3. Remedy to spectral leakage.
4. Remedy to improper use of DFT.
5. Remedy to picket-fence effect.

Now we shall discuss each remedy one by one as follows.

1. **Remedy to Truncation Error:** The truncation is shown in Figure 5.6(c). Here we have used a rectangular window function $w_R(t)$. If we want to perform DFT using FFT on a digital computer, we would be required to perform on a finite piece of the signal $s(t)$ because we have limitations in computer storage memory. Thus truncation is usually necessary (if signal $s(t)$ is not naturally limited) and it results in a smearing effect. This effect is shown in Figure 5.6(c).

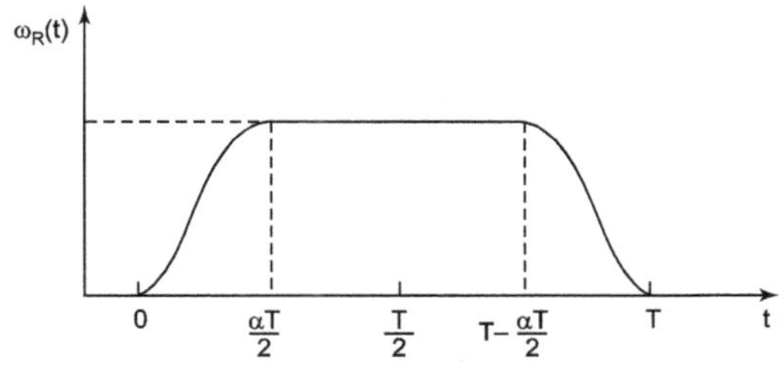

FIGURE 5.8 Data window.

Truncation error can be reduced by selecting a window function $w(t)$ that minimizes the spreading of the spectra into higher frequencies. A raised cosine window function is shown in Figure 5.8. This window function reduces the time function more gradually near the ends of the interval than the rectangular window function, which is shown in Figure 5.6(b). These window functions are called data windows.

By using these window functions, the convergence of series is more rapid and results in less spectral leakage.

These window functions can be expressed mathematically as follows:

$$\begin{aligned} w_R(t) &= 0.50 + 0.50\cos\frac{2\pi t}{aT} && 0 < t \le \frac{aT}{2} \\ &= 1. && \frac{aT}{2} < t < T - \frac{aT}{2} \\ &= 0.50 + 0.50\cos\frac{2\pi t}{aT}. && T - \frac{aT}{2} < t \le T \end{aligned}$$

Typical values of a lie between 0.10 and 0.20.

2. **Remedy to Aliasing Error:** Aliasing error can be reduced by choosing a sampling rate higher than the Nyquist rate to avoid any spectral overlap. There is a need for prior knowledge of the spectrum of the signal so we can choose the appropriate sampling rate. If we have no prior knowledge of the signal then we must filter the signal before sampling using a low pass filter. Prefiltering is done in order to ensure about the highest-frequency

component so as to decide the appropriate sampling rate. By doing so we can reduce aliasing error.

3. **Remedy to Spectral Leakage:** This leakage of energy into other frequencies can be minimized by selecting sampling frequency in such a way that the zero and peak samples of the truncated signal $s(t)$ are always present. Longer bandwidth which is some integral multiple of the bandwidth of the signal will reduce spectral leakage. This is shown in Figure 5.6(e). Spectral leakage can also be reduced by choosing the window function for which the first and the last set of samples are weighted by a raised cosine wave, which is shown in Figure 5.8.
4. **Remedy to Improper Use of DFT:** Improper use of DFT can be taken care of by adding some extra zero samples to the set of non-zero samples so that the N-point DFT produces the correct result.
5. **Remedy to Picket-Fence Effect:** Picket-fencing can be reduced by varying the number of points N in a given time period T and adding zeros at the end of the original record while maintaining the original record intact. This method artificially alters the period and this new period changes the locations of spectral lines without changing the continuous form of the original spectrum. By shifting the sample points, we can easily observe those spectral components which were originally hidden from view.

EXAMPLE 5.8

Find out the DFT of the sequence $s(n)$ illustrated in the figure below for $N = 6$.

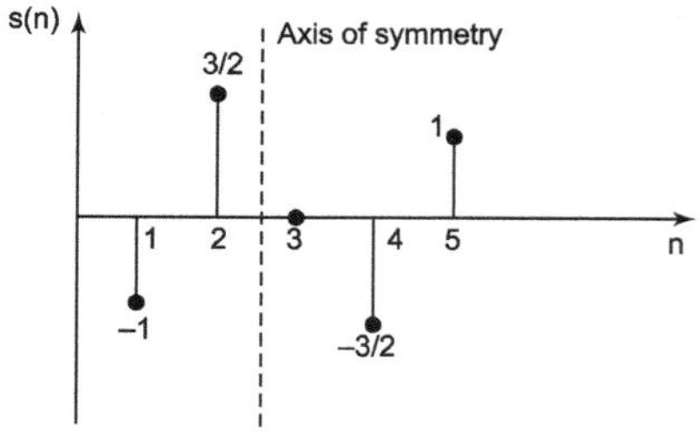

FIGURE 5.9 Graphical representation of sequence $s(n)$ with $N = 6$.

Solution:

For the above sequence $s(n)$

$$s(0) = s\left(\frac{N}{2}\right) = s(3) = 0$$

$$s(1) = -1$$

$$s(2) = -s(4) = \frac{3}{2}$$

$$s(5) = -s(1) = 1$$

Using a formula of DFT as:

$$S(K) = \sum_{n=0}^{5} s(n) W_N^{nK}, \qquad \text{where } W_N = e^{-j2\pi/N}$$

$$= \sum_{n=0}^{5} s(n) e^{-j2\pi K/N}$$

$$= \sum_{n=0}^{5} s(n) e^{-j2\pi K/b} = \sum_{n=0}^{5} s(n) e^{-j2\pi nK/3}$$

$$= s(0)e^{-j2\pi .0K/3} + s(1)e^{-j2\pi .1K/3} + s(2)e^{-j2\pi .2K/3}$$

$$+ s(3)e^{-j2\pi .3K/3} + s(4)e^{-j2\pi .4K/3} + s(5)e^{-j2\pi .5K/3}$$

$$= 0 \times 1(-1)e^{-j\pi K/3} + \frac{3}{2}e^{-j2\pi K/3} + 0e^{-j2\pi 3K/3}$$

$$\left(-\frac{3}{2}\right)e^{-j4\pi K/3} + (1)e^{-j\pi 5K/3}$$

$$= -e^{-j\pi K/3} + \frac{3}{2}e^{-j\pi K/3} - \frac{3}{2}e^{-j\pi 4Kn/3} + e^{-j\pi 5K/3}$$

$$S(K) = -e^{-j\pi K/3} + \frac{3}{2}e^{-j\pi 2K/3} - \frac{3}{2}e^{-j\pi 4K/3} + e^{-j\pi 5K/3}$$

$K = 0, \qquad S(0) = 0$

$K = 1, \qquad S(1) = -j\frac{\sqrt{3}}{2}$

$K = 2, \qquad S(2) = \frac{5}{2}\sqrt{3}$

$K = 3, \qquad S(3) = 0$

$K = 4, \qquad S(4) = -j\frac{\sqrt{3}}{2}$

$K = 5, \qquad S(5) = +j\frac{\sqrt{3}}{2}$

$$S(K) = \left\{0, -j\frac{\sqrt{3}}{2}, j\frac{5}{2}\sqrt{3}, 0, -j\frac{5}{2}\sqrt{3}, j\frac{\sqrt{3}}{2}\right\}$$

EXAMPLE 5.9

Determine the circular convolution of two sequences $s_1(n)$ and $s_2(n)$ given by

$$s_1(n) = s_2(n) = 1, \quad \text{for} \quad 0 \le n \le N-1$$
$$= 0, \text{ otherwise}$$

Solution:

Since both the sequences $s_1(n)$ and $s_2(n)$ are of unity value.

DFTs of the sequences will be

$$S_1(K) = S_2(K) = \sum_{n=0}^{N-1} 1.e^{-j2\pi nK/N}$$

The above computation is applied for $N = 4$

$$S_1(K) = S_2(K) = \sum_{n=0}^{3} 1.e^{-j2\pi nK/4}$$

$$= 1 + e^{-j2\pi 1K/2} + e^{-j\pi K} + e^{-j\pi 3K/2}$$

$K = 0, \qquad S_1(0) = S_2(0) = 4$

$K = 1, \qquad S(1) = S_2(1) = 0$

$K = 2, \qquad S_1(2) = S_2(2) = 0$

$K = 3, \qquad S_1(3) = S_2(n) = 0$, etc.

That is, $S_1(K) = S_2(K) = N$, for $K = 0$

$$= 0, \quad \text{otherwise}$$

Resultant of circular convolution of two sequences

$$s_1(n) = s_2(n) = 1, 0 \le n \le N-1$$
$$= 0, \quad \text{otherwise}$$

is given by

$$S_3(K) = S_1(K).\ S_2(K) = N.N = N^2, \quad \text{for } K = 0$$

Inverse DFT

$$s_3(n) = \frac{1}{N}\sum_{K=0}^{N-1} S_3(K)e^{j2\pi nK/N}$$

$$=\frac{1}{N}.N^2=N \quad \text{for} \quad 0\le n\le N-1$$

This is the result of circular convolution.

EXAMPLE 5.10

Find the symmetry in each case for the periodic sequence $s(n)$ as shown in Figure 5.10.

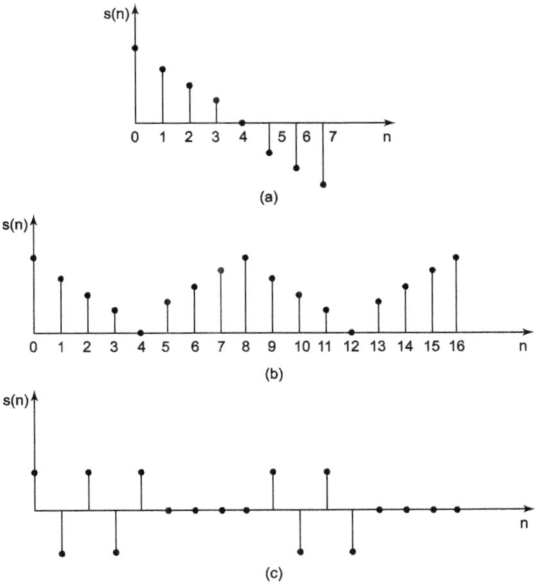

FIGURE 5.10.

Solution:

a. For the sequence $s(n)$ in Figure 5.10(a), we find $N=8$ and that,

$$s(1)=-s(7)$$

$$s(2)=-s(6)$$

$$s(3)=-s(5)$$

except $s(0)$ and $s(4)$.

Therefore, it obeys symmetry

$$s(n)=-s(N-m)$$

Hence the Fourier coefficients $S(K)$ will be imaginary.

b. For the Figure 5.10(b), we find that

$$s(1)=s(7)$$

$$s(2)=s(6)$$

and $\quad s(3)=s(5)$

thus it obeys the symmetry $s(n) = s(N - m)$. Hence the Fourier coefficients $S(K)$ are real.

c. Sequence $s(n)$ of Figure 5.10(c) does not obey any symmetry, thus the Fourier coefficients $S(K)$ will be complex numbers.

EXAMPLE 5.11

Show for an arbitrary K, that

a. $S(-K) = S^*(K)$

b. $S\left(\frac{N}{2} - K\right) = S^*\left(\frac{N}{2} + K\right)$

where $S(K)$ is the DFT of $s(n)$.

$S^*(K)$ is the complex conjugate of $S(K)$. Here $s(n)$ is real and N is even.

Solution:

a. DFT of $s(n)$ is given by

$$S(K) = \sum_{n=0}^{N-1} s(n) W_N^{-nK} \qquad (1)$$

where $\qquad W_N = e^{j2\pi/N}$

Substituting $K = -K$ in Eq. (1), we get

$$S(-K) = \sum_{n=0}^{N-1} s(n) W_N^{-nK} \qquad (2)$$

Taking the complex conjugate of both sides of Eq. (2), we get

$$S*(-K) = \sum_{n=0}^{N-1} s(n) W_N^{-nK} \qquad (3)$$

Comparing Eqs. (1) and (3), we get

$S(K) = S^*(-K)$ Proved.

b. Since $s(n)$ is real, then

$$S(K) = S^*(-K)$$

It holds symmetry in its Fourier coefficients (DFTs)

$$S\left(\frac{N}{2} - K\right) = \sum_{n=0}^{N-1} s\left(\frac{N}{2} - n\right) W_N^{-(N/2-n)K} \qquad (4)$$

$$S\left(\frac{N}{2}+K\right)=\sum_{n=0}^{N-1} s\left(\frac{N}{2}-n\right)W_N^{-(N/2-n)K}$$

or $$S\left(\frac{N}{2}+K\right)=\sum_{n=0}^{N-1} s(N/2-n)W_N^{-(N/2-n)K} \tag{5}$$

Taking the complex conjugate of both sides of Eq. (5), we get

$$S^*\left(\frac{N}{2}+K\right)=\sum_{n=0}^{N-1} s\left(\frac{N}{2}-n\right)W_N^{-(N/2-n)K} \tag{6}$$

Because sequence $s(n)$ is a real sequence.

$$s^*(n)=s(n)$$

Comparing Eqs. (4) and (6), we get

$$S\left(\frac{N}{2}-K\right)=S^*\left(\frac{N}{2}+K\right) \text{ Proved.}$$

EXAMPLE 5.12

If sequence $s_1(n)$ is a periodic sequence with period N and it is also periodic with period $2N$. Assume that $S_1(K)$ is the DFT of $s_1(n)$ with period N and $S_2(K)$ is the DFT of $s_1(n)$ with period $2N$. Determine $S_2(K)$ in terms of $S_1(K)$.

Solution:

DFT of $s_1(n)$ with period N is given by

$$S_1(K)=\sum_{n=0}^{N-1} s_1(n)e^{-(j2\pi/N)nK} \tag{1}$$

Substituting $W_N = e^{-2\leq/N}$ in above Eq. (1), we get

$$S_1(K)=\sum_{n=0}^{N-1} s_1(n)W_N^{nK} \tag{2}$$

Substituting DFT of $s_1(n)$ with period $2N$ is given by

$$S_2(K)=\sum_{n=0}^{2N-1} s_1(n)e^{-(j2\pi/2N)nK} \tag{3}$$

Substituting $W_{2N} = e^{-2\pi/2N}$ in Eq. (3), we get

$$S_1(K)=\sum_{n=0}^{2N-1} s_1(n)W_{2N}^{nK} \tag{4}$$

Now, we want to express $S_2(K)$ in terms of $S_1(K)$, where $s(n)$ is a periodic sequence and it satisfies the condition

$$s(n) = s(n+N) = s(n+2N) \quad (5)$$

$$W_{2N}^{nK} = W_N^{nK/2} = e^{-j(2\pi/2N)nK} = e^{-j(2\pi/N)nK/2} \quad (6)$$

Substituting Eq. (6) in Eq.(4), we get

$$S_2(K) = \sum_{n=0}^{2N-1} s_1(n) W_{2N}^{nK}$$

$$= \sum_{n=0}^{2N-1} s_1(n) W_{2N}^{nK/2} \quad (7)$$

$$= \sum_{n=0}^{2N-1} s_1(n) W_{2N}^{nK/2} + \sum_{n=N}^{2N-1} s(n+N) W_N^{(n+N)K/2}$$

$$= S_1\left(\frac{K}{2}\right) + \left[\sum_{n=N}^{2N-1} s(n+N)\right] W_N^{(n+N)K/2}$$

where $S_1\left(\frac{K}{2}\right) = \sum_{n=N}^{N-1} s_1(n) W_N^{nK/2}$

$$= S_1\left(\frac{K}{2}\right) + \left[\sum_{n=N}^{N-1} s_1(n) W_N^{(n)K/2}\right] W_N^{NK/2}$$

$$= S_1\left(\frac{K}{2}\right) + S_1\left(\frac{K}{2}\right) W_N^{NK/2}$$

or $$S_2(K) = \left(1 + W_N^{NK/2}\right) S_1\left(\frac{K}{2}\right) \quad (8)$$

This is the desired expression.

Note. If K is even then $W_N^{NK/2} = 1$

and if K is odd then $W_N^{NK/2} = (-1)^K$ if, therefore, for K is even,

$$S_2(K) = 2S_1\left(\frac{K}{2}\right)$$

for K is odd,

$$S_2(K) = \left[1 + (1)^K\right] S_1\left(\frac{K}{2}\right) = [1-1] S_1\left(\frac{K}{2}\right)$$

$$= [0] S_1\left(\frac{K}{2}\right) = 0.$$

EXAMPLE 5.13

Consider two periodic sequences, $s(n)$ is periodic with period N and $y(n)$ is periodic with period M. The sequence $z(n)$ is defined as $z(n) - s(n) + y(n)$,

a. Show that $z(n)$ is periodic with period MN.

b. DFTs of $s(n)$ and $y(n)$ will be periodic with period N and M, respectively.

Find $Z(L)$ in the terms $S(L)$ and $Y(L)$, where $Z(L)$, $S(L)$, and $Y(L)$ are the DFTs of $z(n)$, $s(n)$, and $y(n)$, respectively.

Solution:

Given:

$$z((n))_K = s((n))_N + y((n))_M \tag{1}$$

Here double brackets are used as it illustrates a sequence with its time period.

For linearity, both sequences $s(n)$ and $y(n)$ should have the same period.

$$z((n))_K = s((n))_K + y((n))_K \tag{2}$$

In Eq. (2) all the sequences have the same period, that is, K.

a. Now we want to determine the relation between K, M, and N.

$$\text{DFT of } z(n) = Z(L) = \sum_{n=0}^{K-1} z(n) e^{-j(j2\pi K)nL} \tag{3}$$

where L is the discrete-frequency index.

We know that $Z(L)$ is periodic with period K. Hence

$$Z(0) = Z(L)$$

$$Z(1) = Z(L+1)$$

$$Z(2) = Z(L+2) \text{ etc.}$$

Substituting Twiddle factor $WN = e^{-j2\pi/L}$ in Eq. (3), we get

$$Z((L))_K = \sum_{n=0}^{K-1} z(n)W_N^{nL} = \sum_{n=0}^{K-1}\left[s(n)+y(n)\right]W_K^{nL}$$

$$= \sum_{n=0}^{K-1} s((n))W_K^{nL} + \sum_{n=0}^{K-1} y((n))W_K^{nL}$$

or $$Z((L))_K = \sum_{n=0}^{K-1} s((n))e^{-(j2\pi/N)nL} + \sum_{n=0}^{K-1} y((n))e^{-(j2\pi/M)nL}$$

or $$= \sum_{n=0}^{K-1} s((n))e^{-(j2\pi/MN)MnL} + \sum_{n=0}^{K-1} y((n))e^{-(j2\pi/MN)NnL}$$

(We know that $M \times 2\pi = N \times 2\pi = K \times 2\pi = 2\pi$, Hence $K = MN$.)

$$Z((L))_K = \sum_{n=0}^{K-1} s((n))e^{-(j2\pi/K)nL} + \sum_{n=0}^{K-1} y((n))e^{-(j2\pi/K)nL}$$

or $$Z((L))_K = \sum_{n=0}^{K-1} s(n)W_K^{nL} + \sum_{n=0}^{K-1} y(n)W_K^{nL} \quad (4)$$

The relation of Eq. (4) shows that if $K = MN$, then only the sum of equation. $z((n))_k = s((n))_N + y((n))_M$ is possible and relevant, and that LCM of M and N is MN, where $M \neq N$.

b. Now, we are required to determine the relationship between $Z((L))_K$, $S((L))_K$ and $Y((L))_K$, where $K = MN$.

$$Z((L))_K = \sum_{n=0}^{K-1} z((n))W_K^{nL}$$

$$= \sum_{n=0}^{MN-1} s((n))W_{MN}^{nL} + \sum_{n=0}^{MN-1} y((n))W_{MN}^{nL}$$

or $$Z((L))_K = S_2((L))_K + Y_2((L))_K \quad (5)$$

Now, $S_2((L))_K$ can be expressed in terms of $S((L))_N$.

Where, $$S((L)) = \sum_{n=0}^{N-1} s((n))W_N^{nL} \quad (6)$$

Since $s((n))$ is also periodic with period MN

$$s((n)) = s((n+N)) = s((n+2N)) = \ldots = s((n \ +(M-1)N)) \tag{7}$$

Double brackets are used to show the periodicity of a particular sequence. From the above two equations (6) and (7), we can expand $S((L))$ as

$$\sum_{n=0}^{N-1} s((n))W_{NM}^{nL} + \sum_{n=0}^{N-1} s((n))W_{NM}^{(n+N)L} + \sum_{n=0}^{N-1} s((n))W_{NM}^{(n+2N)L} + \ldots.$$

$$+\sum_{n=0}^{N-1} s((n))W_{NM}^{[n+(M-1)N]L} \tag{8}$$

Rewriting Eqn. (8) in a convenient form, we get

$$\sum_{n=0}^{N-1} s((n))W_N^{NL/M} + W_N^{NL/M}\sum_{n=0}^{N-1} s((n))W_N^{NL/M} + \ldots + W_N^{NL/(M-1)}\sum_{n=0}^{N-1} s((n))W_N^{NL/M}$$

Now, we can write an expression for $S_2((L))$ as

$$S_2((L)) = 1 + W_N^{NL/M} + W_N^{N(L+1)/M} + \ldots + W_N^{N(M+1)L/M}\sum_{n=0}^{N-1} W_N^{NL/M}$$

where $$= F.S\left(\left(\frac{L}{M}\right)\right) \tag{9}$$

$$F = 1 + W_N^{NL/M} + W_N^{N(L+1)/M} + W_N^{N(M-2)L/M} + \ldots + W_N^{N(M-1)L/M}$$

$$= \sum_{P=0}^{M-1} W_N^{PN} = \sum_{P=0}^{M-1} 1 = M \tag{10}$$

where P is any integer

$$\left[W_N^{PN} = e^{-j2\pi/N.PN} = 1\right]$$

$$F = M \tag{11}$$

Substituting $F = M$ in Eq. (1), we get

$$S_2((L)) = FS\left(\left(\frac{L}{M}\right)\right) = MS\left(\left(\frac{L}{M}\right)\right) \tag{12}$$

Let $$Y((L)) = \sum_{n=0}^{M-1} y((n))W_M^{nL}$$

Since $y((n))$ is also periodic in NM

$$y((n)) = y((n+M)) = y(n+2m) + \ldots + y[n+(N-1)M] \tag{13}$$

$$Y_2((L)) = \sum_{n=0}^{M-1} y((n))W_{NM}^{nL} + \sum_{n=0}^{M-1} y((n))W_{NM}^{(n+M)L} + \sum_{n=0}^{M-1} y((n))W_{NM}^{(n+M+1)L}$$

$$+ \ldots + \sum_{n=0}^{M-1} y((n))W_{NM}^{(n+(N-1)M)L}$$

$$= \sum_{n=0}^{M-1} y((n))W_{NM}^{nL} + W_{NM}^{nL} \sum_{n=0}^{M-1} y((n))W_{NM}^{nL} + \ldots$$

$$+ W_{NM}^{(N-1)L} \sum_{n=0}^{M-1} y((n))W_{NM}^{nL}$$

$$= \left[1 + W_{NM}^{ML} + W_{NM}^{(M+1)L} + \ldots + W_{NM}^{(N-1)ML}\right] \sum_{n=0}^{M-1} y((n))W_{NM}^{nL}$$

$$\left(\text{Putting } W_{NM}^{NL} = W_{M}^{ML/N} \text{ and } W_{NM}^{nL} = W_{M}^{n/N}\right)$$

$$= \left[1 + W_{M}^{ML/N} + W_{M}^{(M+1/N)L} + \ldots + W_{M}^{(N-1/N)L}\right] \sum_{n=0}^{M-1} y((n))W_{M}^{nL/N}$$

$$= N \sum_{n=0}^{M-1} y((n))W_{M}^{nL/N} = NY\left(\frac{L}{N}\right) \tag{14}$$

where $\left[1 + W_{M}^{ML/N} + W_{M}^{(M+1/N)L} + \ldots + W_{M}^{(N-1/N)ML} = N\right]$ already determined.

$$V((L))_K = S_2((L))_K + Y_2((L))_K \tag{15}$$

$$\boxed{S_2((L)) = MS\left(\left(\frac{L}{F}\right)\right) \text{ and } Y_2((L)) = NY\left(\left(\frac{L}{M}\right)\right)}$$

Substituting the value of $S_2((L))$ and $Y_2((L))$ in Eq. (15), we get

$$V((L))_K = S_2((L))_K + Y_2((L))_K = MS\left(\left(\frac{L}{F}\right)\right) + NY\left(\left(\frac{L}{M}\right)\right)$$

Hence proved.

EXERCISES

1. Differentiate between discrete-time Fourier transform (DTFT) and discrete Fourier transform (DFT) of a sequence.
2. How DTF can be used as a liner transformation tool in signal analysis and digital signal processing?
3. Discuss various properties of the DFT in brief.
4. Differentiate between circular convolution and linear convolution of two discrete-time sequences.
5. Discuss circular convolution of two discrete-time sequences in detail.
6. What is Parseval's theorem for discrete-time sequences.
7. What is the relationship between DFT and z-transform?
8. How lines convolution can be performed using the DFT?
9. In what case resultant of linear convolution and circular convolution will be the same? Discuss.
10. Discuss various problems of pitfalls in using DFT.
11. Write short notes on the following:
    (i) Truncation error. ***(ii)*** Aliasing error.
    (iii) Spectral leakage. ***(iv)*** Improper use of DFT.
    (v) Picket-fence effect.
12. Discuss remedies of various problems of pitfalls in using DFT.

NUMERICAL EXERCISES

1. Evaluate the N-point DFT of the sequences given as follows:

(i) $s(n) = \begin{cases} \frac{1}{2}, & 0 \leq 2 \\ 0, & \text{elsewhere} \end{cases}$ ***(ii)*** $s(n) = (A)^n u(n)$

2. Determine the DFT of the following sequences:

(i) $s(n) = \begin{cases} 2, & 2 \le n \le 6 \\ 0, & \text{for } n = 0,1,7,8,9 \end{cases}$ Given $N = 10$

(ii) $s(n) = \begin{cases} 1, & 3 = 3,7 \\ 0, & \text{for } n = 1,2,4,5,6,8,9 \end{cases}$ Given $N = 10$

(iii) $s(n) = \begin{cases} 1, & 0 \le n \le 5 \\ 2, & 6 \le n \le 9 \end{cases}$ Given $N = 10$

3. Determine the circular convolution of the following sequences using the time-domain formula.

(i) $s_1(n) = \{1, 2, 3, 1\}$ and $s_2(n) = \{4, 3, 2, 2\}$

(ii) $s_1(n) = \{1, 1, 1, 1\}$ and $s_2(n) = \{2, 2, 2, 2\}$

(iii) $s_1(n) = \{1, 2, 3, 1\}$ and $s_2(n) = \{1, 1, 1, 1\}$

4. Determine linear convolution using DFT method

$$s_1(n)\begin{cases} 1, & n = 0 \\ \frac{1}{2}, & n = 1 \\ 0, & \text{otherwise} \end{cases} \quad \text{and } s_2(n) = \begin{cases} \frac{1}{2}, & n = 0 \\ 1, & n = 1 \\ 0, & \text{otherwise} \end{cases}$$

5. Find the circular convolution of two sequences given as follows:

$$s_1(n)\begin{cases} 1, & n = 0 \\ \frac{1}{2}, & n = 1 \\ 0, & \text{otherwise} \end{cases} \quad \text{and } s_2(n) = \begin{cases} \frac{1}{2}, & n = 0 \\ 1, & n = 1 \\ 0, & \text{otherwise} \end{cases}$$

6. Given the two 5-point sequences as

$$s_1(n) = \{\underset{\uparrow}{0},1,2,3,4\}$$

$$s_1(n) = \{\underset{\uparrow}{0},1,0,0,0\}$$

Determine $s(n) = S_1(n) \, \textcircled{N} \, s_2(n)$ by DFT method.

CHAPTER 6

FAST FOURIER TRANSFORM (FFT) ALGORITHMS

6.1 INTRODUCTION

The FFT does not represent a transform different from the discrete Fourier transform (DFT) but they are special algorithms for speedier implementation of DFT. FFT requires a comparatively smaller number of arithmetic operations such as multiplications and additions than DFT. FFT also requires lesser computational time than DFT. The fundamental principle on which all these algorithms are based is that of decomposing the computation of the DFT of a sequence of length N into successively smaller DFTs. The way in which this principle is implemented leads to a variety of different algorithms, all with comparable improvements in computational speed. Thus, we can say that DFT plays an important role in several applications of digital signal processing such as linear filtering, correlation, analysis and spectrum analysis.

Direct Computation of the DFT. DFT for a complex-valued sequence, $s(n)$ may be expressed as

$$S(K) = \sum_{n=0}^{N-1} s(n)e^{-j2\pi nK/N} \tag{6.1}$$

Here both $S(K)$ and $s(n)$ are complex-valued discrete-frequency and discrete-time sequences of N-points, respectively.

$$S(K) = S_R(K) + jS_I(K)$$

$$s(n) = s_R(n) + js_I(n)$$

$$e^{-j2\pi K/N} = \cos\frac{2\pi nK}{N} = j\sin\frac{2\pi nK}{2}$$

Substituting the value of $S(K)$, $s(n)$, and $e–^{j2\pi nK/N}$ in Eq. (6.1), we get

$$S(K) = \sum_{n=0}^{N-1} s(n)e^{-j2\pi nK/N}, \quad K = 0, 1, \ldots, N-1 \tag{6.2}$$

$$S_R(K) + jS_I(K) = \sum_{n=0}^{N-1}[s_R(n) + js_I(n)]\left[\cos\frac{2\pi nK}{N} - j\sin\frac{2\pi nK}{2}\right]$$

$$= \sum_{n=0}^{N-1}\left[s_R(n)\cos\frac{2\pi nK}{N} + s_I(n)\sin\frac{2\pi nK}{2}\right]$$

$$-j\sum_{n=0}^{N-1}\left[s_R(n)\cos\frac{2\pi nK}{N} - s_I(n)\cos\frac{2\pi nK}{2}\right] \tag{6.3}$$

Separating real and imaginary parts of above Eq. (6.3), we get

$$S_R(K) = \sum_{n=0}^{N-1}\left[s_R(n)\cos\frac{2\pi nK}{N} + s_I(n)\sin\frac{2\pi nK}{N}\right] \tag{6.4}$$

and

$$s_I(K) = -\sum_{n=0}^{N-1}\left[s_R(n)\sin\frac{2\pi nK}{N} - s_I(n)\cos\frac{2\pi nK}{N}\right] \tag{6.5}$$

where $s_R(n)$ and $s_I(n)$ are the real and imaginary parts of a discrete-time sequence $s(n)$, respectively.

$S_R(K)$ and $S_I(K)$ are the real and imaginary parts of the DFT of $s(n)$, respectively.

From Eqs. (6.4) and (6.5), it is clear that for each value of K, the direct computation of DFT of $S(n)$, that is, $S(K)$ requires $4N$ real multiplications and $4(N - 1)$ real additions. Since $S(K)$ is computed for N different values of K, the direct computation of the DFT of a sequence $s(n)$ requires $4N \times N = 4N^2$ real multiplications and $4(N - 1) \times N = 4N(N - 1)$ real additions. In other words, we can say that direct DFT computation requires N^2 complex multiplications and $N(N - 1)$ complex additions.

The direct computation of Eqs. (6.4) and (6.5) requires:

i. N^2 Complex multiplications or $4N^2$ Real multiplications.

ii. $N(N - 1)$ Complex additions or $4N(N - 1)$ Real additions

iii. $2N^2$ Evaluations to trigonometric functions such as

$$\sin\frac{2\pi nK}{N} \text{ and } \cos\frac{2\pi nK}{N}$$

6.2 GOERTZEL ALGORITHM

In Goertzel algorithm, the periodicity of the phase factors W_N^K is exploited. This algorithm allows us to express the computation of the DFT as linear filtering.

DFT of sequence $s(n)$ is given by

$$S(K) = \sum_{m=0}^{N-1} s(m) W_N^{Km} \tag{6.6}$$

where $W_N = e-^{j2\pi/N}$ is called phase factor.

Since $$W_N^{-KN} = \left(e^{-j(2\pi/N)}\right)^{-KN} = e^{j2\pi K} = 1, \quad K = 0,1,2,\ldots$$

Now multiplying above Eq. (6.6) by W_N^{-KN}, we get

$$S(K) = W_N^{-KN} \sum_{m=0}^{N-1} s(n) W_N^{Km} = \sum_{m=0}^{N-1} s(m) W_N^{K(m-N)}$$

or $$S(K) = \sum_{m=0}^{N-1} s(m) W_N^{-K(N-m)} \tag{6.7}$$

From the inspection of Eq. (6.7), we note that it is in the form of convolution. Now, if we define the discrete-time sequence $y_K(n)$ as

$$y_K(n) = \sum_{m=0}^{N-1} s(m) W_N^{-K(N-m)} \tag{6.8}$$

Then, $y_K(n)$ is the convolution of a sequence $s(n)$ of length N with a filter that has a unit-simple response

$$H_K(n) = W_N^{-KN}.u(n) \tag{6.9}$$

Comparing Eqs. (6.7) and (6.8), we get

$$S(K) = yk(n)\big|_{n=N} \tag{6.10}$$

System function $H(z)$ of a filter whose unit-sample response, $h_K(n) = W_N^{-KN}.u(n)$.

Taking the z-transform of Eq. (6. 9), we get

$$H_K(z) = \mathcal{Z}[h_K(n)] = \mathcal{Z}\left[W_N^{-KN}\right]$$

$$= \frac{1}{1 - W_N^{-K_z-1}} = \frac{z}{z - W_N^{-K}} \tag{6.11}$$

or $$H_K(z) = \frac{z}{W_N^{-K}} = \frac{N(z)}{D(z)}$$

where $N(z)$ and $D(z)$ are numerator and denominator polynomial of rational z-transform.

The zeros of a z-transform $H_K(z)$ are the values of z for which $H_K(z) = 0$ and poles of a z-transform $H_K(z)$ are the values of z for which $HK(z) = \infty$.

In other words, the roots of numerator polynomial $N(z)$ are called zeros, and roots of denominator polynomial $D(z)$ are called poles of rational z-transform.

In the above z-transform, there is one zero at $z = 0$ and one pole at $z = W_N^{-K}$

These poles and zeros are illustrated graphically by a pole-zero plot as shown in Figure 6.1.

This filter has a pole at $z = W_N^{-K}$

$$e^{j\omega K} = e^{j2\pi K/N}$$

or
$$\omega_K = \frac{2\pi K}{N}$$

It means it has a pole on the unit circle at the frequency $\omega_K = \dfrac{2\pi K}{N}$.

Thus, the entire DFT can be computed by passing the block of input data into a parallel bank of N single-pole filters.

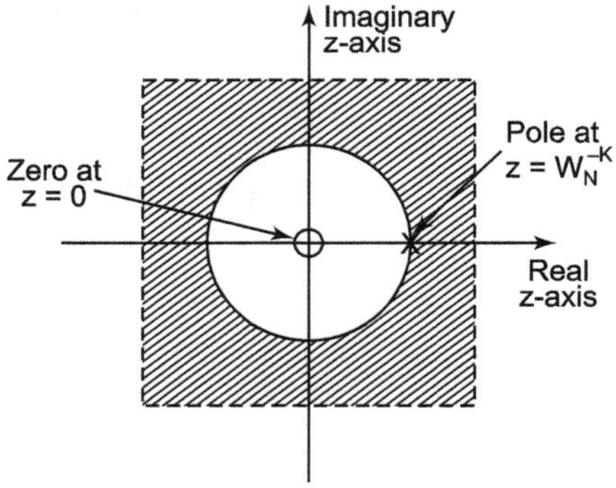

FIGURE 6.1 Pole-zero plot for above z-transform $Hg(z)$.

$$H_K(z) = \frac{1}{1 - W_N^{-K} z^{-1}}$$

or
$$\frac{Y_K(z)}{S_K(z)} = \frac{1}{1 - W_N^{-K} z^{-1}}$$

or
$$Y_K(z)\left[1 - W_N^{-K} z^{-1}\right] = S_K(z)$$

or
$$Y_K(z) - W_N^{-K} z^{-1} Y_K(z) = S_K(z)$$

or
$$Y_K(z) = W_N^{-K} z^{-1} Y_K(z) + S_K(z) \tag{6.12}$$

Taking the inverse z-transform of above Eq. (6.12), we get

$$y_K(n) = W_N^{-K} y_K(n-1) + S_K(n) \tag{6.13}$$

When initial conditions are zero, $y(-1) = 0$

The desired output is $S(K) = y_K(n)|_{n=N} = y_K(N)$ (6.14)

for $K = 0, 1, \ldots, N-1$.

Since both the input $s(n)$ and W_N^{-K} are complex, the computation of each new value of $y_K(n)$ requires four real multiplications and four real additions.

System function of first-order filter* is

$$H_K(z) = \frac{1}{1 - W_N^{-K} z^{-1}}$$

and its difference equation is

$$y_K(n) = W_N^{-K} y_K(n-1) + s_K(n)$$

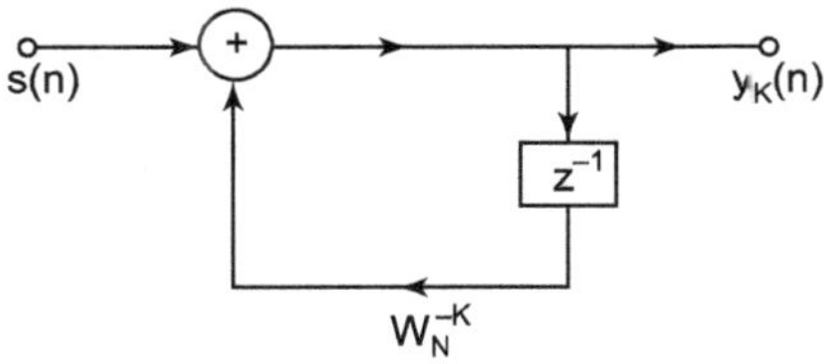

FIGURE 6.2 Flow graph of first-order complex recursive computation of $S(K)$.

The complex multiplications and additions in Eq. (6.13) can be avoided by combining the pairs of single-pole filters as they possess complex-conjugate poles. Complex-conjugate poles are those poles whose imaginary parts have opposite signs. For example, one pole is given by $p_1 = a + jb$ then its complex-conjugate pole will be given by $p_2 - p_1^* = a - jb$.

We can get a second-order filter from a first-order filter by multiplying both the numerator and denominator of a system function of the first-order filter by a factor $\left(1 - W_N^K z^{-1}\right)$.

$$H_K(z) = \frac{1 - W_N^{-K} z^{-1}}{\left(1 - W_N^K z^{-1}\right)\left(1 - W_N^K z^{-1}\right)}$$

$$= \frac{1 - W_N^K z^{-1}}{1 - 2\cos\left(\frac{2\pi K}{N}\right) z^{-1} + z^{-2}} \tag{6.15}$$

This is the system function of a second-order filter. The above system function can be written as follows:

$$H_K(z) = \frac{Y_K(z)}{S(z)} = \frac{1 - W_N^K z^{-1}}{1 - 2\cos\left(\frac{2\pi K}{N}\right) z^{-1} + z^{-2}} \tag{6.16}$$

Eq. (6.16) can also be written as

$$\frac{Y_K(z)}{S(z)} = \frac{1}{1 - 2\cos\left(\frac{2\pi K}{N}\right) z^{-1} + z^{-2}}$$

or

$$\frac{Y_K(z)}{S(z)} \cdot \frac{V_K(z)}{S(z)} = \frac{1}{1 - 2\cos\left(\frac{2\pi K}{N}\right) z^{-1} + z^{-2}}$$

The above equation can be separated into two parts as

$$\frac{Y_K(z)}{V_K(z)} = 1 - W_N^K z^{-1} \tag{6.17}$$

and

$$\frac{V_k(z)}{s(z)} = \frac{1}{1 - 2\cos\left(\frac{2\pi K}{N}\right) z^{-1} + z^{-2}} \tag{6.18}$$

Taking the inverse z-transform of Eqs. (6.17) and (6.18), we get

$$y_K(n) = \upsilon_K(n) - W_N^K \upsilon_K(n-1) \tag{6.19}$$

where all initial conditions are zero.

$$\upsilon_K(n) = 2\cos\frac{2\pi K}{N}\upsilon_K(n-1\} - \upsilon_K(n-2) + s(n) \tag{6.20}$$

where initial conditions are zero.

The Direct Form-II Realization of the system function given by Eq. (6.20) is shown in Figure 6.3.

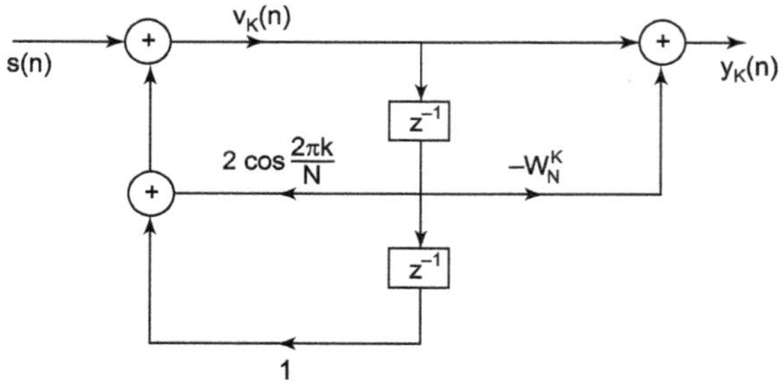

FIGURE 6.3 Flow graph of second-order recursive computation of $S(K)$. *This is an illustration of Goertzel algorithm.*

Only two multiplications are required to implement the poles of the system given in Eq. (6.20). Since the coefficients are real and (–1) is not counted as a multiplication. Four additions are required for implementing the poles. Complex multiplication $(-W_N^K)$ is required to implement zero. This operation is not performed at every iteration of the difference equation.

Goertzel algorithm is especially attractive for DFT computation of a smaller number of points. Number of complex multiplications required for N-point DFT computation using Goertzel algorithm is N_2 and number of additions required is $2N^2$.

From Eq. (6.16), we get

$$\left[1 - 2\cos\left(\frac{2\pi K}{N}\right) z^{-1} + z^{-2}\right] Y_K(z) = \left[1 - W_N^K z^{-1}\right] S(z) \tag{6.21}$$

Taking inverse z-transform of above Eq. (6.21), we get

$$y_K(n) - 2\cos\left(\frac{2\pi K}{N}\right) y_K(n-1) + y_k(n-2) = s(n) - W_N^K s(n-1) \tag{6.22}$$

with initial conditions are zero.

$$y_K(n) = 2\cos\left(\frac{2\pi K}{N}\right)y_K(n-1) - y_k(n-2) + s(n) - W_N^K s(n-1) \qquad (6.23)$$

The signal flow graph of second-order filter is illustrated in Figure 6.3.

6.3 FAST FOURIER TRANSFORM ALGORITHMS

Direct computation of the DFT is less efficient because it does not exploit the properties of symmetry and periodicity of the phase factor $W_N = e-^{2\pi/N}$.

These properties are:

Symmetry property: $W_N^{K+N/2} = -W_N^K$

Periodicity property: $W_N^{K+N} = -W_N^K$

As we already know that all computationally efficient algorithms for DFT are collectively known as FFT algorithms and these algorithms exploit the above two properties of phase factor, W_N.

6.3.1 Classification of FFT Algorithms

According to the storage of the components of the intermediate vector, FFT algorithms are classified into two groups:

1. In-Place FFT algorithms.
2. Natural Input–Output FFT algorithms.

In-Place FFT Algorithms: In this FFT algorithm, component of an intermediate vector can be stored at the same place as the corresponding component of the previous vector.

In-place FFT algorithms reduce the memory space requirement.

Natural Input–Output FFT Algorithms. In this FFT algorithm, both input and output are in natural order. It means both discrete-time sequence $s(n)$ and is DFT $S(K)$ are in natural order. This type of algorithm consumes more memory space for the preservation of natural order of $s(n)$ and $S(K)$.

The disadvantage of an In-place FFT algorithm is that the output appears in an unnatural order necessitating proper shuffling of $s(n)$ or $S(K)$.

In-place FFT algorithms are superior to the natural input–output FFT algorithms although it needs shuffling of $s(n)$ or $S(K)$. This shuffling operation is known as Scrambling.

The scrambled value of an integer I is defined as a new number generated by reversing the order of all bits in the equivalent binary number for that integer.

Another classification of FFT algorithms is based on decimation of $s(n)$ or $S(K)$. Decimation means decomposition into decimal parts.

On the basis of decimation process, FFT algorithms are of two types:

1. Decimation-in-Time FFT algorithms.
2. Decimation-in-Frequency FFT algorithms.

Decimation-in-Time (DIT) FFT Algorithms. In DIT FFT algorithms, the sequence $s(n)$ will be broken up into odd-numbered and even-numbered subsequences.

This algorithm was first proposed by Cooley and Tukey in 1965.

Decimation-in-Frequency (DIP) FFT Algorithms. In DIF FFT algorithms, the sequence $s(n)$ will be broken up into two equal halves. This algorithm was first proposed by Gentlemen and Sande in 1966.

Computation reduction factor of FFT algorithms

$$= \frac{\text{Number of computations required for direct DFT}}{\text{Number of computation required for FFT algorithm}}$$

$$= \frac{N^2}{\frac{N}{2}\log_2(N)}.$$

6.3.2 Number of Stages in DFT Computation using FFT Algorithms

Number of stages in DFT computation using FFT algorithms depends upon the total number of points (N) in a given sequence.

For these algorithms, number of points in a discrete-time sequence,

$$N = 2^r \text{ where } r > 0,$$

r is the number of stages required for DFT computation via FFT algorithms.

Let us have a 8-point discrete-time sequence, $N = 8 = 2^3$. It requries three stages for DFT computations.

In DIT FFT algorithm, input discrete-time sequence $s(n)$ is in Bit-reversed order but output, $S(K)$ is in netural order for in-place computation. In Decimation-in-frequency (DIF) FFT algorithm, input discrete-time sequence $s(n)$ is in natural order but its DFT is in Bit-reversed order for

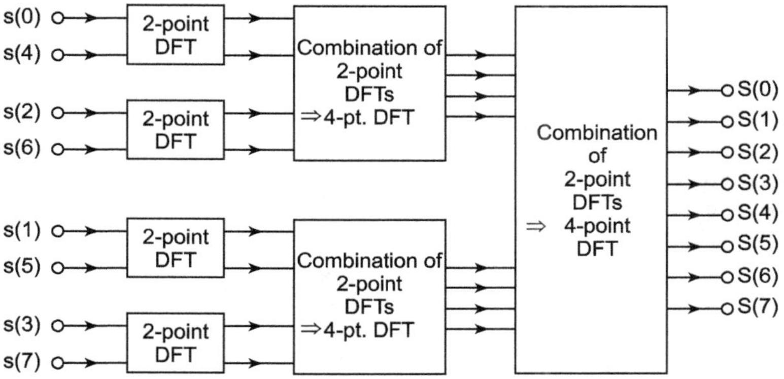

FIGURE 6.4 Three stages in N-point DFT computation via decimation-in-time *FFT* algorithm ($N = 8$).

in-place computation. For in-place computation, smaller memory space is required.

Generally, we use Radix-2 FFT algorithms. In Radix-2 FFT algorithms, the original discrete-time sequence, $s(n)$ is divided into two parts and DFT computation is done on each part separately and the resultant of each part added to get the overall discrete-frequency sequence.

In the DIT FFT algorithm, original sequence $s(n)$ is divided into even-numbered points and odd-numbered points. But in the DIF FFT algorithm, original discrete-time sequence $s(n)$ is divided into two parts as first half and second half. Figure 6.4 illustrates the number of stages required in N-point DFT computation via. DIT FFT algorithm (here $N = 8$).

6.3.3 DIT FFT Algorithm

This FFT algorithm was first proposed by Cooley and Tukey in 1965. Algorithms in which the decomposition is based on breaking the original sequence s(n) into successively smaller subsequences are called DIT FFT algorithms.

The principle of DIT is conveniently explained by considering $N = 2^r$, where $r > 0$. In this algorithm, DFT can be computed by separating original sequence $s(n)$ into two $N/2$-point sequences. These two $N/2$-point sequences consist of Even-numbered and Odd-numbered points in $s(n)$.

DFT of sequence $s(n)$ is given by

$$S(K) = \sum_{n=0}^{N-1} s(n) W_N^{nK}, \qquad K = 0,1,\ldots,N-1 \tag{6.24}$$

where $W_N = e^{-j2\pi/N}$ is called the phase factor.

Now separating $s(n)$ into even- and odd-numbered points, we get

$$S(K) = \sum_{m=0}^{N-1} s(n)W_N^{nK} = \sum_{n \text{ Even}} s(n)W_N^{nK} + \sum_{n \text{ Odd}} s(n)W_N^{nK} \tag{6.25}$$

Substituting $n = 2m$ for even and $n = 2m + 1$ for n odd in Eq. (6.25), we get

$$S(K) = \sum_{m=0}^{\frac{N}{2}-1} s(2m)W_N^{2mK} + \sum_{m=0}^{\frac{N}{2}-1} s(2m+1)W_N^{(2m+1)K}$$

$$= \sum_{m=0}^{\frac{N}{2}-1} s(2m)W_N^{2mK} + \sum_{m=0}^{\frac{N}{2}-1} s(2m+1)W_N^{2mK}W_N^{K}$$

$$= \sum_{m=0}^{\frac{N}{2}-1} s(2m)W_N^{2mK} + W_N^{K}\sum_{m=0}^{\frac{N}{2}-1} s(2m+1)W_N^{2mK}$$

Since $\quad W_N^2 = \left(e^{-j2\pi/N}\right)^2 = e^{(-j2\pi)(N/2)} = W_{N/2}$

Putting the value of W_N^2 in above equation, we get

or $$S(K) = \sum_{m=0}^{\frac{N}{2}-1} s(2m)W_{N/2}^{mK} + W_N^{K}\sum_{m=0}^{\frac{N}{2}-1} s(2m+1)W_{N/2}^{mK}$$

$$= G(K) + W_N^K H(K)$$

or $$S(K) = G(K) + W_N^K H(K) \tag{6.26}$$

where $G(K) = \sum_{m=0}^{\frac{N}{2}-1} s(2m)W_{N/2}^{mK}$ DFT of even-numbered points of sequence $s(n)$.

$H(K) = \sum_{m=0}^{\frac{N}{2}-1} s(2m+1)W_{N/2}^{mK}$ DFT of odd-numbered points of sequence $s(n)$.

Both $s(n)$ and $S(K)$ are periodic with period N, consequently, $G(K)$ and $H(K)$ will be also periodic with period $N/2$.

After computing $N/2$-point DFTs, that is, $G(K)$ and $H(K)$, they are then combined to produce the N-point DFT, $S(K)$. To illustrate the procedure let us consider Eq. (6.26) again

$$S(K) = G(K) + W_N^K H(K), \quad K = 0, 1, \ldots, N-1$$

Now we are considering a case where sequence has 8 points or samples.

$K = 0,\quad S(0) = G(0) + W_N^0 H(0)$

$K = 1,\quad S(1) = G(1) + W_N^1 H(1)$

$K = 2,\quad S(2) = G(2) + W_N^2 H(2)$

$K = 3,\quad S(3) = G(3) + W_N^3 H(3)$

$K = 4,\quad S(4) = G(4) + W_N^4 H(4) = G(0) + W_N^4 H(0)$

Since $G(4) = G(0 + 4) = G(0); H(4) = (0 + 4) = H(0)$

$K = 5,\quad S(5) = G(5) + W_N^5 H(5) = G(1) + W_N^5 H(1)$

Since $G(5) = G(1 + 4) = G(1); H(5) = (1 + 4) = H(1)$

$K = 6,\quad S(6) = G(6) + W_N^6 H(6) = G(2) + W_N^6 H(2)$

Since $G(6) = G(2 + 4) = G(2); H(6) = (2 + 4) = H(2)$

$K = 7,\quad S(7) = G(7) + W_N^7 H(7) = G(3) + W_N^7 H(3)$

Since $G(7) = G(3 + 4) = G(3); H(7) = (3 + 4) = H(3)$

Note. $G(K)$ and $H(K)$ are periodic with period $N/2$. Above sequence has 8 points then its DFT also has 8 points. Therefore $G(K)$ and $H(K)$ are also periodic with period $N/2$, that is, $8/2 = 4$.

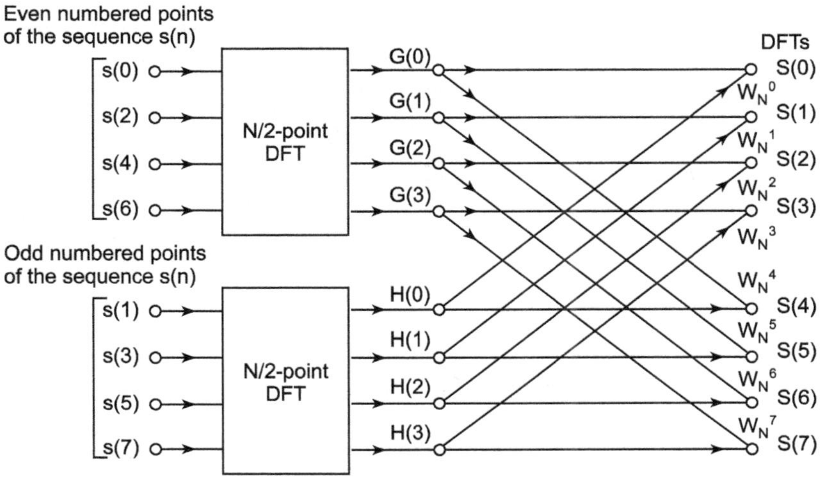

FIGURE 6.5 Flow graph of the DIT decomposition of an N-point DFT computation into two $N/2$-point DFT computation ($N = 8$).

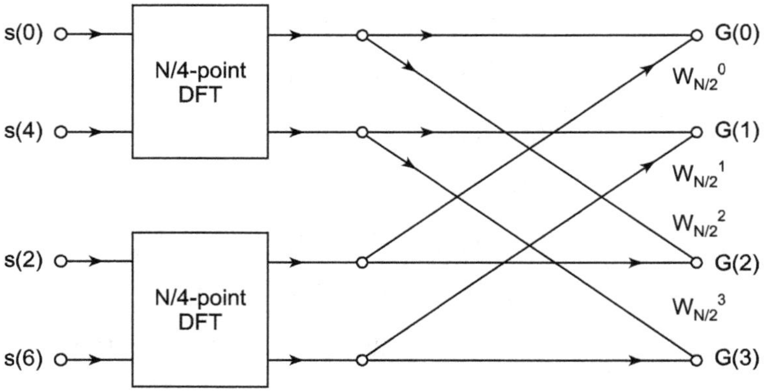

FIGURE 6.6 Flow graph of the decimation-in-time decomposition of an $N/2$-point DFT computation into two $N/4$ point DFT computations (where $N = 8$).

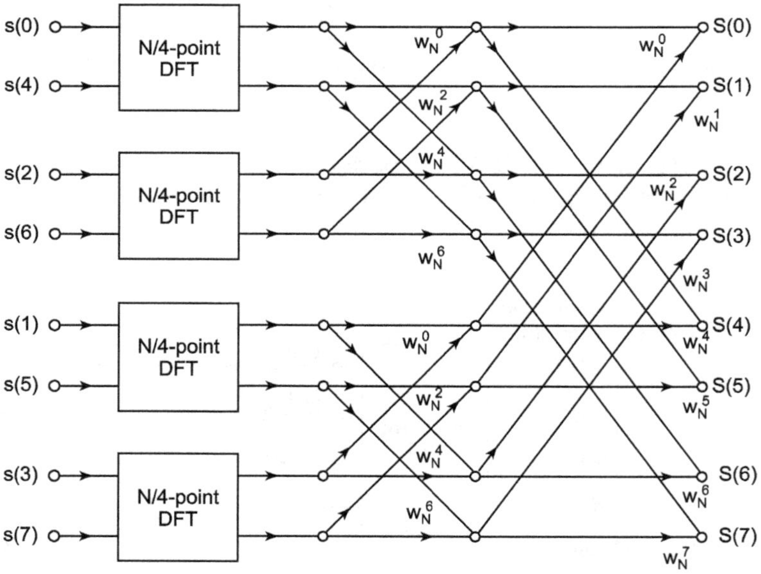

FIGURE 6.7 Result of substituting Figure 6.6 into Figure 6.5.

or

$$G\left(K+\frac{N}{2}\right)=G(K), H\left(K+\frac{N}{2}\right)=H(K)$$

Figure 6.6 illustrates the flow graph of DIT decomposition of N-point DFT computation into two $N/2$-point DFT computation (For $N = 8$)

$G(K)$ and $H(K)$ in Eq. (6.26) can be computed as follows:

$$G(K) = \sum_{m=0}^{N/2-1} g(m)W_{N/2}^{mK} + \sum_{m \text{ Even}} g(m)W_{N/2}^{mK} + \sum_{m \text{ Odd}} g(m)W_{N/2}^{mK}$$

Substituting $m = 2l$ for m even and $m = 2l + 1$ for m odd in above equation, we get

$$G(K) = \sum_{l=0}^{N/4-1} g(2l)W_{N/2}^{2lK} + \sum_{l=0}^{N/4-1} g(2l+1)W_{N/2}^{(2l+1)K}$$

$$= \sum_{l=0}^{N/4-1} g(2l)W_{N/4}^{lK} + W_{N/2}^{K} \sum_{l=0}^{N/4-1} g(2l+1)W_{N/4}^{lK} \quad (6.27)$$

Similarly,

$$H(K) = \sum_{l=0}^{N/4-1} g(2l)W_{N/4}^{lK} + W_{N/2}^{K} \sum_{l=0}^{N/4-1} h(2l+1)W_{N/4}^{lK} \quad (6.28)$$

4-point DFT is computed according to Eqs. (6.27) and (6.28). This computation is shown in Figure 6.6.

After inserting Figure 6.6 into Figure 6.5, we get Figure 6.7.

Two-point DFT is computed $s(0)$ as given in Figure 6.8. It is also known as butterfly computation because of its appearance.

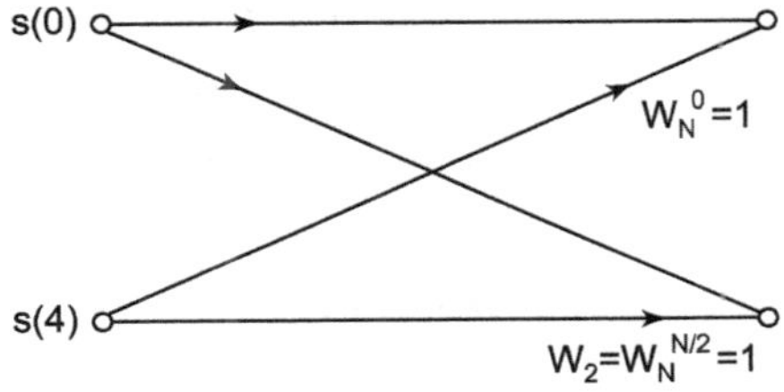

FIGURE 6.8 Flow graph of 2-point DFT.

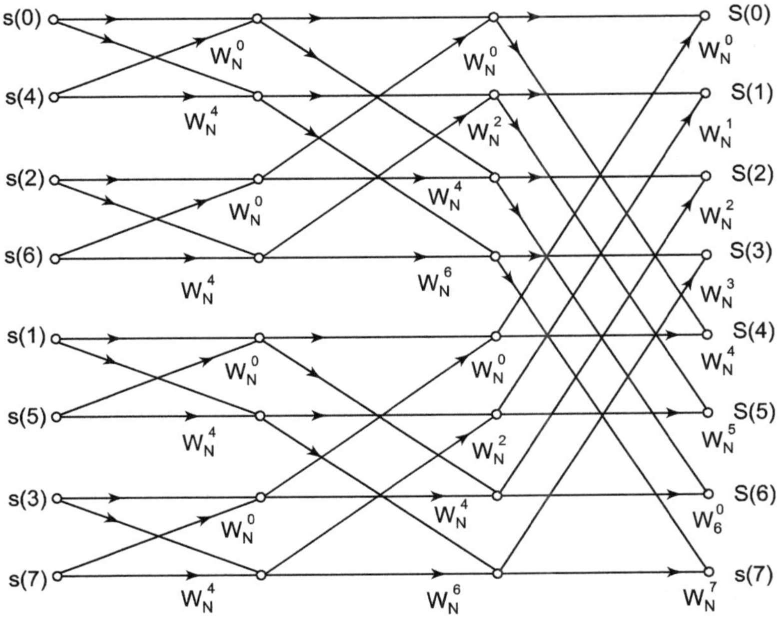

FIGURE 6.9 Flow graph of complete decimation-in-time of an 8-point DFT computation.

Substituting Figure 6.8 into Figure 6.7, we get Figure 6.9 which is a complete DIT decomposition of 8-point DFT computation.

6.3.4 Steps for Computation of DIT FFT Algorithm

Followings are the important steps involved in the computation of DIT FFT algorithms:

1. Data shuffling must be done first. It is performed by Bit Reversal.
2. Number of stages in the calculation is r = log2 N, where N = number points in DFT.
3. 2-point DFT is to be performed at the first stage which involves no multiplication, but it requires some addition and subtraction.
4. General formulation for DIT FFT algorithm provides us a readymade formula for calculation of phase rotation factors required for each butterfly.

EXAMPLE 6.1

Find the DFT of the following discrete-time sequence

$$s(n)=\{1,-1,-1,-1,1,1,1,-1\}$$

using Radix-2 DIT FFT algorithm.

Solution:

The twiddle factor or phase rotation factor $W_N = e^{-j2\pi/N}$ involved in the FFT calculation are found out as follows for $N = 8$.

$$W_N = \text{Phase rotation factor} = e^{-j2\pi/N}$$

$$W_8^0 = e^{-j(2\pi/8)N} = e^0 = 1$$

$$W_8^1 = e^{-j(2\pi/8).1} = e^{-j\pi/4} = \cos\frac{\pi}{4} - j\sin\frac{\pi}{4} = \frac{1-j}{\sqrt{2}}$$

$$W_8^2 = e^{-j(2\pi/8).2} = e^{-j\pi/2} = \cos\frac{\pi}{3} - j\sin\frac{\pi}{2} = -j$$

$$W_8^3 = e^{-j(2\pi/8).3} = e^{-j3\pi/4} = \frac{-1-j}{\sqrt{2}}$$

In Radix-2 DIT FFT algorithm, original sequence $s(n)$ is decomposed into two parts even-numbered point and odd-numbered point sequence. This is done very easily by bit reversal of original discrete-time sequence. Flow

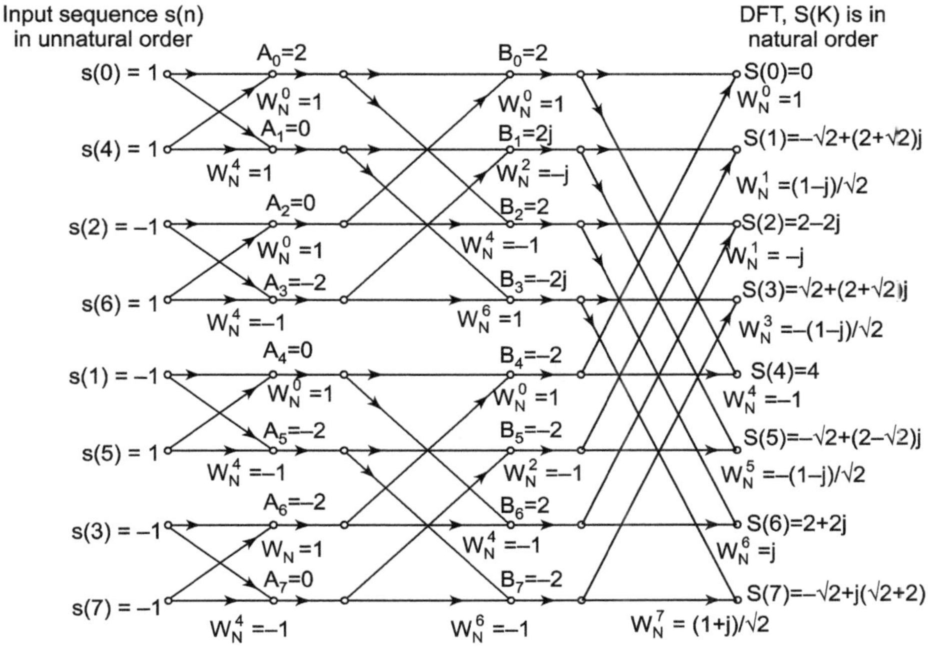

FIGURE 6.10 Flow graph of Radix-2 decimation-in-time FFT algorithm for N = 8.

graph of Radix-2 DIT FFT algorithm is given in Figure 6.10. Determination of DFT using Radix-2 DIF FFT algorithm requires three stages because the number of points in a given sequence is 8, that is, $2^3 = 2r = N = 8$, where r is number of stages required = 3.

Stage I:

$$A_0 = s(0) + s(4)\,W_N^0 = 1 + 1 \times 1 = 2$$

$$A_1 = s(0) + s(4)\,W_N^4 N = 1 + 1 \times (-1) = 0$$

$$A_2 = s(2) + s(6)\,W_N^0 = -1 + 1 \times (1) = 0$$

$$A_3 = s(2) + s(6)\,W_N^4 = -1 + 1 \times (-1) = -2$$

$$A_4 = s(1) + s(5)\,W_N^0 = -1 + 1 \times (1) = 0$$

$$A_5 = s(1) + s(5)\,W_N^4 = -1 + 1 \times (-1) = -2$$

$$A_6 = s(3) + s(7)\,W_N^4 = -1 + (-1) \times 1 = -2$$

$$A_7 = s(3) + s(7)\,W_N^4 = -1 + (-1) \times (-1) = 0$$

Stage II:

$$B_0 = A_0 + A_2 W_N^0 = 2 + 0 \times 1 = 2$$

$$B_1 = A_1 + A_3 W_N^2 = 0 + (-2)(-j) = 2j$$

$$B_2 = A_0 + A_2 W_N^4 N = 2 + 0 \times (-1) = -2$$

$$B_3 = A_1 + A_3 W_N^6 = 0 + (-2) j = -2j$$

$$B_4 = A_4 + A_6 W_N^0 = 0 + (-2)(1) = 2$$

$$B_5 = A_5 + A_7 W_N^2 = -2 + 0 \times (-j) = 2j$$

$$B_6 = A_4 + A_6 W_N^4 = 0 + (-2)(-1) = 2$$

$$B_7 = A_5 + A_7 W_N^6 = -2 + 0 \times (j) = -2j$$

Stage III:

$$S(0) = B_0 + B_4 W_N^0 = 2 + (-2)(1) = 0$$

$$S(1) = B_1 + B_5 W_N^1 = 2j + (-2)\left(\frac{1-j}{\sqrt{2}}\right) = 2j - \sqrt{2} + \sqrt{2j}$$
$$= \sqrt{2} + \left(2 + \sqrt{2}\right) j$$

$$S(2) = B_2 + B_6 W_N^2 = 2 + (2)(-j) = 2 - 2j$$

$$S(3) = B_3 + B_7 W_N^3 = 2j + (-2)\left(\frac{-1-j}{\sqrt{2}}\right)$$
$$= -2j + \sqrt{2} + \sqrt{2j} = \sqrt{2} + \left(-2 + \sqrt{2}\right) j$$

$$S(4) = B_0 + B_4 W_N^4 = 2 + (-2)(-1) = 4$$

$$S(5) = B_1 + B_5 W_N^5 = 2j + (-2)\left(\frac{-(1-j)}{\sqrt{2}}\right)$$
$$= -2j + \sqrt{2} - \sqrt{2j} = \sqrt{2} + \left(2 + \sqrt{2}\right) j$$

$$S(6) = B_2 + B_6 W_N^6 = 2 + 2(j) = 2 + 2j$$

$$S(7) = B_0 + B_7 W_N^5 = -2j + (-2)\frac{(1-j)}{\sqrt{2}} = -2j - \sqrt{2} - \sqrt{2j}$$
$$= \sqrt{2} - \left(2 + \sqrt{2}\right) j$$

Resultant discrete-frequency sequency will be

TABLE 6.1 Phase Rotation Factors for Quick Computation

Number of points in DFT, N	Stage 1	Stage 2	Stage 3	Stage 4	Stage 5
4 No. of stages = 2	Twiddle factor not required	W_4^0, W_4^1	—	—	—
8 No. of stages = 3	Twiddle factor not required	W_8^0, W_8^2	W_8^0, W_8^1 W_8^2, W_8^3	—	—
16 No. of stages = 4	Twiddle factor not required	W_{16}^0, W_{16}^4	W_{16}^0, W_{16}^2 W_{16}^4, W_{16}^6	W_{16}^0, W_{16}^0 W_{16}^2, W_{16}^3 W_{16}^4, W_{16}^5 W_{16}^6, W_{16}^7	—
32 No. of stages = 4	Twiddle factor not required	W_{32}^0, W_{32}^8	W_{32}^0, W_{32}^4 W_{32}^8, W_{32}^{12}	W_{32}^0, W_{32}^2 W_{32}^{14} W_{16}^4, W_{16}^5 W_{16}^6, W_{16}^7	W_{32}^0, W_{32}^1.... W_{32}^{15}

$$
\begin{aligned}
(K) &= \{S(0), S(1), S(2), S(3), S(4), S(5), S(6), S(7)\} \\
&= \left\{0, -\sqrt{2} + \left(2 + \sqrt{2}\right) j, 2 - 2j, \sqrt{2} + \left(-2 + \sqrt{2}\right) j, 4, \right. \\
&\quad \left. \sqrt{2} + \left(2 - \sqrt{2}\right) j, 2 + 2j - \sqrt{2} - \left(2 + \sqrt{2}\right) j \right\}
\end{aligned}
$$

6.3.5 Decimation-in-Frequency FFT Algorithm

In DIF FFT algorithm, the output DFT sequence $S(K)$ is broken into smaller and smaller subsequences. For the derivation of this algorithm, the number of points or samples in a given sequence should be $N - 2^r$ where $r > 0$. For this purpose, we can first divide the input sequence into the first-half and the second-half of the points.

DFT of sequence $s(n)$ is given by

$$S(K) = \sum_{n=0}^{N-1} s(n) W_N^{nK}, \quad K = 0, 1, \ldots N-1 \tag{6.29}$$

where $WN = e{-j}^{2\pi/N}$ is called the phase factor.

$$= \sum_{\text{Ist Half}} s(n) W_N^{nK} + \sum_{\text{IInd Half}} s(n) W_N^{nK}$$

$$= \sum_{m=0}^{N/2-1} s(n)W_N^{nK} + \sum_{n=N/2}^{N-1} s(n)W_N^{nK}$$

Putting $n = n + \frac{N}{2}$ in the second part of the above equation, we get

$$S(K) = \sum_{n=0}^{N/2-1} s(n)W_N^{nK} + \sum_{n=0}^{N/2-1} s\left(n+\frac{N}{2}\right)W_N^{(n+N/2)K}$$

or
$$S(K) = \sum_{n=0}^{N/2-1} s(n)W_N^{nK} + W_N^{(N/2)K}\sum_{n=0}^{N/2-1} s\left(n+\frac{N}{2}\right)W_N^{nK}$$

Since $W_N^{(N/2)K} = (1)^K$. Substituting the value of $W_N^{NK/2}$ in above equation, we get

$$S(K) = \sum_{n=0}^{N/2-1} s(n)W_N^{nK} + (-1)\sum_{n=0}^{N/2-1} s\left(n+\frac{N}{2}\right)W_N^{nK} \tag{6.30}$$

or
$$S(K) = \sum_{n=0}^{N/2-1}\left[s(n)(-1)^K s\left(n+\frac{N}{2}\right)\right]W_N^{nK} \tag{6.31}$$

Putting, $S(2m)$ for K Even — It is for even-numbered points of DFT
$S(2m + 1)$ for K Odd — It is for odd-numbered points of DFT in Eq. (6.31), we get

$$S(2m) = \sum_{n=0}^{N/2-1}\left[s(n)+s\left(n+\frac{N}{2}\right)\right]W_N^{2mn} = \sum_{n=0}^{N/2-1} g(n)W_{N/2}^{mn} \tag{6.32}$$

$$S(2m+1) = \sum_{n=0}^{N/2-1}\left[s(n)-s\left(n+\frac{N}{2}\right)\right]W_N^{(2m+1)n}$$

$$= \sum_{n=0}^{N/2-1}\left[s(n)-s\left(n+\frac{N}{2}\right)\right]W_N^{2mn} = W_{N/2}^{n}$$

or
$$S(2m+1) = \sum_{n=0}^{N/2-1}[h(n)]W_N^{n}W_{N/2}^{mn} \tag{6.33}$$

where
$$g(n) = s(n) + s\left(n+\frac{N}{2}\right)$$

$$h(n) = s(n) - s\left(n+\frac{N}{2}\right)$$

$$W_N^2 = W_{N/2}$$

$$m = 0, 1, \ldots, \left(\frac{N}{2} - 1\right)$$

Eqs. (6.32) and (6.33) can be recognized as $N/2$-point DFTs. The DFT can be computed by first forming the sequences $g(n)$ and $h(n)$, then computing the $N/2$-point DFTs of these two sequences that are called the even-numbered output points and odd-numbered points, respectively.

Eqs. (6.32) and (6.33) are illustrated in Figure 6.11 for 8-point DFT. 8-point DFT is broken into two 4-point DFTs. The first 4-point DFT is for even-numbered points and the second 4-point DFTs are for odd-numbered points of $S(K)$. Now, 4-point DFTs is further broken into two 2-point DFTs in the same manner as given above. It is illustrated in Figure 6.12.

After inserting Figure 6.12 into Figure 6.11, we get Figure 6.13 and this is illustrated as follows. Now 2-point DFT is computed as follows in Figure 6.14.

After inserting Figure 6.14 into Figure 6.13, we get a flow graph of complete DIF decomposition of an N-point DFT computation ($N = 8$) and illustrated in Figure 6.15.

Note. In DIF decomposition, input discrete-time sequence $s(n)$ is in natural order and output DFT sequence $S(K)$ is in Bit-reversed order for an in-place computation. But in DIT decomposition, input discrete-time

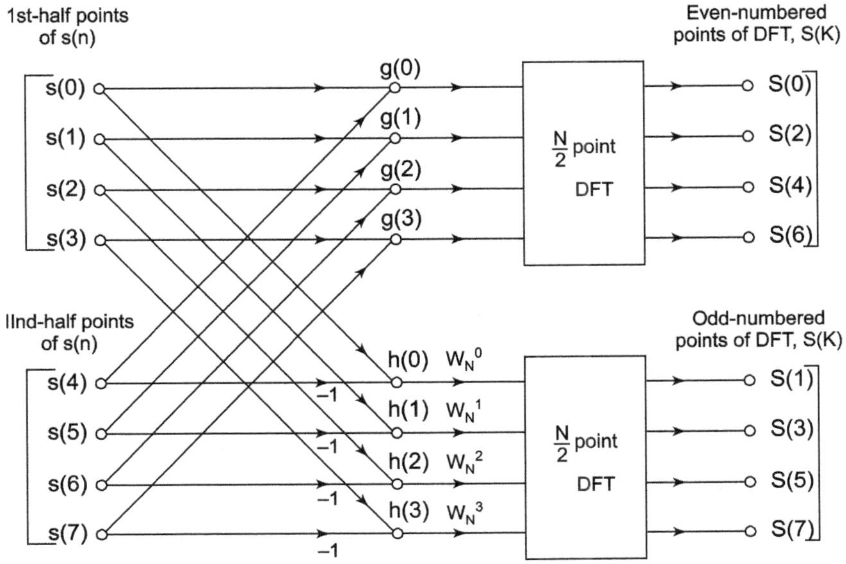

FIGURE 6.11 Flow graph of the decimation-in-frequency decomposition of an N-point DFT compilation into two $N/2$-point. DFT computation. Here $N = 8$.

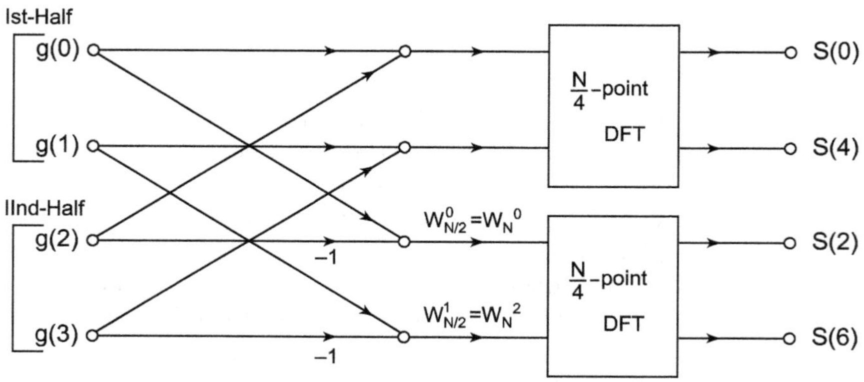

FIGURE 6.12 Flow graph of the decimation-in-frequency decomposition *N*/2-point DFT computation into two *N*/4 point DFT computation Here, *N* = 8.

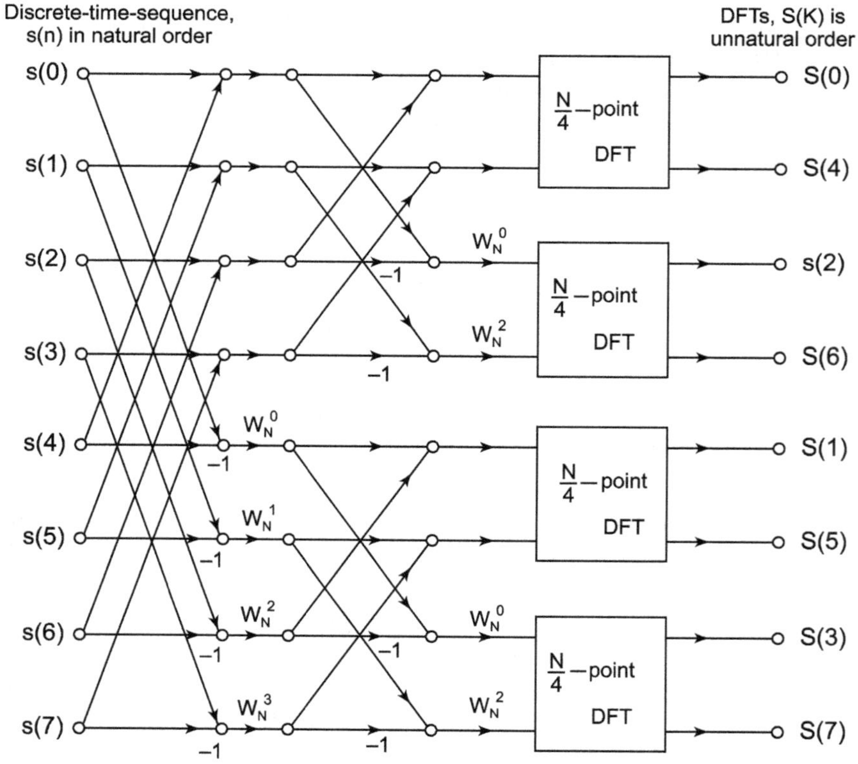

FIGURE 6.13 Flow graph after inserting Figure 6.12 into Figure 6.11.

sequence, $s(n)$ is in Bit-reversed order and output DFT sequence, $S(K)$ is in natural order for an in-place computation.

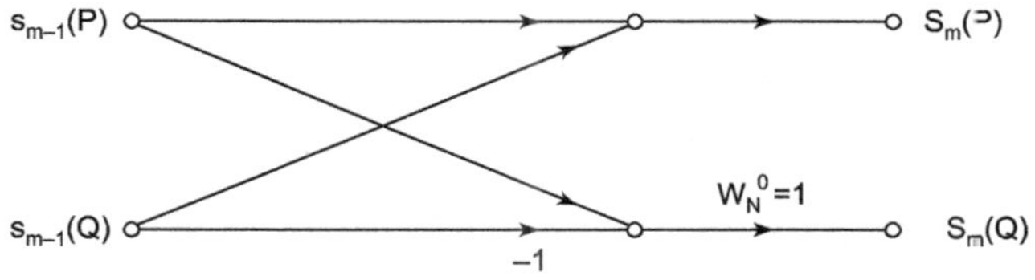

FIGURE 6.14 Flow graph of a 2-point DFT which is required in the last stage of decimation-in-frequency decomposition.

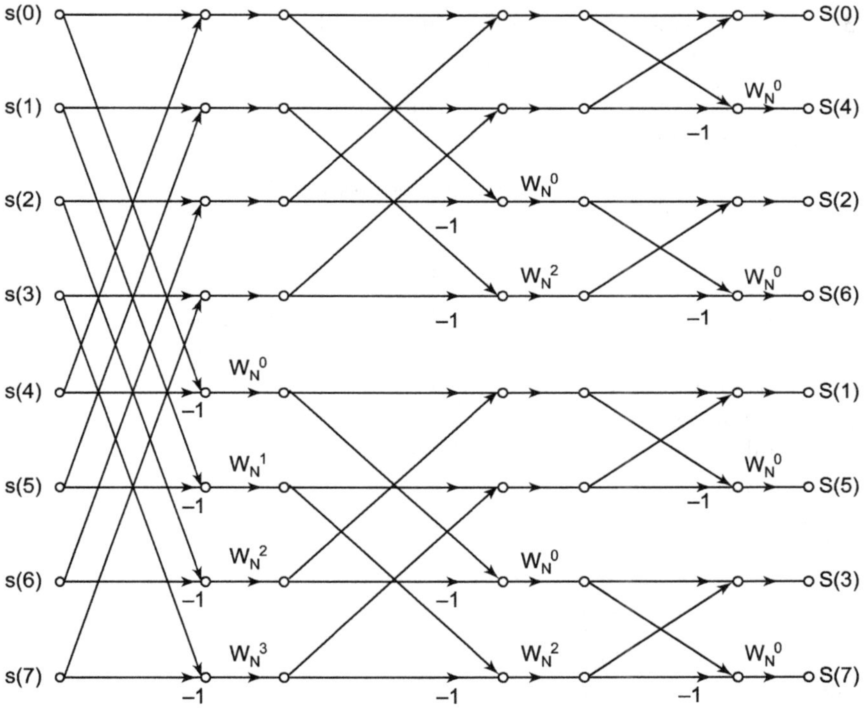

FIGURE 6.15 Flow graph of complete DIF decomposition of an *N*-point DFT computation (*N* = 8).

6.3.6 Steps for Computation of DIF FFT Algorithm

Given below are the important steps for the computation of DIF FFT algorithms.

1. Data shuffling is not required but whole sequence is divided into two parts: first half and second half. From these we calculate $g(n)$ and $h(n)$ as follows:

$$g(n) = s(n) + s\left(n + \frac{N}{2}\right)$$

and $$h(n) = s(n) - s\left(n + \frac{N}{2}\right)$$

where $$n = 0, 1, \ldots, \frac{N}{2} - 1.$$

2. From these two sequences, we compute $N/2$-point DFTs. Before computation of $N/2$-point DFT for $h(n)$, it is multiplied by W_N^n.
3. Number of stages required in the calculation of N-point DFT is $r = \log_2 N$, where N is the number of points in DFT.
4. $N/2$-point DFT is to be performed at the first stage which involves $N/2$ multiplication.
5. In the DIF algorithm, twiddle factors can be found from the previous Table (6.1) in the reverse order.
6. Finally data shuffling is performed. It is also performed by Bit reversal.

Bit Reversal. In bit reversal, MSB (most significant bit) and LSB (least significant bit) of data are exchanged. Bit reversal is helpful in shuffling the data in DIT and DIF FFT algorithms.

If the number of bits required to convert decimal numbers into binary = 3.

Total number of samples = $2^3 = 8$.

We can understand bit reversal very easily with the help following example for an 8 point discrete-time sequence.

Original sequence				Bit-reversed sequence		
Decimal		Original Binary		Bit-reversed Binary		Decimal
$s(0)$	$\Rightarrow$	$s(000)$	$\rightarrow$	$s(000)$	$\Rightarrow$	$s(0)$
$s(1)$	$\Rightarrow$	$s(001)$	$\rightarrow$	$s(100)$	$\Rightarrow$	$s(4)$
$s(2)$	$\Rightarrow$	$s(010)$	$\rightarrow$	$s(010)$	$\Rightarrow$	$s(2)$
$s(3)$	$\Rightarrow$	$s(011)$	$\rightarrow$	$s(110)$	$\Rightarrow$	$s(6)$
$s(4)$	$\Rightarrow$	$s(100)$	$\rightarrow$	$s(001)$	$\Rightarrow$	$s(1)$
$s(5)$	$\Rightarrow$	$s(101)$	$\rightarrow$	$s(101)$	$\Rightarrow$	$s(5)$
$s(6)$	$\Rightarrow$	$s(110)$	$\rightarrow$	$s(011)$	$\Rightarrow$	$s(3)$
$s(7)$	$\Rightarrow$	$s(111)$	$\rightarrow$	$s(111)$	$\Rightarrow$	$s(7)$

EXAMPLE 6.2

Find the DFT of the following discrete-time sequence

$$s(n)=\{1,-1,-1,-1,1,1,1,-1\}$$

using Radix-2 DIF FFT algorithm.

Solution.

The twiddle factor or phase rotation factor $W_N = e^{-j2\pi/N}$ involved in the FFT calculation are found out as follows for $N = 8$.

$$W_N = \text{Phase rotation factor } e^{-j2\pi/N}.$$

$$W_8^0 = e^{-j(2\pi/8)0} = e^0 = 1$$

$$W_8^1 = e^{-j(2\pi/8)1} = e^{-j\pi/4} = \frac{1-j}{\sqrt{2}}$$

$$W_8^2 = e^{-j(2\pi/8)2} = e^{-j\pi/2} = -j$$

$$W_8^3 = e^{-j(2\pi/8)3} = e^{-j3\pi/4} = \frac{-(1+j)}{\sqrt{2}}$$

In Radix-2 DIF FFT algorithm, original sequence $s(n)$ is decomposed into two subsequences as first half and second half of a sequence. There is no need to reordering (shuffling) the original sequence as in Radix-2 DIT FFT algorithm. But resultant discrete-frequency sequence is shuffled (reordered) into natural order because these are obtained in unnatural order. Flow graph of Radix-2 DIF FFT

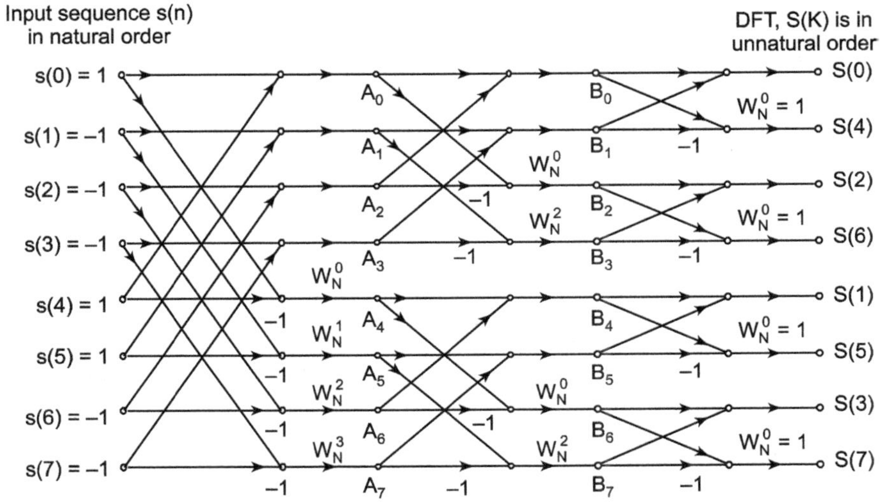

FIGURE 6.16 Flow graph of Radix-2 DIF FFT algorithm $N = 8$.

algorithm for $N = 8$ is shown in Figure 6.11. Determination of DFT using Radix-2 DIF FFT algorithm requires three stages because the number of points in a given sequence is 8, that is, $2^3 = 2^r = N = 8$, where r is number of stages required = 3.

Stage I:

$$A_0 = s(0) + s(4) = 1 + 1 = 2$$
$$A_1 = s(1) + s(5) = -1 + 1 = 0$$
$$A_2 = s(2) + s(6) = -1 + 1 = 0$$
$$A_3 = s(3) + s(7) = -1 - 1 = -2$$
$$A_4 = [s(0) + (-1)\, s(4)]\ W_8^0 = [-1 + (-1)\,(1)] \times 1 = 0$$
$$A_5 = [s(1) + (-1)\, s(5)]\ W_8^1 = [-1 + (-1)\,(1)]\ \frac{(1-j)}{\sqrt{2}} = -\sqrt{2}(1-j)$$
$$A_6 = [s(2) + (-1)\, s(6)]\ W_8^2 = [-1 + (-1) \times 1](-j) = 2j$$
$$A_7 = [s(3) + (-1)\, s(7)]\ W_8^3 = [-1 + (-1)(-1)]\ \left\{\frac{-(1-j)}{\sqrt{2}}\right\} = 0$$

Stage II:

$$B_0 = A_0 + A_2 = 2 + 0 = 2$$
$$B_1 = A_1 + A_3 = 0 + (-2) = -2$$
$$B_2 = [A_0 + (-1)A_2]\ W_8^0 = [2 - 0] \times 1 = 2$$
$$B_3 = [A_1 + (-1)A_3]\ W_8^2 = [0 + (-1)\,(-2)] \times (-1) = -2j$$
$$B_4 = A_4 + A_6 = 0 + 2j = 2j$$
$$B_5 = A_5 + A_7 = [-\sqrt{2}\,(1-j)] + 0 = -\sqrt{2}\,(1-j)$$
$$B_6 = [A_4 + (-1)\,A_6]\ W_8^0 = [0 + (-1)\,2j] \times 1 = -2j$$
$$B_7 = [A5 + (-1)\,A_7]\ W_8^2 = [-\sqrt{2}\,(1-j)$$
$$+ (-1) \times 0] \times (-j) = -\sqrt{2}\,(1+j)$$

Stage III:

$$S(0) = B_0 + B_1 = 2 + (-2) = 0$$
$$S(4) = B_0(-1)\,B_1 = 2 + (-1)\,(-2) = 4S$$
$$(2) = B_2 + B_3 = 2 + (-2j) = 2 - 2j$$
$$S(6) = B_2 + (-1)B_3 = 2 + (-1)\,(-2j) = 2 + 2j$$
$$S(1) = B_4 + B_5 = 2j + [-\sqrt{2}(1-j)] = 2j - \sqrt{2} + \sqrt{2}\,j = \sqrt{2} + (2 + \sqrt{2})j$$
$$S(5) = B_4 + (-1)B_5 = 2j + (-1)\,[-\sqrt{2}(1-j)] = 2j - \sqrt{2} + \sqrt{2}\,j$$
$$= \sqrt{2} + (2 + 2\ )j$$
$$S(3) = B_6 + B_7 = -2j + \sqrt{2}(1+j) = -2j + \sqrt{2} + \sqrt{2}\,j = \sqrt{2} + (-2 + \sqrt{2})j$$
$$S(7) = B_6 + (-1)B_7 = -2j + (-1)\,\sqrt{2}(1+j) = -2j - \sqrt{2} + \sqrt{2}\,j$$
$$= -\sqrt{2} - (2 + \sqrt{2}\ )j$$

These are in unnatural order. Resultant discrete-frequency sequence can be written as

$$S(K) = \{S(0),\ S(1),\ S(2),\ S(3),\ S(4),\ S(5),\ S(6),\ S(6),\ S(7)\}$$

or

$$S(K) = \left[0, -\sqrt{2} + \left(2 + \sqrt{2}\right) j, 2 - 2j, \sqrt{2} + \left(-1 + \sqrt{2}\right) j, 4,\right.$$

$$\left.\sqrt{2} + \left(2 - \sqrt{2}\right) j, 2 + 2j, -\sqrt{2} - \left(2\sqrt{2}\right) j\right\}.$$

6.3.7 Number of Complex Multiplications Required in DIF FFT Algorithm

Number of complex multiplications required in DIF FFT algorithm are the same as that required in DIT FFT algorithm. Number of complex multiplication required in these FFT algorithms are $N/2 \log_2 N$, where $N = 2^r$, $r > 0$ and

TABLE 6.2

No. of points (or samples) in a sequence $s(n)$, N	Complex multiplications in direct computation of DFT = $N \times N = A$	Complex multiplications in FFT algorithms $N/2 \log^2 N = B$	Speed improvement Factor = A/B
$2^2 = 4$	16	4	$\frac{16}{4} = 4.0$
$2^3 = 8$	64	12	$\frac{64}{12} = 5.3$
$2^4 = 8$	256	32	$\frac{256}{32} = 8.0$
$2^5 = 32$	1024	80	$\frac{1024}{80} = 12.8$
$2^6 = 64$	4096	192	$\frac{4096}{192} = 21.3$
$2^7 = 128$	16384	448	$\frac{16384}{448} = 36.6$
$2^8 = 256$	65536	1024	$\frac{65536}{1024} = 64.0$
$2^9 = 512$	262144	2304	$\frac{262144}{2304} = 113.8$

N is the total number of points (or samples) in a discrete-time sequence. Thus the total computations (number of multiplication and addition operations) are the same in both FFT algorithms.

Now we will compare the computational complexity for the direct computation of the DFT and FFT algorithm. This comparison is given in Table 6.2.

EXERCISES

1. What is Fast Fourier Transformation (FFT)?
2. Explain Goertzel algorithm.
3. Define following terms:
   (*a*) Decimation-in-time (*b*) Decimation-in-frequency
   (**c**) Phase rotation factor or twiddle factor.
4. Distinguish between decimation-in-time and decimation-in-frequency algorithms.
5. Give classification of FFT algorithms.
6. Give various steps required for the computation of decimation-in-time (DIT) FFT algorithms and decimation-in-frequency (DIF) FFT algorithms.
7. Discuss the number of complex multiplication required in FFT algorithms.
8. Give flow graph of complete decimation-in-time (DIT) FFT algorithm for $N = 8$.
9. Give flow graph of complete decimation-in-frequency (DIF) FFT algorithm for $N = 8$, where N is the total number of samples in DFT.
10. Find the DFT of the following sequences using decimation-in-time (DIT) and decimation-in-frequency (DIF) FFT algorithms

    (*a*) $s(n) = \{1, 1, 1, 1, 1, 1, 1, 1\}$ (*b*) $s(n) = \{1, 0, 0, 0, 1, 1, 1, 0\}$
    (**c**) $s(n) = \{1, 0, 0, 1, -1, 1\}$ (*d*) $s(n) = \{1, 1, 1, 1, 0, 0, 0, 0\}$.

CHAPTER 7

IMPLEMENTATION OF DISCRETE-TIME SYSTEMS

7.1 INTRODUCTION

In this chapter, we will study the realization of linear-time invariant (LTI) discrete-time systems in either software or hardware.

There are various structures for the realization of finite-duration impulse response (FIR) and IIR discrete-time systems. LTI discrete-time systems are characterized by the general linear constant-coefficient difference equation,

$$y(n) = -\sum_{k=1}^{N} A_k y(n-k) + \sum_{k=0}^{M} B_k s(n-k) \tag{7.1}$$

where $s(n)$ and $y(n)$ are the input and output of a discrete-time system, respectively. A_k and B_k are the system coefficients.

Taking the z-transform of Eq. (7.1), we get

$$Y(z) = -\sum_{k=1}^{N} A_k z^{-k} Y(z) + \sum_{k=0}^{M} B_k z^{-k} S(z)$$

or

$$\frac{Y(z)}{S(z)} = \frac{\sum_{k=0}^{M} B_k z^{-k}}{1 + \sum_{k=0}^{N} A_k z^{-k}}$$

But the system function is given by

$$H(z) = \frac{z\text{-tranform of output } y(n)}{z\text{-tranform of input } s(n)} = \frac{Y(z)}{S(z)}$$

or
$$\boxed{H(z)=\frac{Y(z)}{S(z)}=\frac{\sum_{k=0}^{M-1} B_k z^{-k}}{1+\sum_{k=1}^{N-1} A_k z^{-k}}} \quad (7.2)$$

where $H(z)$ is the system function of the LTI discrete-time system.

Zeros and poles of the system function $H(z)$ depend on the choice of the system parameters (B_k) and (A_k) and these parameters determine the frequency response characteristics of the system.

Network structure in block diagram form implies a hardware configuration for the realization of the discrete-time system.

EXAMPLE 7.1

Find out the system function of the system for which output and input is given as

$$y(n)=\frac{5}{6}y(n-1)-\frac{1}{6}y(n-2)+s(n)$$

Also, draw a pole-zero plot for the above system function.

Solution:

The system function is defined as the ratio of z-transform of output and z-transform of input of a system keeping all initial conditions zero.

$$H(z)=\frac{Y(z)}{S(z)}=\frac{\mathcal{Z}[y(n)]}{\mathcal{Z}[s(n)]}$$

where, $Y(z) = z$-transform of $y(n)$

$S(z) = z$-transform of $s(n)$

Given: $$y(n)=\frac{5}{6}y(n-1)-\frac{1}{6}y(n-2)+s(n)$$

Taking the z-transform of both sides of above difference equation, we get

$$Y(z)=\frac{5}{6}z^{-1}Y(z)-\frac{1}{6}z^{-2}Y(z)+S(z)$$

or $$\frac{Y(z)}{S(z)}=\frac{1}{1-\frac{5}{6}z^{-1}+\frac{1}{6}z^{-2}}$$

or $$H(z)=\frac{Y(z)}{S(z)}=\frac{z^2}{z^2-\frac{5}{6}z+\frac{1}{6}}=\frac{6z^2}{6z^2-5z+1}$$

$$= \frac{6z^2}{6z^2 - 2z - 3z + 1} = \frac{6z^2}{6z(3z-1) - 1(3z-1)}$$

or
$$H(z) = \frac{Y(z)}{S(z)} = \frac{6z^2}{(2z-1)(3z-1)}$$

This is the system function of the given difference equation.

$H(z)$ can be written in terms of two polynomials one is called *numerator polynomial* $N(z)$ and the other is called *denominator polynomial* $D(z)$.

$$H(z) = \frac{N(z)}{D(z)} = \frac{6z^2}{(2z-1)(3z-1)}$$

The roots of the numerator polynomial of a system function $H(z)$ is called zeros of $H(z)$ and the roots of the denominator polynomial of a system function $H(z)$ is called poles of $H(z)$.

In this problem

$$H(z) = \frac{N(z)}{D(z)} = \frac{6z^2}{(2z-1)(3z-1)}$$

Zeros of $H(z) : N(z) = 0$

$$6z^2 = 0$$

$$z = 0$$

$H(z)$ will have one zeros at $z = 0$

Poles of $H(z) : D(z) = 0$

$$(2z-1)\,(3z-1) = 0$$

or
$$2z - 1 = 0$$

$$z = \frac{1}{2}$$

and
$$3z - 1 = 0$$

$$z = \frac{1}{3}$$

$H(z)$ will have two poles, one at

$$z = \frac{1}{2} \text{ and other at } z = \frac{1}{3}.$$

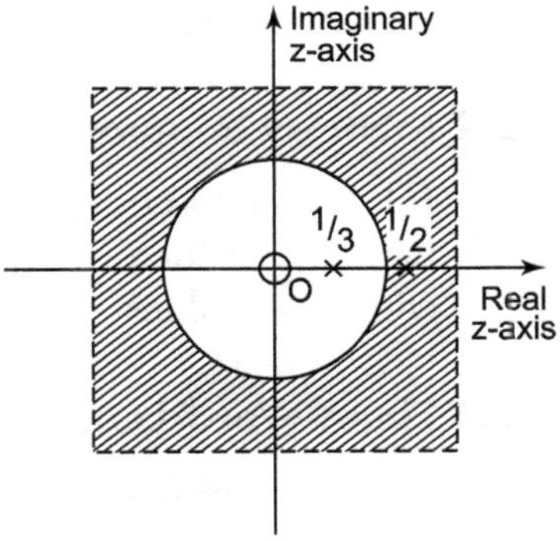

FIGURE 7.1 Pole-zero plot for $H(z) = \dfrac{6z^2}{(2z-1)(3z-1)}$

There poles and zeros of $H(z)$ can be shown by a pole-zero plot. It is shown in Figure 7.1.

Note. The location of zeros is shown by circle (O) and location of poles are shown by cross ($\times$) in the pole-zero plot.

7.2 MAJOR FACTORS INFLUENCING OUR CHOICE OF SPECIFIC REALIZATION

There are three major factors that influence our choice of specific realization. There are

1. Computation complexity.
2. Memory requirements.
3. Finite-word-length effects in the computations.

7.2.1 Computational Complexity

It is defined as the number of arithmetic operations required to compute an output value $y(n)$ for the system. Arithmetic operations are multiplications, divisions, additions and subtractions. Nowadays some other operations are also included for the measurement of computational complexity. These operations are how many times a fetch from memory is performed or how many times a comparison between two numbers is performed per output sample.

7.2.2 Memory Requirements

It is defined as the number of memory locations required to store the system parameters such as previous inputs, previous outputs and any intermediate computed values.

7.2.3 Finite-Word-Length Effects in the Computations

These are also known as finite precision effects. These are defined as the rounding-off effects that occur in any digital implementation of the system, either in hardware or software. The computations that are performed in the process of determining an output from the system must be rounded off to fit within the limited precision constraints of the hardware used in the implementation. We select a realization that is not very sensitive to finite-word-length effects.

7.3 NETWORK STRUCTURES FOR IIR SYSTEMS

Infinite-duration Impulse Response (IIR) systems are described by the system function given by Eq. (7.2), as

$$\boxed{H(z)=\frac{Y(z)}{S(z)}=\frac{\sum_{k=0}^{M-1} B_k z^{-k}}{1+\sum_{k=1}^{N-1} A_k z^{-k}}}$$

where M and N are integer numbers.

This system can be realized by several types of network structures. These are given as follows:

1. Direct-form network structures
2. Transposed-form network structures
3. Cascade-form network structures
4. Parallel-form network structures.

These are discussed one by one in detail in the following subsections.

7.3.1 Direct-Form Network Structures

As already stated that the rational system function of Eq. (7.2) characterizes an IIR system. This system can be viewed as cascading of two systems $H_1(z)$ and $H_2(z)$,

$$H(z)=\frac{\sum_{k=0}^{M-1} B_k z^{-k}}{1+\sum_{k=1}^{N-1} A_k z^{-k}}=H_1(z).H_2(z) \tag{7.3}$$

where $$H_1(z)=\sum_{k=0}^{M-1} B_k z^{-k} \tag{7.4}$$

and $$H_2(z)=\frac{1}{1+\sum_{k=1}^{N-1} A_k z^{-k}}=\left[1+\sum_{k=1}^{N} A_k z^{-k}\right]^{-1}$$

$$=1-\sum_{k=1}^{N} A_k z^{-k}=1+\sum_{k=1}^{N}(-A_k)z^{-k} \tag{7.5}$$

There will be two different direct-form realizations, characterized by whether $H_1(z)$ proceeds $H_2(z)$ or whether $H_2(z)$ proceeds $H_1(z)$. Since $H_1(z)$ is an FIR system because it has a finite number of filter coefficients B_K. Figure 7.2 illustrates the block diagram of two types of direct-form realizations.

Direct-Form I Realization: By attaching the all-pole system $H_2(z)$ in cascade (series) with All-zero system $H_1(z)$, we obtain the direct-form I realization as shown in Figure 7.3(*b*). This realization requires $M + N - 1$ multiplications and $M + N$ additions. It requires $M + N - 1$ memory locations. All-pole system is a system that has only poles in its system function but does not possess any zeros in it. All-zero system is a system that has only zeros in its system function but does possess any pole in its system function.

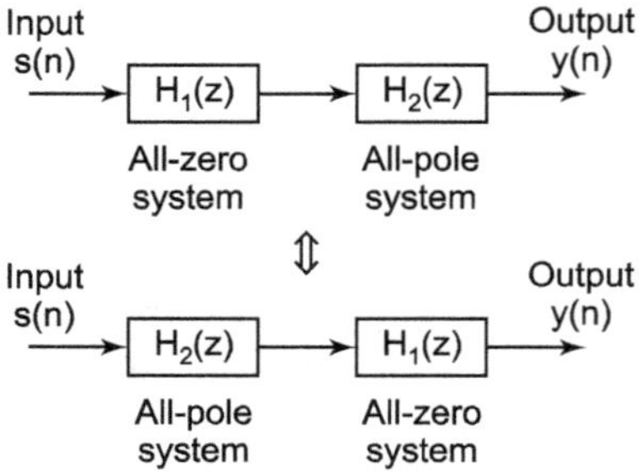

FIGURE 7.2 Block diagram of two types of direct-form realizations.

Direct-Form II Realization: If the All-pole system $H_2(z)$ is placed before the All-zero system $H_1(z)$ in cascade then we obtain a more compact network structure. This is called Direct-Form II Realization. Figures 7.4 and 7.5 illustrate the Direct-Form II network structures.

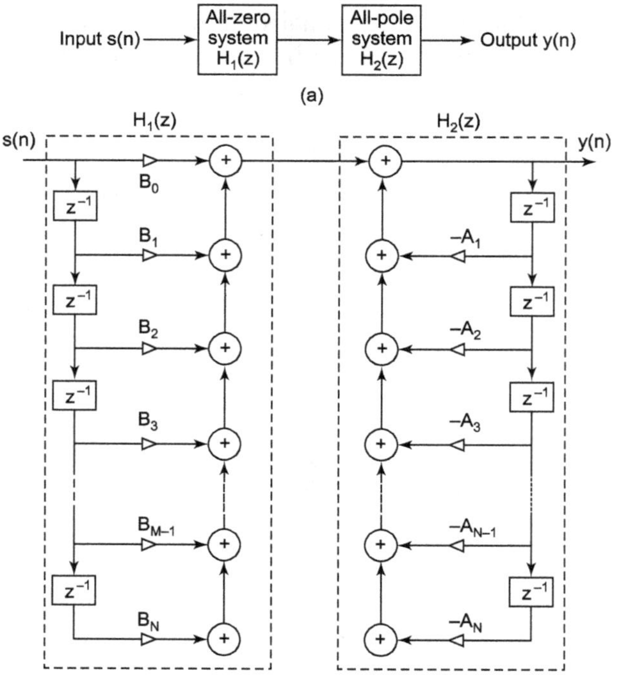

(a) Block diagram of direct form-I network structure.
(b) Direct form-I Realization.

FIGURE 7.3.

This network structure requires $M + N - 1$ multiplications and $M + N$ additions. It requires a maximum of M or N memory locations. Since the direct-form II realization reduces the number of memory locations by using common delay elements for both all-zero and all-pole systems.

These forms of network structure are obtained directly from system function $H(z)$ without any rearrangement of $H(z)$. But both the direct-form network structures are very sensitive to parameter quantization. Therefore, they are not recommended in practical applications.

In practical applications, we prefer direct-form II instead of direct-form I because it requires lesser number of memory locations. Figure 7.4 can be reconfigured in a compact form as given in Figure 7.5.

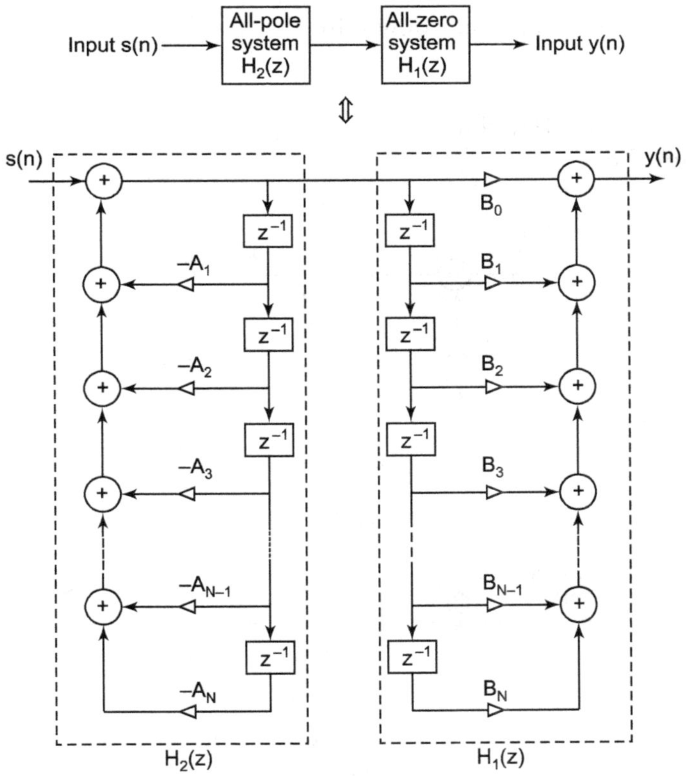

FIGURE 7.4 Direct-form realization with $H_2(z)$ cascaded with $H_1(z)$.

7.3.2 Transposed-Form Network Structures

Before studying the above topic, we should know the following terms and definitions.

Signal Flow Graph: It provides an alternative but equivalent graphical representation of a block diagram network structure. Signal flow graphs are comprised of branches and nodes. A signal flow graph is a collection of directed branches that connect at nodes. The sum of incoming signals at a node is equal to the sum of outgoing signals at the same node. A signal flow graph is shown in Figure 7.6.

For example. Consider a second-order IIR filter given in Figure 7.6(a)

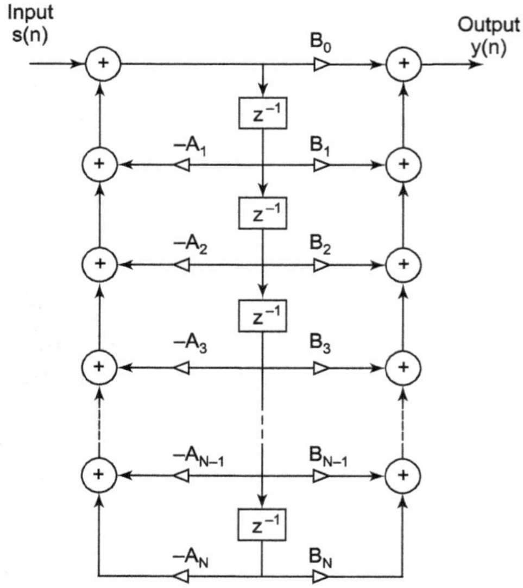

FIGURE 7.5 Direct-form II realization ($N = M$).

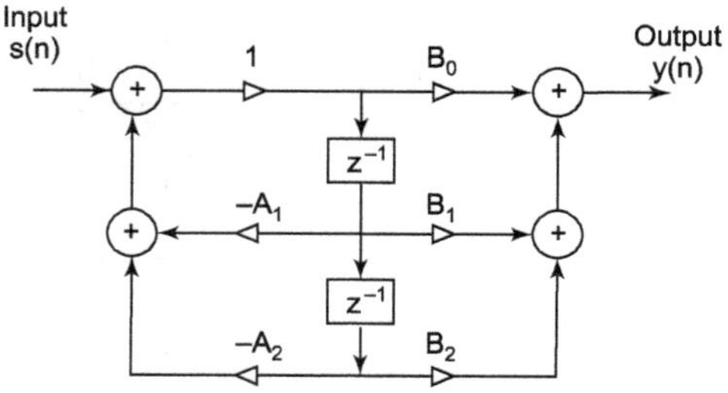

(a) Second order IIR filter structure.

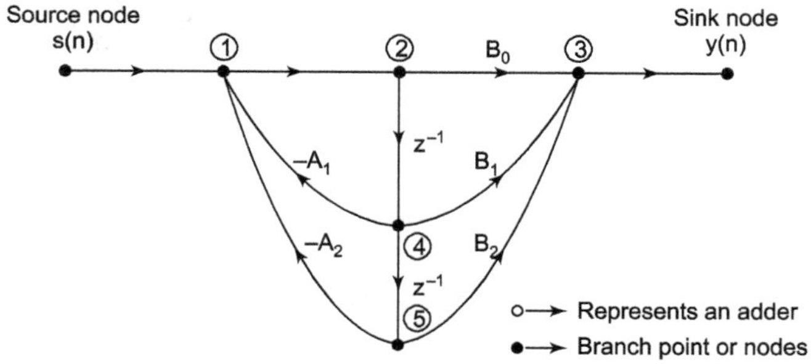

(b) Signal flow graph of the above filter structure.

FIGURE 7.6.

A signal flow graph contains the same information as the network structure of the system. The difference in both, branch points and adders in the block diagram are represented by nodes in the signal flow graph.

One useful technique in deriving a new system structure for FIR and IIR systems is stemmed by transposition theorem. This theorem is also called the *flow graph reversal theorem.*

Transposition Theorem: This theorem states that if we reverse the direction of all the transmittances, interchange input and output and also interchange branch points and summers or adders in the flow graph, the system function remains unchanged. This theorem resulting a network structure which is called a *transposed network structure.*

Transposition of the network structure of Figure 7.6 is performed as follows:

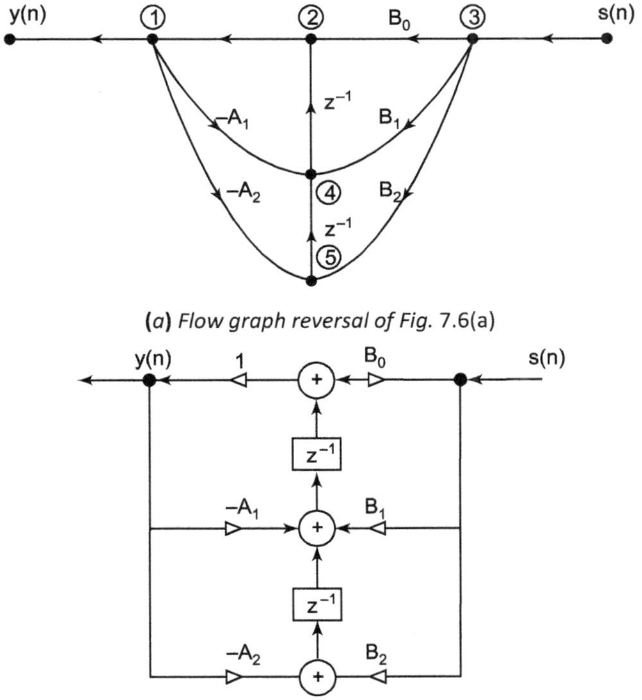

(*b*) *Transposed network structure of Fig. 7.6(b)*

FIGURE 7.7.

Transposed direct-form II network structure requires the same number of multiplications, additions and memory locations as required in the original direct-form II network structure.

7.3.3 Cascade-Form Network Structure

Cascade-form network structure is implemented by attaching all the factors of a system function $H(z)$ in series, that is, in cascade. Consider a high-order IIR system with system function given by Eq. (7.2). This system function $H(z)$ can be factored into a cascade of second-order subsystems. This network structure is shown in Figure 7.8:

$$H(z)=\frac{Y(z)}{S(z)}=\prod_{k=1}^{N'}H_k(z)=H_1(z).H_2(z).H_3(z),...,H_{N'}(z) \tag{7.6}$$

Here ν is used for showing the product of various parts of $H(z)$, that is,

$$H(z)=\prod_{k=1}^{N'}H_k(z)=H_1(z).H_2(z).H_3(z).....H_{N'}(z)$$

where N' is the integer part of $\left(\frac{N+1}{2}\right)$

$H_k(z)$ has the general form

$$H_k(z)=\frac{B_{k_0}+B_{k_1}z^{-1}+B_{k_2}z^{-2}}{1+A_{k_1}z^{-1}+A_{k_2}z^{-2}} \tag{7.7}$$

The coefficients A_{ki} and B_{ki} are real coefficients.

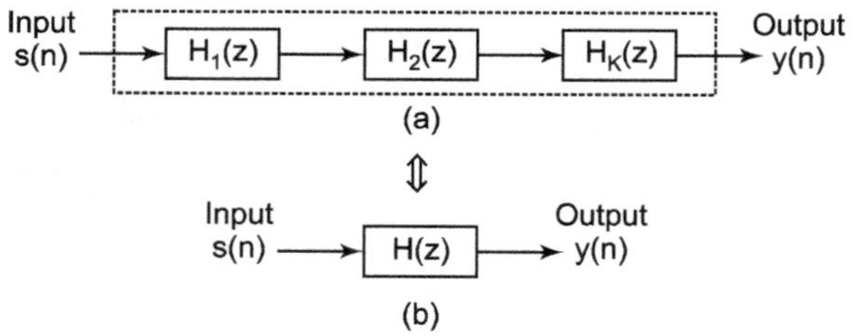

FIGURE 7.8 (a) Cascade-form network structure of second-order subsystems. (b) Equivalent network structure of above cascade-form network structure.

7.3.4 Parallel-Form Network Structure

A parallel-form realization of an IIR system can be obtained by performing a partial-fraction expansion of $H(z)$.

System function $H(z)$ of Eq. (7.2) can be written in partial fractions as

$$H(z)=C+\sum_{k=1}^{N}\frac{a_k}{1-P_k z^{-1}} \tag{7.8}$$

where P_k are the poles and α_k are the coefficients or residues in the partial-fraction expansion and constant C is defined as the ratio of B_N/A_N.

This network structure consists of a parallel bank single-pole filter. Generally, some of the poles of $H(z)$ may be complex-valued and coefficients are also complex-valued. For avoiding multiplications by complex numbers, we can combine pairs of complex-conjugate poles to form two-pole subsystems.

Two real-valued poles can also be combined to form two-pole subsystems. Real-valued poles have some real values and there is no imaginary part.

These subsystems have the system function in the form

$$H_k(z) = \frac{B_{k0} + B_{k1}z^{-1}}{1 + Ak_1 zh - 1 + A_{k_2} z^{-2}} \tag{7.9}$$

Coefficients B_{ki} ard A_{ki} real-valued system parameters.

Parallel-form network structures are shown in Figures 7.9 and 7.10.

$H(z)$ can be implemented in either direct forms or in transposed direct forms.

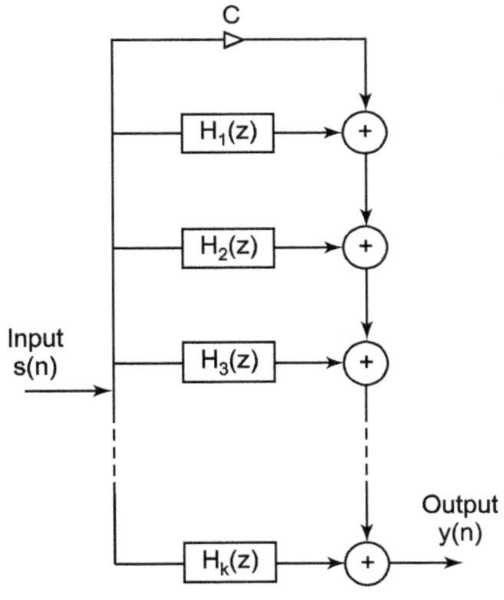

FIGURE 7.9 Paralled-form network structure of IIR system.

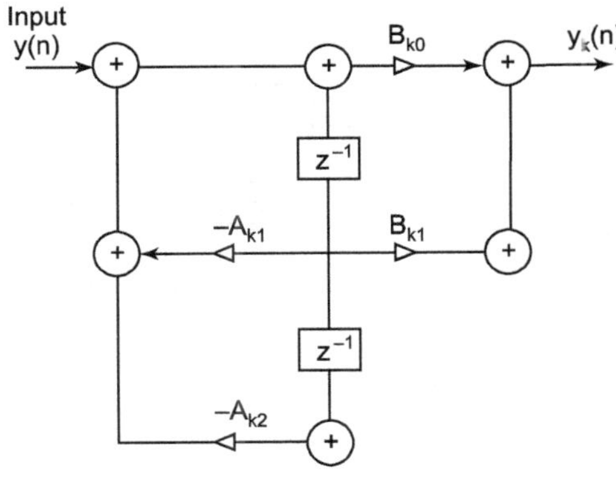

FIGURE 7.10 Structure of second-order section in a parallel-form network structure realization.

EXAMPLE 7.2

Sketch the direct-form I, direct-form II, cascade and parallel-form network structures for the system characterized by following difference equations.

$$y(z) = \frac{3}{4}y(n-1) - \frac{1}{8}y(n-2) + s(n) + \frac{1}{3}s(n-1).$$

Solution:

For computing transfer function of the above system, we take z-transform of both sides of above difference equation, we get

$$Y(z) = \frac{3}{4}z^{-1}Y(z) - \frac{1}{8}z^{-2}Y(z) + S(z) + \frac{1}{3}z^{-1}S(z)$$

or
$$H(z) = \frac{Y(z)}{S(z)} = \frac{1+\frac{1}{3}z^{-1}}{1-\frac{3}{4}z^{-1}+\frac{1}{8}z^{-2}} = H_1(z).H_2(z)$$

Direct-form I. $H_1(z) = 1 + \frac{1}{3}z^{-1}$. It is an all-zero system because it has only zeros in it.

$$H_2(z) = \frac{1}{1-\frac{3}{4}z^{-1}+\frac{1}{8}z^{-2}}$$

It is an all-pole system because it has only poles in it.

Figure 7.10(*a*) and (*b*) illustrates the direct-form I network structure of the system function $H(z)$ given for the above problem.

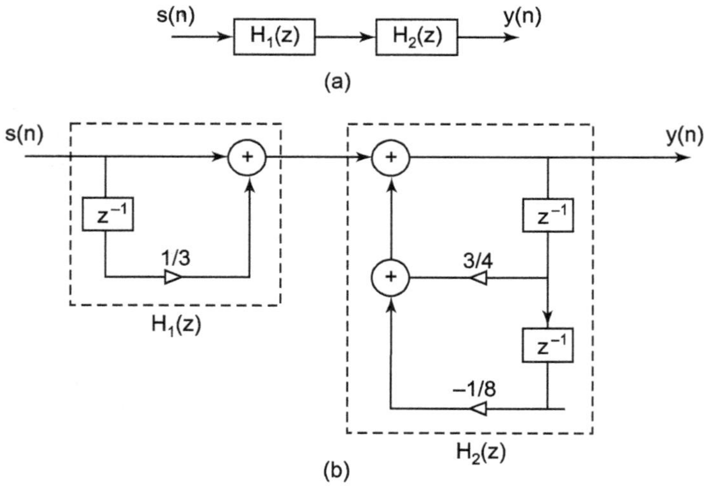

(a) Block diagram of direct-form I of above problem.

(b) Direct-form I, network structure of above filter.

FIGURE 7.11.

Direct-form II. Direct-form II network structure is shown in Figure 7.12.

Cascade-form network structure. Cascade-form network structure for $H(z)$ is illustrated in Figure 7.13.

$$H(z) = \frac{1+\frac{1}{3}z^{-1}}{1-\frac{3}{4}z^{-1}+\frac{1}{8}z^{-2}} = \frac{\left(1+\frac{1}{3}z^{-1}\right)}{\left(1-\frac{1}{2}z^{-1}\right)\left(1-\frac{1}{8}z^{-1}\right)}$$

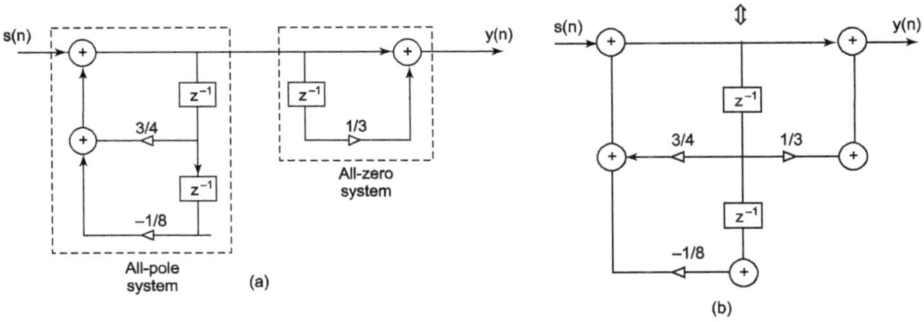

FIGURE 7.12 Direct-form II realization of above filter.

$$= \underbrace{\left\{\frac{1+\frac{1}{3}z^{-1}}{1-\frac{1}{2}z^{-1}}\right\}}_{H_1(z)\text{ direct form-II}} = \underbrace{\frac{1}{\left(1-\frac{1}{4}z^{-1}\right)}}_{H_2(z)\text{ cascaded with direct form-II}}$$

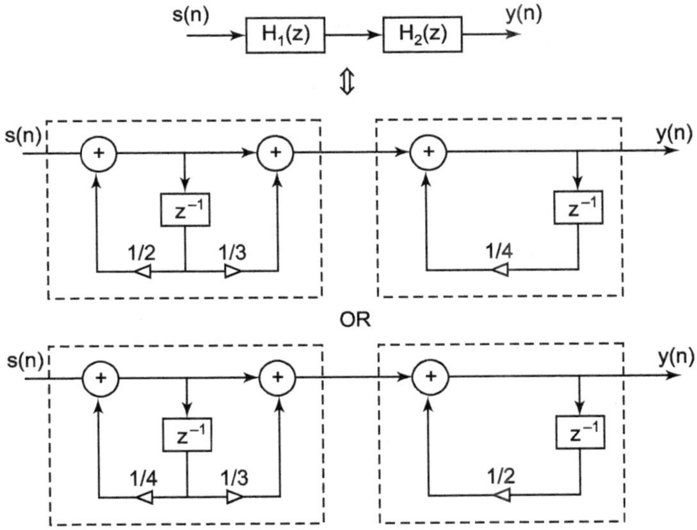

FIGURE 7.13 Cascade-form network structure of above filter.

or

$$= \underbrace{\left\{\frac{1+\frac{1}{3}z^{-1}}{1-\frac{1}{2}z^{-1}}\right\}}_{H_1(z)\text{ direct form-II}} = \underbrace{\frac{1}{\left(1-\frac{1}{4}z^{-1}\right)}}_{H_2(z)\text{ cascaded with direct form-II}}$$

Parallel-form network structure. Parallel-form network structure for $H(z)$ is shown in Figure 7.14.

$$H(z)=\frac{1+\frac{1}{3}z^{-1}}{\left(1-\frac{1}{2}z^{-1}\right)\left(1-\frac{1}{4}z^{-1}\right)}=\frac{A}{\left(1-\frac{1}{2}z^{-1}\right)}+\frac{B}{\left(1-\frac{1}{4}z^{-1}\right)}$$

$$=\frac{\frac{10}{3}}{1-\frac{1}{2}z^{-1}}+\frac{\left(-\frac{7}{3}\right)}{1-\frac{1}{4}z^{-1}}=\frac{10}{3}H_1(z)+\left(-\frac{7}{4}\right)H_2(z)$$

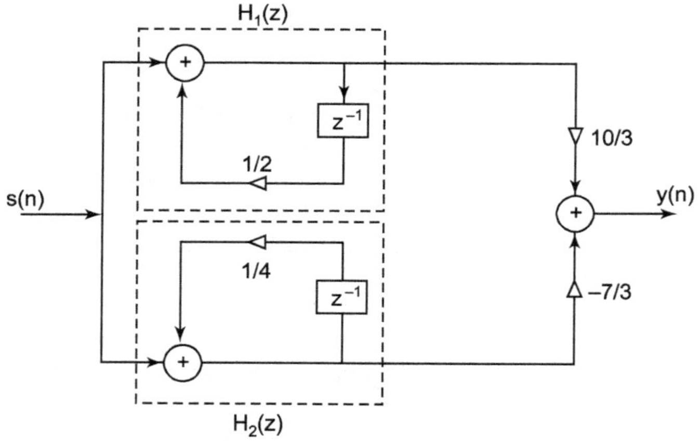

FIGURE 7.14 Parallel-form network structure of the above filter.

EXAMPLE 7.3

Sketch the block diagram representation of direct-form I and direct-form II network structures of a system with system function.

$$H(z)=\frac{1.2-4.8z^{-2}}{1-0.9z^{-1}+0.28z^{-2}-0.16z^{-3}}$$

Solution:

Given system function

$$H(z)=\frac{Y(z)}{S(z)}=\frac{1.2-4.8z^{-2}}{1-0.9z^{-1}+0.28z^{-2}-0.16z^{-3}}=H_1(z)H_2(z)$$

where $\quad H_1(z)=1.2-4.8z^{-2}$ is an all-zero system

$$H_2(z) = \frac{1.2 - 4.8z^{-2}}{1 - 0.9z^{-1} + 0.28z^{-2} - 0.16z^{-3}} \text{ is an all-pole system.}$$

Figures 7.15 and 7.16 illustrate the direct-form I and direct-form II network structures of the above system function $H(z)$, respectively.

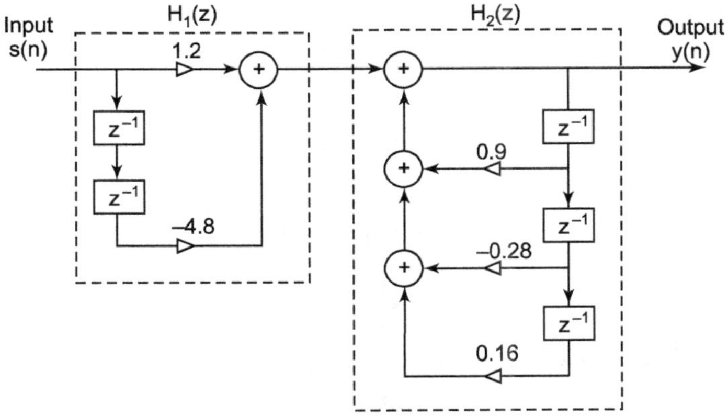

FIGURE 7.15 Direct-form I realization of the above system.

EXAMPLE 7.4

Sketch a cascade network structure of the system characterized by the transfer function

$$H(z) = \frac{2(z+2)}{z(z-0.1)(z+0.5)(z+0.4)}$$

Solution:

Given the system function

$$H(z) = \frac{2(z+2)}{z(z-0.1)(z+0.5)(z+0.4)}$$

$$= \frac{2z^{-3}\left(1+2z^{-1}\right)}{\left(1-0.1z^{-1}\right)\left(1-0.5z^{-1}\right)\left(1-0.4z^{-1}\right)}$$

Parallel-form network structure for system function $H(z)$ is shown in Figure 7.17.

EXAMPLE 7.5

Sketch the parallel-form network structure of the system with transfer function.

$$H(z) = \frac{2z(z+3)}{z^2 + 0.3z + 0.02}$$

Solution:

Given, the transfer function

$$H(z) = \frac{2z(z+3)}{z^2 + 0.3z + 0.02} = \frac{2z(z+3)}{(z+0.1)(z+0.2)}$$

$$= \frac{A}{z+0.1} + \frac{B}{z+0.2} = \frac{-5.8}{z+0.1} + \frac{11.2}{z+0.2}$$

$$= \frac{-5.8z^{-1}}{1+0.1z^{-1}} + \frac{11.2z^{-1}}{1+0.2z^{-1}} = H_1(z) + H_2(z) \qquad (1)$$

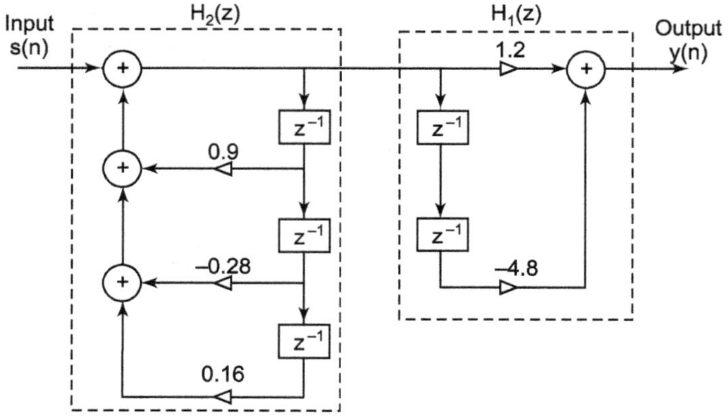

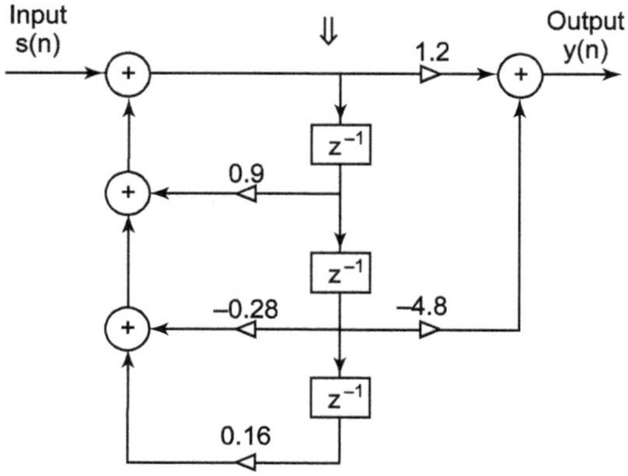

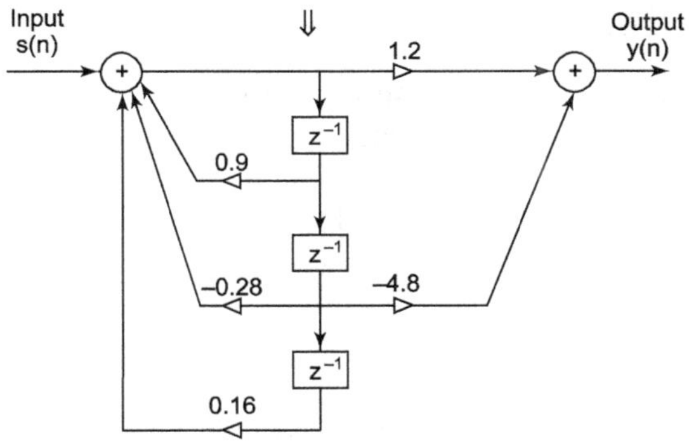

FIGURE 7.16 Direct-form II realization of the above system (Ex. 7.3).

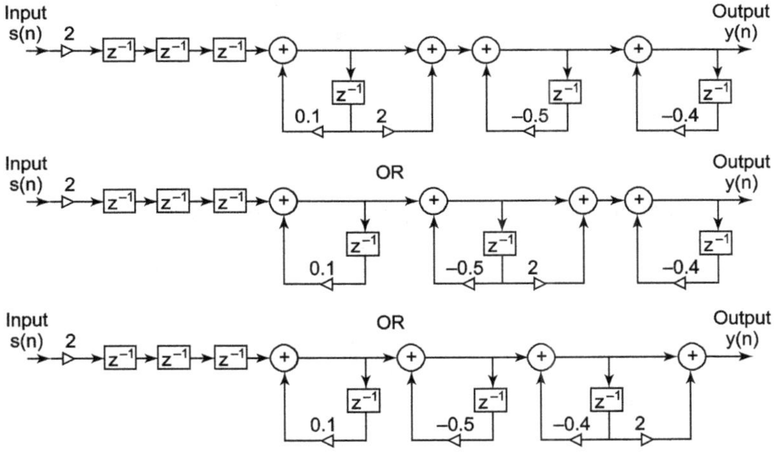

FIGURE 7.17 Parallel-form network structure of the above system (Ex. 7.4).

Figure 7.18 illustrates the parallel-form network structure for system function $H(z)$ given by Eq. (1).

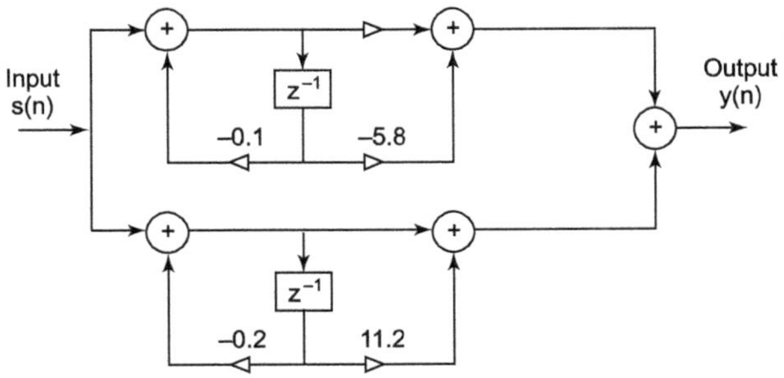

FIGURE 7.18 Parallel-form network structure of the above system (Ex. 7.5).

7.4 NETWORK STRUCTURE FOR FIR SYSTEMS

FIR systems are systems whose impulse response $h(n)$ is determined for the finite number of sample points. Impulse response $h(n)$ is the response of a system to a particular input, that is, unit-sample sequence $\delta(n)$. It is defined as

$$\delta(n) = \begin{cases} 1, & n \geq 0 \\ 0, & n < 0 \end{cases}$$

z-transform of impulse response $h(n)$ is called system function or transfer function $H(z)$ of the system.

System function of an LTI discrete-time system is characterized by Eq. (7.2) as

$$H(z) = \frac{\sum_{k=0}^{M-1} B_k z^{-k}}{1 + \sum_{k=1}^{N-1} A_k z^{-k}}$$

If $A_k = 0$ then,

$$H(z) = \sum_{k=0}^{M-1} B_k z^{-k} \tag{7.10}$$

Eq. (7.10) is the system function of FIR system

$$H(z) = \frac{Y(z)}{S(z)} = \sum_{k=0}^{M-1} B_k z^{-k}$$

or

$$y(z) = \sum_{k=0}^{M-1} B_k z^{-k} S(z) \tag{7.11}$$

Taking the inverse z-transform of Eq. (7.11), we get

$$y(n) = \sum_{k=0}^{M-1} B_k s(n-k) \tag{7.12}$$

Unit-sample response of the FIR system is

$$h(n) = \begin{cases} B_n, & 0 \leq n \leq M-1 \\ 0, & \text{otherwise} \end{cases} \tag{7.13}$$

where M is the length of the FIR filter.

There are several types of network structures for FIR filters/systems given as follows:

1. Direct-form network structures.
2. Cascade-form network structures.
3. Frequency-sampling network structures.

We will discuss them one by one.

7.4.1 Direct-Form Network Structures

The direct-form realization of the FIR system follows immediately from the non-recursive realization. Direct-form realization is shown in Figure 7.19.

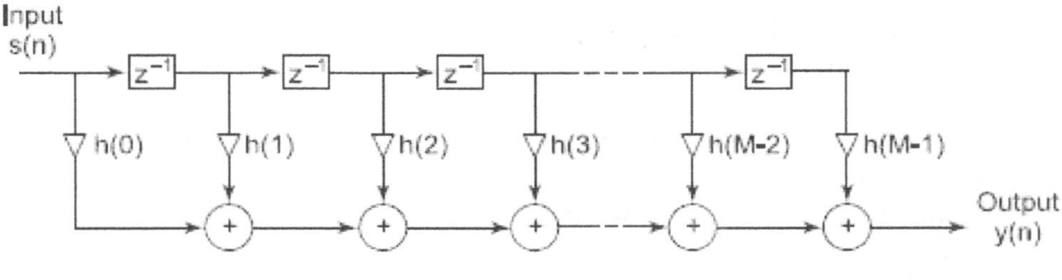

FIGURE 7.19 Direct-form realization of FIR system.

Non-recursive realization is given by Eq. (7.12)

$$y(n) = \sum_{k=0}^{M-1} B_k s(n-k)$$

or by convolution summation/linear convolution of $h(n)$ and $s(n)$, given by

$$y(n) = \sum_{k=0}^{M-1} h(k)s(n-k) \tag{7.14}$$

This network structure requires $M - 1$ memory locations for storing $M - 1$ previous inputs. The complexity of this network structure has M number of multiplications and $M - 1$ number of additions per output sample. This structure resembles a Tapped-Delay-Line system. Therefore, this network structure is also called *Tapped-Delay-Line Filter*.

Unit-sample response $h(n)$ of linear-phase FIR filter satisfies either symmetry or asymmetry condition.

Symmetry condition $\Rightarrow h(n) = h(M - 1 - n)$

Asymmetry condition $\Rightarrow h(n) = -h(M - 1 - n)$

For a system that satisfies either of the two conditions the number of multiplications is reduced by half.

7.4.2 Cascade-Form Network Structures

The cascade-form realization follows from Eq. (7.2). Here system function $H(z)$ is being factored into second-order FIR system.

$$H(z) = \prod_{k=1}^{N} H_k(z) \tag{7.15}$$

where

$$H_k(z) = B_{k0} + B_{k1}z^{-1} + B_{kz}z^{-2}, \quad k = 1,2,\ldots,N' \tag{7.16}$$

where N' is the integer part of $\left(\frac{M+1}{2}\right)$. The zeros of system function $H(z)$ are grouped in pairs to get the second-order FIR systems of the form given in Eq. (7.16). Coefficients B_{ki} in Eq. (7.16) should be real-valued.

Real coefficient B_{ki} can be achieved by forming pairs of complex-conjugate roots. Complex-conjugate roots are roots whose value is complex and each one is a complex conjugate of the others.

For example, one root is given by $2 + j3$. It is a complex root because it has one real part and one imaginary part. Its complex-conjugate root is given by just reversing the sign of the imaginary part of the root. The second root will be $2 - j3$.

The basic fourth-order FIR filter structure is shown in Figure 7.20.

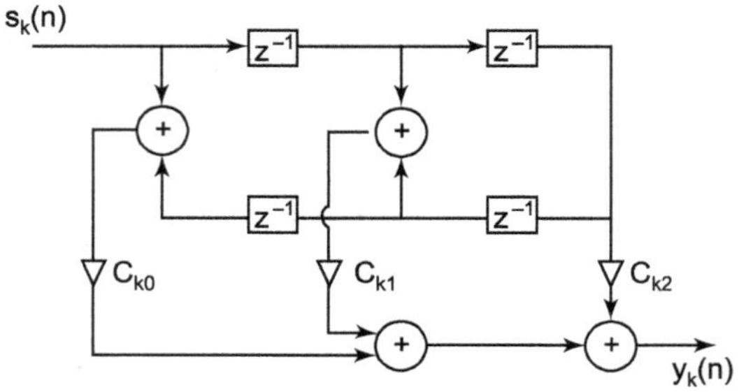

FIGURE 7.20 Fourth-order structure of FIR system.

A comparison of two pairs of pole results in fourth-order FIR filter. Its cascade-form realization is shown in Figure 7.21.

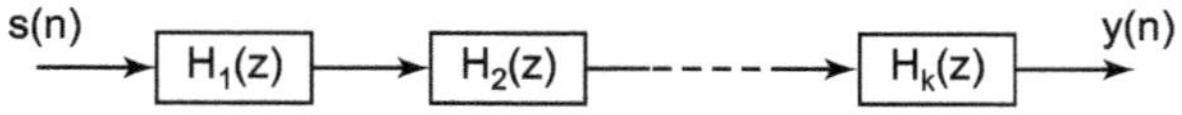

FIGURE 7.21 Cascade-form realization of FIR filter.

In linear-phase FIR filters, symmetry in $h(n)$ implies that the zeros of $H(z)$ also have symmetry. If z_k and z_k^* are the pair of complex-conjugate zeros then $1/z_k$ and $1/z_k^*$ are also a pair of complex-conjugate zeros. By doing so we can form fourth-order sections of the FIR systems.

This formation can be given as

$$H_k(z) = C_{k0}\left(1 - z_k z^{-1}\right)\left(1 - z_k^* z^{-1}\right)\left(1 - \frac{z^{-1}}{z_k}\right)\left(1 - \frac{z^{-1}}{z_k^*}\right)$$

$$= C_{k0} + C_{k1}z^{-1} + C_{k1}z^{-2} + C_{k1}z^{-3} + z^{-4} \qquad (7.17)$$

Coefficients C_{k1} and C_{k2} are the functions of zk.

The combination of two pairs of poles resulting in the fourth-order filter section reduces the number of multiplications from 6 to 3. Figure 7.22 illustrates the cascade-form realization of a FIR system.

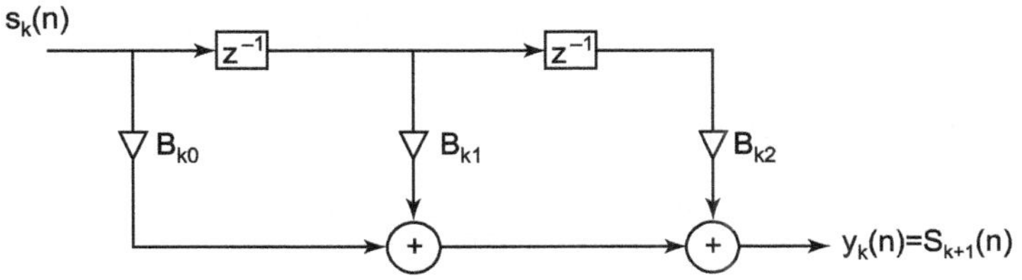

FIGURE 7.22 Cascade-form realization of an FIR system.

7.4.3 Frequency-Sampling Network Structures

The frequency-sampling realization is an alternative network structure for an FIR system. In this network structure, parameters of the system are characterized by frequency response instead of the impulse response $h(n)$.

Derivation: Frequency response of FIR system is given by Discrete-Time Fourier Transform (DTFT) of $h(n)$, as follows:

$$H(\omega)=\sum_{n=0}^{M-1} h(n)e^{-j\omega n} \tag{7.18}$$

Now for deriving the frequency-sampling structure, frequency sampling of DTFT of $h(n)$ is done.

The values of $H(\omega)$ at frequencies $\omega_K = \frac{2\pi}{M}(K+a)$.

$$H(K+a)=H\left(\frac{2\pi}{M}(K+a)\right)$$

$$=\sum_{n=0}^{M-1} h(n)e^{-j(2\pi/M)(K+a)n} \tag{7.19}$$

The set of values $\{H(K + \alpha)\}$ are called the frequency samples of $H(\omega)$.

If $\alpha = 0$, then it called DFT of $\{h(n)\}$.

Impulse response $h(n)$ in terms of the frequency samples $H(K + \alpha)$ is given by

$$h(n)=\frac{1}{M}\sum_{n=0}^{M-1} H(K+a)\, e^{-j(2\pi/M)(K+a)n} \tag{7.20}$$

$n = 0, 1,\ldots, M - 1$

If $\alpha = 0$, then it is called inverse DFT of $\{H(K)\}$.

System function $H(z)$ is given by

$$H(z)=\sum_{n=0}^{M-1}h(n)z^{-n} \tag{7.21}$$

Substituting the value of $h(n)$ in Eq. (7.21), we get

$$H(z)=\sum_{n=0}^{M-1}\left[\frac{1}{M}\sum_{n=0}^{M-1}h(K+a)\,e^{-j(2\pi/M)(K+a)n}\right]z^{-n} \tag{7.21a}$$

Now interchanging the order of summations in Eq. (7.21*a*), we get

$$H(z)=\sum_{K=0}^{M-1}H(K+a)\underbrace{\left[\frac{1}{M}\sum_{n=0}^{M-1}\left\{e^{-j(2\pi/M)\,(K+a)}\,z^{-1}\right\}^{n}\right]}_{\text{This is a geometric progression}}$$

$$=\sum_{K=0}^{M-1}H(K+a\,)\frac{1}{M}\left\{\frac{1-e^{-j2\pi(K+a)}z^{-M}}{1-e^{-[j2\pi(K+a)]/M}z^{-1}}\right\}$$

or

$$H(z)=\sum_{K=0}^{M-1}H(K+a)\frac{1}{M}\left\{\frac{1-e^{j2\pi K}e^{j2\pi a}z^{-M}}{1-e^{[j2\pi(K+a)]/M}z^{-1}}\right\}$$

Hence $\quad e^{j2\pi K}=1,\quad K=0,1,2,$

$$K=0,\ e^{j2\pi.0}=e^{0}=1$$

$$K=1,\ e^{j2\pi.1}=e^{j2\pi}=1 \text{ and so on}$$

$\therefore \quad e^{j2\pi K}=1$

or

$$H(z)=\sum_{K=0}^{M-1}H(K+a)\frac{1}{M}\left\{\frac{1-e^{j2\pi a}z^{-M}}{1-e^{[j2\pi(K+a)]/M}z^{-1}}\right\}$$

$$=\frac{1-e^{j2\pi a}z^{-M}}{M}\sum_{K=0}^{M-1}\frac{H(K+a)}{1-e^{j(2\pi/M)(K+a)}z^{-1}}$$

or

$$H(z)=\frac{1-e^{j2\pi a}z^{-M}}{M}\sum_{K=0}^{M-1}\frac{H(K+a)}{1-e^{j(2\pi/M)(K+a)}z^{-1}} \tag{7.22}$$

This system function is characterized by a set of frequency samples $H(K+\alpha)$.

System function $H(z)$ consists of two subsystem functions in cascade and is given by

$$H(z)=H_1(z)\cdot H_2(z) \tag{7.23}$$

This FIR system realization is viewed as a cascade of two subsystems $H_1(z)$ and $H_2(z)$. $H_1(z)$ is an all-zero filter. It is also called *comb filter with system function.*

$$H_1(z)=\frac{1}{M}\left[1-e^{j2\pi a}z^{-M}\right] \tag{7.24}$$

The second subsystem with system function

$$H_2(z) = \sum_{K=0}^{M-1} \frac{H(K+a)}{1 - e^{j(2\pi/M)(K+a)} z^{-1}} \tag{7.25}$$

Pole locations are identical to the zero locations and both occur at $\omega_K = (2\pi/M)\ (K + \alpha)$, which are the frequencies at which the desired frequency response is specified. Frequency-sampling realization of the FIR system given by Eq. (7.22) is shown in Figure 7.23:

$$\begin{aligned} H_2(z) &= \sum_{K=0}^{M-1} \frac{H(K+a)}{1 - e^{j(2\pi/M)(K+a)} z^{-1}} \\ &= \frac{H(a)}{1 - e^{2\pi/M} z^{-1}} + \frac{H(K+a)}{1 - e^{j(2\pi/M)(K+a)} z^{-1}} \\ &+ \frac{H(2+a)}{1 - e^{j(2\pi/M)(2+a)} z^{-1}} + \ldots + \frac{H(M-1+a)}{1 - e^{j(2\pi/M)(K-1+a)} z^{-1}} \end{aligned} \tag{7.26}$$

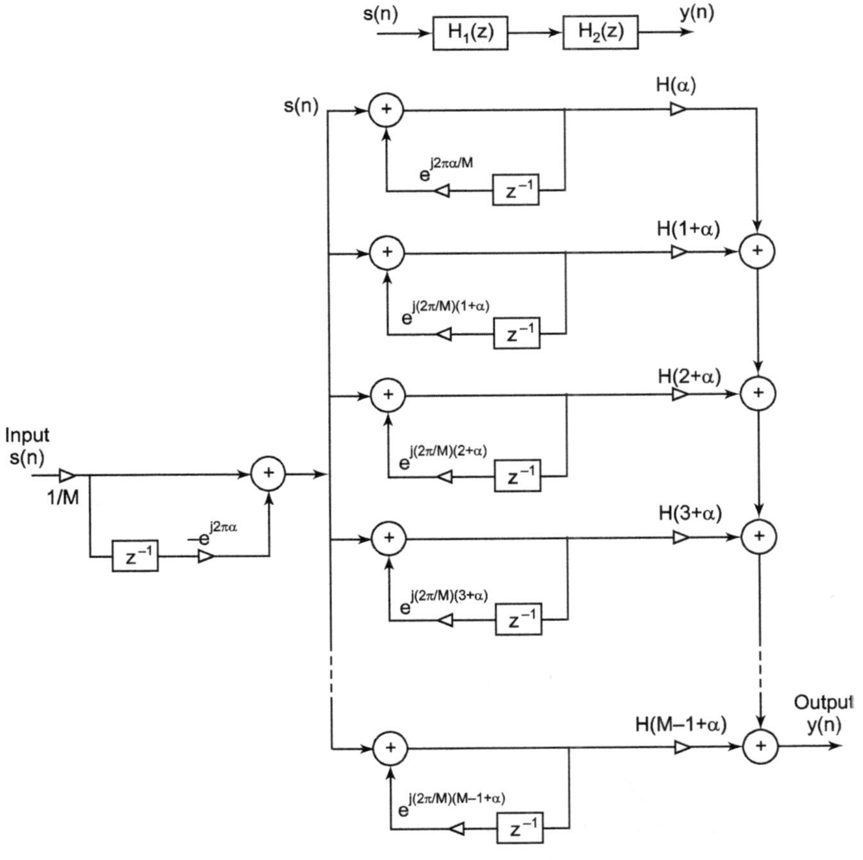

FIGURE 7.23 Frequency-sampling realization of FIR system.

When the desired frequency response characteristic of the FIR system is narrow band, its most of the gain parameters $H(K + a)$ are zero. Consequently, the corresponding resonant filters can be eliminated as the gain parameters are zero for them. Now only non-zero-gain parameter resonant filters will be retained.

In other words, we can say this network structure requires fewer computations than the corresponding direct-form realization.

The frequency-sampling network structure can be simplified further by exploiting the property of symmetry in $H(K + \alpha)$ which is

$$H(K) = H*(M-K) \text{ for } a = 0$$

and

$$H\left(K+\frac{1}{2}\right) = H*\left(M-K-\frac{1}{2}\right) \text{ for } a = \frac{1}{2} \tag{7.27}$$

Thus for $\alpha = 0$, the system function $H_2(z)$ is simplified as

$$H_2(z) = \sum_{K=0}^{M-1} \frac{H(M-1a)}{1-e^{j(2\pi/M)(K+a)}z^{-1}}$$

$$= \sum_{K=0}^{M-1} \frac{H(M-1+a)}{1-e^{j(2\pi/M)(K+0)}z^{-1}}$$

$$= \sum_{K=0}^{M-1} \frac{H(M-1+a)}{1-e^{j(2\pi/M)(K)}z^{-1}}$$

$$= \frac{H(0)}{1-z^{-1}} + \sum_{K=0}^{\frac{M-1}{2}} \frac{A'(K)+B'(K)z^{-1}}{1-2\cos\left(\frac{2\pi K}{M}\right)z^{-1}+z^{-2}}$$

$$M \text{ is odd} \tag{7.28}$$

$$H_2(z) = \frac{H(0)}{1-z^{-1}} + \frac{H(M/2)}{1+z^{-1}} + \sum_{K=1}^{\frac{M-1}{2}} \frac{A'(K)+B'(K)z^{-1}}{1-2\cos\left(\frac{2\pi K}{M}\right)z^{-1}+z^{-2}}$$

$$M \text{ is even} \tag{7.29}$$

where

$$A'(K) = H(K) - H(M-K)$$

$$B'(K) = H(K)e^{-j2\pi K/M} + H(M-K)e^{-j2\pi K/M} \tag{7.30}$$

EXAMPLE 7.6

Sketch a direct-form realization for the following linear-phase FIR filter. Its impulse response is given by

$$h(n) = (1, 2, 3, 4, 3, 2, 1\}.$$

Solution:

System function $H(z)$ of the above FIR filter is computed by

$$\begin{aligned} H(z) &= z\text{-transform of } h(n) = \mathcal{Z}[h(n)] \\ &= \sum_{n=0}^{M-1} h(n)z^{-n} = \sum_{n=0}^{6} h(n)z^{-n} \\ &= h(0)z^{-0} + h(1)z^{-1} + h(2)z^{-2} + h(3)z^{-3} + h(4)z^{-4} \\ &\qquad + h(5)z^{-5} + h(6)z^{-6} \\ &= 1 + 2z^{-1} + 3z^{-2} + 4z^{-3} + 3z^{-4} + 2z^{-5} + 1z^{-6} \end{aligned}$$

or

$$H(z) = \frac{Y(z)}{S(z)} = 1 + 2z^{-1} + 3z^{-2} + 4z^{-3} + 3z^{-4} + 2z^{-5} + 1z^{-6}$$

Direct-form realization of the system function $H(z)$ given in this problem is illustrated in Figure 7.24.

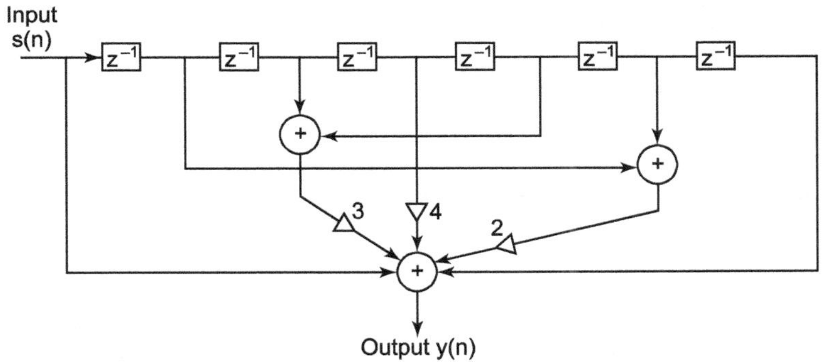

FIGURE 7.24 Direct-form realization for the above linear-phase filter.

EXAMPLE 7.7

Sketch the block diagram for the direct-form realization and frequency-sampling realization of $M = 16$, $\alpha = 0$, symmetric linear-phase FIR filter. This filter has the following frequency samples

$$H\left(\frac{2\pi K}{16}\right) = \begin{cases} 1, & K = 0,1,2 \\ \frac{1}{2}, & K = 3 \\ 0, & K = 4,5,6,7 \end{cases}$$

Also, compare the computational complexity of these structures.

Solution:

Since the given FIR filter is symmetric. By exploiting the property of symmetry, the number of multiplications per output point can be reduced from 16 to 8 in the direct-form realization. The number of additions per output point is 16 – 1= 15. The block diagram of direct-form realization is shown in Figure 7.25.

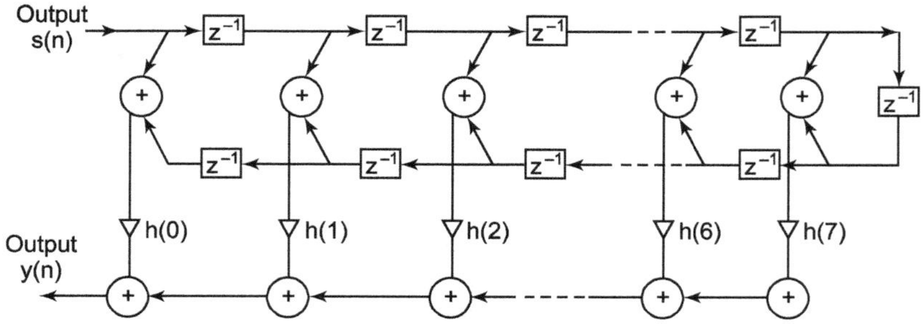

FIGURE 7.25 Block diagram of direct-form realization.

Frequency-sampling Network Structure. For this network structure, we use

$$H_1(z)=\frac{1}{M}\left(1-z^{-M}e^{j2\pi a}\right) \tag{1}$$

and

$$H_2(z)=\sum_{K=0}^{M-1}\frac{H(K+a)}{1-e^{j(2\pi/M)(K+a)}z^{-1}}$$

$$H_2(z)=\frac{H(0)}{1-z^{-1}}+\frac{H(M/2)}{1+z^{-1}}\sum_{K=0}^{\frac{M}{2}-1}\frac{A'(K)+B'(K)z^{-1}}{1-2\cos\left(\frac{2\pi K}{M}\right)z^{-1}+z^{-2}} \tag{2}$$

For M even

$$H_2(z)=\frac{H(0)}{1-z^{-1}}+\sum_{K=1}^{\frac{M}{2}-1}\frac{A'K+B'(K)z^{-1}}{1-2\cos\left(\frac{2\pi K}{M}\right)z^{-1}+z^{-2}} \qquad \text{For } M \text{ odd}$$

The frequency-sampling network structure can be simplified further by exploiting the property of symmetry in $H(K+\alpha)$ as

$$H(K)=H*(M-K) \text{ for } a=0$$

$$H\left(K+\frac{1}{2}\right)=H*\left(M-K-\frac{1}{2}\right) \text{ for } a=\frac{1}{2}$$

We use Eqs. (1) and (2) for the frequency-sampling realization and drop all terms which have zero-gain coefficients $\{H(K)\}$.

The non-zero coefficients are $H(K)$ and corresponding pairs are $H(M - K)$ for $K = 0, 1, 2$.

Since $H(0) = 1$, the single-pole filter requires no multiplication. The two double-pole filter sections require three multiplications each for a total of six multiplications. Total number of addition is 9. Therefore, the frequency-sampling realization of this FIR system is computationally more efficient than the direct-form realization. Frequency-sampling realization of $H(z)$ given in this problem is shown in Figure 7.26.

EXAMPLE 7.8

Sketch a cascade-form network realization in such a way that it requires only three delay elements for a system characterized by the following transfer function

$$H(z) = \frac{z^{-1} - \alpha}{1 - \alpha z^{-1}} \cdot \frac{z^{-1} - \beta}{1 - \beta z^{-1}}$$

where α are β constants.

Solution:

For cascade realization, we break the original transfer function in this manner

$$H(z) = H_1(z) H_2(z) = \left(\frac{z^{-1} - \alpha}{1 - \alpha z^{-1}} \right) \left(\frac{z^{-1} - \beta}{1 - \beta z^{-1}} \right) \tag{1}$$

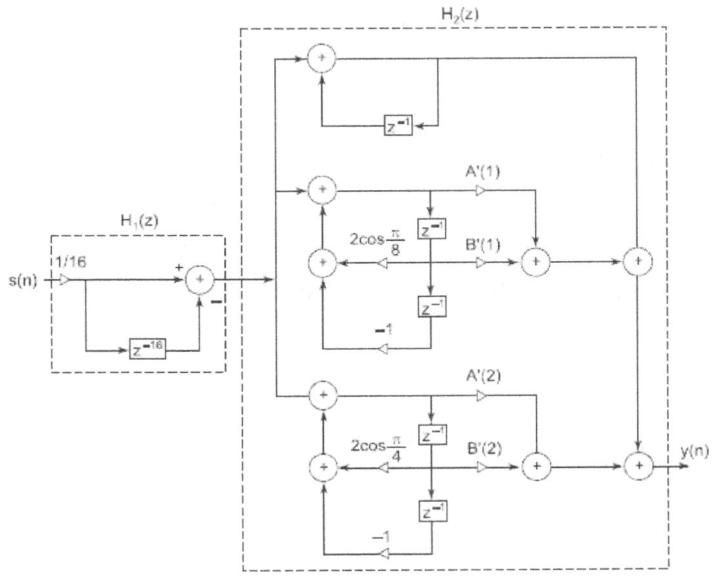

FIGURE 7.26 Frequency-sampling realization for FIR filter (Ex. 7.7).

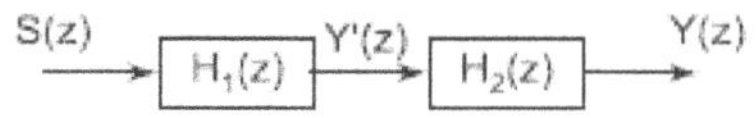

FIGURE 7.27 $H(z)$ is realized in cascading of $H_1(z)$ and $H_2(z)$.

or
$$H_1(z) = \frac{Y'(s)}{S(z)} = \frac{z^{-1} - a}{1 - az^{-1}} \tag{2}$$

and
$$H_2(z) = \frac{Y(z)}{Y'(z)} = \frac{z^{-1} - \beta}{1 - \beta z^{-1}} \tag{3}$$

From Eq. (2), we get

$$\frac{Y'(s)}{S(z)} = \frac{z^{-1} - a}{1 - az^{-1}}$$

or
$$Y'(z)\left[1 - az^{-1}\right] = S(z)\left[z^{-1} - a\right]$$

or
$$Y'(z) - az^{-1}Y'(z) = z^{-1}S(z) - aS(z)$$

On taking inverse z-transform, we get

$$y'(n) - ay(n-1) = s(n-1) - as(n)$$

or
$$y'(n) - ay(n-1) = s(n-1) - as(n) \tag{4}$$

From Eq. (3), we get

$$\frac{Y(z)}{Y'(z)} = \frac{z^{-1} - \beta}{1 - \beta z^{-1}}$$

or
$$Y(z)\left[1 - \beta z^{-1}\right] = Y'(z)\left[z^{-1} - \beta\right]$$

or
$$Y(z) - \beta z^{-1}Y(z) = z^{-1}Y'(z) - \beta Y'(z)$$

On taking inverse z-transform, we get

$$y(n) - \beta y(n-1) = y'(n-1) - \beta y'(n)$$

or
$$y(n) = \beta y(n-1) + y'(n-1) - \beta y'(n) \tag{5}$$

Figure 7.28 shows the cascade realization. It contains only three delay elements as per our requirement.

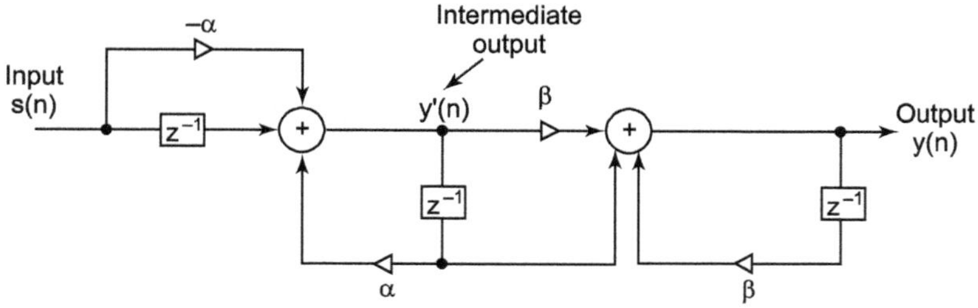

FIGURE 7.28 Cascade-form realization of transfer function given.

EXAMPLE 7.9

Find out the parallel-form realization for a system which is characterized by the transfer function

$$H(z)=\frac{1}{1-z^{-1}-z^{-2}}.$$

Solution:

Given transfer function $H(z)$ can be modified as follows:

$$H(z)=\frac{1}{1-z^{-1}-z^{-2}}$$

or
$$H(z)=\frac{1}{z^{-2}-z-1}$$

Factorization of $z^2 - z - 1$ results in

$$z^2-z-1=\left[z-\left(\frac{1+\sqrt{5}}{2}\right)\right]\left[z-\left(\frac{1-\sqrt{5}}{2}\right)\right]$$

Also substitute $a=\dfrac{1+\sqrt{5}}{2}$ and $\beta=\dfrac{1+\sqrt{5}}{2}$

then
$$z^2-z-1=(z-a)\,(z-\beta)$$

and
$$a-\beta=\left(\frac{1+\sqrt{5}}{2}\right)-\left(\frac{1-\sqrt{5}}{2}\right)=\sqrt{5}$$

or
$$H(z)=\frac{z^2}{(z-a)\,(z-\beta)}=\frac{1}{\left(1-az^{-1}\right)\left(1-\beta z^{-1}\right)}$$

$$=\frac{A}{\left(1-a\,z^{-1}\right)}+\frac{B}{\left(1-\beta z^{-1}\right)}\qquad\text{(By partial fraction)}$$

Determination of constants A and B.

$1-\alpha z^{-1}=0$

or $\alpha z^{-1}=1$,

or $z=\alpha$

$$A=\frac{1}{1-\beta z^{-1}}=\frac{1}{1-\dfrac{\beta}{a}}=\frac{a}{a-\beta}=\frac{a}{\sqrt{5}}$$

$1-\beta z^{-1}=0$

or $\beta z^{-1}=1$,

or $z=\beta$

$$B=\frac{1}{1-az^{-1}}=\frac{1}{1-\dfrac{a}{\beta}}=\frac{\beta}{\beta-a}=\frac{\beta}{-\sqrt{5}}$$

or
$$H(z) = \frac{A}{1-\alpha z^{-1}} + \frac{B}{1-\beta z^{-1}}$$
$$= \frac{\alpha/\sqrt{5}}{1-\alpha z^{-1}} + \frac{-\beta/\sqrt{5}}{1-\beta z^{-1}} = H_1(z) + H_2(z)$$

Figure 7.29(a) shows the realization of $H_1(z)$ and $H_2(z)$ separately and Figure 7.29(b) shows parallel realization of $H_1(z)$ and $H_2(z)$.

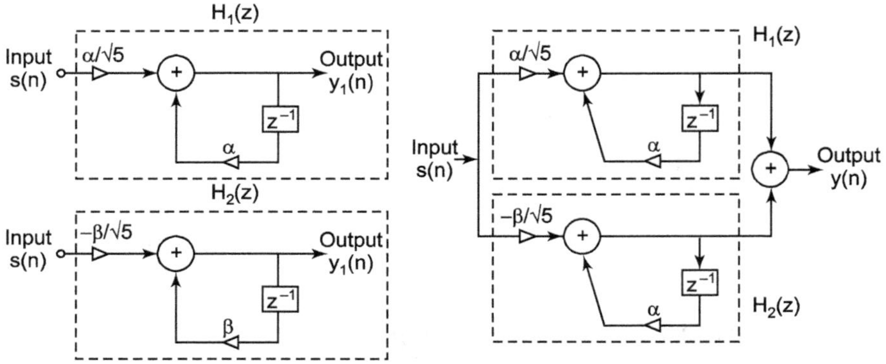

FIGURE 7.29 (a) Realization of $H_1(z)$ and $H_2(z)$ separately.
(b) Parallel-form realization of above transfer function.

$$H(z) = H_1(z) + H_2(z)$$

EXAMPLE 7.10

Draw the network structure for the following transfer function.
$$H(z) = \frac{Y(z)}{S(z)} = \frac{z^2}{z^2 - zr\cos\theta \cdot z + r^2}$$

Solution.

Given transfer function
$$H(z) = \frac{Y(z)}{S(z)} = \frac{z^2}{z^2 - zr\cos\theta . z + r^2}$$
$$= \frac{1}{1 - 2r\cos\theta . z^{-1} + r^2 z^{-2}}$$
where $Y(z)$ and $S(z)$ are the z-transforms of output the input, respectively.
$$Y(z)\left(1 - 2r\cos\theta\ z - 1 + r^2 z^{-2}\right) = S(z) \qquad (1)$$

Taking the inverse z-transform of Eq. (1), we get

$$y(n) \; -2r\cos\theta\, y(n-1) + r^2 y(n-2) = s(n)$$

or
$$y(n) - 2r\cos\theta\, y(n-1) + r^2 y(n-2) + s(n) \qquad (2)$$

Network structure for transfer function $H(z)$ is shown in Figure 7.30.

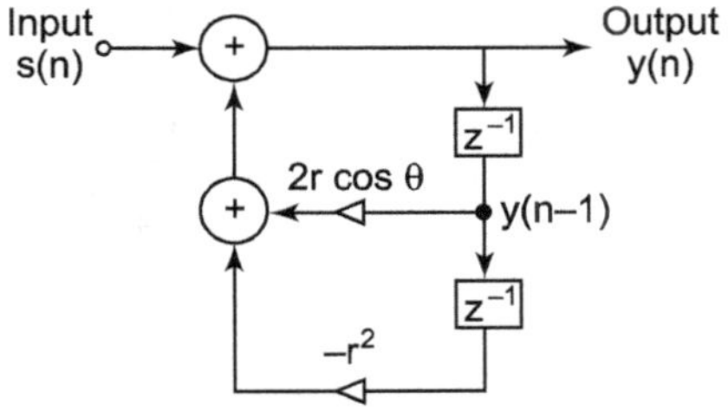

FIGURE 7.30 Network structure for the transfer function $H(z)$.

EXAMPLE 7.11

Find the transfer function $H(z)$ of the system shown in Figure 7.31. Also determine the difference equation which characterize it.

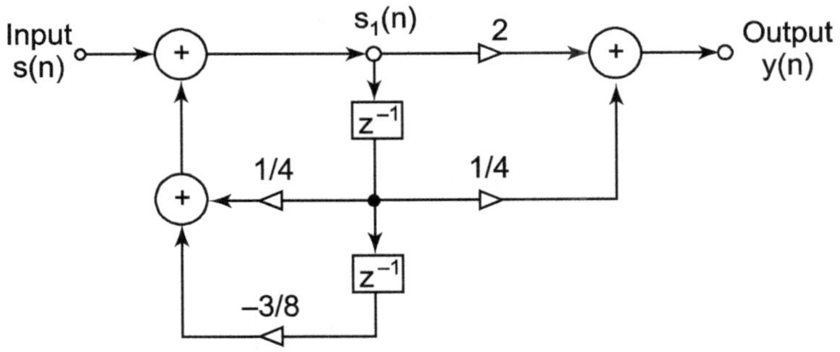

FIGURE 7.31 Network structure of a discrete-time system with input $s(n)$ and output $y(n)$.

Solution:

From the network structure of Figure 7.31.

$$s_1(n) = s(n) = \left(\frac{1}{4}\right)s_1(n-1) - \frac{3}{8}s_1(n-2) \qquad (1)$$

And
$$y(n) = 2s_1(n) + \frac{1}{4}s_1(n-1) \qquad (2)$$

Eqs. (1) and (2) are the difference equations of above network structure. Taking the z-transform of equations (1) and (2), we get

$$S_1(z) = S(z) + \left(\frac{1}{4}\right)z^{-1}S_1(z) - \frac{3}{8}z^{-2}S_1(z) \qquad (3)$$

and

$$Y(z) = 2S_1(z) + \frac{1}{4}z^{-1}S_1(z) \qquad (4)$$

From Eq. (3)

$$S_1(z) = \frac{1}{\left(1 - \frac{1}{4}z^{-1} + \frac{3}{8}z^{-3}\right)}S(z) \qquad (5)$$

From Eq. (4)

$$Y(z) = \left(1 + \frac{1}{4}z^{-1}\right)S_1(z) \qquad (6)$$

Substituting Eq. (5) in Eq. (6), we get

$$Y(z) = \left(z + \frac{1}{4}z^{-1}\right)S_1(z)$$

$$= \left(2 + \frac{1}{4}z^{-1}\right)\frac{1}{\left(1 - \frac{1}{4}z^{-1} + \frac{3}{8}z^{-3}\right)}S(z)$$

or

$$\frac{Y(z)}{S(z)} = \frac{\left(2 + \frac{1}{4}z^{-1}\right)}{\left(1 - \frac{1}{4}z^{-1} + \frac{3}{8}z^{-3}\right)} \qquad (7)$$

But transfer function of a system is defined as the ratio of z-transform of output, $Y(z)$ to the z-transform of input, $S(z)$ and is given by

$$H(\mathrm{z}) = \frac{Y(z)}{S(z)} = \frac{\left(2 + \frac{1}{4}z^{-1}\right)}{\left(1 - \frac{1}{4}z^{-1} + \frac{3}{8}z^{-2}\right)}$$

This is the transfer function of above system given in Figure 7.31.

EXAMPLE 7.12

Find and draw a parallel-form of realization for the system which is characterized by the following transfer function

$$H(z) = \frac{z\left(2z^2 + z + 1\right)}{\left(z^2 + \frac{1}{4}z - \frac{1}{8}\right)\left(z + \frac{1}{8}\right)}$$

Draw the realization with only first-order systems.

Solution:

Given transfer function $H(z)$ can be modified as

$$H(z) = \frac{z(2z^2 + z(1)}{\left(z^2 + \frac{1}{4}z - \frac{1}{8}\right)\left(z + \frac{1}{8}\right)}$$

$$= \frac{2 + z^{-1} + z^{-2}}{\left(1 + \frac{1}{4}z^{-1} - \frac{1}{8}z^{-2}\right)\left(1 + \frac{1}{8}z^{-1}\right)}$$

(Dividing both numerator and denominator by z^n)

or $$H(z) = \frac{2 + z^{-1} + z^{-2}}{\left(1 + \frac{1}{4}z^{-1}\right)\left(1 - \frac{1}{4}z^{-1}\right)\left(1 + \frac{1}{8}z^{-1}\right)}$$

$$= \frac{A}{\left(1 + \frac{1}{2}z^{-1}\right)} + \frac{B}{\left(1 - \frac{1}{4}z^{-1}\right)} + \frac{C}{\left(1 + \frac{1}{8}z^{-1}\right)}$$

(By partial-fraction method)

$$= \frac{32/9}{\left(1 + \frac{1}{2}z^{-1}\right)} + \frac{44/9}{\left(1 - \frac{1}{4}z^{-1}\right)} + \frac{29/38}{\left(1 + \frac{1}{8}z^{-1}\right)}$$

$$= H_1(z) + H_2(z) + H_3(z)$$

The network realization for the system given in this problem is shown in Figure 7.32. It is a parallel-form of network structure.

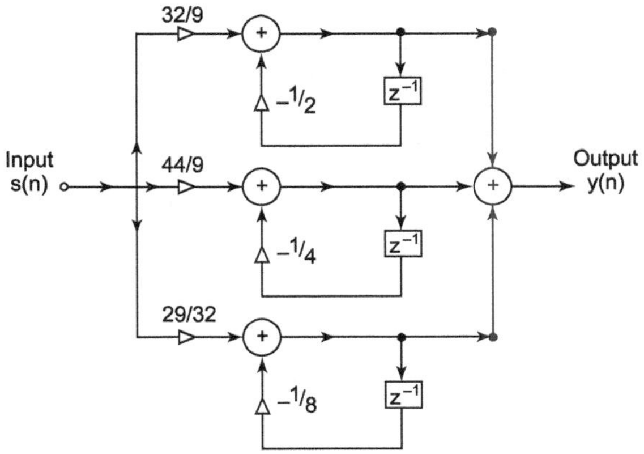

FIGURE 7.32 Parallel-form of network realization of transfer function $H(z)$.

EXERCISES

1. What are the major factors that influence our choice of specific realization? Discuss.
2. Discuss network structures for IIR systems given as follows:
   a. Direct-form I and Direct-form II network structures.
   b. Transposed-form network structures.
   c. Cascade-form network structures.
   d. Parallel-form network structures.
3. Define transposition theorem and network flow graph with an example.
4. Discuss network structures for FIR systems given as follows:
   a. Direct-form network structure.
   b. Cascade-form network structure.
   c. Frequency-sampling network structures.

NUMERICAL EXERCISES

1. Sketch the direct-form I and direct-form II network structures of the systems with transfer function

   a. $H(z) = \dfrac{1+z^{-1}}{1-0.5z^{-1}+0.06z^{-2}}$ **b.** $H(z) = \dfrac{z^{-1}-3z^{-2}}{\left(10-z^{-1}\right)\left(1+0.5z^{-1}+0.5z^{-2}\right)}$

   c. $H(z) = \dfrac{3z^2+6z}{(z-0.2)\left(z^2+0.5z-0.5\right)}$ **d.** $H(z) = \dfrac{2+3z^{-2}+1.5z^{-4}}{10-0.2z^{-1}+0.35z^{-2}}$

   e. $H(z) = \dfrac{6\left(1+3z^{-1}\right)}{3-0.9z^{-1}+0.006z^{-2}}.$
2. Sketch a cascade-form network structures of the systems with transfer function

   a. $H(z) = \dfrac{2z(z+3)}{z^2+0.3z+0.02}$ **b.** $H(z) = \dfrac{z^{-1}+4z^{-2}}{5-2z^{-1}+0.15z^{-2}}$

c. $H(z) = \dfrac{2 - 10z^{-2}}{1 - 0.8z^{-1} + 0.15z^{-2}}$ **d.** $H(z) = \dfrac{3x^2 - 12}{z^2 - 0.4 + 0.03}$

e. $H(z) = \dfrac{4z^3 + 16z^2 + 4z - 24}{2z^4 + 1.6z^3 + 0.5z^2 + 0.1z}$.

3. Find and sketch parallel-form network structure of the systems with transfer function

a. $H(z) = \dfrac{2z(z+3)}{z^2 + 0.3z + 0.02}$

b. $H(z) = \dfrac{z^{-1} + 4z^{-2}}{5 - 2z^{-1} + 0.15z^{-2}}$

c. $H(z) = \dfrac{z^2 - 1}{z^2 + 0.7z + 0.1}$

4. System function $H(z)$ of an FIR filter is given by

$$H(z) = 1 + 3z^{-1} + 3.5z^{-2} + 2z^{-3} + 2^{-4}$$

Sketch the direct-form network structure of this system (filter).

5. Sketch the direct-form network structure and frequency-sampling network structure of the $M = 32$, $\alpha = 0$, symmetric linear-phase FIR filter which has frequency samples.

$$H\left(\frac{2\pi K}{32}\right) = \begin{cases} 1, & K = 0,1,2 \\ \frac{1}{3} & K = 3 \\ 0, & K = 4,5,\ldots,15 \end{cases}$$

Also compute of the complexity of these two structures.

6. Find the direct-form I, direct-form II, cascade and parallel-form network structures for the following systems

a. $y(n) = \frac{1}{2}y(n-1) + \frac{1}{4}y(n-2) + s(n-1)$

b. $y(n) = y(n-1) - \frac{1}{2}y(n-2) + s(n-1) + s(n-2)$

where $y(n)$ and $s(n)$ are outputs and inputs of the system, respectively.

CHAPTER 8

DIGITAL FILTERS

8.1 INTRODUCTION

Filtering is a process by which the frequency spectrum of a signal can be modified, reshaped, or manipulated to achieve some desired objectives. These objectives are:

1. To eliminate noise contaminated in signal.
2. To remove signal distortion due to an imperfect transmission channel.
3. To separate two or more distinct signals which were purposely mixed for maximizing channel utilization.
4. To resolve signals into their frequency components.
5. To demodulate the signals which were modulated at the transmitter end.
6. To convert digital (discrete-time) signals into analog signals.
7. To limit the bandwidth of the signals.

Filters are of two types depending upon the type of signal to be processed.

Analog Filters: Analog filter is a system in which both the input and the output are continuous-time signals. Block diagram of analog filter is shown in Figure 8.1.

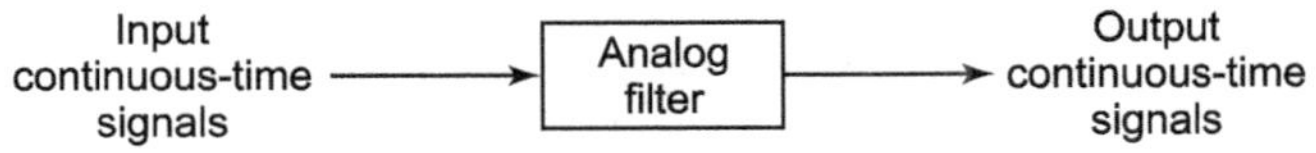

FIGURE 8.1 Analog filter.

Digital Filters: Digital filter is a system in which both the input and the output are discrete-time signals. Block diagram of the digital filter is shown in Figure 8.2.

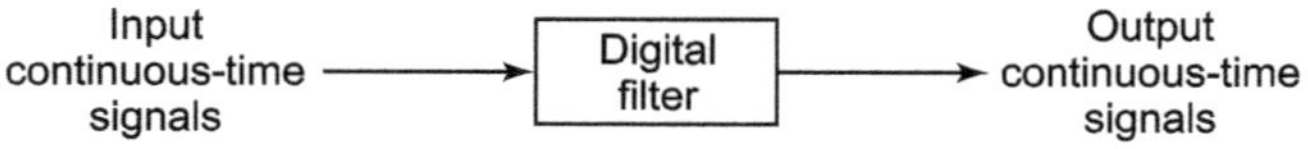

FIGURE 8.2 Digital filter.

Review of Analog Filter Design

There are five kinds of analog filters given as:

1. low-pass analog filter,
2. high-pass analog filter,
3. band-pass analog filter,
4. band-stop analog filter, and
5. all-pass analog filter.

Pole-zero configuration and corresponding frequency responses of various kinds of analog filters are given in Table 8.1.

TABLE 8.1 Pole-zero Configuration and Corresponding Frequency Responses of Various Kinds of Analog Filters

	Name of the analog filter	Its transfer function $H(s)$	Frequency response $H(j\omega)$	Pole-zero locations of $\lvert H(j\omega>)\rvert$
1.	Low-pass filter (LPF)	$\dfrac{r_0^2}{s^2\left(\dfrac{r_0}{Q}\right)s+r_0^2}$	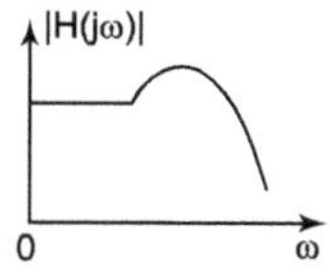	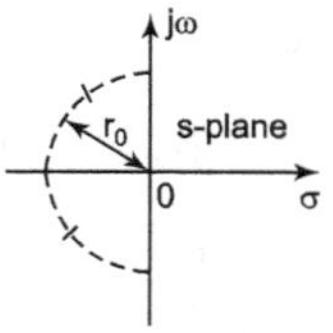
2.	High-pass filter (HPF)	$\dfrac{s^2}{s^2+\left(\dfrac{r_0}{Q}\right)s+r_0^2}$	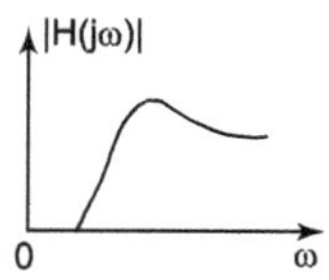	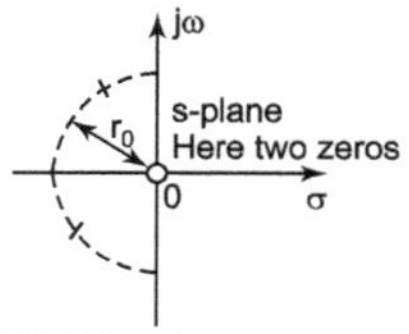

(*Continued*)

	Name of the analog filter	Its transfer function $H(s)$	Frequency response $H(j\omega)$	Pole-zero locations of $\lvert H(j\omega>)\rvert$
3.	Band-pass filter (BPF)	$\dfrac{(r_0/Q)s}{s^2+\left(\dfrac{r_0}{Q}\right)s+r_0^2}$	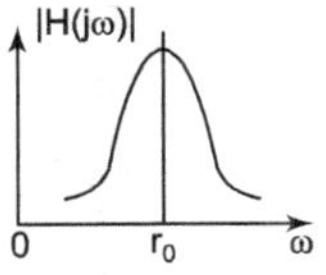	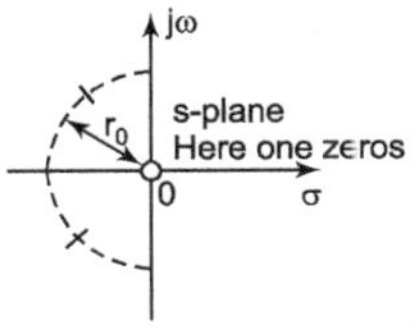
4.	Band-stop filter (BSF)	$\dfrac{s_0+r_0^2}{s^2+\left(\dfrac{r_0}{Q}\right)s+r_0^2}$	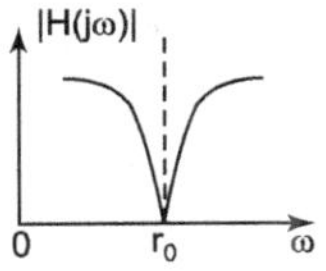	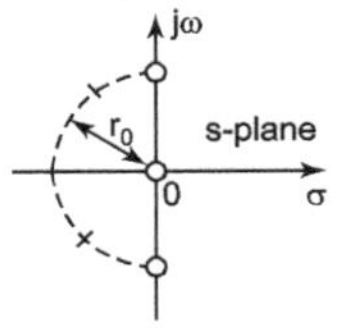
5.	All-pass filter (APF)	$\dfrac{s^2-(r_0/Q)s+r_0^2}{s^2+\left(\dfrac{r_0}{Q}\right)s+r_0^2}$	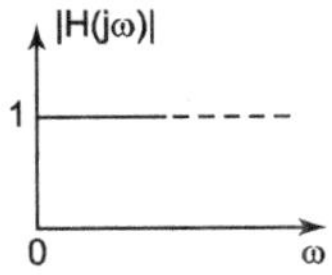	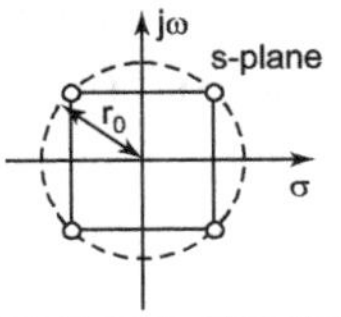

In Table 8.1, all the poles and zeros are seen to be located on a circle of radius r_0. This design was given by Butterworth therefore, is called *Butterworth Analog filter design.* Here Q is the quality factor of the circuit at resonance. In the pole-zero plot given in Table 8.1 location of poles and zeros are shown by cross (×) and circle (O), respectively.

All poles are located on the circle of radius, it is one kind of similarity that exists between the analog and digital filters. Another similarity is that the design of all types of analog filters is done from the LPFs using frequency transformation. Digital filters are also designed in the same manner as analog filters. Pole-zero characteristics are very important in both types of filters design (analog and digital filters). Pole-zero characterization of analog filters is summarized in Table 8.2 given below.

TABLE 8.2 Summary of Pole-zero Characterization of Analog Filter

	Name of the Analog Filter	Pole-zero Characterization
1.	Low-pass filter (LPF)	There are no zeros at the origin $(\sigma, \omega) = (0, 0)$ and poles are on the circle of radius r_0 in a complex conjugate.
2.	High-pass filter (HPF)	There are two zeros at the origin for two complex-conjugate poles on the circle of radius r_0.

(Continued)

	Name of the Analog Filter	Pole-zero Characterization
3.	Band-pass filter (BPF)	There is one zero at the origin and two complex-conjugate poles on a circle of radius r_0.
4.	Band-stop filter (BSF)	There are two zeros on the imaginary axis of s-plane on a circle of radius r_0 and two complex-conjugate poles on the same circle. Zeros on the imaginary axis of s-plane ensure that there is no output of the filter at frequency $\omega = \omega_0$. Band-stop filter (BSF) is also called Notch filters.
5.	All-pass filter (APF)	Two complex poles in the left half-plane and two complex zeros in the right half-plane lie on the circle of radius r_0. These pairs of complex poles and zeros lie in the quadrantal symmetry. For all-pass filters (APF), magnitude is unity for all values of ω.

A first-order analog filter is one that has only one pole on the real axis (s-axis) of s-plane. Its transfer function is given by

$$H(s) = \frac{1}{s + A}$$

and its impulse response will be given by

$$h(t) = \text{Inverse Laplace transform of } H(s)$$
$$= \pounds^{-1}[H(s)] = \pounds^{-1}\left[\frac{1}{s+A}\right] = e^{-At}$$

In other words, we can say that a first-order low-pass analog filter has one pole in the frequency-domain and its impulse response (time-domain response) is exponentially decreasing with time in the range $0 \le t \le \infty$.

Similarly, a first-order digital filter has one real zero on the real axis of z-plane, and this real zero lies inside the unit circle $|z| = 1$ or $r_0 = 1$. The transfer function of the first-order digital filter is given by

$$H(z) = \frac{1}{1 - Az^{-1}}$$

and its impulse (unit-sample) response will be given by $h(n)$ = Inverse z-transform of $H(z)$

$$\mathcal{Z}^{-1}\left[H(z) = \mathcal{Z}^{-1}\right]\left[\frac{1}{1 - Az^{-1}}\right] = A^n$$

Note. It is worth noting here that digital filters have a similarity with the expression of analog filters. Impulse responses (time-domain responses) for both the filters (analog and digital) decrease exponentially to zero as n or t approaches infinity.

The impulse response is used to characterize both types of filters. The impulse response of a digital filter is given in the form $h(n) = An$. It is called an infinite impulse response (IIR) digital filter.

Figure 8.3 shows the LPF specifications for the analog case.

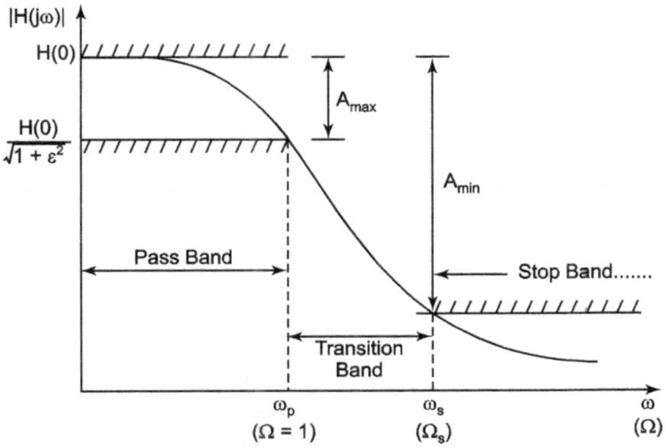

FIGURE 8.3 Butterworth approximation of magnitude response of an Nth order filter [Tolerance limits: A_{max} and A_{min}].

A_{max} is the maximum allowable attenuation in the pass band ($0 \le \Omega \le 1$) and A_{min} is the minimum allowable attenuation in the stop band ($\Omega = \Omega_s$).

There will exist a transition band in the range of frequencies ($1 \le \Omega \le \Omega_s$).

There are four specifications of a LPF; A_{max}, A_{min}, 1, and Ω_s. These specifications depend upon the tolerance limits along with corresponding frequencies.

If we divide the frequency ω by ω_p and this division is denoted by $\Omega = \dfrac{\omega}{\omega_p}$, then the cut-off frequency is normalized to unity.

Also $\Omega = \dfrac{\omega}{\omega_p}$, where ω_s is the stop-band frequency.

$$\boxed{|H(j\Omega)|^2 = \frac{|H(0)|^2}{1+\varepsilon^n \Omega^{2n}}} \tag{8.1}$$

where $H(0)$ is the maximum magnitude at zero frequency which is equal to unity and $\varepsilon = \sqrt{(10^{0.1A\max} - 1}$. It is illustrated in Figure 8.3.

Here at $\omega = \omega_p$ and $\Omega = \dfrac{\omega}{\omega_p} = \dfrac{\omega_p}{\omega_p} = 1$

Substituting $\Omega = 1$ in Eq. (8.1)

$$|H(j\Omega)|^2 \text{ at } \omega = \omega_p = \frac{|H(0)|^2}{1+\varepsilon^2} \tag{8.2}$$

Writing Eq. (8.1) in decibel (dB) as

$$H(j\Omega)_{dB} = 20\log|H(j\Omega)|$$
$$= 20\log H(0) - 10\log\left(1+\varepsilon^2\Omega^{2n}\right) \quad (8.3)$$

At $\Omega = 1$, $H(0) = 1$, attenuation loss will be maximum and is given by

$$A_{max} = 10\log\left[1+\varepsilon^2\Omega^{2n}\right] \quad (8.4)$$

or

$$\varepsilon = [10^{0.1}A_{max-1}]^{1/2} \quad (8.5)$$

At $\Omega = \Omega_s$, the minimum allowable attenuation loss is given by

$$A_{min} = 20\log\left(1+\varepsilon^2\Omega_s^{2n}\right)^{1/2} \quad (8.6)$$

or

$$\Omega_s^n = \left(10^{0.1A\min} - 1\right)^{1/2}/\varepsilon \quad (8.7)$$

Taking the log of both sides of Eq. (8.7) yields

$$n\log_{10}\Omega_s = \frac{1}{2}\log_{10}\left(10^{0.1A_{min}} - 1\right) + \log_{10}(\varepsilon)$$
$$= \log_{10}\sqrt{10^{0.1A_{min}} - 1} - \log_{10}\sqrt{10^{0.1A_{max}} - 1}$$
$$= \log_{10}\frac{\sqrt{10^{0.1A_{min}} - 1}}{\sqrt{10^{0.1A_{max}} - 1}}$$

or

$$\boxed{n = \log_{10}\frac{\sqrt{10^{0.1A_{min}} - 1}}{\sqrt{10^{0.1A_{max}} - 1}}\cdot\frac{1}{\log_{10}(\Omega_s)}}$$

Note: If $\varepsilon = 1$ then $\varepsilon = \left[10^{0.1A_{max}} - 1\right]^{1/2}$

or

$$1 = \left[10^{0.1A_{max}} - 1\right]^{1/2}$$

or

$$10^{0.1A_{max}} = 2$$

$$\log_{10}\left[10^{0.1A_{max}}\right]\log_{10}(2)0.1A_{max} = 0.3010$$
$$A_{max} = 3 \text{ at } \Omega = 1$$

For Chebyshev approximation, the order of filter can be determined as

$$\boxed{n = \cosh^{-1}\left(\frac{10^{0.1A_{min}} - 1}{10^{0.1A\max} - 1}\right)^{1/2} - \frac{1}{\cosh^{-1}(\Omega_s)}} \quad (8.8)$$

where

$$\cosh^{-1}x = \log e[x + (x^2 - 1)^{1/2}]$$

First of all, we determine the order of the filter n, then we determine the expression for transfer function $H(s)$ for Butterworth filter or for Chebyshev filter. From the above expressions, we can design a LPF. To find the transfer function $H(s)$ for high-pass, band-pass, or band-stop filters, we use the low-pass expression. Here we use frequency transformation given in Table 8.3.

TABLE 8.3 Frequency Transformations from LPF to HPF, BPF, and BSF

Type of Filter	Filter Specifications		Frequency variable Transformation
	Pass Band	Stop Band	
Low-pass filter (LPF)	$0 \to \Omega_1$	$\Omega_2 \to \infty$	Transition band = $\Omega_2 - \Omega_1$
High-pass filter (HPF)	$\Omega_2 \to \infty$	$0 \to \Omega_1$	$q = \frac{\Omega_2}{s}$
Band-pass filter (BPF)	$\Omega_1 \to \Omega_2$ $\Omega_0 \to \sqrt{\Omega_1\Omega_2}$	$0 \to \Omega_1$ $\Omega_4 \to \infty$	Transition band $\Omega_3 \to \Omega_1$ $\Omega_2 \to \Omega_4$ $q = Q\left[\left(\frac{s}{\Omega_0}\right) + \left(\frac{\Omega_0}{s}\right)\right]$
Band-stop filter (BSF)	$0 \to \Omega_1$ $\Omega_2 \to \infty$ $\Omega_0 \to \sqrt{\Omega_1\Omega_2}$	$\Omega_3 \to \Omega_4$	Transition band $\Omega_3 \to \Omega_4$ $\Omega_4 \to \Omega_2$ $q = \frac{1}{Q\left[\left(\frac{s}{\Omega_0}\right) + \left(\frac{\Omega_0}{s}\right)\right]}$

8.2 MAJOR CONSIDERATIONS IN USING DIGITAL FILTERS

The procedure for the implementation of a digital filter has the following steps in order

1. selection of filter,
2. specification of the frequency response characteristic of the filter,
3. phase response specification,
4. filter design,
5. filter realization, and
6. filter implementation.

8.2.1 Selection of Filter

The selection of filter depends on the market demand and intended applications.

8.2.2 Specification of the Frequency Response Characteristics of the Filter

After deciding on the choice of filter type to be used, analog or digital, we decide about the specifications of the frequency response characteristics of the filter.

Digital filters are classified into four types similar to analog filter on the basis of frequency response.

These are:

1. low-pass (LP) filter,
2. high-pass (HP) filter,
3. band-pass (BP) filter and
4. band-stop (BS) filters.

- Low-pass (LP) filters are those which pass low frequencies from zero to a cut-off frequency ω_P with approximately unity gain.
- High-pass (HP) filter passes frequencies from cut-off frequency $\omega_{P1} = \omega_P$ to $\omega_{P2} = \pi$ with a unity gain.
- Band-pass (BP) filter passes frequencies in a chosen range from ω_{P1} to ω_{P2} with a unity gain.
- Band-stop (BS) filter stops frequencies in the chosen range from ω_{S1} to ω_{S2}.

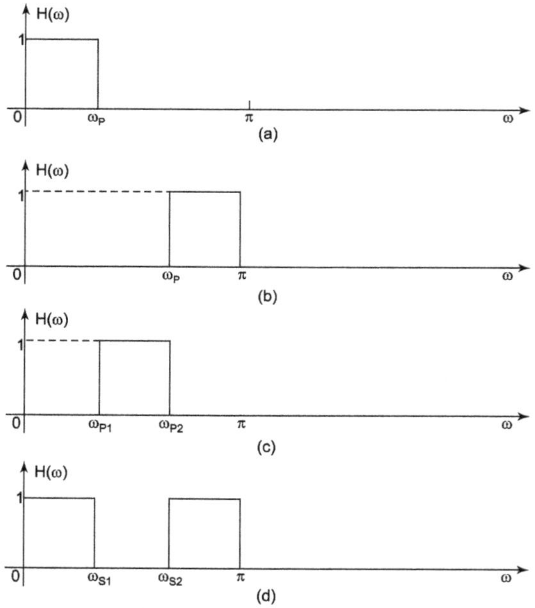

FIGURE 8.4 Frequency response of ideal filters (a) Low-pass filter, (b) high-pass filter, (c) band-pass filter, and (d) band-stop filter.

Each class of filters is specified in terms of pass-band frequency ω_P, stop-band frequency (ω_S, transition band $\omega_P \leq \omega \leq \omega_S$, pass-band tolerance δ_P, stop-band tolerance δ_S, positive tolerance δ^+ of magnitude response, negative tolerance δ^- of magnitude response, pass-band ripple (δ^+, δ^-) defined in dB and stop-band attenuation.

Pass-band ripple defined in dB is given as

$$A_p = \{20\ \log_{10}(1+\delta^+) - 20\ \log_{10}(1\ -\delta^-)\}_{\max} \tag{8.9}$$

Stop-band attenuation defined in dB is given as

$$As = -20\log_{10}(\delta_S) \tag{8.10}$$

Low-pass filter: Specifications of a LPF is given in terms of its frequency response $H(\omega)$ as

$$\begin{cases} 1-\delta^- \le |H(\omega)| \le +\delta^+, & 0 \le \omega \le \omega_p \\ \quad 0 \le |H(\omega)| \le \delta_S, & \omega_S \le \omega \le \pi \end{cases} \tag{8.11}$$

LPF specification is shown in Figure 8.5.

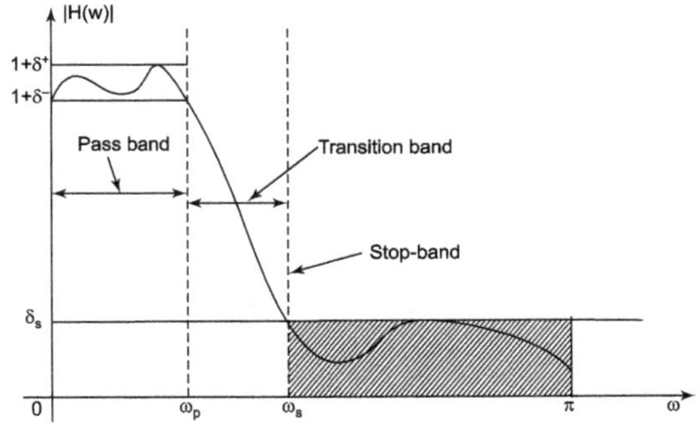

FIGURE 8.5 Low-pass filter specifications.

High-pass Filter: Specification of a high-pass filter is given in terms of its frequency response $H(\omega)$ as

$$\begin{cases} \quad 0 \le |H(\omega)| \le \delta_S, & 0 \le \omega \le \omega_S \\ 1-\delta^- \le |H(\omega)| \le 1+\delta^+, & \omega_P \le \omega \le \pi \end{cases} \tag{8.12}$$

Its specifications are shown in Figure 8.6.

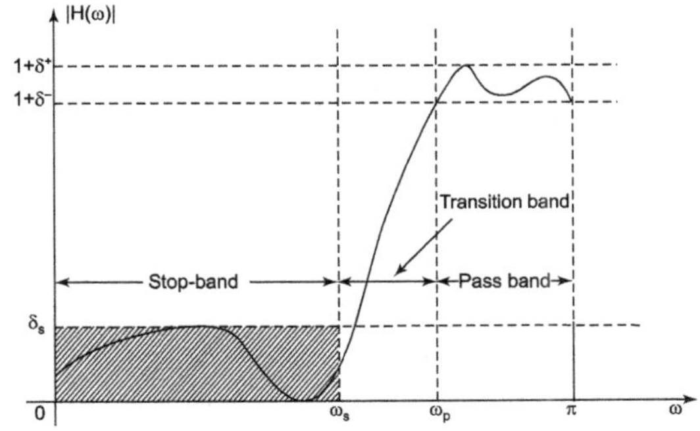

FIGURE 8.6 High-pass filter specifications.

Band-pass filter: Specifications of a band-pass filter is given in terms of its frequency response $H(\omega)$ as

$$\begin{cases} 0 \le |H(\omega)| \le \delta_{S_1}, & 0 \le \omega \le \omega_{S_1} \\ 1-\delta^{-1} \le |H(\omega)| \le 1+\delta^{+}, & \omega_{P_1} \le \omega \le \omega_{P_2} \\ 0 \le |H(\omega)| \le \delta_{S_2}, & \omega_{s_2} \le \omega \le \pi \end{cases} \tag{8.13}$$

Its specifications are shown in Figure 8.7.

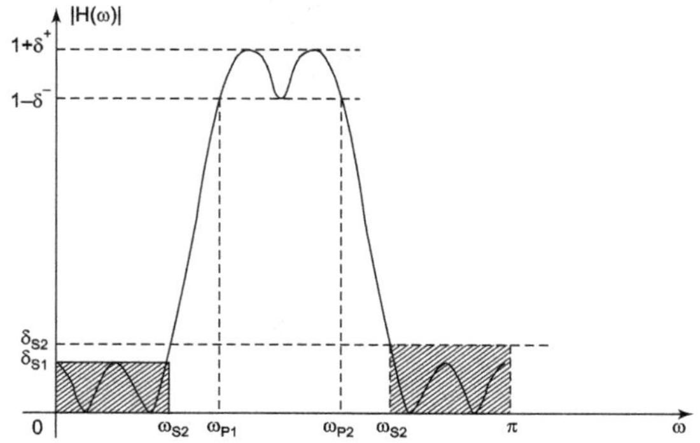

FIGURE 8.7 Band-stop filter specifications.

Band-stop filter: Specifications of a band-stop filter is given in terms of its frequency response $H(\omega)$ as

$$\begin{cases} 1-\delta_1^{-} \le |H(\omega)| \le 1+\delta_1^{+}, & 0 \le \omega \le \omega_{P_1} \\ 0 \le |H(\omega)| \le \delta_S, & \omega_{P_1} \le \omega \le \omega_{P_2} \\ 1-\delta_2^{-} \le |H(\omega)| \le 1+\delta_2^{+}, & \omega_{s_2} \le \omega \le \pi \end{cases} \tag{8.14}$$

Its specifications are shown in Figure 8.8.

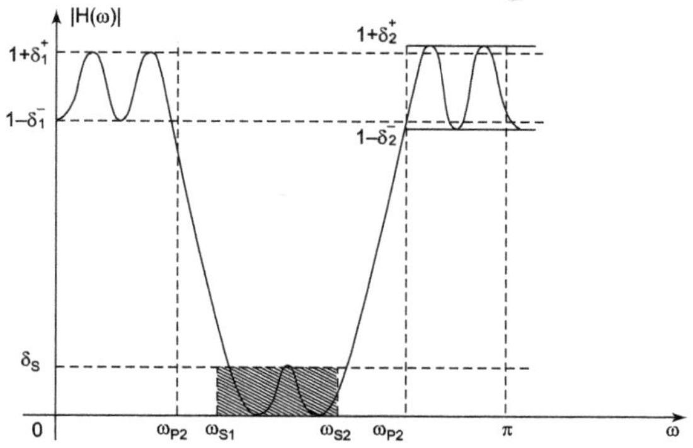

FIGURE 8.8 Band-stop filter specifications.

8.2.3 Phase Response Specifications

The frequency response of a filter can be written as

$$\begin{aligned} H(\omega) &= H_R(\omega) + jH_I(\omega) \\ &= |H(\omega)|e^{j\theta(\omega)} \end{aligned} \tag{8.15}$$

in terms of the real and imaginary part $H_R(\omega)$ and $H_I(\omega)$, respectively and also in terms of magnitude response $|H(\omega)|$ and phase response (ω).

Where magnitude response

$$A(\omega) = |H(\omega)| = \sqrt{H_R^2(\omega) + H_I^2(\omega)} \tag{8.16}$$

and phase response

$$\theta(\omega) = \tan^{-1}\left[\frac{H_1(\omega)}{H_R(\omega)}\right] \tag{8.17}$$

$H(\omega)$ is a continuous function of ω.

A digital filter whose frequency response can be expressed in continuous-phase from as

$$H(\omega) = A(\omega)e_P^{j\omega\pi} \tag{8.18}$$

is said to have a linear phase.

Where τ_P is the phase delay (measured in samples). Linear phase filters preserve input signal shape with minimum distortion.

A digital filter with frequency response $H(\omega) = A(\omega)ej^{(\theta_0 - \omega\tau g)}$ is called generalized-linear phase filter.

Where τ_g is group delay measured in samples. In such cases, the envelope of the output signal is approximately a distortionless version of that of the input signal.

Generalized-linear phase real digital filters are of four types. Real digital filters are filters that work on the real sequences and not on the imaginary sequences. For these filters, both input and output are real sequences. The sequences on which these filters are used will have no imaginary part in them.

Type I. These have integral group delay and their initial phase delay is zero.

Type II. These have fractional group delay and their integral phase delay is zero.

Type III. These have integral group delay and their initial phase delay is $\frac{\pi}{2}$.

Type IV. These have fractional group delay and their initial phase delay is $\frac{\pi}{2}$.

Types I and II have constant phase delay while types III and IV have varying phase delay.

Minimum-phase filter is a filter whose all zeros are located inside or on the unit circle.

8.2.4 Filter Design

Now we determine unit-sample response $h(n)$ or system function $H(z)$ with targeted specifications for minimum complexity.

FIR filters are designed either by the windowing design method or by least-square design method. These filters are discussed in Chapter 10.

An analog IIR filter is first designed and converted into an equivalent digital IIR filter. This conversion required four steps:

Step I. A given analog filter is converted into a digital filter that has approximately identical frequency response.

Step II. Specifications of the above digital IIR filter are transformed to those of the analog IIR filter.

Step III. Now analog IIR filter design has been completed.

Step IV. The digital design is extracted from the analog design.

This approach of designing the digital filter from analog filter easy and reliable but there is a lack of generality.

Flow chart of digital filter design (digital IIR filter) is shown in Figure 8.9.

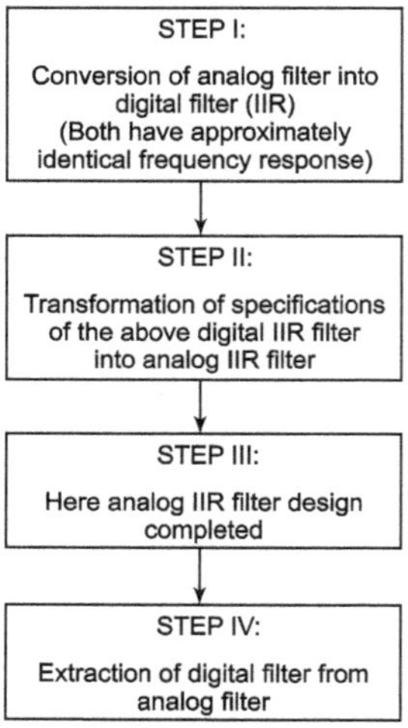
STEP I:
Conversion of analog filter into digital filter (IIR) (Both have approximately identical frequency response)
STEP II:
Transformation of specifications of the above digital IIR filter into analog IIR filter
STEP III:
Here analog IIR filter design completed
STEP IV:
Extraction of digital filter from analog filter

FIGURE 8.9 Flowchart of digital filter design (digital IIR filter).

8.2.5 Filter Realization

Construction of a block diagram of the filter comprising of elementary components such as adders, multipliers, and delay elements is called *filter realization*. A digital filter can be realized by direct-forms (I and II), parallel, cascade, and transposed forms.

8.2.6 Filter Implementation

Filter implementation is carried out by building the filter either in software or in hardware. Filters are also implemented by a combination of software and hardware.

8.3 COMPARISON BETWEEN DIGITAL AND ANALOG FILTERS

A comparative study between digital filter and analog filter is given in Table 8.4.

TABLE 8.4

S. No.	Analog filters	Digital filter
1.	In analog filters, both inputs and outputs are continuous-time signals.	In digital filters, both inputs and outputs are discrete-time signals.
2.	Implementation of such filters is carried out using passive components (resistors, capacitors, inductors) and active components (transistors, operational amplifiers).	These are implemented on a digital computer or microcomputer using DSP integrated circuits. Three basic elements such as adder, multiplier, and delay elements are used for implementing digital filters.
3.	These operate in the infinite frequency range theoretically but are limited in practice by the finite maximum operating frequencies of the semiconductor devices used. For example, OP AMPs function upto 100 MHz, and higher frequencies are handled by microwave devices.	The frequency range is restricted to half of the sampling rate. It is also restricted by the maximum computational speed available in a particular application. This is a drawback of a digital filter.
4.	Main disadvantages of analog filters are their higher noise sensitivity, nonlinearities, dynamic range limitations, lack of flexibility in designing and reproductivity, errors generated due to drift, and variations in the value of active and passive components used in circuits.	Digital filters require additional *A/D* and *D/A* converter sections for connecting to the physical analog world.
5.	These have a higher frequency range of operation as well as they can interact directly with the real analog world.	The main advantages of digital filters are that they are insensitive to noise, have higher linearity, unlimited dynamic range, flexibility in software design, high accuracy, reliability is higher.

8.4 COMPARISON BETWEEN IIR AND FIR DIGITAL FILTERS

A comparison between IIR and FIR digital filters is given in Table 8.5.

TABLE 8.5

S. No.	IIR Digital filters	IIR Digital filters
1.	IIR digital filters are characterized by rational system function as $H(z) = \dfrac{\sum_{K=0}^{M} B_K z^{-K}}{1 + \sum_{K=1}^{N} A_k z^{-K}}$	FIR digital filters are characterized by system function which is not rational as $H(z) = \sum_{K=0}^{M-1} B_k z^{-K}$
2.	The impulse response of these digital filters is computed for infinite number of samples (points), that is, $h(n) \neq 0$, $0 \leq n < \infty$ where $h(n)$ is the impulse response of the above filter. Hence these are called Infinite Impulse Response (IIR) digital filters.	The impulse response of these digital filters is computed for finite number of samples (points), that is, $h(n) \neq 0,\ 0 \leq n \leq M-1$ $= 0$, elsewhere Hence these are called Finite Impulse Response (FIR) digital filters.
3.	These filters do not have linear phase and these are used where some phase distortion is tolerable.	These filters have linear phase characteristics. These filters are used in speech processing to eliminate the adverse effects of frequency dispersion due to the non-linearity of phase.
4.	Theoretically, these filters are stable. After truncation their coefficients become unstable.	These filters are realized by direct convolution which is why these are stable.
5.	These filters have less flexibility for obtaining non-standard frequency responses or for those filters for which analog filter design techniques are not available.	These filters have greater flexibility to control the shape of their magnitude response and realization efficiency.
6.	These filters are usually realized by a recursive method. The present output of these filters also depends on previous outputs as well as present and past inputs. It is a feedback system.	These filters are generally realized nonrecursively or by direct convolution. These are not feedback systems. The present output of these filters does not depend on previous outputs.

(Continued)

S. No.	IIR Digital filters	IIR Digital filters
7.	IIR filters are more susceptible to round-off noise associated with finite precision arithmetic, quantization error, and coefficient in accuracies.	These effects are less severe in FIR digital filters.
8.	There is a shorter time delay in these filters.	Time delay increases with increase in the order of the filter.
9.	These, require a lesser number of arithmetic operations and these have lower computational complexity and smaller memory requirements.	For sharp amplitude response, we require higher-order FIR digital filter. This is the main drawback of an FIR filter.
10.	IIR filters have resemblance with analog filters. The common method for IIR digital filter design is to design an IIR analog filter followed by analog-to-digital transformation by either method given below: invariant impulse response method, bilinear transformation method, etc.	FIR filters are unique to the discrete-time domain. These cannot be derived from analog filters.

8.5 REALIZATION PROCEDURES FOR DIGITAL FILTERS

Digital filters can be realized in various ways. Realization depends upon the relationship between input sequence and output sequence of the digital filter. These realization procedures can be classified into three major categories given as:

1. recursive realization,
2. non-recursive realization, and
3. FFT realization.

8.5.1 Recursive Realization

A digital filter is said to be recursively realized if its present output $y(n)$ depends both on the previous outputs as well as on the present and previous inputs. It requires the feedback of output. Recursive realization is shown in Figures 8.10 and 8.11:

$$\text{System function} \quad H(z) = \frac{Y(z)}{S(z)} = \frac{B_0 + B_1^{z-1} + B_2^{z-2} + B_3^{z-3}}{1 + A_1^{z-1} + A_2^{z-2} + A_3^{z-3}} \tag{8.19}$$

is a recursively realized digital filter.

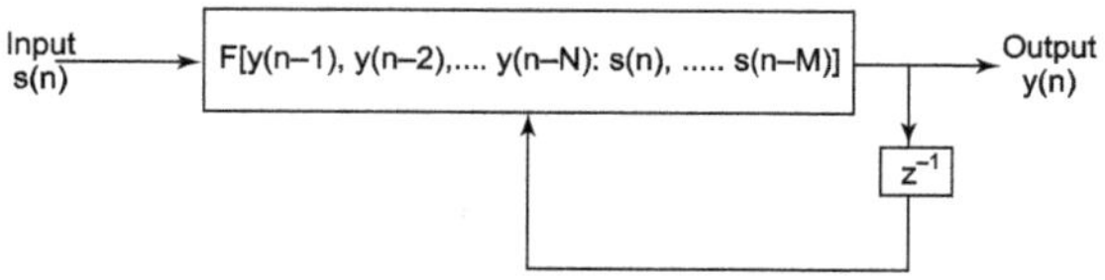

FIGURE 8.10

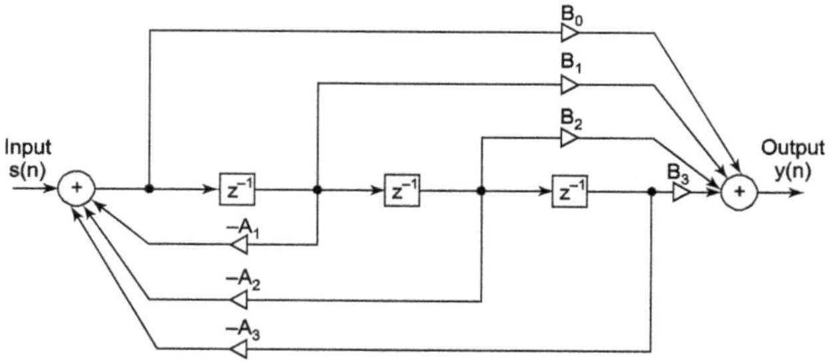

FIGURE 8.11 Block diagram of a recursively realized digital filters.

8.5.2 Non-recursive Realization

Non-recursive realization is one in which present output $y(n)$ depends only on present and past values of inputs but not on previous values of outputs. There is no feedback of outputs. Non-recursive realization is shown in Figure 8.12.

$$\text{System function} \quad H(z) = B_0 + B_1^{z-1} + B_2^{z-2} + B_3^{z-3} \tag{8.20}$$

where B_0, B_1, B_2, B_3 are the filter coefficients.

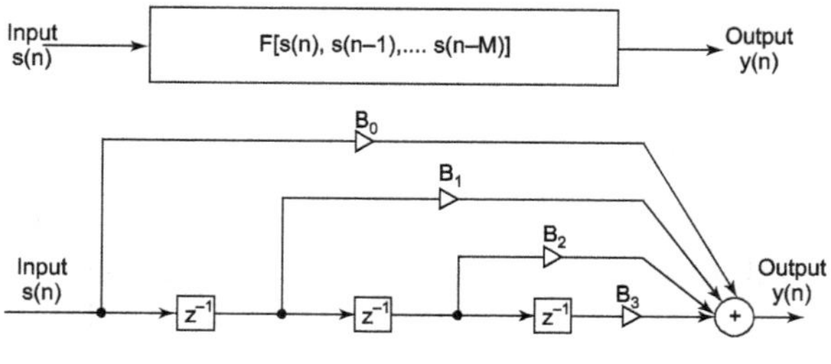

FIGURE 8.12 Block diagram of a nonrecursively realized digital filter.

8.5.3 Fast Fourier Transform (FFT) Realization

In this realization procedure, the input signal $s(n)$ is transformed by the FFT algorithms thereby increasing the computational speed considerably. After filtering the spectrum of the signal, an inverse FFT transformation is performed. FFT realization is shown in Figure 8.13.

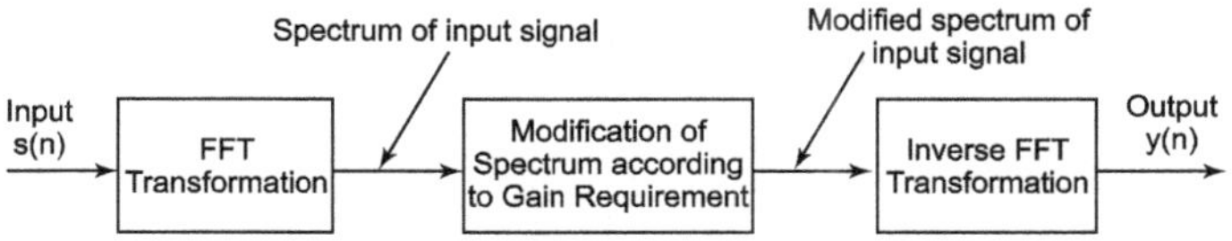

FIGURE 8.13 Block diagram of FFT realization.

From the spectrum of input signal, we can choose any frequency content either low-frequency, high-frequency or other frequencies.

Note. IIR digital filters are easily realized using the recursive realization procedure. FIR digital filters are easily realized using either non-recursive realization or FFT realization procedures.

8.6 NOTCH FILTERS

A notch filter is a filter that contains one or more notches or ideally perfect nulls in its frequency response characteristics. Figure 8.14 shows the frequency response characteristics of a notch filter.

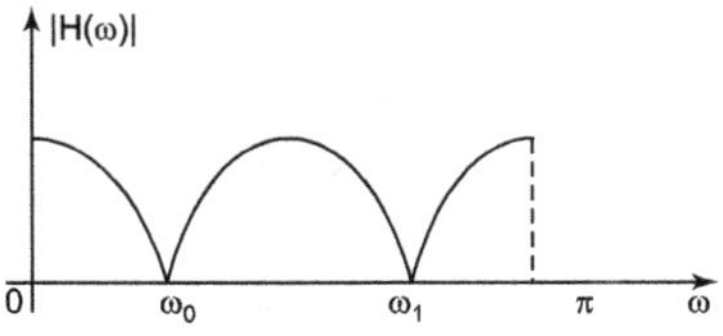

FIGURE 8.14 Frequency response characteristics of a notch filter.

There are two nulls at frequencies ω_0 and ω_1. Null points are frequency points where amplitudes are zero.

These filters are useful in many applications where some specific frequency components are to be eliminated. These are used in instrumentation and recording systems that have a power-line frequency of 60 Hz or 50 Hz and their harmonics are to be eliminated.

To insert a null in the frequency response of a filter at a frequency ω_0, there is a need to introduce a pair of complex-conjugate zeros on the unit circle at an angle.

$$z_1 = e^{j\omega 0}$$

$$z_2 = z_1^* = e^{-j\omega 0}$$

Thus the system function for an FIR notch filter is given by

$$\begin{aligned} H(z) &= B_0\left(1 - z_1 z^{-1}\right)\left(1 - z_2 z^{-1}\right) \\ &= B_0(1 - e^{j\omega 0_z -1})(1 - e^{j\omega 0_z -1}) \\ &= B_0\left[(1 - e^{j\omega 0})(1 - e^{j\omega 0})z^{-1} + z^{-2}\right] \\ &= B_0\left[1 - 2\left(\frac{e^{j\omega_0} + e^{-j\omega_0}}{2}\right)z^{-1} + z^{-2}\right] \end{aligned}$$

or $$H(z) = B_0\left[1 - 2\cos\omega_0 z^{-1} + z^{-2}\right] \tag{8.21}$$

where B_0 is a constant.

Figure 8.15 shows the frequency response of a notch FIR filter which has a null at $\omega = \dfrac{\pi}{4}$.

There is a problem associated with the FIR notch filter. The problem is that the notch has a relatively large bandwidth. Due to the large bandwidth, other frequency components around the desired null are severely attenuated. For reducing the bandwidth of the null, we resort to a more sophisticated longer FIR filter.

Now we want to improve the frequency response characteristics by inserting poles in the system function $H(z)$ of Eq. (8.21).

We place a pair of complex-conjugate poles at

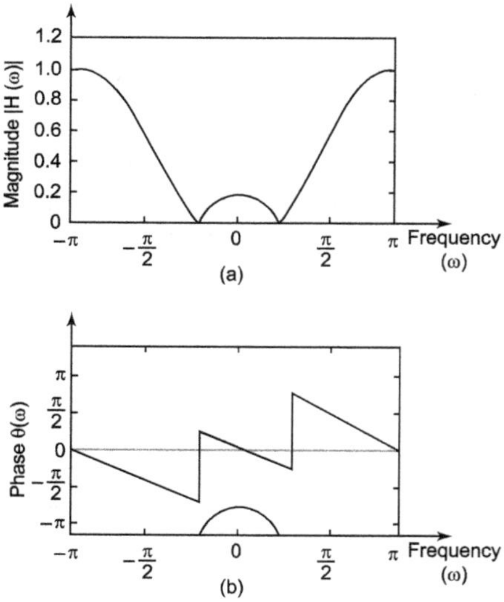

FIGURE 8.15 Frequency response characteristics of a notch filter without poles $H(z) = B_0\,[1 - 2\cos\omega_0 z^{-1} + z^{-2}]$ where B_0 is a constant, (a) magnitude response characteristics of a notch filter (null at $\omega = \pi/4$) and (b) phase response characteristics of a notch filter (null at $\omega = \pi/4$).

$$p_1 = re^{j\omega 0} \text{ and } p_2 = p_1^* = re^{-j\omega 0}$$

These poles introduce a resonance in the vicinity of the null and reduce the bandwidth of the notch. The resulting system function of a notch filter with one pair of the complex conjugate of zeros and one pair of the complex conjugate of poles is given by

$$H(z) = \frac{B_0\left(1 - z_1 z^{-1}\right)\left(1 - z_2 z^{-1}\right)}{\left(1 - p_z z^{-1}\right)\left(1 - p_1 z^{-1}\right)}$$

$$= \frac{B_0\left(1 - e^{j\omega 0} z^{-1}\right)\left(1 - e^{j\omega_0} z^{-1}\right)}{\left(1 - re^{j\omega_0} z^{-1}\right)\left(1 - re^{-j\omega_0} z^{-1}\right)}$$

or
$$H_z = B_0 \frac{1 - 2\cos\omega_0 z^{-1} + z^{-2}}{1 - 2r\cos\omega_0 z^{-1} + r^2 z^{-2}} \tag{8.22}$$

The frequency response characteristics of a notch filter with a pole at $r = 0.85$ and null at $\omega = \frac{\pi}{4}$ given in Eq. (8.22) are plotted in Figure 8.16.

By comparing the frequency response of the FIR filter shown in Figures 8.15 and 8.16, we see that the introduction of poles reduces the bandwidth of the notch. The introduction of a pole in the vicinity of the null reduces the bandwidth of the notch and also results in a small ripple in the pass-band of the filter due to resonance produced by the pole. The effect of the ripple can be reduced by inserting additional poles and/or zeros in the system function of the notch FIR filter.

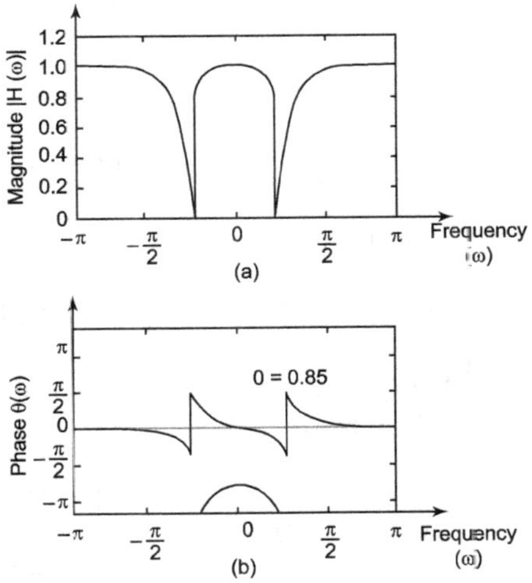

FIGURE 8.16 Frequency response characteristics of a notch filter with pole $r = 0.85$ and null at $\omega = \pi/4$. $H_z = B_0 \frac{1 - 2\cos\omega_0 z^{-1} + z^{-2}}{1 - 2r\cos\omega_0 z^{-1} + r^2 z^{-2}}$

(a) Magnitude response characteristics of a notch filter with pole at $r - 0.85$ and null at $\omega = \pi/4$.
(b) Phase response characteristics of a notch filter with pole at $r = 0.85$ and null at $\omega = \pi/4$.

8.7 COMB FILTERS

The simplest form of a comb filter can be viewed as a notch filter in which the nulls occur periodically across the frequency band. It has an analogy with an ordinary comb filter which has periodically spaced teeth. These types of filters have applications in a wide range of practical systems such as in the rejection of power-line harmonics, suppression of clutter from fixed objects in moving-target-indication radars.

A simple form of comb filter is illustrated by considering a moving-average FIR filter. It is described by the difference equation given as

$$y(n) = \frac{1}{M+1}\sum_{m=0}^{M} s(n-m) \tag{8.23}$$

Transfer function of the above filter of Eq. (8.23) can be determined by taking z-transform of both sides of Eq. (8.23),

$$Y(z) = \mathcal{Z}\left[\frac{1}{M+1}\sum_{m=0}^{M} s(n-m)\right]$$

$$= \frac{1}{M+1}\left[\sum_{m=0}^{M} s(n-m)\right] = \frac{1}{M+1}\sum_{m=0}^{M} z^{-m}S(z)$$

or

$$\frac{Y(z)}{S(z)} = \frac{1}{M+1}\sum_{m=0}^{M} z^{-m}$$

where $\sum_{m=0}^{M} z^{-m}$ is a geometric progression with first term equal to 1 and common ratio is equal to z^{-1}.

or

$$H(z) = \frac{Y(z)}{S(z)} = \frac{1}{M+1}\left[\frac{1-(z^{-1})^{M+1}}{1-z^{-1}}\right] \tag{8.24}$$

Frequency response of FIR filter given in Eq. (8.24) is determined by substituting $z = e^{j\omega}$,

or

$$H(\omega) = \frac{1}{M+1}\left[\frac{1-\left(e^{j\omega}\right)^{-M-1}}{1-\left(e^{j\omega}\right)^{-1}}\right]$$

$$= \frac{1}{M+1}\left[\frac{1-e^{-j\omega(M+1)}}{1-e^{-j\omega}}\right]$$

$$= \frac{1}{M+1}\left[\frac{1-e^{j\omega(M+1)/2}-e^{-j\omega(M+1)/2}}{e^{-j\omega/2}-e^{-j\omega/2}}\right]e^{-j\omega M/2}$$

$$= \frac{1}{M+1}\left[\frac{\left(\frac{e^{j\omega(M+1)/2}-e^{j\omega(M+1)/2}}{2j}\right)}{\left(\frac{e^{j\omega/2}-e^{-j\omega/2}}{2j}\right)}\right]e^{-j\omega M/2}$$

or $$H(\omega)=\frac{1}{M+1}e^{-j\omega M/2}\left[\frac{\sin\omega[M+1]/2}{\sin(\omega/2)}\right] \tag{8.25}$$

From observation of Eq. (8.25), we see that the filter has zeros on the unit circle at

$$z=e^{j2\pi m/(M+1)}, m=0,\ 1,\ \ldots,M \tag{8.26}$$

Note. Pole at $z = 1$ is cancelled by the zero at $z = 1$. So that the FIR filter does not contain poles outside $z = 0$.

A plot of mangitude characteristics of frequency response of Eq. (8.25) shows the existence of the periodically spaced zeros in frequency response at $\omega_m = \frac{2\pi m}{M+1}$ for $m = 1, 2, 3,\ldots M$. Figure 8.17 shows magnitude response characteristics of filter given in Eq. (8.25).

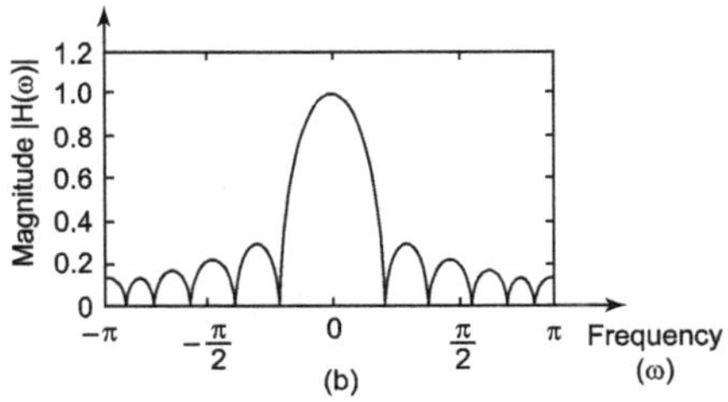

FIGURE 8.17 Magnitude response characteristics for comb filter is given by Eq. (7.23) with $M = 10$.

In the general case, we can develop a comb filter by taking an FIR filter. It is given by

$$H(z)=\sum_{m=0}^{M}h(m)z^{-m} \tag{8.27}$$

Replacing z by z^P in Eq. (8.27), where P is the positive integer, we get

$$H_P(z)=\sum_{m=0}^{M}h(m)\left(z^P\right)^{-m}=\sum_{m=0}^{M}h(m)z^{-Pm} \tag{8.28}$$

If $H(\omega)$ is the frequency response of the original FIR filter then frequency response of the filter of Eq. (8.28) is given by

$$H_P(\omega)=\sum_{m=0}^{M}h(m)e^{-j\omega Pm}=H(P\omega) \tag{8.29}$$

Consequently, frequency response characteristics $HP(\omega)$ is a P-order repetition of frequency response of original filter $H(\omega)$ in the range

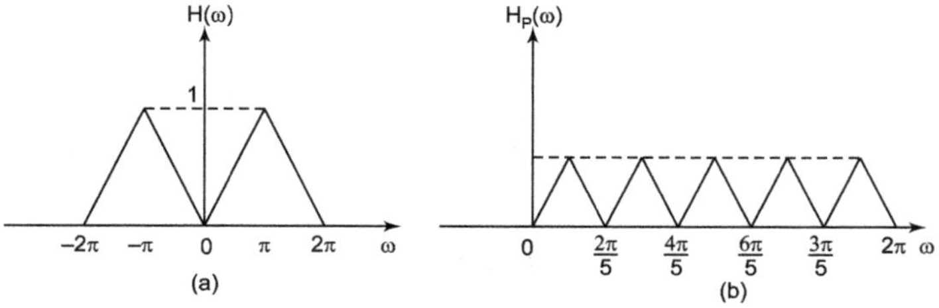

FIGURE 8.18 Comb filter with frequency response $H_p(\omega)$ obtained from $H(\omega)$. (a) Frequency response $H(\omega)$. (b) Frequency response $H_p(\omega)$ for $P = 5$.

$0 \le \omega < 2\omega$. Figure 8.18 shows the relationship between $H_P(\omega)$ and $H(\omega)$ for $P = 5$.

Now, let the original FIR filter with transfer function $H(z)$ have a spectral null at some frequency. Then the filter with transfer function $H_P(z)$ has periodically spaced spectral null at

$$\omega_m = \omega_0 + \frac{2\pi m}{P}, m = 0,1,2,3,...P-1$$

Figure 8.19 shows an FIR comb filter with $M = 3$ and $P – 3$. This filter

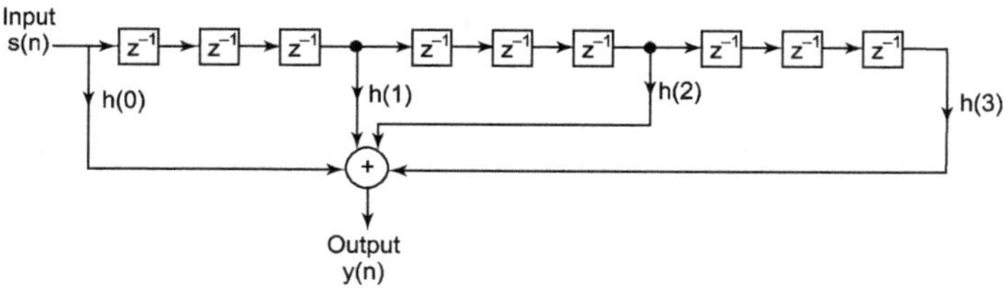

FIGURE 8.19 Network structure of an FIR digital comb filter with $M = 3$ and $P = 3$.

can be viewed as an FIR filter of length 10 with only 4 of the 10 filter coefficients non-zero. We know that the transfer function of the original FIR filter is given by

$$H(z) = \frac{1}{M+1}\left[\frac{1-z^{-(M+1)}}{1-z^{-1}}\right]$$

But we also know that

$$H_P(\omega) = H\left(z^P\right) = \frac{1}{M+1}\left[\frac{1-z^{-P(M+1)}}{1-z^{-P}}\right] \tag{8.30}$$

The frequency response of the original FIR filter is given by

$$H(\omega) = \frac{e^{-j\omega M/2}}{M+1}\frac{\sin\omega[(M+1)/2]}{\sin(\omega/2)}$$

But we know that $H_p(\omega) = H(P\omega)$

$$\therefore \qquad H_P(\omega) = \frac{e^{-j\omega M/2}}{M+1} \frac{\sin P_\omega[(M+1)/2]}{\sin(P\omega/2)} \tag{8.31}$$

The filter has zeros on the unit circle at $z = e^{j2\pi m/L(M+1)}$, for all integer values of *in* except *in* $m = 0, P, 2P, \ldots, MP$.

Figure 8.20 shows magnitude response for $P = 5$ and $M = 10$.

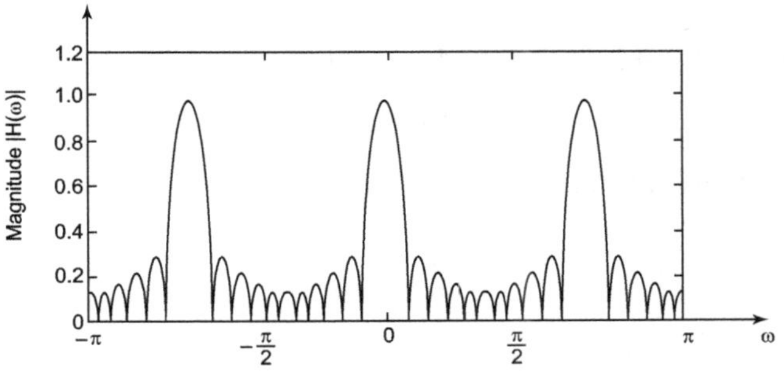

FIGURE 8.20 Magnitude response characteristics for a comb filter given by Eq. (8.31) with $P = 5$ and $M = 10$.

8.8 ALL-PASS FILTERS

An all-pass filter is defined as a system that has a constant magnitude response at all frequencies.

$$|H(\omega)| = 1, \quad 0 \le \omega \le \pi \tag{8.32}$$

The pure delay system is an example of an all-pass filter and its system function is given by

$$H(z) = z^{-m} \tag{8.33}$$

This system passes all signals without modification except for a delay of m samples. All-pass filters have application as phase equalizers. When these are placed in a cascade with a system that has an undesired phase response, a phase equalizer compensates for the poor phase characteristics of the system and produces an overall linear phase response.

8.9 DIGITAL SINUSOIDAL OSCILLATORS

It can be viewed as a limiting form of a two-pole resonator. This resonator has complex-conjugate poles which lie on the unit circle.

The transfer function of second-order systems is given by

$$H(z) = \frac{B_0}{1 + A_1 z^{-1} + A_2 z^{-2}} \tag{8.34}$$

where A_1 and A_2 are parameters and are given by

$$A_1 = -2r\cos\omega_0 \text{ and } A_2 = r^2$$

This has complex-conjugate poles at $p = re^{\pm j\omega 0}$

$$H(z) = \frac{B_0}{1 - 2r\cos\omega_0 z^{-1} + r^2 z^{-2}} \tag{8.35}$$

Unit-sample response,

$$h(n) = \mathcal{Z}^{-1}[H(z)] = \mathcal{Z}^{-1}\left[\frac{B_0}{1 - 2r\cos\omega_0 z^{-1} + r^2 z^{-2}}\right]$$

or

$$h(n) = \frac{B_0 r^n}{\sin\omega_0}\sin(n+1)\omega_0 . u(n) \tag{8.36}$$

If the poles are located on the unit circle $|z| = 1$ or $r = 1$ and B_0 is set to A_0 $\sin\omega_0$, then

$$h(n) = A_0 \sin(n+1)\omega_0 u(n) \tag{8.37}$$

Impulse response of the second-order system with complex-conjugate poles on the unit circle $|z|$ is sinusoidal and this system is called a **digital sinusoidal oscillator**. It is also called a digital sinusoidal generator. It is a basic component of a digital frequency synthesizer.

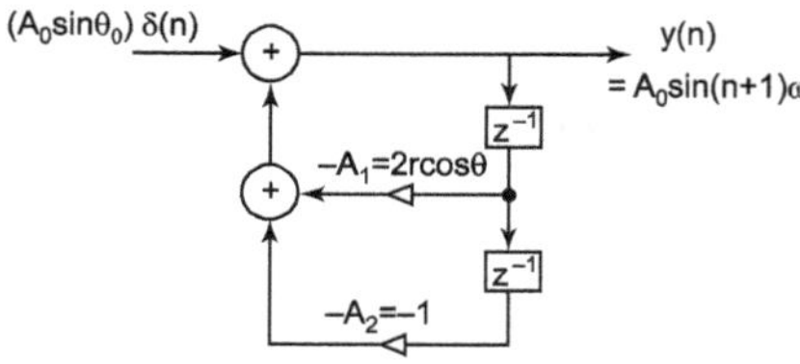

FIGURE 8.21 Digital sinusoidal oscillator.

The difference equation of system of Figure 7.21 is given by

$$y(n) = A_1(n-1) - y(n-2) + B_0\delta(n) \tag{8.38}$$

where parameters $A_1 = -2\cos\omega_0$ and $B_0 = A\sin\omega_0$

The initial conditions are

$$y(-1) = 0 \text{ and } y(-2) = 0$$
$$y(n) = -A_1 y(n-1) - y(n-2) + B_0 \delta(n)$$

By iteration method using initial conditions given above

$n = 0,$ $\quad y(0) = +2\cos\omega_0 y(-1) - y(-2) + A\sin\omega_0\delta(0)$
$\quad = +2\cos\omega_0(0) - (0) + A\sin\omega_0(l) = A\sin\omega_0$

$n = 1,$ $\quad y(1) = +2\cos\omega_0 y(0) - y(-1) + A\sin\omega_0\delta(l)$
$\quad = +2\cos\omega_0(A\sin\omega_0) - (0) + A\sin\omega_0(0)$
$\quad = +A\sin 2\omega_0$

$n = 2,$ $\quad y(2) = 2\cos\omega_0 y(l) - y(0) + A\sin\omega_0\delta(2)$
$\quad = 2\cos\omega_0(A\sin 2\omega_0) - A\sin\omega_0 + A\sin\omega_0(0)$
$\quad = A2\cos\omega_0 \sin 2\omega_0 - A\sin\omega_0 + 0$
$\quad = 2A\cos\omega_0[2\sin\omega_0\cos\omega_0] - A\sin\omega_0$
$\quad = A\sin\omega_0[4\cos^2\omega_0 - l]$
$\quad = 3A\sin\omega_0 - 4\sin^3\omega_0 = A\sin 3\omega_0$
$\quad \vdots$

and so forth.

Note. Application of the impulse at $n = 0$ serves the purpose of "beginning of sinusoidal oscillation."

In some practical applications, we require modulation of two sinusoidal carrier signals in phase quadrature, one is A sin (ω0*n* and second is A cos ω0*n*. These signals can be generated by coupled-form oscillator. These signals can be obtained by using trigonometric formulas given below

$$\cos(C + D) = \cos C\cos D - \sin C\sin D$$
$$\sin(C + D) = \sin C\cos D + \cos C\sin D$$

where from definition $C = \omega_0$, $D = \omega_0$

and $\quad y_c(n) = \cos n\omega_0 u(n) \quad (8.39)$

$\quad y_s(n) = \sin n\omega_0 u(n) \quad (8.40)$

Thus we find two coupled difference equations

$$y_c(n) = (\cos\omega_0)y_c(n-1) - (\sin\omega_0)y_s(n-1) \quad (8.41)$$

$$y_s(n) = (\sin\omega_0)y_c(n-1) + (\cos\omega_0)y_s(n-1) \quad (8.42)$$

Eqs. (8.41) and (8.42) can be written in the matrix form as

$$\begin{bmatrix} y_c(n) \\ y_s(n) \end{bmatrix} = \begin{bmatrix} \cos\omega_0 & -\sin\omega_0 \\ \sin\omega_0 & \cos\omega_0 \end{bmatrix} \begin{bmatrix} y_c(n-1) \\ y_s(n-1) \end{bmatrix} \quad (8.43)$$

The structure of the realization of Eq. (8.43) is shown in Figure 8.22.

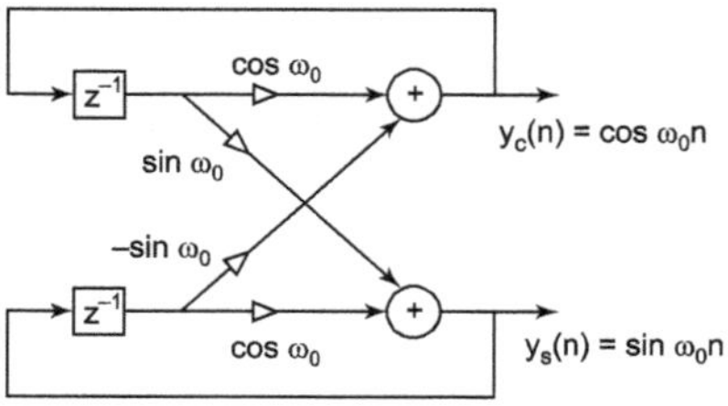

FIGURE 8.22 Realization of the coupled-form oscillator.

8.10 DIGITAL RESONATORS

It is a special two-pole band-pass filter with the pair of complex-conjugate poles placed near the unit circle $|z| = 1$ or $r = 1$. Here resonator means that the filter has a large magnitude response in the vicinity of the pole location. The resonant frequency of the filter is determined by determining the angular position of the pole. These are useful in many applications such as in band-pass filtering and speech generation.

Figure 8.23 illustrates the pole-zero pattern of a digital resonator without zeros.

Figure 8.24 shows the magnitude response of a digital resonator without zeros.

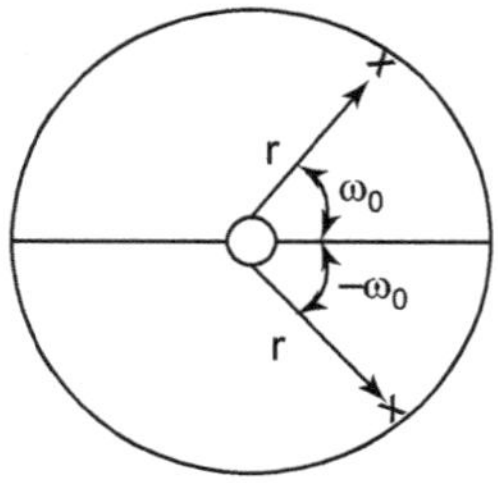

FIGURE 8.23 Pole-zero pattern of digital resonator without zeros.

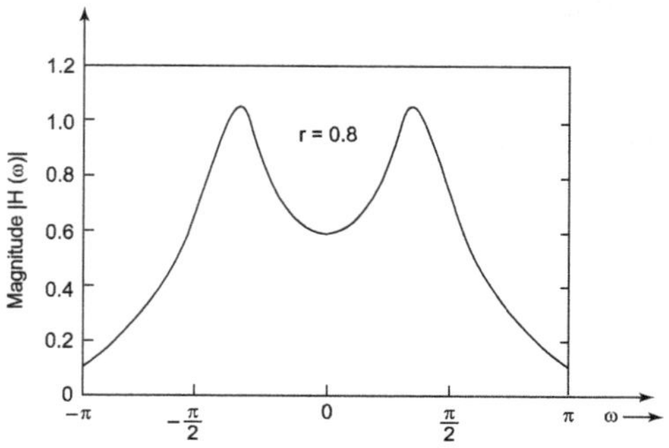

FIGURE 8.24 Magnitude response of a digital resonator without zeros.

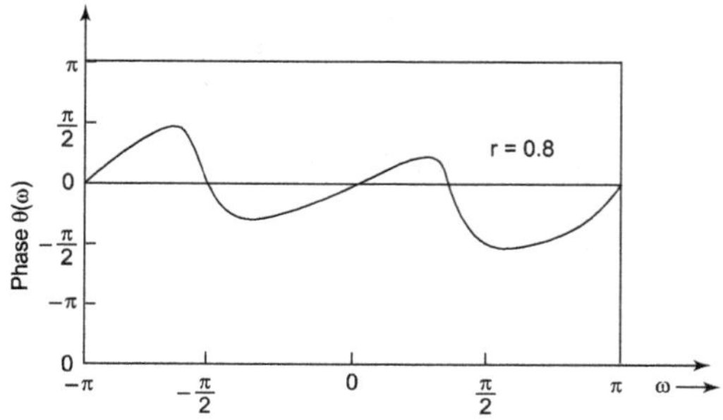

FIGURE 8.25 Phase response of a digital resonator without zeros.

Figure 8.25 shows the phase response of a digital resonator without zeros. Digital resonators are designed by selecting the complex-conjugate poles at

$$p_1 = re^{j\omega 0}$$

and $$p_2 = p_1^* = re^{j\omega 0}, \ 0 < r < 1$$

In addition to one pair of complex-conjugate poles, we can select up to two zeros. There are many possible choices but we have two cases of special interest.

Case I. One choice is to locate the zeros at the origin ($z = 0$).
Case II. Another choice is to locate a zero at $z = 1$ and other at $z = 1$.

Choice of case II completely removes the response of the filter at frequencies $\omega = 0$ and $\omega = \pi$. It is useful in many DSP applications.

The transfer function of a digital resonator with zeros at the origin $z = 0$ is given by

$$H(z) = \frac{B_0}{\left(1 - re^{j\omega_0} z^{-1}\right)\left(1 - re^{j\omega_0} z^{-1}\right)} \tag{8.45}$$

or $$H(z) = \frac{B_0}{1 - 2r\cos\omega_0 z^{-1} + r^2 z^{-2}} \tag{8.46}$$

The frequency response of digital resonator given in Eq. (8.45) is.

$$H(z) = \frac{B_0}{\left(1 - re^{j\omega_0} e^{-j\omega}\right)\left(1 - re^{j\omega_0} e^{-j\omega}\right)}$$

$$H(\omega_0) = H(\omega)\big|_{\omega=\omega_0} = \frac{B_0}{\left(1 - re^{j\omega_0} e^{-j\omega}\right)\left(1 - re^{j\omega_0} e^{-j\omega_0}\right)}$$

$$= \frac{B_0}{(1 - r)\left(1 - re^{-j2\omega_0}\right)}$$

Magnitude of $H(\omega_0)$

$$|H(\omega_0)| = \frac{B_0}{(1 - r)\sqrt{1 + r^2 - 2r\cos 2\omega_0}} \tag{8.47}$$

Now we want to determine the value of B_0 for unity magnitude,

$$|H(\omega_0)| = 1 \tag{8.48}$$

or $$\frac{B_0}{(1 - r)\sqrt{1 + r^2 - 2r\cos 2\omega_0}} = 1$$

or $$B_0 = (1-r)\sqrt{1+r^2-2r\cos 2\omega_0}$$

where B_0 is the desired normalization factor.

Frequency response of the resonator

$$H(\omega) = \frac{B_0}{\left(1-re^{j\omega_0}z^{-1}\right)\left(1-re^{j\omega_0}z^{-1}\right)}$$

can be expressed as

$$|H(\omega)| = \frac{B_0}{\sqrt{1+r^2-2r\cos(\omega_0-\omega)}\sqrt{1+r^2-2r\cos(\omega_0-\omega)}}$$

$$= \frac{B_0}{V_1(\omega)V_2(\omega)}$$

where $V_1(\omega)$ and $V_2(\omega)$ are the magnitudes of the vectors from p_1 and p_2 to the point ω in the unit circle $|z| = 1$ or $r = 1$.

$$V_1(\omega) = \sqrt{1-r^2-2r\cos(\omega_0-\omega)} \quad (8.50)$$

$$V_2(\omega) = \sqrt{1+r^2-2r\cos(\omega_0-\omega)} \quad (8.51)$$

$\theta_1(\omega)$ and $\beta_2(\omega)$ are angles of vectors $V_1(\omega)$ and $V_1(\omega)$, respectively.

$$\theta(\omega) = 2\omega - \theta_1(\omega) - \theta_2(\omega) \quad (8.52)$$

$V(\omega)$ takes its minimum value $(1 - r)$ at $\omega = \omega_0$ for any value r.

Product $V_1(\omega)$, $V_2(\omega)$ reaches a minimum value at the frequency.

$$\omega_r = \cos^{-1}\left(\frac{1+r^2}{2r}\cos\omega_0\right) \quad (8.53)$$

It is the resonant frequency of the filter. We have observed that when r is very close to unity, the resonant frequency ω will be approximately equal to ω_0. ω_0 is the angular position of the pole. It is also observed that as r approaches unity, the resonance peak becomes sharper because $V_1(\omega)$ changes more rapidly in relative size in the vicinity of ω_0. A quantitative measure of the sharpness of the resonance is given by the 3-dB bandwidth of the filter.

3-dB bandwidth is given by

$$\Delta\omega \approx 2(1-r) \quad (8.54)$$

r is close to unity.

Figures 8.24 and 8.25 show the magnitude and phase response of a digital resonator for $\omega = \pi/3$.

Phase response changes very rapidly and suddenly near resonant frequency.

System function of the digital resonator when the zeros of the digital resonator are placed at $z = 1$ and $z = -1$ is given by

$$H(z) = \frac{(1 - z^{-1})(1 + z^{-1})}{\left(1 - re^{j\omega_0} z^{-1}\right)\left(1 - re^{j\omega_0} z^{-1}\right)} \tag{8.55}$$

$$= E \frac{1 - z^2}{1 - 2r\cos\omega_0 z^{-1} + r^2 z^{-2}}$$

where E is a constant.

Frequency response is determined by substituting $z = e^{j\omega}$ in Eq. (8.56),

$$H(\omega) = E \frac{1 - e^{-2j\omega}}{\left[1 - 2re^{j(\omega_0 - \omega)}\right]\left[1 - 2re^{j(\omega_0 - \omega)}\right]} \tag{8.57}$$

Here we have observed that the zeros at $z = 1$ and $z = -1$ affect both the magnitude and phase response of digital resonator.

Magnitude response is given by

$$|H(\omega)| = E \frac{N(\omega)}{V_1(\omega)V_2(\omega)} \tag{8.58}$$

where $N(\omega)$ is given by

$$N(\omega) = \sqrt{2(1 - \cos 2\omega)} \tag{8.59}$$

Due to the presence of the zero factor, the resonant frequency changes from that given by

$$\omega_r = \cos^{-1}\left(\frac{1 + r^2}{2r}\cos\omega_0\right)$$

The bandwidth of the filter also changes. Figure 8.26 shows the magnitude and phase characteristics for $\omega_0 = \pi/3$ and $r = 0.8$. Here we observe that this filter has a slightly smaller bandwidth than the resonator which has zeros at the origin, $|z| = 0$.

There is a very small shift in the resonant frequency due to the presence of the zeros.

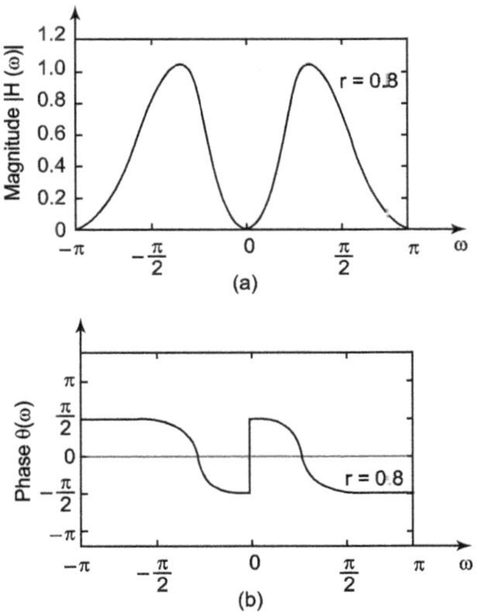

FIGURE 8.26 Frequency response characteristics of a digital resonator with zero at $\omega = 0$ and $\omega = \pi$ and $\omega_0 = \pi/3$ and $r = 0.8$.
(a) Magnitude response characteristics of a digital resonator with zeros at $\omega = 0$ and $\omega = \pi$ and $\omega_0 = \pi/3$ and $r = 0.8$.
(*b*) Phase response characteristics of a digital resonator with zeros at $\omega = 0$ and $\omega = \pi$ and $\pi = \pi/3$ and $r = 0.8$.

EXERCISES

1. What is filtering? Give some objectives of the filtering process.
2. Define analog and digital filters.
3. Give a review of analog filter design in brief.
4. What are the various considerations in using digital filters?
5. Compare digital and analog filters.
6. Compare IIR and FIR digital filters.
7. Describe various methods of filter realization such as recursive, non-recursive, and FFT realization.
8. Sketch a flow chart of digital filter design.
9. Write a summary of pole-zero characteristics of analog filters.
10. Discuss the frequency response characteristics of frequency-selective filters such as low-pass, high-pass, band-pass, and band-reject filters.
11. Describe the phase response of a filter.

CHAPTER 9

DESIGN AND ANALYSIS OF INFINITE IMPULSE RESPONSE (IIR) DIGITAL FILTERS

9.1 INTRODUCTION

Digital filters are of two types depending upon the number of sample points used to determine the unit-sample (impulse) response of an LTI discrete-time system. If infinite number of sample points are used to determine the unit-sample response then these digital filters are called Infinite-duration Impulse Response (IIR) digital filters.

IIR digital filter design procedures are extensions of those originally developed for analog filters. In fact, IIR digital filters are commonly used to replace existing analog filters.

9.2 APPROXIMATION OF IIR DIGITAL FILTERS FROM ANALOG FILTERS

The corresponding discrete-time transfer function $H_D(z)$ of an IIR digital filter is obtained from the continuous-time transfer function $H_A(s)$ of an analog filter. This approximation is obtained using one of the following methods:

1. impluse response invariance method, and
2. bilinear transformation method.

All these methods will be discussed in the next sections.

9.2.1 Realizability Constraints Imposed on Transfer Function $H_D(z)$

For a realizable IIR digital filter, its transfer function $H_D(z)$ must satisfy the following constraints:

1. It must be a rational function of z with real coefficients.
2. Poles of $H_D(z)$ must lie within the unit-circle of the z-plane, that is, $|z| = 1$.
3. The degree of numerator polynomial, $N(z)$ must be equal or less than that of the denominator polynomial, $D(z)$.

The first constraint is imposed by the assumption that the signals are real. The second and third constraints are due to the assumption of a stable and causal filter, respectively.

9.2.2 Impulse-Response Invariance Method

In this method, the impulse response of a derived digital filter is the same as that of the given reference analog filter at sampling intervals. Now we have a reference analog filter and its transfer function, $H_A(s)$. It is converted into a digital filter with the transfer function $H_D(z)$. Figure 9.1 illustrates the impulse response of an analog filter and its corresponding derived digital filter. From Figure 9.1, it is seen that the impulse response for both the filters is approximately the same at some given sampling intervals.

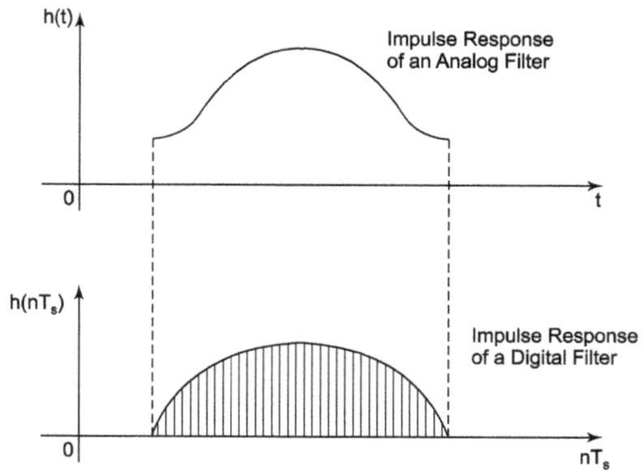

FIGURE 9.1 Impulse response for analog filter and derived digital filter is approximately the same.

This method has the following three design steps:

Step I: Deduce impulse response $h_A(t)$ of the analog filter, that is,

$$h_A(t) = \text{Inverse Laplace Transform of } H_A(s)$$
$$= \pounds^{-1}[H_A(s)].$$

Step II: Deduce the discrete-time version of $h_A(t)$. It can be achieved by putting $t = nT_s$.

$$h_D(nT_s) = h_A(t)|t = nTs, n = 0, 1, \ldots$$

where t = Continuous-time independent variable

n = Discrete-time index

T_s = Sampling interval $\frac{1}{f_s}$, in seconds

f_s = Sampling rate in H_z.

Step III: Determination of $H_D(z)$ by taking z-transform of $h_D(nT_s)$

$$H_D(z) = z\text{-transform of } h_D(nTs)$$
$$= \mathcal{Z}\,[h_D(nT_s)]$$

$H_D(z)$ is the required transfer function of digital filter.

EXAMPLE 9.1

Design an IIR digital filter from second-order analog filter whose transfer function is given as $H_A(s) = b/\{(s + a)^2 + b^2\}$ using impulse-response invariance method.

Solution:

We have the transfer function of second-order analog filter,

$$H_Z(s) = \frac{b}{(s+a)^2 + b^2} \text{ where } s = \sigma = j\Omega.$$

Step I. Impulse response $h_A(t)$ of an analog filter is determined by taking inverse Laplace transformation of $H_A(s)$.

$$h_A(t) = \text{Inverse Laplace transform of } H_A(s)$$
$$= \pounds^{-1}\left[H_A(s)\right] = \pounds^{-1}\left[\frac{b}{(s+a)^2 + b^2}\right]$$
$$= \pounds^{-1}\left[\frac{b}{(s+a+jb)(s+a-jb)}\right]$$
$$= \pounds^{-1}\left[\frac{A}{(s+a+jb)} + \frac{B}{(s+a-jb)}\right]$$

where A and B are constant (from partial fraction expansion)

$$= \pounds^{-1}\left[\frac{(-1/2j)}{(s+a+jb)} + \frac{(1/2j)}{(s+a-jb)}\right]$$
$$= -\frac{1}{2j}e^{-at}e^{jbt} + \frac{1}{2j}e^{-at}e^{jbt}$$

$$= e^{-at}\left[\frac{e^{jbt} - e^{-jbt}}{2j}\right] e^{-at} \sin bt$$

or $$h_A(t) = \begin{cases} e^{-at} \sin bt, & t \geq 0 \\ 0, & \text{otherwise} \end{cases}$$

Step II. $$h_D(nTs) = h_A(t)\big|_{t=nTs}$$

$$= \left[e^{-at} \sin bt\right]_{t=nT_s} = e^{-anT_s} \sin bnT_s$$

where T_s is the sampling period.

or $$h_D(nT_s) = \begin{cases} e^{-anT_s \sin bnT_s}, & n \geq 0 \\ 0, & \text{otherwise} \end{cases}$$

where $h_D(nT_s)$ is the impulse response of digital filter.

Step III. Determination of transfer function of digital filter $H_D(z)$ from $h_D(nTs)$.

$$H_D(z) = z\text{-transform of } h_D(nT_s)$$

$$= \mathcal{Z}\left[h_D(nT_s)\right] = \mathcal{Z}\left[e^{-anT_s} \sin bnT_s\right)$$

$$= \frac{ze^{-aT_s} \sin bT_s}{z^2 - 2ze^{-aT_s} \cos bT_s + e^{-2aT_s}}$$

This is the transfer function of derived digital FIR filter.

Poles of analog filter which are complex conjugate in the s-plane and the corresponding frequency response are shown in Figure 9.2(a). Its frequency response is seen to have a peak shape. Figure 9.2(b) illustrates the pole-zero location and frequency response of digital filters.

The magnitude has been scaled by the value $|H_A(j0)|$, so that at zero frequency, both the responses commence at zero dB.

When the analog filter is band limited the impulse-response invariance method produces a digital filter whose frequency response is given by

$$H_D(z)\big|_{z=e^{j\omega}} = H_D(e^{j\omega}) = \frac{1}{T_s} H_A(j\Omega) \quad (1)$$

Transfer function $H(s)$ can be written in the general partial fraction form

$$H_A(s) = \frac{b}{(s+a)^2 + b^2} = \sum_{m=1}^{M} \frac{R_m}{s - s_m} \quad (2)$$

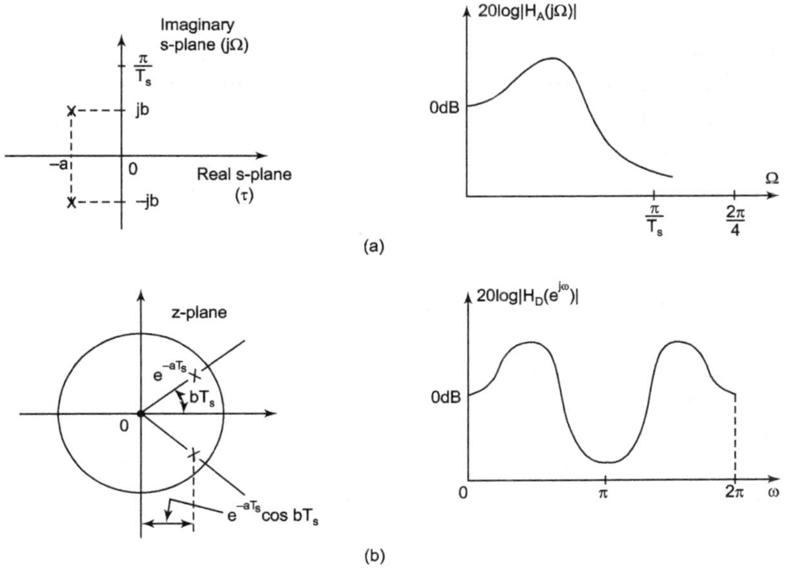

FIGURE 9.2 (a) Pole locations and frequency response of the analog filter, (b) Pole-zero locations and frequency response of the digital filter.

where R_m is the mth residue at the mth pole $s = s_m$ of the transfer function $H_A(s)$. The poles may be real or complex conjugate.

Corresponding transfer function for digital filter will be given by

$$H_D(z) = \sum_{m=1}^{M} \frac{R_m}{1 - e^{-T_s s_m} z^{-1}} \tag{3}$$

R_m and s_m have their usual meaning.

A digital filter has an extremely high gain due to high sampling rate. For this reason, it is advised to multiply Eq. (3) by T_s and then use it.

$$H_D(z) = \sum_{m=1}^{M} \frac{T_s R_m}{1 - e^{T_s s_m} z^{-1}} \tag{4}$$

Taking the inverse z-transform of above Eq. (4), we get

$$h_D(z) = \mathcal{Z}^{-1}\left[H_D(z)\right] = \mathcal{Z}^{-1}\left[\sum_{m=1}^{M} \frac{T_s R_m}{1 - e^{-T_s s_m} z^{-1}}\right]$$

or

$$h_D(n) = T_s h_A\left(nT_s\right) \tag{5}$$

It is the basis of impulse-response invariance method that we choose a unit-sample response for the digital filter that is similar to the impulse response of the analog filter.

Distortion in the frequency response is introduced due to aliasing in the impulse-response invariance method of deriving the digital IIR filter.

The relationship between analog and digital frequency is linear and consequently, the shape of frequency response is preserved except for aliasing.

This method can only be used for band-limited filters. So first of all, all filters are band limited to avoid severe aliasing distortion.

9.2.3 Bilinear Transformation Method

In impulse-response invariance method, the derived IIR digital filter has exactly the same impulse response as the original analog filter for continuous time $t = nT_s$, where T_s is the sampling time.

Now, we have another approximation technique of IIR digital filters from analog filters. It is called *bilinear transformation method.* Digital filter derived from this method has approximately the same time-domain response as the original analog filter for any value of input.

By bilinear transformation method, digital filter is derived from the analog filter as

$$H_D(z) = H_A(s)\big|_{s=(2/Ts)(z-1/z+1)} \tag{9.1}$$

where
$H_D(z)$ = Transfer function of digital filter
$H_A(s)$ = Transfer function of analog filter
T_s = Sampling period.

Bilinear transformation method is one of the best currently available methods for designing IIR digital filters from reference analog filters due to simplicity and similarity of the frequency response of IIR digital filters to that of reference analog filters. This method produces true frequency-to-frequency transformation.

The bilinear transformation method is applicable to all types of filters.

9.2.3.1 Derivations of Formula for Bilinear Transformation Method

Figure 9.3 shows a simple block diagram of an analog filter. In this figure, the analog filter is an analog integrator and its response is shown.

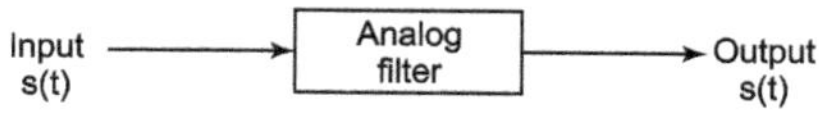

FIGURE 9.3 Analog filter is an analog integrator.

$$H_D(z) = H_A(s)\big|_{s=(2/T_s)(z-1/z+1)}$$

For this derivation, we consider a transfer function of an analog integrator given by

$$H_A(s) = \frac{1}{s} \tag{9.2}$$

Impulse response of the above analog filter is determined as

$$h_A(t) = \text{Inverse Laplace transform of } H_A(s)$$
$$= £^{-1}\left[H_A(s)\right] = £^{-1}\left[\frac{1}{s}\right] = u(t) \tag{9.3}$$

where $u(t)$ is called the unit-step function and it is defined as

$$u(t) = \begin{cases} 1, & \text{for } t \geq 0^+ \\ 0, & \text{for } t \leq 0^- \end{cases} \tag{9.4}$$

Response to an arbitrary input (excitation) $s(t)$ is determined by the convolution integral as

$$y(t) = \int_0^t s(\tau)h_A(t-\tau)d\tau \tag{9.5}$$

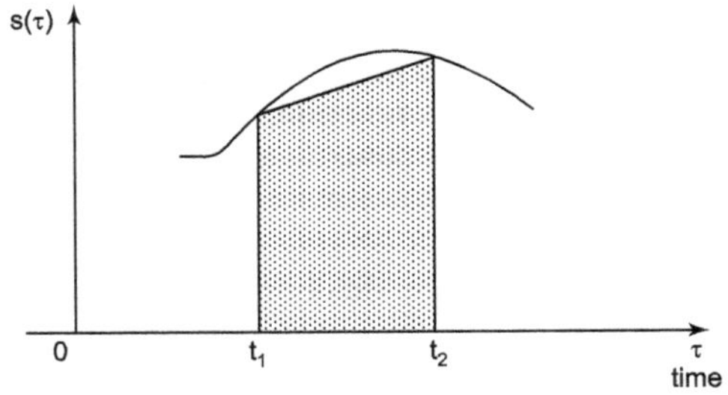

FIGURE 9.4 Response of above analog filter (analog integrator)

If $0 \pm < t < t_2$, Eq. (9.5), can be written as

$$y(t_2) = \int_0^{t_2} s(t)h_A(t_2-\tau)d\tau \tag{9.6}$$

$$y(t_1) = \int_0^{t_1} s(t)h_A(t_1-\tau)d\tau \tag{9.7}$$

Subtracting Eq. (9.7) from Eq. (9.6), we get

$$y(t_2) - y(t_1) = \int_0^{t_2} s(t)h_A(t_2-\tau)d\tau - \int_0^{t_1} s(t)h_A(t_1-\tau)d\tau \tag{9.8}$$

We know that $h_A(t_2 - \tau) = h_A(t_1 - \tau) = 1$

$$\text{for } 0^+ < \tau < t_1$$
$$0^+ < \tau \le t_2$$

Substituting the values $h_A(t_2 - \tau) = h_A(t_1 - \tau) = 1$ in Eq. (9.8), we get

$$y(t_2) - y(t_1) = \int_0^{t_2} s(\tau)d\tau - \int_0^{t_1} s(\tau)d\tau = \int_{t_1}^{t_2} s(\tau)d\tau \qquad (9.9)$$

As $t_1 \to t_2$, from Figure 9.4

$$y(t_2) - y(t_1) \equiv \frac{t_2 - t_1}{2}\left[s(t_1) + s(t_2)\right] \qquad (9.10)$$

Putting $t_1 = nT_s - T_s$ and $t_2 = nT_s$ in above Eq. (9.10), we get

or $$y(nT_s) - y(nT_s - T_s) = \frac{nT_s - nT_s + T_s}{2}\left[s(nT_s - T_s) + s(nT_s)\right]$$

or $$y(nT_B) - y(nT_s - T_s) = \frac{T_s}{2}\left[s(nT_s - T_s) + s(nT_s)\right] \qquad (9.11)$$

This is the difference equation of derived digital filter (integrator) from analog filter (integrator).

Transfer function of digital integrator can be determined by z-transformation of the above difference Eq. (9.11). By z-transforming, we get

$$Y(z) - Z^{-1}Y(z) = \frac{T_s}{2}\left[z^{-1}S(z) + S(z)\right]$$

or $$Y(z)\left[1 - z^{-1}\right] = \frac{T_s}{2}\left[S(z)S(1 + z^{-1})\right]$$

or $$\frac{Y(z)}{S(z)} = \frac{T_s}{2}\left[\frac{1 + z^{-1}}{1 - z^{-1}}\right] = \frac{T_s}{2}\left[\frac{z + 1}{z - 1}\right]$$

or $$H_D(z) = \frac{Y(z)}{S(z)} = \frac{T_s}{2}\left[\frac{z + 1}{z - 1}\right] \qquad (9.12)$$

$H_D(z)$ is the transfer function of the digital filter and is defined as the ratio of z-transform of output sequence to z-transform of the applied input sequence, keeping all initial conditions zero:

$$HD(z) = \frac{z\text{-transform of output sequence}}{z\text{-transform of input sequence}}$$

$$= \frac{Z[y(n)]}{Z[s(n)]} = \frac{Y(z)}{S(z)}$$

Eq. (9.12) gives the transfer function of the digital integrator.

From $\quad H_A(s) = \dfrac{1}{s}$ and $H_D(z) = \dfrac{T_s}{2}\left(\dfrac{z+1}{z-1}\right)$, we get

$$H_D(z) = H_A(s)\big|_{s=(2/T_s)(z-1/z+1)} \tag{9.13}$$

that is, the transfer function of a digital integrator can be obtained by applying the bilinear transformation

$$s = \frac{2}{T_s}\left(\frac{z-1}{z+1}\right) \text{ to } H_A(s)$$

Here the digital filter has been assumed to have approximately the same response (time-domain) as the analog filter for any value of input signal.

9.2.3.2 Properties of Mapping of Bilinear Transformation

Here mapping properties of bilinear transformation will be studied. The relation between the frequency response of the derived digital filter and that of the original analog filter can be established by examining these mapping properties.

We know that for bilinear transformation method

$$s = \frac{2}{T_s}\left(\frac{z-1}{z+1}\right) \tag{9.14}$$

or $\quad sT_s(z+1) = 2(z-1)$

or $\quad sT_sz + sT_s = 2z - 2$

or $\quad sT_sz - 2z = sT_s - 2$

or $\quad z(sT_s - 2) = -(sT_s + 2)$

or $\quad z = \dfrac{-(sT_s+2)}{sT_s - 2}$

or $\quad z = \dfrac{2+sT_s}{2-sT_s}$

or
$$z = \frac{2/sT_s + s}{2/sT_s - s} \tag{9.15}$$

$s = \sigma + j\omega$ is a complex variable that is used in the Laplace transform of the analog system and $z = re^{j\theta}$ is another complex variable that is used for z-transform of digital systems:

$$re^{j\theta} = \frac{2/T_s + \sigma + j\omega}{2/T_s - \sigma - j\omega} = \frac{(2T_s + \sigma) + j\omega}{(2T_s - \sigma) - j\omega}$$

$$= \sqrt{\frac{(2/T_s+\sigma)^2+\omega^2}{(2/T_s-\sigma)^2-\omega^2}}\, e^{j\left[\tan^{-1}\omega(2Ts+\sigma)+\tan^{-1}\omega/2/T_s-\sigma\right]} \quad (9.16)$$

Comparing both sides of Eq. (9.16), we get

$$r = \sqrt{\frac{(2/T_s+\sigma)^2+\omega^2}{(2/T_s-\sigma)^2+\omega^2}} \quad (9.17)$$

and $$\theta = \tan^{-1}\frac{\omega}{(2/T_s+\sigma)} + \tan^{-1}\frac{\omega}{(2/T_s-\sigma)} \quad (9.18)$$

Let us consider Eq. (9.17):

Case I: if $\sigma > 0$, then $r > 1$, that is, the bilinear transformation maps the open right-half s-plane onto the region exterior to the unit-circle $|z| = 1$, that is, $r = 1$ of the z-plane.

Case II: if $\sigma < 0$, then $r < 1$, that is, the bilinear transformation maps the open left-half s-plane onto the interior of the unit-circle $|z| = 1$ of the z-plane.

Case III: if $\sigma = 0$, then $r = 1$, that is, the bilinear transformation maps the imaginary axis of the s-plane onto the unit-circle $|z| = 1$ of the z-plane.

Now for Case III where $\sigma = 0$ and $r = 1$. Substituting $\sigma = 0$ in Eq. (9.18), we get

$$\theta = \tan^{-1}\frac{\omega}{(2/T_s+\sigma)} + \tan^{-1}\frac{\omega}{(2/T_s-\sigma)} = 2\tan^{-1}\left(\frac{\omega T_s}{2}\right)$$

or $$\theta = 2\tan^{-1}\left(\frac{\omega T_s}{2}\right) \quad (9.19)$$

Hence from Eq. (9.19), we get

$$\theta = 2\tan^{-1}\left(\frac{\omega T_s}{2}\right)$$

if $$\omega = 0 \text{ then } \theta = 2\tan^{-1}\left(\frac{0T_s}{2}\right) = 2\tan^{-1}(0) = 0$$

if $$\omega = +\infty \text{ then } \theta = 2\tan^{-1}\left(\frac{+\infty T_s}{2}\right) = 2\tan^{-1}(+\infty) = +\pi$$

1. The origin $(\sigma, \omega) = (0, 0)$ of s-plane maps onto points $(r, \theta) = (1, 0)$ of the z-plane.
2. Positive imaginary axis $(\sigma, \omega) = (0, +\infty)$ of s-plane maps onto point $(r, \theta) = (1, +\pi)$, that is, upper semicircle $|z| = 1$.
3. Negative imaginary axis $(\sigma, \omega) = (0, -\infty)$ of s-plane maps onto point $(r, \theta) = (1, -\pi)$, that is, lower semicircle $|z| = 1$.

This transformation is shown in Figure 9.5.

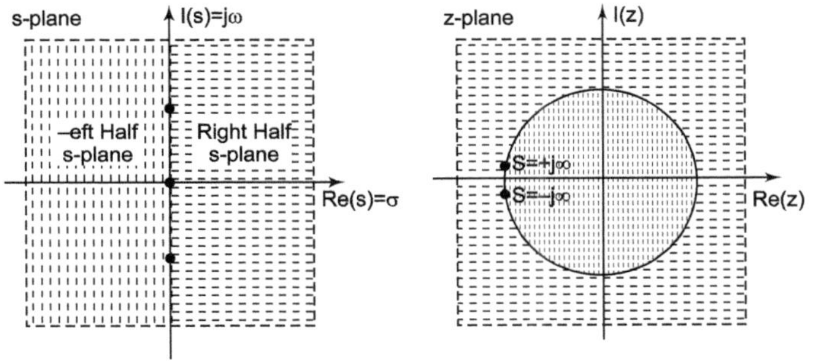

FIGURE 9.5 Illustration of bilinear transformation from s-plane onto z-plane.

9.2.3.3 Warping Effect

At low frequencies, the derived digital filter has the same frequency response as the reference analog filter. For higher frequencies, the relation between analog frequency w and digital frequency W is highly nonlinear and a distortion is introduced in the frequency scale of the digital filter relative to the analog filter. This distortion is known as warping effect:

$$\theta = 2\tan^{-1}\left(\frac{\omega T_s}{2}\right) \tag{9.20}$$

or

$$\frac{\theta}{2} = \tan^{-1}\left(\frac{\omega T_s}{2}\right)$$

or

$$\frac{\Omega T_s}{2} = \tan^{-1}\left(\frac{\omega T_s}{2}\right), \text{ since } \theta = \Omega T_s$$

or

$$\Omega = \boxed{\Omega = \frac{2}{7}\tan^{-1}\left(\frac{\omega T_s}{2}\right)} \tag{9.21}$$

Eq. (9.21) relates the frequency of reference analog filter (ω) and frequency of the derived digital filter (Ω). Warping effect influences both amplitude response phase response of a digital filter at higher frequencies.

9.2.3.4 Influence of the Warping Effect on the Amplitude Response of a Digital Filter

Demonstration of influence of warping effect on the amplitude response of a derived digital filter from reference analog filter is performed by

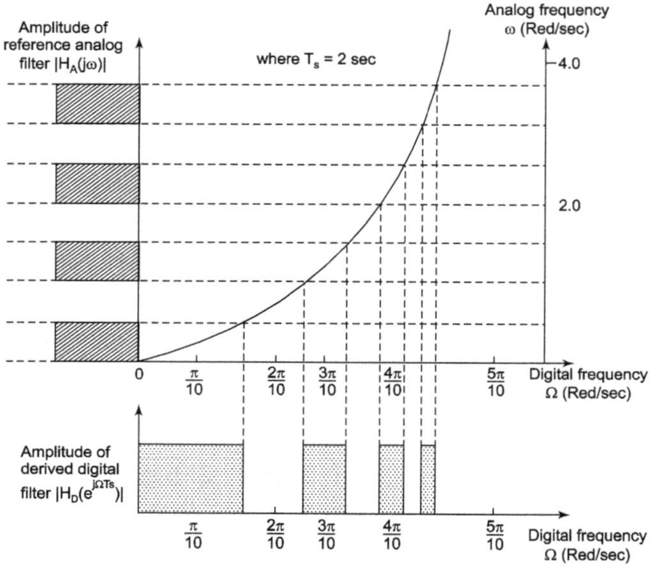

FIGURE 9.6 Demonstration of influence of the warping effect on the amplitude response of a derived digital filter from reference analog filter in bilinear transformation method.

Considering an analog filter with a number of passbands centered at fixed intervals is shown in Figure 9.6. The derived digital filter has the same number of passbands. But its center frequencies and bandwidths of passbands is reducing disproportionately. This is shown in Figure 9.6.

9.2.3.5 Influence of the Warping Effect on the Phase Response of a Derived Digital Filter

For the purpose of demonstrating the influence of warping effect on the phase response of a derived digital filter from a reference analog filter, we consider an analog filter with linear phase response. As we have shown in Figure 9.7, the phase response of a derived digital filter is nonlinear.

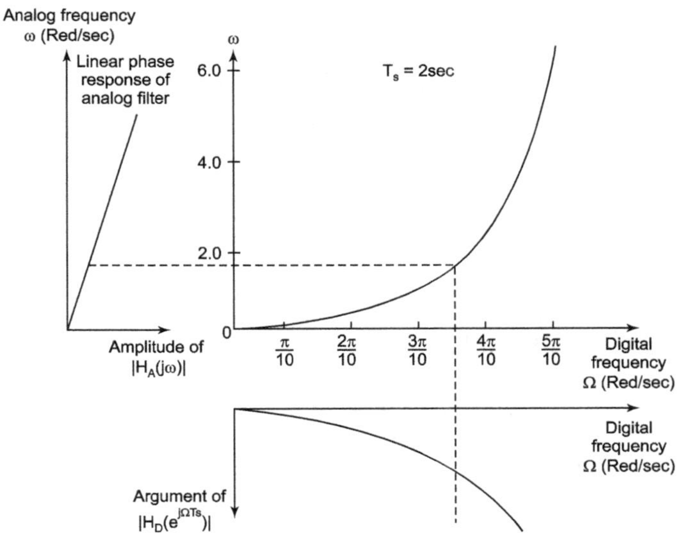

FIGURE 9.7 Demonstration of influence of the warping effect on the phase response of a derived digital filter from reference analog filter in bilinear transformation method.

EXAMPLE 9.2

Obtain an IIR digital filter from the second-order analog filter whose transfer function is given as $H_A(s) = s + a/(s + a)^2 + b^2$ using impulse-response invariance method.

Solution:

Step I: Impulse response $h_A(t)$ of an analog filter is determined by inverse Laplace transformation of $H_A(s)$.

$$h_A(t) = \text{Inverse Laplace transformation of } H_A(s)$$

$$= £^{-1}[H_A(s)] = £^{-1}\left[\frac{s+a}{(s+a)^2+b^2}\right]$$

$$= £^{-1}\left[\frac{s+1}{(s+a+jb)(s+a-jb)}\right]$$

$$= £^{-1}\left[\frac{A}{(s+a+jb)} + \frac{B}{(s+a-jb)}\right]$$

$$= £^{-1}\left[\frac{1/2}{s+a+jb} + \frac{1/2}{s+a-jb}\right] \text{(by partial fraction)}$$

$$= \frac{1}{2}e^{-(a+jb)t} + \frac{1}{2}e^{-(a-jb)t}$$

$$= e^{-at}\left[\frac{e^{jbt} + e^{-jbt}}{2}\right] e^{-at} \cos bt$$

or $$h_A(t) = \begin{cases} e^{-at} & \cos(bt), t \geq 0 \\ 0, & \text{otherwise} \end{cases}$$

Step II: $h(nT_s) = h_A(t)|_{t=nTs} = [e^{-at} \cos bt]_{t=nTs} = e^{-anTs} \cos bnT_s$

where T_s = Sampling period

or $$h(nT_s) = \begin{cases} e^{-anT_s} \cos bnT_s, & n \geq 0 \\ 0, & \text{otherwise} \end{cases}$$

Step III: Determination of $H(z)$:

$$H(z) = z\text{-transform of } h(nT_s)$$

$$= \mathcal{Z}\left[h(nT_s)\right] = \mathcal{Z}\left[e^{-anT_s} \cos bnT_s\right]$$

$$= \frac{1 - e^{-aT_s \cos bT_s} z^{-1}}{\left(1 - e^{-(a+jb)T_s} z^{-1}\right)\left(1 - e^{-(a+jb)T_s} z^{-1}\right)}$$

or $$H(z) = \frac{1 - e^{-aT_s} \cos bT_s z^{-1}}{1 - 2e^{-aT_s} \cos bT_s z^{-1} + e^{-2aT_s} z^{-2}}$$

This is transfer function of digital IIR filter and its network structure is shown in Figure 9.8.

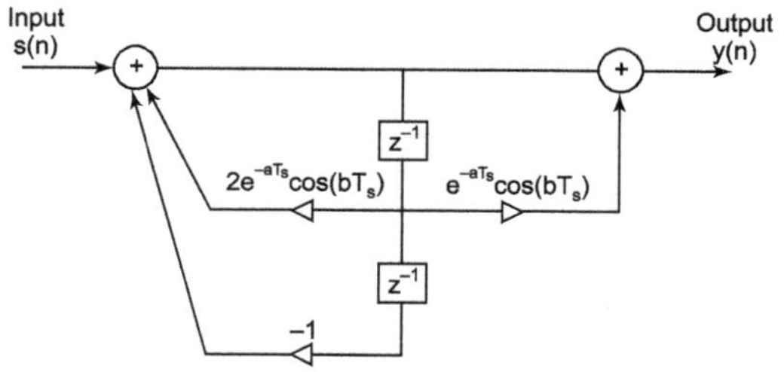

FIGURE 9.8 Network structure of above IIR digital filter.

EXAMPLE 9.3

Convert following analog filter into digital filter using impulse-response invariance method.

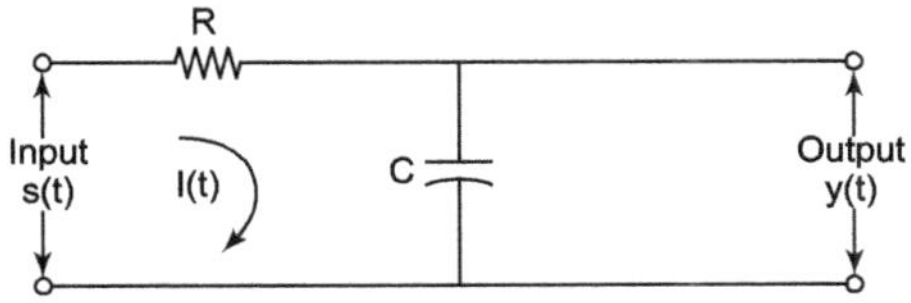

FIGURE 9.9 Analog filter.

Solution:

Determination of $H_A(s)$:

$$s(t) = RI(t) + \frac{1}{C}\int_0^t I(t)dt \tag{1}$$

$$y(t) = \frac{1}{C}\int_0^t dt \tag{2}$$

Taking the Laplace transform above Eqs. (1) and (2), we get

$$S(s) = RI(s) + \frac{1}{Cs}I(s) = \left(R + \frac{1}{Cs}\right)I(s) \tag{3}$$

$$Y(s) = \frac{1}{Cs}I(s) \tag{4}$$

From Eqs. (3) and (4), we get

$$S(s) = \left(R + \frac{1}{Cs}\right)Y(s)Cs = (RCs+1)Y(s)$$

or
$$H_A(s) = \frac{Y(s)}{S(s)} = \frac{1}{1+RCs} \tag{5}$$

Determination of $h_A(t)$ from $H_A(s)$:

$$h_A(t) = \text{Inverse Laplace transform of } H_A(s)$$

$$= \pounds^{-1}\left[H_A(s)\right] = \pounds^{-1}\left[\frac{1}{1+RCs}\right]$$

$$= \pounds^{-1}\left[\frac{\frac{1}{RC}}{1+\frac{1}{RC}}\right] = \frac{1}{RC}e^{-t/RC}$$

or
$$h_A(t) = \begin{cases} \frac{1}{RC}e^{-t/RC}, & t \geq 0 \\ 0, & \text{otherwise} \end{cases}$$

Determination of $h(nT_s)$:

$$h(nT_s) = h_A(t)\,|_{t=nT_s} = \frac{1}{RC}e^{-nT_s/RC}$$

$$= \tau e^{-\tau} nT_s \qquad \text{Assuming } \tau = \frac{1}{RC}$$

or
$$h(nT_s) = \begin{cases} \tau e^{-\tau nT_s}, & \text{for } n \geq 0 \\ 0, & \text{otherwise} \end{cases}$$

Determination of $H_D(z)$ from $h(nT_s)$

$$H_D(z) = z\text{-transform of } h(nT_s) = Z\left[\tau e^{-\tau nT_s}\right]$$

or
$$H_D(z) = \frac{\tau}{1 - e^{-\tau T_s} z^{-1}}$$

This is the transfer function of digital filter. Figure 9.10 illustrates the network structure of the above digital filter.

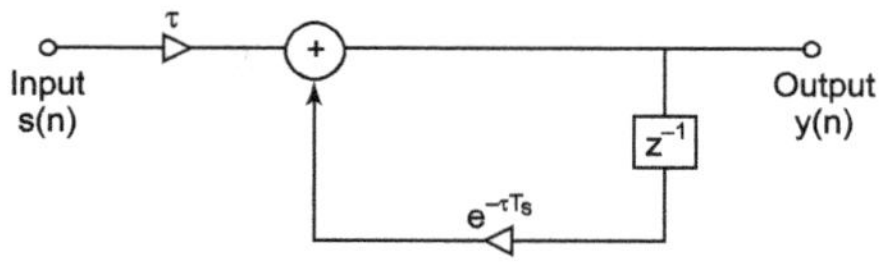

FIGURE 9.10 Network structure of above digital filter.

EXAMPLE 9.4

Convert the analog filter with transfer function $H(s) = s + 0.1/(s + 0.1)^2 + (3)^2$ into a digital filter using bilinear transformation method. Given sampling point $T_s = 2\ s$.

Solution:

From bilinear transformation method

$$H_D(z) = H_A(s)\,|_{s=(2/T_s)(z-1/z+1)}$$

This is the transfer function of a derived digital filter.

$$= \left[\frac{s+0.1}{(s+0.1)^2+9}\right]_{s=(2/T_s)(z-1/z+1)}$$

$$= \frac{\left(\dfrac{z-1}{z+1}\right)+0.1}{\left[\left(\dfrac{z-1}{z+1}\right)+0.1\right]^2+9}$$

$$= \frac{\dfrac{z-1+0.1z+0.1}{(z+1)}}{\dfrac{(z-1+0.1z+0.1)^2+9(z+1)^2}{(z+1)^2}}$$

$$= \frac{(z+1)+(1.1z+0.9)}{(1.1z-0.9)^2+9(z+1)^2} = \frac{1.1z^2+0.2z-0.9}{10.21z^2+16.02z+9.81}$$

or $$H_D(z) = \frac{0.1066\left(1+0.18z^{-1}-0.82z^{-2}\right)}{\left(1+1.569z^{-1}-0.96z^{2}\right)}$$

EXAMPLE 9.5

Design a first order low-pass IIR digital filter with a 3-dB bandwidth of 0.2π by using bilinear transformation method. Transfer function of analog filter is given as

$$H_A(s) = \frac{\Omega_c}{s+\Omega_C}$$

where Ω_C is the 3-dB bandwidth of analog filter.

Solution:

The digital filter is specified to have $\omega_C = 0.2\pi$ for 3-dB bandwidth.

$$\Omega_C = \frac{2}{T_s}\tan\frac{\omega_C}{2} = \frac{2}{T_s}\tan\frac{0.2\pi}{2}$$

$$= \frac{2}{T_s}\tan 0.1\pi = \frac{0.65}{T_s}$$

$$H_A(s) = \frac{\Omega_C}{s+\Omega_C}$$

Substituting the $\Omega_C = 0.65/T_s$ in the transfer function of above analog filter.

$$H_A(s) = \frac{\Omega_C}{s+\Omega_C} = \frac{0.65/T_s}{s+0.65/T_s} = \frac{0.65}{sT_s+0.65}$$

From the bilinear transformation method

$$H_D(z) = H_a(s)\big|_{s=(2/T_s)(z-1/z+1)}$$

$$H_D(z) = \frac{0.65}{\dfrac{2}{T}\left(\dfrac{z-1}{z+1}\right)T_s+0.65}$$

$$= \frac{0.65(z+1)}{2(z-1)+0.65(z+1)} = \frac{0.65(z+1)}{2.65z+(1-1.35)}$$

$$= \frac{0.65(z+1)}{2.65z-1.35}$$

or $$H_D(z) = \frac{0.245\left(1+z^{-1}\right)}{1-0.509z^{-1}}$$

This the transfer function of a derived digital filter.

9.2.4 Digital Butterworth Filter

Butterworth method for analog filter design plays a very important role because of its simplicity and also because the magnitude characteristics are very nearly ideal near the cut-off frequency of high order filter. Figure 9.11(a) shows the magnitude characteristics of the Butterworth filter for different orders. Figure 9.11(b) illustrates the pole locations in the s-plane for a Butterworth analog filter ($n = 3$).

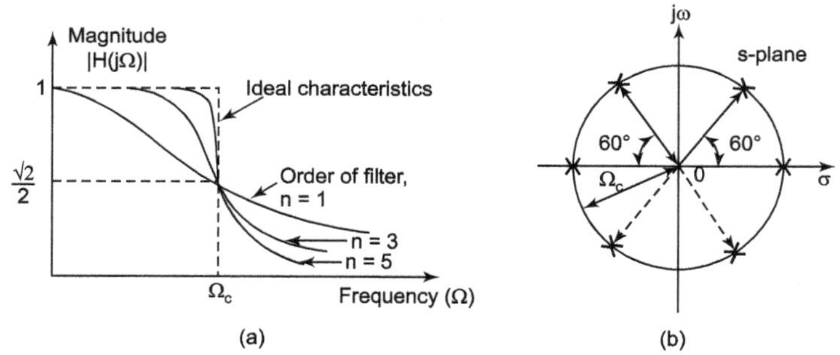

FIGURE 9.11 (a) Magnitude characteristics of a Butterworth analog filter for different orders of the filter. (b) Pole locations in the s-plane for the Butterworth analog filter. [Here order of the filter $n = 3$]

As the order of the Butterworth filter increases, the magnitude characteristic's role-off becomes much faster and transition band decreases. All the characteristic curves pass through a common point at $|H(j\Omega_C)| = \sqrt{2}/2$ (magnitude at cut-off frequency $\Omega = \Omega_C$).

Transfer function of a Butterworth analog filter is given as

$$\boxed{H_A(s)H_A(-s) = \frac{1}{1+\left(\frac{a}{j\Omega_C}\right)^{2n}}} \qquad (9.22)$$

where Ω_C is the cut-off frequency and n is the order of the Butterworth analog filter. The roots of the denominator polynomial are the poles of the squared magnitude function

$$1+\left(\frac{s}{j\Omega_C}\right)^{2n}=0$$

or

$$\left(\frac{s}{j\Omega_C}\right)^{2n}=-1$$

or

$$\left(\frac{s}{j\Omega_C}\right)=(-1)^{1/2n}$$

or

$$\boxed{s_m=(-1)^{1/2n}\,j\Omega_C} \tag{9.23}$$

Thus we see that there are $2n$ poles spaced on a circle of radius Ω_C in the s-plane.

Here $H_A(s)$ is a rational transfer function in s with positive real coefficients and poles of $H(s)$ must appear in complex conjugate.

For a third-order Butterworth analog filter ($n = 3$), the poles will be equally spaced around the circle. It is shown in Figure 9.11(b). These poles are spaced by $\frac{2\pi}{6}=60°$. For stable causal filter, the poles must lie in the left half of the s-plane on the Butterworth circle.

Transfer function of a third-order Butterworth analog filter is constructed by using the complex poles at the angles $\frac{2\pi}{3},\frac{3\pi}{3}$ and $\frac{4\pi}{3}$ and this transfer function is given by

$$H(s)=\frac{1}{(s+1)\left(s+e^{j\pi/3}\right)\left(s+e^{j\pi/3}\right)}=\frac{1}{(s+1)(s^2+s+1)}$$

$$=\frac{1}{s^3+2s^2+2s+1} \tag{9.24}$$

$H(j0)$ is always equal to unity for Butterworth filter and its normalized cut-off frequency Ω_C is also equal to unity.

To obtain a digital Butterworth filter from an analog Butterworth filter we map pole pattern from s-plane to z-plane using bilinear transformation. Hence corresponding $2n$ zeros of magnitude squared function lie at $z = -1$.

Bilinear transformation is a conformal mapping because the Butterworth circle in the s-plane maps into a circle in the z-plane. Butterworth circle in the z-plane is not centered at the origin of the z-plane, that is, ($z = 0$).

Now we will derive an expression to the Butterworth circle in the z-plane.

Let $s = \sigma + j\Omega$, where s is the complex variable used in Laplace transformation, σ and Ω are the real and imaginary parts of the complex variable s.

Let the cut-off frequency be Ω_1. In Figure 9.11(b) cut-off frequency is Ω_1 denoted by Ω_C. Equation of the Butterworth circle in the s-plane centered at the origin $(\sigma, \Omega) = (0, 0)$ is given by

$$\Omega_1^2 = \sigma^2 + \Omega^2 \tag{9.25}$$

We know from bilinear transformation of analog filter into IIR digital filter that s-plane is mapped into z-plane.

1. Left-half of s-plane maps into interior of the unit-circle, $|z| = 1$.
2. Right-half of the s-plane maps into exterior of the unit-circle, $|z| = 1$.
3. A circle maps into a circle.
4. σ-axis maps into c-axis in the z-plane.

From Bilinear transformation, we also know that

$$s = \frac{2}{T_s}\left(\frac{z-1}{z+1}\right) \tag{9.26}$$

It can be written as

$$z = \frac{1 + \left(\frac{T_s}{2}\right)s}{1 - \left(\frac{T_s}{2}\right)s} \tag{9.27}$$

where T_s = sampling period.

Substituting $s = \sigma + j\Omega$ in Eq. (9.27), we get

$$z = \frac{1 + \left(\frac{T_s}{2}\right)(\sigma + j\Omega)}{1 - \left(\frac{T_s}{2}\right)(\sigma + j\Omega)}$$

$$= \frac{\left(1 + \frac{T_s}{2}\sigma\right) + j\left(\frac{\Omega T_s}{2}\right)}{\left(1 - \frac{T_s}{2}\sigma\right) + j\left(\frac{\Omega T_s}{2}\right)}$$

$$= \frac{A + jB}{A' - jB} \quad \text{where } A = 1 + \frac{T_s}{2}\sigma, A' = 1 - \frac{T_s}{2}\sigma, B = \frac{\Omega T_s}{2}$$

$$= \frac{(A + jB)(A' + jB)}{(A' - jB)(A' + jB)}$$

$$= \frac{AA' + jA'B + jAB + j^2B^2}{(A')^2 + B^2}$$

$$= \frac{\left(AA' + B^2\right) + j(AB + A'B)}{(A')^2 + B^2}$$

$$= \frac{\left(AA' + B^2\right) + jB(A + A')}{(A')^2 + B^2} \qquad [\text{But } A + A' = 2]$$

or

$$= \frac{(AA' = B^2) + j2B}{(A')^2 + B^2} \tag{9.28}$$

Now our aim is to find the equation of the Butterworth circle in the z-plane. Here z is a complex variable and is given by $z = C + jD$, where C and D are real and imaginary parts of z. We know that the center of the Butterworth circle in the s-plane cannot be mapped to the z-plane using bilinear transformation but only the points on the circle in s-plane can be mapped onto the points in the z-plane.

From Eq. (9 28)

$$z = C + jD = \frac{\left(AA' - B^2\right) + j(2B)}{(A')^2 + (B)^2}$$

Substituting

$$I = 1 + \frac{T_s}{2}\tau$$

$$A' = 1 + \frac{T_s}{2}\tau$$

$$B = \frac{\Omega T_s}{2}$$

$$B_1 = \frac{\Omega_1 T_s}{2} = \frac{\left(\sqrt{\sigma^2 + \Omega^2}\right)T_s}{2}$$

In Eq. (9.28), we get

$$z = C + jD = \frac{1 + B_1^2 + j2B}{1 + B_1^2 - 2\sqrt{B_1^2 - B^2}} \tag{9.29}$$

Comparing real and imaginary parts of the previous equation, we get

$$C = \frac{1 - B_1^2}{1 + B_1^2 - 2\sqrt{B_1^2 - B^2}} \tag{9.30}$$

and

$$D = \frac{2B}{1 + B_1^2 - 2\sqrt{B_1^2 - B^2}} \tag{9.31}$$

Both C and D have the same denominator. If $B = 0$ then D is also zero.

From $B = \frac{\Omega T_s}{2}$ if $B = 0$, then $\Omega = 0$ as $T_s = 0$.

The frequency $\Omega = 0$ corresponds to σ-axis (real axis) of the s-plane. It means c-axis of the z-plane maps into the σ-axis of the s-plane and vice-versa.

Now we need to find two extreme points C_1 and C_2 on the z-plane Butterworth circle and whose diameter will be $(C_1 - C_2)$. The center of the Butterworth circle on the z-plane will be given by $(C_1 - C_2/2)$.

From the above discussion, we substitute $\sigma = +\Omega_1$ or $B_1 = \Omega_1 T_s/2$ and $B = 0$ and $C = C_1$ in the Eq. (9.30), we get

$$C = \frac{1 - B_1^2}{1 + B_1^2 - 2\sqrt{B_1^2 - B^2}}$$

or

$$C_1 = \frac{1 + B_1^2}{1 + B_1^2 - 2B_1} = \frac{(1 + B_1)(1 - B_1)}{(1 - B_1)} = \frac{1 + B_1}{1 - B_1}$$

or

$$C_1 = \frac{1 + \frac{\Omega_1 T_s}{2}}{1 - \frac{\Omega_1 T_s}{2}} \quad \left(\text{Substituting } B_1 + \frac{\Omega_1 T_s}{2}\right) \tag{9.32}$$

Now we need to find another extreme point $C = C_2$ corresponding to $\sigma = \Omega_1$ of the s-plane Butterworth circle. Substituting $B = 0$ and $C = C_2$ and $\sigma = -\Omega_1$ or $B_1 = \Omega_1 T_s/2$ in Eq. (9.30):

$$C = \frac{1 - B_1^2}{1 + B_1^2 - 2\sqrt{B_1^2 - B^2}}$$

$$C_2 = \frac{1 - \left(\frac{\Omega_1 T_s}{2}\right)^2}{1 + \left(\frac{\Omega_1 T_s}{2}\right) + 2\left(\frac{\Omega_1 T_s}{2}\right)}$$

or

$$C_1 = \frac{\left(1 - \frac{\Omega_1 T_s}{2}\right)\left(1 + \frac{\Omega_1 T_s}{2}\right)}{\left(1 + \frac{\Omega_1 T_s}{2}\right)} \tag{9.33}$$

or

$$C_2 = \frac{\left(1 - \frac{\Omega_1 T_s}{2}\right)}{\left(1 + \frac{\Omega_1 T_s}{2}\right)}$$

Therefore the two extreme points are

$$C_1 = \frac{1 + \frac{\Omega_1 T_s}{2}}{1 - \frac{\Omega_1 T_s}{2}} = \frac{1+B_1}{1-B_1} \quad \text{and} \quad C_2 = \frac{1 + \frac{\Omega_1 T_s}{2}}{1 - \frac{\Omega_1 T_s}{2}} = \frac{1+B_1}{1-B_1}$$

Equation to the circle in the z-plane is given by

$$(C - C_0)^2 + D^2 = R^2 \tag{9.34}$$

R is the radius of circle and is determined by

$$R = \left(\frac{C_1 - C_2}{2}\right) = \frac{1}{2}\left[\frac{1+B_1}{1-B_1} - \frac{1-B_1}{1+B_1}\right] = \frac{1}{2}\left[\frac{4B_1}{1-B_1^2}\right] = \frac{2B_1}{1-B_1^2} \tag{9.35}$$

Center of the circle will be determined by

$$\begin{aligned} C_0 &= C_2 + R \\ &= \frac{1-B_1}{1+B_2} + \frac{2B_1}{1-B_1^2} = \frac{(1-B_1)^2 + 2B_1}{1-B_1^2} \\ &= \frac{1+B_1^2 - 2B_1 + 2B_1}{1-B_1^2} = \frac{1+B_1^2}{1-B_1^2} \end{aligned} \tag{9.36}$$

Substituting the values of R and C_0 in the equation of circle

$$(C - C_0)^2 + D^2 = R^2$$

we get,

$$\left[C - \frac{(1+B_1^2)}{(1-B_1^2)}\right]^2 + D^2 = \left[\frac{2B_1}{1-B_1^2}\right]^2 \tag{9.37}$$

Both the unit-circle and the Butterworth circle are illustrated in Figure 9.12.

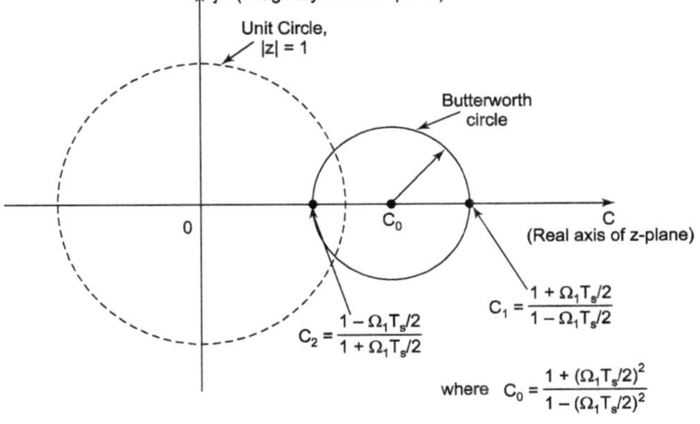

FIGURE 9.12 Butterworth circle in the z-plane. (It is transformed from the s-plane using bilinear transformation method.)

EXAMPLE 9.6

Find the Butterworth circle in the z-plane and the corresponding pole locations for a third-order Butterworth analog filter. Given $B_1 = 1/4$ and $\Omega_1 = 1$ in the s-plane.

Solution:

First, we will determine the two extreme points on the real axis of z-plane. From the theory of the Butterworth filter, we know that

$$C_1 = \frac{1+B_1}{1-B_1} = \frac{1+1/4}{1-1/4} = \frac{5/4}{3/4} = \frac{5}{3}$$

and

$$C_2 = \frac{1+B_1}{1-B_1} = \frac{1+1/4}{1-1/4} = \frac{3/4}{5/4} = \frac{5}{3}$$

The radius of the Butterworth circle in the z-plane will be determined by

$$R = \frac{2B_1}{1-B_1^2} = \frac{1\times 1/4}{1-(1/4)^2} = \frac{1/2}{15/16} = \frac{8}{15}$$

Center of the Butterworth circle

$$C_0 = C_2 + R = \frac{3}{5} + \frac{8}{15} = \frac{9+8}{15} = \frac{17}{15}$$

It is given that it is a third-order filter (n = 3) and there will be 6 poles located on the Butterworth circle in the s-plane. The angle between two consecutive poles will be $\frac{2\pi}{2n} = \frac{2\pi}{2\times 3} = 60°$

Complex conjugate poles will be determined as

$$s = -(\cos\ 60° \pm j\sin\ 60°)$$

$$= -\left(\frac{1}{2} \pm \frac{\sqrt{3}}{2}\right)$$

Since

$$s = \sigma + j\Omega = -\left(\frac{1}{2} \pm j\frac{\sqrt{3}}{2}\right)$$

$\therefore$

$$\sigma = -\frac{1}{2} \quad \text{and} \quad \Omega = +\frac{\sqrt{3}}{2}$$

Given $\Omega_1 = 1$, but $B_1 = \dfrac{\Omega_1 T_s}{2}$

or

$$\frac{1}{4} = 1\times\frac{T_s}{2} \quad \text{or} \quad T_s = \frac{1}{2}$$

Substituting the above values in Eq. (1),

$$z = \frac{1+\left(\frac{T_s}{2}\right)s}{1-\left(\frac{T_s}{2}\right)s} \tag{1}$$

$$= \frac{1+\left(\frac{1}{2}\times\frac{1}{2}\right)-\left\{\left(\frac{1}{2}+\frac{j\sqrt{3}}{2}\right)\right\}}{1-\left(\frac{1}{2}\times\frac{1}{2}\right)\left\{-\left(\frac{1}{2}+\frac{j\sqrt{3}}{2}\right)\right\}}$$

$$z = \frac{1-\frac{1}{8}+\frac{j\sqrt{3}}{8}}{1+\frac{1}{8}-\frac{j\sqrt{3}}{8}} = \frac{\frac{7}{8}+\frac{j\sqrt{3}}{8}}{\frac{9}{8}-\frac{j\sqrt{3}}{8}} = \frac{7+j\sqrt{3}}{9-j\sqrt{3}}$$

$$\text{Magnitude of } z = |z| = \frac{\sqrt{(7)^2+\left(\sqrt{3}\right)^2}}{(9)^2+\left(-\sqrt{3}\right)^2} = \sqrt{\frac{52}{84}} = 0.787$$

$$\text{Phase of } z = \angle z = \tan^{-1}\left(\frac{\sqrt{3}}{7}\right) - \tan^{-1} = \left(-\frac{\sqrt{3}}{9}\right)$$

$$= \tan^{-1}\left(\frac{\sqrt{3}}{7}\right) + \tan^{-1} = \left(\frac{\sqrt{3}}{9}\right)$$

$$= 13.9 + 10.89 = 24.78°$$

Transfer function of a third-order Buttterworth filter is given by

$$H_A(s) = \frac{1}{\left(s+e^{j0}\right)\left(s+e^{j\pi/3}\right)\left(s+e^{-j\pi/3}\right)}$$

$$= \frac{1}{(s+1)^2\left(s^2+s+1\right)}$$

We now determine the transfer function of the digital filter using the bilinear transformation method

$$H_D(z) = H_A(s)\,|_{s=(s/T_s)(z-1/z+1)} = H_A(s)\,|_{s=(2/1/2)(z-1/z+1)}$$

$$= H_A(s)\,|_{s=4(z-1/z+1)},\ \text{Given } T_s = \frac{1}{2}\,\text{sec}$$

$$= \left.\frac{1}{(s+1)(s^2+s+1)}\right|_{s=4(z-1/z+1)}$$

$$= \frac{1}{\left[4\left(\frac{z-1}{1+1}\right)+1\right]\left[\left\{4\left(\frac{z-1}{z+1}\right)\right\}^2+4\left(\frac{z-1}{z+1}\right)+1\right]}$$

$$= \frac{1}{\left(\frac{4z-4+z+1}{z+1}\right)\left(\frac{16(z-1)^2+4(z-1)(z+1)+(z+1)^2}{(z+1)^2}\right)}$$

$$= \frac{(z+1)^2}{(5z-3)\left(16z^2-32z+4z^2-4+z^2+1+2z\right)}$$

$$= \frac{(z+1)^3}{(5z-3)\left(22z^2-30z+13\right)}$$

or $$H_D(z) = \frac{(z+1)^3}{(5z-3)\left(21z^2-30z+13\right)}$$

$$= \frac{0.2(z+1)^3}{(z-0.6)(z-0.7143-j-.3299)(z-0.7143+j0.3299)}$$

$$= \frac{0.2(z+1)^3}{(z-z_0)(z-z_1)(z-z_1^*)}$$

where z_1^* is the complex conjugate of z_1 and z_0, z_1, z_1^* are the poles of the third-order Butterworth filter.

Now let us find where the poles will be

$$z_0 = 0.6$$
$$z_1 = 0.7143 + j0.3299 = 0.787\angle 24.78°$$
$$z_1^* = 0.7143 - j0.3299 = 0.787\angle 24.78°$$

poles z_1 and z_1^* both have magnitude of 0.787 and argument $0 = 24.78°$.

The corresponding right-hand plane poles are obtained by taking inverse of the left-hand plane poles. So that C_1 is a point in the exterior of unit-circle $|z| = 1$ on the real z-axis which is inverse of C_2 which is located in the interior of the unit-circle $|z| = 1$. Poles and zeros of third-order. Butterworth digital filter in the z-plane is shown in Figure 9.13.

Note. This transfer function $H_D(z)$ has three zeros at $z = -1$ which correspond to three zeros at $s = -\infty$ for the transfer function of analog.

9.2.5 Digital Chebyshev Filter

Magnitude response of Butterworth filters is monotonic in both pass-band and stop-band. To get a higher roll-off in the transition band, a Butterworth polynomial of higher order is required. As higher-order Butterworth filters require more stages and are therefore much more expensive.

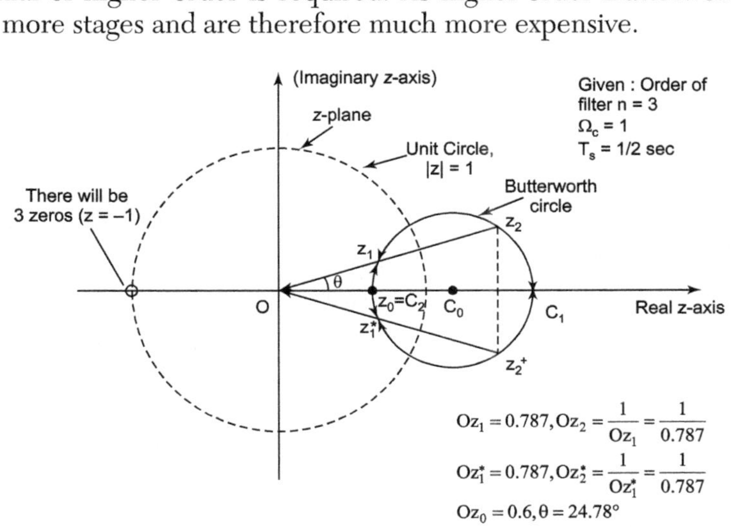

FIGURE 9.13 Demonstration of the poles and zeros of third-order Butterworth digital filter in the *z*-plane.

If the response is allowed to ripple either in the pass band or in the stop band, a lower order filter can be achieved in comparison with the Butterworth design method. This method is also called *equiripple method* and is obtained by using the Chebyshev polynomial. So these filters are also called *Chebyshev filters*.

The magnitude response function has the equiripple behavior in the pass band. It is illustrated in Figure 9.14.

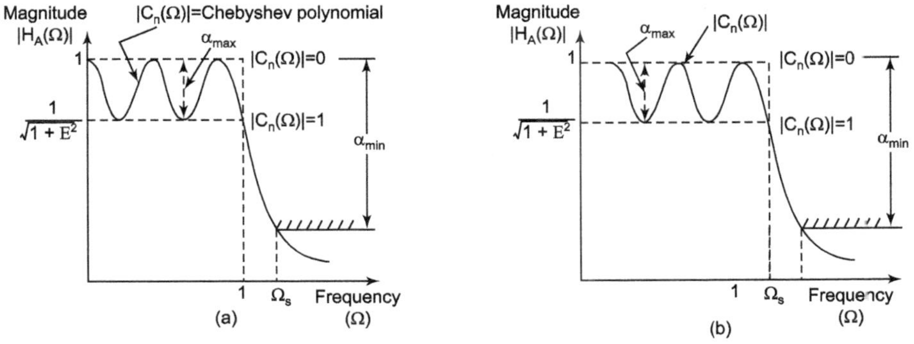

FIGURE 9.14 Magnitude response function approximation using Chebyshev polynominal (a) For $n = 5$ (b) for $n = 6$.

Transfer function of Chebyshev filter is given by

$$\boxed{|H(\Omega)|^2 = \frac{[H(0)]^2}{1+E^2C_n(\Omega)}} \tag{9.38}$$

where $E = \sqrt{10^{A_{\max}-1}}$

$A_{\max} = 0.1a_{\max}$

$\alpha_{\max}$ is the maximum allowable attenuation and $C_n(\Omega)$ is the nth order Chebyshev polynomial. It is expressed by

$$C_n(\Omega) = \begin{cases} \cos\left(n\cos^{-1}\Omega\right) & \text{for } \Omega \le 1 \\ \cosh\left(n\cosh^{-1}\Omega\right) & \text{for } \Omega > 1 \end{cases} \tag{9.39}$$

For Chebyshev filter $H(0) \neq 1$

and $H_A(s) = H_A(j\Omega) = 1$ at $s = j\Omega = j\infty = 0$

Figure 9.14 shows the equiripple behavior for n odd and for n even.

Case I: At $\Omega = 0, C_n(0) = \cos\left(n\cos^{-1}\right) = \cos\left(\frac{nk\pi}{2}\right)$

where $k = 1, 2, \ldots$

For odd values of n, $\frac{nk\pi}{2}$ is also odd, so that

$$C_n(0) = \cos\left(\frac{nk\pi}{2}\right) = 0$$

For these values of k, $|H_A(\Omega)| = 1$

Case II. For even values of k, $\cos\left(\frac{nk\pi}{2}\right) = \pm 1$

Hence, $C_n(0) = \cos\left(\frac{nk\pi}{2}\right) = \pm 1$

For these, values of k, $|H_A(\Omega)| = \frac{1}{\sqrt{1+E^2}}$

For $\Omega = 1, |C_n(\Omega) = 1$, therefore,

$$|H_A(\Omega)|^2 = \frac{1}{\sqrt{1+E^2}} \tag{9.40}$$

For $\Omega > 1$, the hyperbolic function is monotonic and hence beyond the pass band, the curves are monotonic in behavior shape.

From Figure 9.14, pass-band attenuation is given by

$$a_{\max} = 10\log_{10}\left[1+E^2\right] \tag{9.41}$$

All these measurements are for $|H(0)| = 1$, Similarly, stop-band attenuation α_{min} is also specified with respect to $|H_A(0)| = 1$ and this value of attenuation occurs at $\Omega = \Omega s$, where Ωs is the edge of the stop-band frequency. Stop-band attenuation is given by.

$$a_{\max} = 10\log_{10}[1 + E^2 \cosh^2\left(n\cosh^{-1}\Omega s\right) \tag{9.42}$$

Taking antilog of two Eqs. (9.41) and (9.42), we get
From Eq. (8.41),

$$a_{\max} = 10\log_{10}\left[1 + E^2\right]$$

or
$$\frac{a_{\max}}{10} = \log_{10}\left[1 + E^2\right]$$

or
$$A_{\max} = \log_{10}\left[1 + E^2\right]$$

or
$$10A_{\max} = 1 + E^2$$

$$E = \sqrt{10A_{\max} - 1} \tag{9.43}$$

From Eq. (9.42),

$$a_{\min} = 10\log_{10}\left[1 + E^2 \cosh^2\left(n\cos^{-1}\Omega_s\right)\right]$$

or
$$\frac{a_{\max}}{10} = \log_{10}\left[1 + E^2 \cosh^2\left(n\cosh^{-1}\Omega_s\right)\right]$$

or
$$A_{\max} = \log_{10}\left[1 + E^2 \cosh^2\left(n\cosh^{-1}\Omega_s\right)\right]$$

or
$$10A_{\min} = 1 + E^2 \cosh^2\left(n\cosh^{-1}\Omega_s\right)$$

or
$$10A_{\min} - 1 = E^2 \cosh 2\left(n\cosh^{-1}\Omega_s\right)$$

or
$$n\cosh^{-1}\Omega_s = \cosh^{-1}\left[\frac{\sqrt{10A_{\min} - 1}}{E}\right]$$

or
$$n = \cosh^{-1}\left[\frac{\sqrt{10A_{\max} - 1}}{\sqrt{10A_{\max} - 1}}\right]\frac{1}{\cosh^{-1}\Omega_s}$$

Substituting $E = \sqrt{10A_{\max} - 1}$, we get

or
$$n = \cosh^{-1}\left[\frac{\sqrt{10A_{\max} - 1}}{\sqrt{10A_{\max} - 1}}\right]\frac{1}{\cosh^{-1}\Omega_s} \tag{9.44}$$

where,
$$\cos^{-1}(z) = \log e\left(z + \sqrt{z^2 - 1}\right) \tag{9.45}$$

Equation (9.45) can be used to simplify the calculations of $\cosh^{-1}(z)$. A comparison of two relations is given below in the Eqs. (9.46) and (9.47)

$$n = 10_{10}\sqrt{\frac{P}{E}}\frac{1}{\log_{10}\Omega_s} \tag{9.46}$$

where $P = 10^{A}{}_{\min} - 1$.

This is an expression of the order of filter by the Butterworth filter design method:

$$n = \cosh^{-1}\sqrt{\frac{10^{A_{\min}}-1}{10^{A_{\min}}-1}}\frac{1}{\cos^{-1}\Omega_s} \tag{9.47}$$

This is an expression of the order of filter by the Chebyshev filter design method.

We observe that the equations are similar in form except for the difference in the logarithm and $\cosh^{-1}$ functions in the Butterworth and Chebyshev methods respectively. In both cases, we can easily calculate the order of filter (n) from the knowledge of the following parameters:

$\alpha_{\max}$ = maximum allowable attenuation in the pass-band

$\alpha_{\min}$ = minimum allowable attenuation in the stop-band

$$\omega_1 = 2\tan^{-1}\left(\frac{\Omega_1}{2}\right)$$

$$\omega_2 = 2\tan^{-1}\left(\frac{\Omega_2}{2}\right)$$

Pole Locations for Chebyshev Filters

Poles for the Chebyshev transfer function are located on an ellipse. Geometrically an ellipse can be constructed from two circles of radii R and r and these two radii form the major and minor axis respectively of the ellipse. Therefore, in order to determine the pole locations of Chebyshev filters, we need to know the following parameters:

1. $$M = E^{-1} + \sqrt{(1+E^{-2})} \tag{9.48}$$

   where $$E = \sqrt{(10^{A_{\max}}-1)}$$

2. Radius of minor circle is determined by

$$r = \left(\frac{M^{1/n} - M^{-1/n}}{2}\right) \tag{9.49}$$

Radius of major circle is determined by

$$R = \left(\frac{M^{1/n} - M^{-1/n}}{2} \right) \tag{9.50}$$

where n is determined from the equation

$$\boxed{n = \cosh^{-1} \sqrt{\left[\frac{\sqrt{10^{A_{\min}} - 1}}{\sqrt{10^{A_{\max}} - 1}} \right] \frac{1}{\cosh^{-1} \Omega_s}}} \tag{9.51}$$

Figure 9.15 shows the method of construction and shows the locations of three poles in the left half s-plane for a third-order filter. One of these three poles is located on the real axis (σ-axis) and it is denoted by p_1 (its length from the center of the circle is $0 - r$).

Now, we want to find the complex poles p_2 and p_2^*. The method of construction of these complex poles is explained as follows:

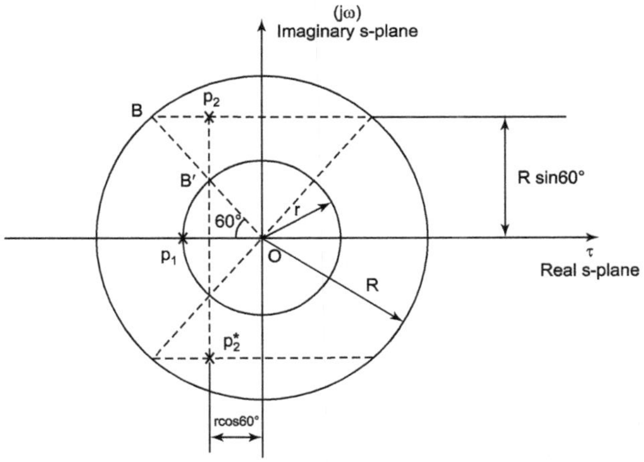

FIGURE 9.15 Determination of pole locations on an ellipse with major axis *R* and minor axis *r*.

1. First of all draw a line OB with an angle $\frac{360}{2n} = \frac{360}{2 \times 3} = 60°$. The line OB intersects the smaller circle at B' and the large circle at B.

2. Now draw a horizontal line Bp_2 from point B and a vertical line p_2B'. Intersection of these two lines (Bp_2 and p_2B') gives us the pole p_2.

3. The complex conjugate of p_2, that is, p_2^* can be determined in the same manner explained above.

Poles are determined as
Pole, $p_1 = -r$
where r is the radius of the minor circle.
Pole, $p_2 = -r \cos 60° \pm jR \sin 60°$

Note. The value of r and R depends upon the value of α_{max}. So r and R are fixed for a given α_{max} and E. The number of poles in an ellipse depends upon the order of the filter n. For example, if the order of the filter n, is 4, the angles are the same as the Butterworth filters and its pole pairs are determined as

$$p_1 p_1^* = r\cos 22.5° \pm jR\sin 22.5°$$

$$p_2 p_2^* = r\cos 67.5° \pm jR\sin 67.5°$$

EXAMPLE 9.7

Design a low-pass digital filter with the following specifications: Maximum pass-band attenuation,

$$a_{\text{max}} = -3 \text{ dB, for } 0 \le \omega \le \frac{2\pi}{10}$$

Minimum stop-band attenuation,

$$a_{\text{min}} = -15 \text{ dB, for } \frac{3\pi}{10} \le \omega \le \pi$$

Using the following two methods,

1. impulse-response invariance method, and
2. bilinear transformation method.

Solution:

We will design the above filter using both methods mentioned above one by one:

1. *Design Using Impulse-Response Invariance Method*

Maximum allowable ripple should be 1 dB. So the given maximum allowable ripple, that is, –3 dB is inadequate for this design. Here we need to calculate E and M separately. Order of the filter for the Chebyshev method is given by

$$n = \cosh^{-1}\sqrt{\frac{10^{A_{\text{min}}} - 1}{10^{A_{\text{min}}} - 1}} \frac{1}{\cos^{-1}\Omega_s}$$

$$= \cosh^{-1}\sqrt{\frac{10^{a_{\text{max}/10}} - 1}{10^{a_{\text{min}/10}} - 1}} \cdot \frac{1}{\cos^{-1}\Omega_s}$$

$$\Omega_s = \frac{\Omega_2}{\Omega_1} = \frac{\omega_2}{\omega_1}$$

Given $\omega_1 = \frac{2}{10}\pi$

and $\omega_2 = \frac{3}{10}\pi$

∴ $\Omega_s = \dfrac{\frac{10}{2}}{\frac{2}{10}\pi} = 1.5$

Determination of the order of the filter:

$$n = \cosh^{-1}\sqrt{\frac{10^{A_{\min}}/10-1}{10^{A_{\min}}/10-1}}\cdot\frac{1}{\cos^{-1}(\Omega_s)}$$

$$= \cosh^{-1}\sqrt{\frac{10^{-3/10}-1}{10^{-15/10}-1}}\cdot\frac{1}{\cos^{-1}(1.5)}$$

$$= \cosh^{-1}\sqrt{\frac{10^{-0.3}-1}{10^{-1.5}-1}}\cdot\frac{1}{\cos^{-1}(1.5)}$$

$$= 2.5 \approx 3$$

Now we will recalculate the value of Ω_1:

$$\Omega_1 = 2\tan\left(\frac{\omega_1}{2}\right) = 2\tan\left(\frac{1}{2}\times\frac{2}{10}\pi\right)$$

$$= 2\tan^{-1}(0.5\pi) = 0.705 = 0.244\pi$$

Determination of E:

$$E = \sqrt{10^{a_{\max}/10}-1} = \sqrt{10^{-3/10}-1}$$

$$= \sqrt{10^{-0.3}-1} = 0.9976$$

Determination of M:

$$M = \frac{1}{E} + \sqrt{1+\frac{1}{E^2}}$$

$$= \frac{1}{0.9976} + \sqrt{1+\frac{1}{(0.9976)^2}} = 2.4183$$

Determination of the values of r and R for the order of the filter $n = 3$

$$r = \left(\frac{M^{1/n} - M^{1/n}}{2}\right) = \left[\frac{(2.4183)^{1/3} - (2.4183)^{-1/3}}{2}\right]$$

$$= 0.29863$$

$$R = \left(\frac{M^{1/n} - M^{-1/n}}{2}\right) = \left[\frac{(2.4183)^{1/3} - (2.4183)^{-1/3}}{2}\right]$$

$$= 1.04363$$

Here the order of the filter is 3 so this filter requires three poles given as follows.

$$p_1 = -\Omega_1 r = -0.705 \times 0.29863 = -0.21053$$

$$p_2 = \Omega_2(-r\cos 60° + jR\sin 60°) = -0.1053 + j0.6372$$

$$p_2^* = \Omega_1(-r\cos 60° + jR\sin 60°) = -0.1053 - j0.6372$$

The transfer function of the analog filter is given by

where $$H_A(s) = \frac{H_0}{\prod_{i=1}^{n}(s - p_i)}$$

It is the general form of any analog filter.

where $$H_0(s) = \begin{cases} 10^{-0.5^{a_{max}}} & \prod_{i=1}^{n}(-p_i) \text{ for } n \text{ even} \\ & \prod_{i=1}^{n}(-p_i) \text{ for } n \text{ odd} \end{cases}$$

Here n is odd, therefore,

$$H_0 = \prod_{i=1}^{n}(-p_i)$$

$$= p_i - p_2 \times -p_3 \qquad \left(\text{but } p_3 = p_2^*\right)$$

$$H_0 = -p_1 \times -p_2 \times -p_2^*$$

$$= -(-0.21053) \times -(-0.1053 + j0.6372) \times -(-0.1053 - j0.6372)$$

$$= (0.21053\left[(-0.1053)^2 + (0.6372)^2\right]$$

$$H_A(s) = \frac{H_0}{(s - p_1)(s - p_2)(s - p_3)}$$

$$= \frac{H_0}{(s - p_1)(s - p_2)(s - p_2^*)}$$

$$= \frac{0.0878}{(s + 0.21053)(s + 0.1053 - j0.6372) \times (s + 0.1053 + j0.6372)}$$

$$\text{or} \quad H_A(s) = \frac{0.0878}{(s+0.21053)\left(s^2+0.2106s+0.4171\right)}$$

It is the transfer function of an analog filter.

Now we have the transfer function for the analog filter and it can be modified into the transfer function of the digital filter in the following manner.

$$H_A(s) = \frac{0.0878}{(s+0.21053)\left(s^2+0.2106s+0.4171\right)}$$

$$= \frac{0.0878}{(s+0.21053)(s+0.1053-j0.6372)(s+0.1053+j0.6372)}$$

$$= \frac{A}{(s+0.21053)} + \frac{B}{(s+0.1053-j0.6372)} + \frac{C}{(s+0.1053+j0.6372)}$$

$$= \frac{0.2115}{(s+0.21053)} + \frac{(0.10524+j0.01738)}{(s+0.1053-j0.6372)} + \frac{(0.10524-j0.01738)}{(s+0.1053+j0.6372)}$$

Determination of impulse response of analog filter:

$$h(t) = \text{Inverse Laplace transform of } H(s)$$

$$= \pounds^{-1}\left[H_A(s)\right]$$

$$= 0.2105\pounds^{-1}\left[\frac{1}{s+0.21053}\right]$$

$$-(0.10524+j0.01738)\pounds^{-1}\left[\frac{1}{s+0.1053-j0.6372}\right]$$

$$-(0.10524+j0.01738)\pounds^{-1}\left[\frac{1}{s+0.1053-j0.6372}\right]$$

$$= 0.2105^{e-0.21053t} - (0.10524+j0.01738)e^{-(0.1053-j0.6372)t}$$

$$-(0.10524+j0.01738)e^{-(0.1053-j0.6372)t}$$

Determination of impulse response (unit-sample response) of a digital filter:

$$h(n) = h(t)\big|_{t=nT_s}$$

where T_s is the sampling period and it is assumed to be equal to 1 s:

$$\text{or} \quad h(n) = 0.2105^{e-0.21053n} - (0.10524+j0.01738)^{e-(0.1053-j0.6372)n}$$

$$-(0.10524+j0.01738)e^{-(0.1053-j0.6372)n}$$

Determination of transfer function of digital filter:

$$
\begin{aligned}
H_{(D)} &= (z)\text{-transform of } h(n) = \mathcal{Z}[h(n)] \\
&= 0.2015z[e^{-0.21053n}) \\
&\quad -(0.10524 + j0.01738)\mathcal{Z}\left[e^{-(0.1053-j0.6372)n}\right] \\
&\quad -(0.10524 + j0.01738)\mathcal{Z}\left[e^{-(0.1053+j0.6372)n}\right] \\
&= 0.2105\left[\frac{1}{1-e^{-0.21053}z^{-1}}\right] \\
&\quad -(0.10524 + j0.01738)\left[\frac{1}{1-e^{-(0.1053-j0.6372)z^{-1}}}\right] \\
&\quad -(0.10524 - j0.01738)\left[\frac{1}{1-e^{-(0.1053+j0.6372)z^{-1}}}\right] \\
&= \frac{0.2105}{1-0.81z^{-1}} - \frac{(0.10524 + j0.01738)}{1-(0.899 + j0.0099)z^{-1}} \\
&\quad - \frac{(0.10524 + j0.01738)}{1-(0.899 + j0.0099)z^{-1}}
\end{aligned}
$$

or
$$
H_D(z) = \frac{0.2105}{1-0.81z^{-1}} + \frac{-0.2105 + 0.1897z^{-1}}{1-1.7998z^{-1} + 0.81z^{-2}}
$$

This is the transfer function of digital filter which is derived from analog filter (Chebyshev filter) using impulse-response invariance method of *IIR* digital filter.

2. Design using Bilinear Transformation Method

Step I. First, the prewarping is required in this method. Prewarping is done in this manner.

$$
\Omega_1 = 2\tan\left(\frac{\omega_1}{2}\right) = 2\tan\left(\frac{1}{2}\times\frac{2}{10}\pi\right) = 0.65
$$

$$
\Omega_2 = 2\tan\left(\frac{\omega_2}{2}\right) = 2\tan\left(\frac{1}{2}\times\frac{2}{10}\pi\right) = 1.02
$$

Ωs is determined as, $\Omega_s = \dfrac{\Omega_2}{\Omega_1} = \dfrac{1.02}{0.65} = 1.569$

And $\cosh^{-1}(\Omega s) = \cosh^{-1}(1.569) = 1.022$

Now the order of the filter is determined by

$$
n = \cosh^{-1}\sqrt{\frac{10^{A_{\min}}-1}{10^{A_{\max}}-1}}\cdot\frac{1}{\cos^{-1}\Omega_s}
$$

$$= \cosh^{-1}\sqrt{\frac{10^{a_{\min}/10}-1}{10^{a_{\max}/10}-1}}\cdot\frac{1}{\cos^{-1}\Omega_s}$$

$$= \cosh^{-1}\sqrt{\frac{10^{15/10}-1}{10^{15/10}-1}}\cdot\frac{1}{\cos^{-1}(1.569)}$$

$$= 2.34 \approx 3$$

Recalculation of Ωs for $n = 3$.

$$n = \cosh^{-1}\left\{\sqrt{\frac{10^{a_{\min}/10}-1}{10^{a_{\min}/10}-1}}\right\}\cdot\frac{1}{\cos^{-1}\Omega_s}$$

or
$$3 = \cosh^{-1}\sqrt{\frac{10^{15/10}-1}{103^{3/10}-1}}\cdot\frac{1}{\cosh^{-1}\Omega_s}$$

or
$$3 = 2.4\frac{1}{\cosh^{-1}\Omega_s}$$

or
$$\cosh^{-1}\Omega_s = \frac{24}{3} = 0.8$$

or
$$\Omega_s = \cosh(0.8) = 1.3374$$

Keeping Ω_2 fixed and Ω_1 is allowed to vary, we get $\Omega_1 = 0.224\pi$.

It is worth noting here that Ω_1 is the same as found by impulse-response invariance method. Therefore, transfer function $H_A(s)$ will also be the same as found by impulse-response invariance method.

$$H_A(s) = \frac{0.0878}{(s+0.21053)\left(s^2+0.2106s+0.4171\right)} \quad (1)$$

We know from the bilinear transformation method $s = \frac{2}{T_s}\left(\frac{z-1}{z+1}\right)$ substituting this value in above Eq. (1), we get

$$H_D(z) = H_A(s)\big|_{s=(2T_s)(z-1/z+1)} \qquad \left(T_s = 1\text{ sec}\right)$$

$$= \frac{0.0878}{\left[2\left(\frac{z-1}{z+1}\right)+0.21053\right]\left[\left\{2\left(\frac{z-1}{z+1}\right)\right\}^2+0.2106\left\{2\left(\frac{z-1}{z+1}\right)\right\}+0.4171\right]}$$

or
$$H_D(z) = \frac{0.0878(z+1)^3}{(z-0.8096)(z^2-1.481z+0.8267)} \quad (2)$$

Determination of frequency response:

Substituting $z = e^{j\Omega}$ in Eq. (2), we get

$$H_D^{(ej\Omega)} = \frac{0.0878\left(e^{j\Omega}+1\right)^3}{\left(e^{j\omega}-0.8096\right)\left(e^{2j\Omega}-1.481e^{j\Omega}+0.8267\right)}$$

Given, $\Omega_1 = \frac{2}{10}\pi$ and $\Omega_2 = \frac{3}{10}\pi$

$$H\left(e^{j\Omega_1}\right) = H\left(e^{j02.\pi}\right) = 0.942\underline{|-164°}$$

and $$\left|H\left(e^{j0.2\pi}\right) = -0.5\text{dB}\right|$$

$$H\left(e^{j0.244\pi}\right) = 0.5865\underline{|-202°}$$

and $$\left|H\left(e^{j0.244\pi}\right)\right| = -4.63\text{ dB}$$

$$H\left(e^{j0.3\pi}\right) = 0.1287\underline{|-97°}$$

and $$\left|H\left(e^{j0.3\pi}\right)\right| = -18\text{ dB}$$

These are the required specifications and will be met by the bilinear transformation.

9.2.6 Inverse Chebyshev Filters

Inverse Chebyshev Filters are also known as type II Chebyshev filters. The magnitude response of a low-pass inverse Chebyshev digital filter is given by

$$\boxed{|H(j\Omega)| = \frac{EC_N\left(\frac{\Omega_2}{\Omega}\right)}{\left[1+E^2C_N^2\left(\frac{\Omega_2}{\Omega}\right)\right]^{1/2}}} \tag{9.52}$$

where E is a constant and Ω_C is the 3 dB cut of frequency.

The Chebyshev polynomial $C_N(x)$ is given by

$$C_N(x) = \begin{cases} \cos\left(N\cos^{-1}x\right), & \text{for } |x| \leq 1 \\ \cos\left(N\cosh^{-1}x\right), & \text{for } |x| > 1 \end{cases}$$

The magnitude response of the inverse Chebyshev filter is shown in Figure (9.16).

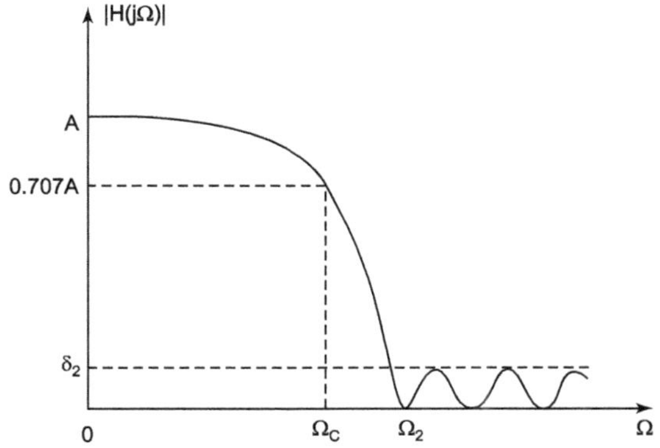

FIGURE 9.16 Magnitude response of the low-pass inverse Chebyshev digital filter.

The magnitude response has maximally flat pass-band and equiripple stop-band, just opposite of the Chebyshev filter's response. That is why the type II Cheybyshev filters are also called inverse Chebyshev filters.

The parameters of the inverse Chebyshev filter are obtained by considering the low-pass filter with the desired specifications as given below.

$$0.707 \le |H(j\Omega)| \le 1,\ 0 \le \Omega \le \Omega_C \tag{9.54}$$

$$|H(j\Omega)| \le \delta_0,\ \Omega \ge \Omega_2 \tag{9.55}$$

From Eqs. (9.52), (9.54), and (9.55), we get

$$(0.707)^2 \le \frac{E^2 C_N^2\left(\dfrac{\Omega_2}{\Omega}\right)}{1+E^2 C_N^2\left(\dfrac{\Omega_2}{\Omega}\right)} \le 1, 0 \le \Omega \le \Omega_C \tag{9.56}$$

$$= \frac{E^2 C_N^2\left(\dfrac{\Omega_2}{\Omega}\right)}{1+E^2 C_N^2\left(\dfrac{\Omega_2}{\Omega}\right)} \le \delta_2^2, \Omega \le \Omega_C \tag{9.57}$$

when $\Omega = \Omega_2$, Eq. (9.57) can be written as

$$\delta_2^2 = \frac{E^2}{1+E^2}$$

or

$$E = \frac{\delta_2}{\sqrt{1+\delta_2^2}} \tag{9.58}$$

When $\Omega = \Omega_2$, Eq. (9.56) becomes

$$0.5 = \frac{E^2 C_N^2\left(\frac{\Omega_2}{\Omega_C}\right)}{1 + E^2 C_N^2\left(\frac{\Omega_2}{\Omega_C}\right)}$$

or $$0.5 + 0.5E^2 C_N^2\left(\frac{\Omega_2}{\Omega_C}\right) = E^2 C_N^2\left(\frac{\Omega_2}{\Omega_C}\right)$$

or $$C_N\left(\frac{\Omega_2}{\Omega_C}\right) = \frac{1}{E} \tag{9.59}$$

Using Eq. (9.53)

$$\cosh\left[N\cosh^{-1}\left(\frac{\Omega_2}{\Omega_C}\right)\right] = \frac{1}{E} \tag{9.60}$$

From Eqs. (9.59) and (9.60), we can get the order of the filter (N).

$$N = \frac{\cosh^{-1}\left(\frac{1}{E}\right)}{\cosh^{-1}\left(\frac{\Omega_2}{\Omega_C}\right)} = \frac{\cosh^{-1}\left[\frac{1}{\delta_2^2} - 1\right]^{1/2}}{\cosh^{-1}\left(\frac{\Omega_2}{\Omega_C}\right)} \tag{9.61}$$

The value of N is chosen to be the nearest integer greater than the value given by Eq. (9.61)

9.2.7 Elliptic Filters

The elliptic filter is sometimes called the Cauer filter. This filter has equiripple pass-band and stop-band. Among the filter types such as Butterworth filter, Chebyshev filter, and inverse Chebyshev filter, for a given filter order, pass-band and stop-band deviations, elliptic filters have the minimum transition bandwidth. The magnitude response of an odd-ordered elliptic filter is shown in Figure (9.17). The magnitude squared response is given by

$$\boxed{|H(j\Omega)|^2 = \frac{1}{1 + E^2 J_N\left(\frac{\Omega}{\Omega_C}\right)}} \tag{9.62}$$

where $J_N(x)$ is the Jacobian elliptical function of order N and E is a constant related to the pass-band ripple.

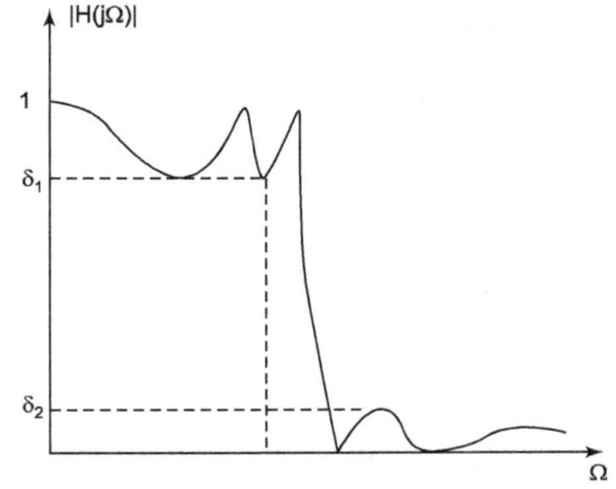

FIGURE 9.17 Magnitude response of a low-pass elliptic filter.

9.3 FREQUENCY TRANSFORMATION

There are basically four types of frequency-selective filters, viz. low-pass filter (LPF), high-pass filter (HPF), band-pass filter (BPF), and band-stop filter (BSF). In the design techniques, we have discussed only LPFs. This LPF can be considered as a prototype filter and its system function can be obtained. Then, if a HPF or BSF or BPF is to be designed, it can be easily obtained by using frequency transformation.

Frequency transformation can be accomplished in two ways. These two ways are:

1. analog-frequency transformation, and
2. digital frequency transformation.

In the analog-frequency transformation the analog system function $H_p(s)$ of the prototype filter is converted into another analog system function $H(s)$ of the desired filter.

Then using any of the mapping techniques, it is converted into the digital filter having a system function $H(z)$.

In the digital frequency transformation, the analog prototype filter is first transformed to the digital domain, to have a system function $H_p(z)$. Then using frequency transformation, it can be converted into the desired "digital filter."

9.3.1 Analog-Frequency Transformation

The frequency transformation formulae used to convert a prototype LPF into a low-pass (with a different cut-off frequency) high-pass, band-pass or band-stop are given as follows:

i. Analog-Frequency Transformation of LPF with cut-off frequency Ω_C to LPF with a new cut-off frequency Ω_C^*:

$$s \longrightarrow \frac{\Omega_C}{\Omega_C^*} s \qquad (9.63)$$

Thus, if the system function of the prototype filter is $H_p(s)$, the system function of the new LPF will be

$$H(s) = H_p\left(\frac{\Omega_C}{\Omega_C^*} s\right) \qquad (9.64)$$

ii. Analog-Frequency transformation of LPF with cut-off frequency Ω_C to HPF with cut-off frequency Ω_C^*:

$$s \longrightarrow \frac{\Omega_C \Omega_C^*}{s} \qquad (9.65)$$

The system function of the HPF is then given by

$$H(s) = H_p\left(\frac{\Omega_C \Omega_C^*}{s}\right) \qquad (9.66)$$

iii. Analog-frequency transformation of LPF with cut-off frequency Ω_C to BPF with low cut-off frequency Ω_1 and higher cut-off frequency Ω_2

$$s \longrightarrow \Omega_C \frac{s(\Omega_2 - \Omega_1)}{s^2 + \Omega_1 - \Omega_2} \qquad (9.67)$$

The system function of the BPF is then given by

$$H(s) = H_p\left(\Omega_c \frac{s^2 + \Omega_C \Omega_C}{s(\Omega_2 - \Omega_1)}\right) \qquad (9.68)$$

iv. Analog-frequency transformation of LPF with cut-off frequency Ω_C to BSF with lower cut-off frequency Ω_1 and higher cut-off frequency Ω_2

$$s \longrightarrow \Omega_C \frac{s(\Omega_2 - \Omega_1)}{s^2 + \Omega_1 - \Omega_2} \qquad (9.69)$$

The system function of BSF is then given by

$$H(s) = H_p\left(\Omega_C \frac{s(\Omega_2 - \Omega_1)}{s^2 \Omega_1 \Omega_2}\right) \qquad (9.70)$$

Analog-frequency transformation formulae are given in Table 9.1

TABLE 9.1 Analog-Frequency Transformation Formulae

	Type of Filter	Transformation
1.	Low-pass filter (LPF)	$s \longrightarrow \frac{\Omega_c}{\Omega_C^*} s$
2.	High-pass filter (HPF)	$s \longrightarrow \left(\frac{\Omega_C \Omega_C^*}{s}\right)$
3.	Band-pass filter (BPF)	$s \longrightarrow \Omega_S \frac{s^2 + \Omega_1 \Omega_2}{s(\Omega_2 - \Omega_1)}$
4.	Band-stop filter (BSF)	$s \longrightarrow \Omega_C \frac{s(\Omega_2 - \Omega_1)}{s^2 + \Omega_1 - \Omega_2}$

EXAMPLE 9.8

A prototype LPF has the system function $H_p(s) = 1/s^2 + 2s + 1$. Find a BPF with $\Omega_0 = 2$ radian/second and $Q = 10\Omega_0^2, = \Omega_1\ \Omega_2$ and $Q = \Omega_0/\Omega_2 - \Omega_1$.

Solution:

From Table 9.1 the required analog-frequency transformation of LPF to BPF is

$$s \longrightarrow \frac{s^2 + \Omega_1 \Omega_2}{s(\Omega_2 - \Omega_1)}$$

that is,

$$s = \Omega_C \frac{s^2 + \Omega_0^2}{s\left(\frac{\Omega_0}{Q}\right)}$$

$$= \Omega_C \frac{s^2 + (2)^2}{s\left(\frac{2}{10}\right)}$$

$$= 5\Omega_C \left(\frac{s^2 + 4}{s}\right)$$

Therefore, system function of BPF will be

$$H_{BPF}(s) = H_p(s)\Big|_{3 = 5\Omega_C\left(\frac{s^2+4}{s}\right)}$$

$$= \left[\frac{1}{s^2 + 2s + 1} \right]_{3=5\Omega_C\left(\frac{s^2+4}{s}\right)}$$

$$= \frac{0.04s^2}{\Omega_C^2 s^4 + 0.4\Omega_C S^3 + \left(8\Omega_C^2 + 0.01\right)s^2 + 1.6\Omega_c s + 16\Omega_C^2}.$$

EXAMPLE 9.9

Transform the prototype LPF with system function

$$H_p(s) = \frac{\Omega_C}{s + \Omega_C}$$

into an HPF with cut-off frequency Ω_C^*.

Solution:

From Table 9.1, the analog-frequency transformation of LPF to HPF if given as

$$s \longrightarrow \frac{\Omega_C \Omega_C^*}{s}$$

Thus, the system function of HPF is given as

$$H_{HPF}(s) = H_p(s)\Big|_{s=\frac{\Omega_C\Omega_C^*}{s}}$$

$$= \left[\frac{\Omega_C}{s + \Omega_C} \right]\Bigg|_{s=\frac{\Omega_C\Omega_C^*}{s}}$$

$$= \frac{\Omega_C}{\frac{\Omega_C \Omega_C^*}{s} + \Omega_C}$$

$$= \frac{\Omega_C s}{\Omega_C s + \Omega_C \Omega_C^*}$$

$$= \frac{\Omega_C s}{\Omega_C s + \left(s + \Omega_C^*\right)} = \frac{s}{s + \Omega_C^*}.$$

9.3.2 Digital Frequency Transformation

Frequency transformation is also possible in the digital domain. The frequency transformation is done in the digital domain by replacing the variable z^{-1} with a function of z^{-1}, that is, $f(z^{-1})$. This mapping must take into account the stability criterion. All the poles lying within the unit-circle that is, $|z| = 1$ must map onto itself and the unit-circle must also map onto itself. For the unit-circle to map onto itself, the implication is that for $|z| = r = 1$

$$e^{-j\omega} = f\left(e^{-j\omega}\right)$$
$$= \left|f\left(e^{-j\omega}\right)\right| e^{jArg[f(e^{-j\omega})]}$$

Hence, we must have $\left|f\left(e^{-j\omega}\right)\right| = 1$ for all frequencies. So, the mapping is that of an all-pass filter and of the form.

$$\boxed{f\left(z^{-1}\right) = \pm\prod_{k=1}^{n}\left[\frac{z^{-1} - A_k}{1 - A_k z^{-1}}\right]} \tag{9.71}$$

For obtaining a stable filter from the stable prototype filter, we must have $|A_k| \leq 1$. The transformation formulae can be obtained from Eq. (9.71) for converting the prototype low-pass digital filter into a digital LPF, digital HPF, digital BPF, or digital BSF. Digital frequency transformation is given in Table 9.2.

TABLE 9.2 Digital Frequency Transformation

Types of Digital Filter	Transformation	Design Parameters
1. Digital LPF	$z^{-1} \to \left[\frac{z^{-1} - A}{1 - Az^{-1}}\right]$	$A = \frac{\sin\left[\left(\omega_c - \omega_c^*\right)/2\right]}{\sin\left[\left(\omega_c + \omega_c^*\right)/2\right]}$
2. Digital HPF	$z^{-1} \to \left[\frac{z^{-1} + A}{1 + Az^{-1}}\right]$	$A = \frac{-\cos\left[\left(\omega_c - \omega_c^*\right)/2\right]}{\cos\left[\left(\omega_c + \omega_c^*\right)/2\right]}$
3. Digital BPF	$z^{-1} \to \left[\frac{z^{-2} - A_1 z^{-1} + A_2}{A_z z^{-2} - A_1 z^{-1} + 1}\right]$	$A_1 = -\frac{2a\,K}{K+1}$ $A_2 = -\frac{(K-1)}{(K+1)}$ $A = \frac{\cos\left[\left(\omega_2 + \omega_1\right)/2\right]}{\cos\left[\left(\omega_2 - \omega_1\right)/2\right]}$ $K = \cot\left(\frac{\omega_2 - \omega_1}{2}\right)\tan\left(\frac{\omega_c}{2}\right)$

(Continued)

Types of Digital Filter	Transformation	Design Parameters
4. Digital BSF	$z^{-1} \rightarrow \left[\dfrac{z^{-2} - A_1 z^{-1} + A_2}{A_z z^{-2} - A_1 z^{-1} + 1}\right]$	$A_1 = -\dfrac{2aK}{K+1}$ $A_2 = -\dfrac{(1-K)}{(1+K)}$ $A = \dfrac{\cos\left[(\omega_2 + \omega_1)/2\right]}{\cos\left[(\omega_2 - \omega_1)/2\right]}$ $K = \tan\left(\dfrac{\omega_2 - \omega_1}{2}\right)\tan\left(\dfrac{\omega_c}{2}\right)$

The frequency transformation may be accomplished in any of the available two techniques, however, caution must be taken to which technique to use.

For example, the impulse-response invariant transformation method is not suitable for HPF_S or BPF_S whose resonant frequencies are higher. In such a case, suppose a low-pass prototype filter is converted into an HPF using analog-frequency transformation and transformed later to a digital filter using impulse-response invariant technique. This will result in aliasing problems. However, if the same prototype LPF is first transformed into a digital filter using impulse-response invariant technique and later converted into an HPF using digital frequency transformation will not have any aliasing problem. Whenever, the bilinear transformation is used, it is of no significance whether the analog-frequency transformation method is used or digital frequency transformation. In this case, both analog and digital frequency transformation techniques will give the same result.

EXERCISES

1. What is an IIR digital filter?
2. How are IIR digital filters realized?
3. What are the various realizability constraints imposed on transfer function of an IIR digital filter?
4. Discuss impulse-response invariance method of deriving IIR digital filter from the corresponding analog filter.

5. Discuss bilinear transformation method of deriving IIR digital filter from the corresponding analog filter.
6. Derive a relationship between complex variables used in Laplace transform (for analog filters) and complex variable z used in z-transform (for digital filters), that is, derive

$$s = \frac{2}{T_s}\left(\frac{z-1}{z+1}\right)$$

where T_s is the sampling period.
7. Discuss various properties of the bilinear transformation method.
8. What is Warping effect? Discuss the influence of Warping effect on amplitude response and phase response of a derived digital filter from a corresponding analog filter.
9. Discuss Digital Butterworth filter.
10. Discuss the magnitude characteristics of an analog Butterworth filter and give its pole locations.
11. Explain the method of constructing Butterworth circle in the z-plane using the bilinear transformation method.
12. Discuss digital Chebyshev filters.
13. Discuss the pole locations for the digital Chebyshev filters.

NUMERICAL EXERCISES

1. Convert the following analog filter with transfer function

$$H_A(s) = \frac{s+0.2}{(s+0.2)^2 + 16}$$

into a digital IIR filter using the following methods:
   i. Impulse-response invariance method.
   ii. Bilinear transformation method.
2. Convert the following analog filters with transfer function

$$H_A(s) = \frac{s+0.2}{(s+0.1)^2 + 9}$$

into a digital IIR filter by using the bilinear transformation method. The digital IIR filter is having a resonant frequency of $\omega_r = \pi/2$.

3. Design a single-pole low-pass IIR digital filter with a 3 dB bandwidth of 0.3π, using bilinear transformation method. The transfer function of the analog filter is given by

$$H_A(s) = \frac{\Omega_C}{s + \Omega_C}$$

where Ω_c is the 3 dB bandwidth of the analog filter.

4. Find the order and poles of a low-pass Butterworth filter that has a-3 dB bandwidth of 500 Hz and an attenuation of 40 dB at 1 kHz.

5. An analog integrator is described by a transfer function $H_A(s) = 1/s$.

   i. Obtain a digital integrator using the bilinear transformation method.

   ii. Obtain the difference equation for the digital integrator relating input $s(n)$ to the output $y(n)$.

CHAPTER 10

DESIGN AND ANALYSIS OF FINITE IMPULSE RESPONSE (FIR) DIGITAL FILTERS

10.1 INTRODUCTION

If the impulse response of a digital filter is determined for some finite number of sample points then these filters are called FIR digital filters. FIR digital filters can readily be designed to have a constant delay as well as prescribed loss specifications. These filters are designed by Fourier Series Method or Numerical-Analysis Methods. Window functions are also used in the designing of FIR digital filters. These functions reduce Gibb's oscillations*.

10.2 PROPERTIES OF FIR DIGITAL FILTERS

A causal FIR digital filter can be characterized by the transfer function

$$H(z) = \sum_{n=0}^{N-1} h(nT_s) z^{-n} \tag{10.1}$$

The frequency response of the above FIR digital filter can be determined by substituting $z = e^{-j\omega T_s}$ in Eq. (10.1).

*See Art. 10.3.1 for details.

$$H\left(e^{j\omega T_s}\right) = \sum_{n=0}^{N-1} h\left(nT_s\right)\left(e^{j\omega T_s}\right)^{-n}$$

$$= \sum_{n=0}^{N-1} h\left(nT_s\right)e^{-j\omega T_s n} = A(\omega)e^{j\phi(\omega)} \tag{10.2}$$

where $A(\omega) = \left|H\left(e^{j\omega T_s}\right)\right|$ is called amplitude response

and $\phi(\omega) = \underline{\left|H\left(e^{j\omega T_s}\right)\right.}$, argument of $H\left(e^{j\omega T_s}\right)$ is called phase response.

Phase Delay: It is the negative ratio of phase $\phi(\omega)$ and frequency w of a filter.

$$\boxed{\tau_g = -\frac{\phi(\omega)}{\omega}} \tag{10.3}$$

Group Delay: It is the negative differentiation of $\phi(\omega)$ with respect to w.

$$\boxed{\tau_g = -\frac{d\phi(\omega)}{d\omega}} \tag{10.4}$$

For linear-phase FIR digital filter both phase delay as well as group delay are constant. If both phase delay and group delay are constant then

$$\phi(\omega) = -\tau\omega \tag{10.5}$$

From Eq. (10.1),

$$H\left(e^{j\omega T_s}\right) = \sum_{n=0}^{N-1} h\left(nT_s\right)e^{-j\omega nT_s}$$

$$= \sum_{n=0}^{N-1} h\left(nT_s\right)\left[\cos\omega nT_s - j\sin\omega nT_s\right]$$

$$= \sum_{n=0}^{N-1} h\left(nT_s\right)\cos\omega T_s - j\sum_{n=0}^{N-1} h\left(nT_s\right)\sin\omega nT_s$$

$$= A + jB$$

$$\phi(\omega) = -\tau\omega = \tan^{-1}\left[\frac{-\sum_{n=0}^{N-1} h\left(nT_s\right)\sin\omega nT_s}{\sum_{n=0}^{N-1} h\left(nT_s\right)\cos\omega nT_s}\right]$$

Thus from Eqs. (10.5) and (10.6)

$$\phi(\omega) = \tan^{-1}\left(\frac{B}{A}\right) = \tan^{-1}\left[\frac{-\sum_{n=0}^{N-1} h(nT_s)\sin\omega nT_s}{\sum_{n=0}^{N-1} h(nT_s)\cos\omega nT_s}\right]$$

or $$\tan(\tau\omega) = \left[\frac{-\sum_{n=0}^{N-1} h(nT_s)\sin\omega nT_s}{\sum_{n=0}^{N-1} h(nT_s)\cos\omega nT_s}\right]$$

or $$\frac{\sin(\tau\omega)}{\cos(\tau\omega)} = \left[\frac{\sum_{n=0}^{N-1} h(nT_s)\sin\omega nT_s}{\sum_{n=0}^{N-1} h(nT_s)\cos\omega nT_s}\right]$$

or $$\sum_{n=0}^{N-1} h(nT_s)\left(\cos\omega nT_s \sin\omega\tau - \sin\omega nT_s \cos\omega\tau\right) = 0$$

or $$\sum_{n=0}^{N-1} h(nT_s)\sin\left(\omega\tau - \omega nT_s\right) = 0$$

or $$\boxed{\phi(\omega) = \sum_{n=0}^{N-1} h(nT_s)\sin(\omega\tau - \omega nT_s)} \tag{10.7}$$

Eq. (10.7) gives the expression for phase response of an FIR digital filter.

10.2.1 Frequency Response

Before discussing frequency response, let us first discuss what symmetrical and anti-symmetrical impulse responses are:

Symmetrical impulse response is the impulse response that satisfies the condition given by

$$\boxed{h(nT_s) = h\left[(N-1-n)T_s \text{ for } 0 \le n \le N-1\right]}$$

It means that impulse response has symmetry about the midpoint between sample points $(N - 2)/2$ and $2N/2$ for even number of samples points or about sample point $(N - 1)/2$ for the odd number of sample points.

Anti-symmetrical impulse response is the impulse response that satisfies the condition given by

$$\boxed{h(nT_s) = -h\left[(N-1-n)T_s \text{ for } 0 \le n \le N-1\right]}$$

It means that impulse response has an anti-symmetry about the midpoint between sample points $(N-2)/2$ and $N/2$ for even number of sample points or about sample point $(N-1)/2$ for the odd number of sample points.

There are four cases of frequency response in constant-delay FIR digital filters given as:

1. Frequency response for symmetrical impulse response with N odd.
2. Frequency response for symmetrical impulse response with N even.
3. Frequency response for anti-symmetrical impulse response with N odd.
4. Frequency response for anti-symmetrical impulse response with N even.

Frequency Response for Symmetrical Impulse Response with N Odd

In this case, the frequency response $H\left(e^{j\omega T_s}\right)=\sum_{n=0}^{N-1}h\left(nT_s\right)e^{-j\omega T_s}$ can be expressed as

$$H\left(e^{j\omega T_s}\right)=\underbrace{\sum_{n=0}^{2}h\left(nT_s\right)e^{-j\omega T_s}}_{\text{I}}+\underbrace{h\left[\left(\frac{N-3}{2}\right)T_s\right]}_{\text{II}}e^{-j\omega}[(N-1)/2]T_s$$

$$+\underbrace{\sum_{n=\left(\frac{N+1}{2}\right)}^{N-1}h\left(nT_s\right)e^{-j\omega nT_s}}_{\text{III}} \tag{10.8}$$

Using equation $h(nT_s) = h[(N-1-n)T_s]$ and substituting $N-1-n=m$ and them $m=n$ in the III summation Eq. (10.8), we get

$$\sum_{n=\left(\frac{n+1}{2}\right)}^{N-1}h[(N-1-n)T_s]e^{-j\omega nT_s}=\sum_{m=\frac{N-3}{2}}^{N-1}h\left(mT_s\right)e^{-j\omega(N-1-m)T_s}$$

$$=\sum_{m=0}^{\left(\frac{N-3}{2}\right)}h\left(mT_s\right)e^{-j\omega(N-1-m)T_s}$$

Now substituting $n=m$ in above Eq. (10.8), we get

$$\sum_{n=\left(\frac{N+1}{2}\right)}^{N-1}h\left(nT_s\right)e^{-j\omega nT_s}=\sum_{n=0}^{\left(\frac{N-3}{2}\right)}h\left(nT_s\right)e^{-j\omega(N-1-m)T_s} \tag{10.9}$$

Substituting Eq. (10.9) into Eq. (10.8), we get

$$H\left(e^{j\omega T_s}\right)=\sum_{n=0}^{\left(\frac{N-3}{2}\right)} h\left(nT_s\right)e^{-j\omega nT_s}+h\left[\left(\frac{N-1}{2}\right)T_s\right]e^{-j\omega[(N-1)/2]T_s}+\sum_{n=\left(\frac{N+1}{2}\right)}^{n-1} h\left(nT_s\right)e^{-j\omega nT_s}$$

$$=\sum_{n=0}^{\left(\frac{N-3}{2}\right)} h\left(nT_s\right)e^{-j\omega nT_s}+h\left[\left(\frac{N-1}{2}\right)T_s\right]e^{-j\omega[(N-1)/2]T_s}+\sum_{n=0}^{\left(\frac{N-3}{2}\right)} h\left(nT_s\right)e^{-j\omega[(N-1)/2]T_s}$$

$$=\sum_{n=0}^{\left(\frac{N-3}{2}\right)} h\left(nT_s\right)e^{-j\omega nT_s}+h\left[\left(\frac{N-1}{2}\right)T_s\right]e^{-j\omega[(N-1)/2]T_s}+\left[\sum_{n=0}^{\left(\frac{N-3}{2}\right)} h\left(nT_s\right)^{e^{j\omega nT_s}}\right]e^{-j\omega(N-1)/2T_s}$$

or

$$H\left(e^{j\omega T_s}\right)=\sum_{n=0}^{\left(\frac{N-3}{2}\right)} h\left(nT_s\right)\left[e^{-j\omega nT_s}+e^{-j\omega(N-1-n)T_s}\right]+h\left[\left(\frac{N-1}{2}\right)T_s\right]e^{-j\omega[(N-1)/2]T_s}$$

$$=e^{-j\omega[(N-1)/2]T_s}\sum_{n=0}^{\left(\frac{N-3}{2}\right)} h\left(nT_s\right)\left[e^{-j\omega[\{(N-1)/2\}-n]T_s}+e^{-j\omega[\{(N-1)/2\}-n]T_s}\right]+h\left[\left(\frac{N-1}{2}\right)T_s\right]e^{-j\omega[(N-1)/2]T_s}$$

$$=e^{-j\omega[(N-1)/2]T_s}\sum_{n=0}^{\frac{N-3}{2}} 2h\left(nT_s\right)\left[e^{-j\omega[\{(N-1)/2\}-n]T_s}+e^{-j\omega[\{(N-1)/2\}-n]T_s}\right]+h\left[\left(\frac{N-1}{2}\right)T_s\right]e^{-j\omega[(N-1)/2]T_s}$$

$$= e^{-j\omega[(N-1)/2]T_s} \sum_{n=02}^{\frac{N-3}{2}} 2h(nT_s)\cos\left[\omega\left(\frac{N-1}{2}-n\right)T_s\right] + h\left[\left(\frac{N-1}{2}\right)T_s\right]e^{-j\omega[(N-1)/2]T_s}$$

or
$$H^{\left(e^{j\omega T_s}\right)} = e^{-j\omega[(N-1)/2]T_s}\left\{\sum_{n=0}^{\frac{N-3}{2}} 2h(nT_s)\cos\left[\omega\left(\frac{N-1}{2}-n\right)T_s\right] + h\left[\left(\frac{N-1}{2}\right)T_s\right]\right\} \quad (10.10)$$

Substituting $k = \dfrac{N-1}{2} - n$ in Eq. (10.10), we get

$$H^{\left(e^{j\omega T_s}\right)} = e^{-j\omega[(N-1)/2]T_s}\left\{\sum_{n=0}^{\frac{N-3}{2}} 2h(nT_s)\cos\left[\omega\left(\frac{N-1}{2}-n\right)T_s\right] + h\left[\left(\frac{N-1}{2}\right)T_s\right]\right\} \quad (10.11)$$

or
$$= e^{-j\omega[(N-1)/2]T_s}\left\{\sum_{k=\frac{N-1}{2}}^{1} 2h\left[\left(\frac{N-1}{2}-k\right)T_s\right]\cos\omega kT_s + h\left[\left(\frac{N-1}{2}\right)T_s\right]\right\}$$

or
$$= e^{-j\omega[(N-1)/2]T_s}\left\{\sum_{k=1}^{\frac{N-1}{2}}[A_k\cos\omega T_s] + A_0\right\}$$

or
$$H\left(e^{j\omega T_s}\right) = e^{-j\omega[(N-1)/2]T_s}\sum_{k=0}^{\frac{N-1}{2}}[A_k\cos\omega T_s] \quad (10.12)$$

where
$$A_0 = h\left[\left(\frac{N-1}{2}\right)T_s\right] \quad (10.13)$$

And
$$A_k = 2h\left[\left(\frac{N-1}{2}-k\right)T_s\right] \quad (10.14)$$

Eq. (10.12) is the frequency response for symmetrical impulse response with N odd.

Note. In a similar manner, we can simplify the frequency responses for the case of symmetrical impulse response with N even anti-symmetrical impulse response with N odd and anti-symmetrical impulse response with N even. The frequency response for these four cases are summarized in Table 10.1.

TABLE 10.1 Frequency Response of Constant-Delay FIR Digital Filters

Impulse Response $h(nTs)$	Number of Samples in Impulse Response N	Frequency Response $H(e^{j\omega T}{}_s)$
Symmetrical	Old	$e^{-j\omega[(N-1)/2]T_s} \sum_{k=0}^{\left(\frac{N-1}{2}\right)} \bullet$ $A_k \cos \omega T_s$
Symmetrical	Even	$e^{-j\omega[(N-1)/2]T_s} \sum_{k=1}^{\left(\frac{N}{2}\right)} \bullet$ $B_k \cos\left[\omega\left(k-\frac{1}{2}\right)T_s\right]$
Anti-symmetrical	Odd	$e^{-j[\omega(N-1/2)]T_s-\pi/2} \sum_{k=1}^{\left(\frac{N}{2}\right)} \bullet$ $A_k \sin \omega k T_s$
Anti-symmetrical	Even	$e^{-j[\omega(N-1/2)T_s-\pi/2]} \sum_{k=1}^{\left(\frac{N-1}{2}\right)} \bullet$ $B_k \sin\left[\omega\left(k-\frac{1}{2}\right)T_s\right]$

where $A_0 = h\left[\left(\frac{N-1}{2}\right)T_s\right], \quad A_k = 2h\left[\left(\frac{N-1}{2}-k\right)T_s\right]$

and $B_k = 2h\left[\left(\frac{N}{2}-k\right)T_s\right]$

The solution of Eq. (10.7) is

$$\tau = \left(\frac{N}{2}-k\right)T_s = \text{Constant phase delay and group delay} \tag{10.15}$$

$$h(nT_s) = h[(N-\ 1\ -n)T_s] = \text{Impulse response, for } 0 < n < N-1 \tag{10.16}$$

Therefore, FIR digital filters have constant phase and group delays over the entire baseband. These delays are constant only for symmetrical impulse response. Impulse response is called symmetrical when $h(nT_s) = h[(N-1-n)T_s]$.

Impulse response symmetry is about midpoint between $[(N/2) - 1]$ and $N/2$ for even N or about $[(N - 1)/2]$ for odd N. This symmetry is shown in Figure 10.1(a) and (b) for $N = 6$ and 7, respectively.

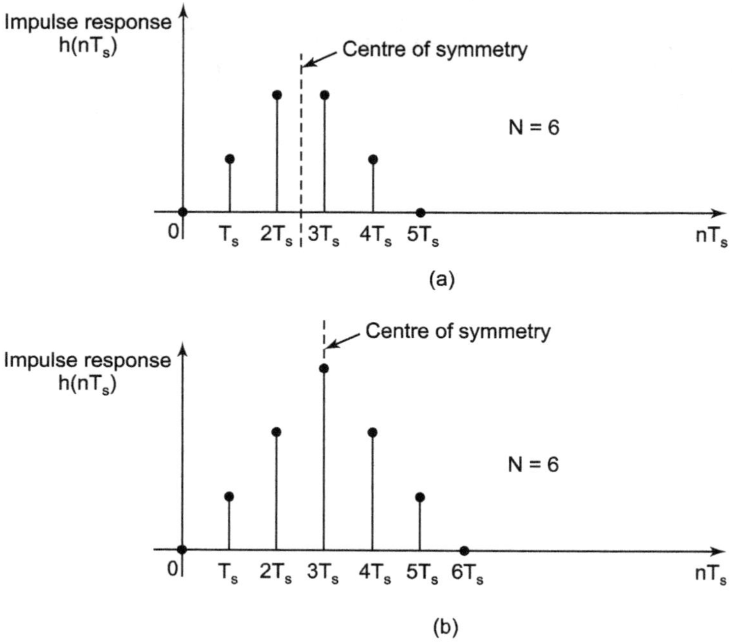

FIGURE 10.1 (a) Impulse response for constant phase and group and delays for even N. (b) Impulse response for constant phase and group delays for odd N.

In some applications, only the group delay is constant, in these cases, the phase response is equal to $\phi(\omega) = \phi_0 - \tau\omega$ where ϕ_0 is a constant. By using this method, we can obtain another class of constant delay FIR digital filter. Putting $\theta_0 = +\pi/2$, the solution is $\tau = (N - 1/2)T_s$ and impulse response $h(nT_s) = -h[N - 1 - n)T_s]$.

In this case, the impulse response $h(nT_s)$ is asymmetrical about the midpoint between samples $\left(\frac{N}{2} - 1\right)$ and $\frac{N}{2}$ for even N or about sample $\left(\frac{N-1}{2}\right)$ for odd N.

This anti-symmetry is shown in Figure 10.2(a) and (b) for $N = 6$ and 7, respectively.

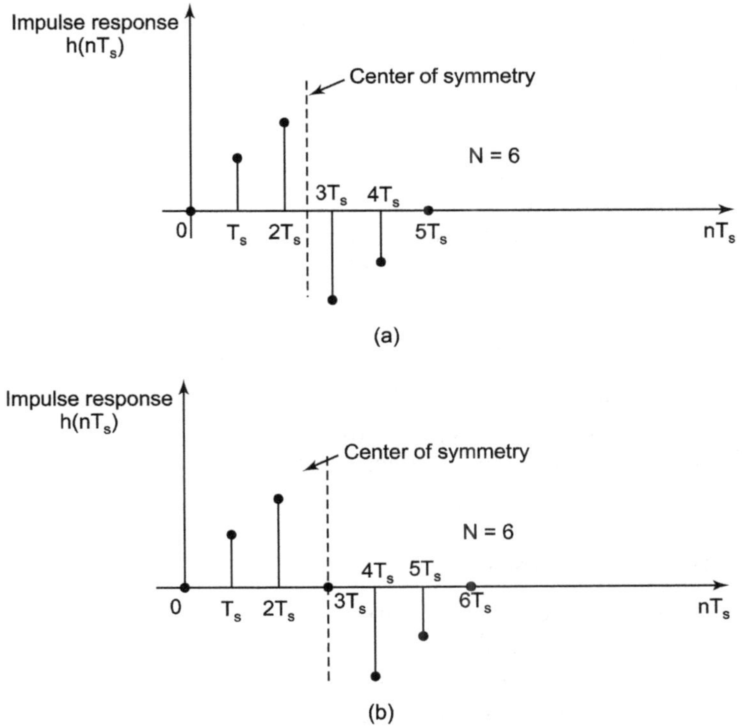

FIGURE 10.2 (a) Impulse response for constant and group delays for even *N*. (b) Impulse response for constant group delays for odd *N*.

10.3 DESIGN OF FIR DIGITAL FILTERS USING FOURIER SERIES METHOD

The frequency response of an FIR digital filter is a periodic function of ω with sampling frequency ω_s. It can be expressed by a Fourier series. Discrete-time Fourier Transform of impulse response $h(nT_s)$ is called frequency response of FIR digital filter.

$$H\left(e^{j\omega T_s}\right) = \sum_{n=-\infty}^{\infty} h\left(nT_s\right)e^{-j\omega nT_s} \tag{10.17}$$

where

$$h\left(nT_s\right) = \frac{1}{\omega_s}\int_{-\omega_s/2}^{\omega_s/2} H\left(e^{j\omega T_s}\right)e^{j\omega nT_s}\,d\omega \tag{10.18}$$

Transfer function $H(z)$ can be determined by substituting $z = e^{j\omega T_s}$ in Eq. (10.17), as

$$H(z) = \sum_{n=-\infty}^{\infty} h(nT_s)z^{-n} \tag{10.19}$$

This is the derived transfer function of a given frequency response. It is a non-causal and infinite order transfer function.

For changing this infinite order transfer function into a finite order transfer function, the series of Eq. (10.19) can be truncated as

$$h(nT_s) = 0, \qquad \text{for} \qquad -\left(\frac{N-1}{2}\right) > n > \left(\frac{N-1}{2}\right)$$

From the above Eq. $h(nT_s)$ is assumed zero outside the some defined sample points. Here $h(nT_s)$ is zero outside the $-(N - 1)/2$ and $(N - 1)/2$ sample points. This is called *truncation of impulse response*. In FIR digital filter design we first truncate impulse response.

10.3.1 Gibb's Oscillations

Slow convergence in the Fourier series results in passband and stopband oscillations. These oscillations are caused by the discontinuity at the passband edge. These oscillations are known as Gibb's oscillations. The frequency of these oscillations increases with the increase of the order of the filter (N).

The amplitude of the last passband ripple and the first stopband ripple remains unchanged. This type of performance is objectionable in practice.

Ways to reduce Gibb's oscillations

There are two methods to reduce Gibb's oscillations:

1. The discontinuity between passband and stopband in the frequency response is avoided by introducing a transition band between passbands and stopbands.
2. Another technique used for the reduction of Gidd's oscillations is to precondition the impulse response $h(nT_s)$. This precondition is imposed by using window functions. Window functions are a class of time domain functions.

10.3.2 Use of Window Functions in the Designing of FIR Digital Filters

Window functions are a class of time domain functions. Gibb's oscillations are reduced using the appropriate window function. Window functions are used to precondition the impulse response $h(nT_s)$.

$$H(z) = z\text{-transform of } h(nT_s) = \mathcal{Z}\left[h(nT_s)\right] = \sum_{n=-\infty}^{\infty} h(nT_s) z^{-n} \tag{10.20}$$

$$H(z) = z\text{-transform of } h(nT_s) = \mathcal{Z}\left[w(nT_s)\right] = \sum_{n=-\infty}^{\infty} w(nT_s) z^{-n} \tag{10.21}$$

$$\begin{aligned} H_\omega(z) &= z\text{-transform of}\left[w(nT_s)\ h(nT_s)\right] = \mathcal{Z}\left[w(nT_s)h(nT_s)\right] \\ &= \sum_{n=-\infty}^{\infty} h(nT_s)h(nT_s)z^{-n} \end{aligned} \tag{10.22}$$

where $w(nT_s)$ is called a window function.

The most frequently used window functions are:

1. Rectangular Window Function.
2. Hann Window Function and Hamming Window Function.
3. Blackman Window Function.
4. Kaiser Window Function.

These window functions will be discussed in the next sections.

10.3.2.1. Rectangular Window Function

It is the simplest window function and is given by

$$\boxed{w(nT_s) = \begin{cases} 1, & \text{for } -\left(\dfrac{N-1}{2}\right) \le n \le \left(\dfrac{N-1}{2}\right) \\ 0, & \text{otherwise} \end{cases}} \tag{10.23}$$

It corresponds to the direct truncation of the Fourier series.

Spectrum of Rectangular Window Function

Frequency response of any window function is given by

$$W\left(e^{j\omega T_s}\right) = \sum_{n=-\infty}^{\infty} w(nT_s) e^{-j\omega n T_s} \tag{10.24}$$

Frequency response of the rectangular window is given by

$$W_R\left(e^{j\omega T_s}\right) = \sum_{n=-\left(\frac{N-1}{2}\right)}^{\left(\frac{N-1}{2}\right)} 1.e^{-j\omega n T_s} = \sum_{n=-\left(\frac{N-1}{2}\right)}^{\left(\frac{N-1}{2}\right)} 1.e^{-j\omega n T_s}$$

This is a Geometric Progression and its sum is given by the formula

$$\frac{A\left[1 - R^{M+1}\right]}{\left[1 - R\right]}$$

where A = First term of Geometric Progression (G.P.)
R = Common Ratio
M = Total number of terms in G.P.

In above G.P., $A = ej\omega T_s[(N-1)/2]$

$$R = e^{j\omega T_s}$$

$$M = \left[\frac{N-1}{2}\right] - \left[-\left(\frac{N-1}{2}\right)\right] = N-1$$

$$W_R\left(e^{j\omega T_s}\right) = e^{j\omega T_s[(N-1)/2]}\frac{\left[1-\left(e^{j\omega T_s}\right)^{N-1+1}\right]}{\left[1-e^{-j\omega T_s}\right]}$$

$$= \frac{e^{j\omega T_s[(N-1)/2]} = e^{j\omega T_s[(N-1)/2]}}{1-e^{-j\omega T_s}}$$

$$= \frac{e^{j\omega T_s[(N-1)/2]} = e^{j\omega T_s[(N+2)/2]}}{1-e^{-j\omega T_s}} = \frac{e^{j\omega NT_s/2} - e^{-j\omega NT_s/2}}{e^{j\omega T_s/2} - e^{-j\omega T_s/2}}$$

$$= \frac{\left(\dfrac{e^{j\omega NT_s/2} - e^{-j\omega nT_s}/2}{2j}\right)}{\left(\dfrac{e^{j\omega T_s/2} - e^{-j\omega T_s/2}}{2j}\right)} = \frac{\sin\dfrac{\omega NT_s}{2}}{\sin\dfrac{\omega T_s}{2}}$$

The frequency response of a rectangular window function is

$$\boxed{W_R\left(e^{j\omega T_s}\right) = \frac{\sin\dfrac{\omega NT_s}{2}}{\sin\dfrac{\omega T_s}{2}}} \tag{10.25}$$

Figure 10.3 illustrates the frequency spectrum of rectangular window function.

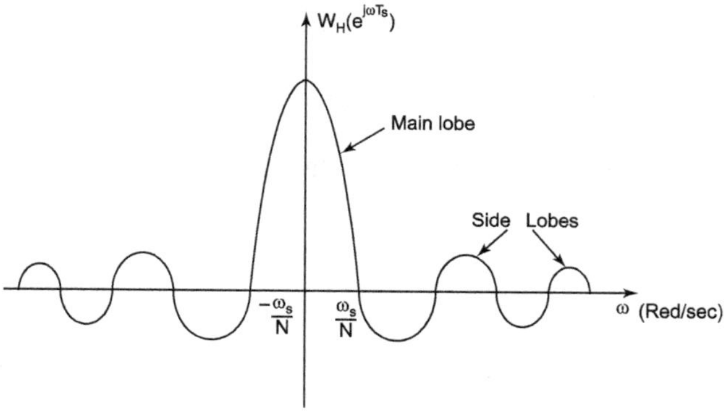

FIGURE 10.3 Spectrum of the rectangular window function

Main lobe width is $\frac{2\omega_s}{N}$ for rectangular window function.

The ripple ratio is defined as

Percentage of ripple ratio is given by,

$$\% \text{ R.R.} = 100\frac{(\text{Maximum side} - \text{lobe Amplitude})}{(\text{Main-lob Amplitude})}$$

RR is 22.34% for $N = 11$ and decreases as N is increased.

10.3.2.2. Hann and Hamming Window Functions

Combined Hann and Hamming window function is called generalized hamming window function and it is given by the following expression

$$w_H(nT_s) = \begin{cases} \alpha + (1-\alpha)\cos\frac{2\pi n}{N-1}, & \text{for} -\left(\frac{N-1}{2}\right) \le n \le \left(\frac{N-1}{2}\right) \\ 0, & \text{otherwise} \end{cases} \quad (10.26)$$

$\alpha = 0.50$ for Hann window function and

$\alpha = 0.54$ for Hamming window function.

Spectrum of Hann and Hamming Window Functions,

Spectrum of Hann and Hamming window functions $w_H(nT_s)$ can be related to that of the rectangular window functions $w_R(nT_s)$.

Rectangular window function $w_R(nT_s) = 1$

but for Hann and Hamming window functions

$$\begin{aligned} w_H(nT_s) &= w_R(nT_s)\left[\alpha + (1-\alpha)\cos\frac{2\pi n}{N-1}\right] \\ &= \alpha w_R(nT_s) + (1-\alpha)\left[\cos\frac{2\pi n}{N-1}\right] w_R(nT_s) \\ &= \alpha w_R(nT_s) + (1-\alpha)\left[\frac{e^{j2\pi n/(N-1)} + e^{-j2\pi n/(N-1)}}{2}\right] w_R(nT_s) \\ &= \alpha w_R(nT_s)\left(\frac{1-\alpha}{2}\right) e^{-j2\pi n/(N-1) w_R(nT_s)} \\ &\quad + \left(\frac{1-\alpha}{2}\right) e^{-j2\pi n/(N-1) w_R(nT_s)} \end{aligned} \quad (10.27)$$

Putting $e^{-j2\pi n/(N-1)} = A$ in Eq. (10.27), we get

$$\begin{aligned} w_H(nT_s) = \alpha w_R(nT_s) + \left(\frac{1-\alpha}{2}\right)(A^{-1})^{-n} w_R(nT_s) \\ + \left(\frac{1-\alpha}{2}\right)(A)^{-n} w_R(nT_s) \end{aligned}$$

[using the theorem of z-transform in which complex scale of z changes, that is,

$$\mathcal{Z}\left[(A)^{-n} w_R(nT_s)\right] = W_R(Az)$$

Taking the z-transform of Eq. (10.27), we get

or $$\mathcal{Z}\left[w_H(nT_s)\right] = \mathcal{Z}\left[\alpha w_R(nT_s)\right] + \mathcal{Z}\left[\left(\frac{1-\alpha}{2}\right)(A^{-1}) w_R(nT_s)\right] + \mathcal{Z}\left[\left(\frac{1-\alpha}{2}\right)(A)^{-n} w_R(nT_s)\right]$$

or $$W_H\left[w_H(z)\right] = \alpha W_R(z) + \left(\frac{1-\alpha}{2}\right) W_R\left(A^{-1}z\right) + \left(\frac{1-\alpha}{2}\right) W_R(Az) \qquad (10.28)$$

Substituting $z = e^{j\omega T_s}$ and $A = e^{j2\pi/(N-1)}$ in Eq. (10.28), we get

$$W_H\left(e^{j\omega T_s}\right) = \alpha W_R\left(e^{j\omega T_s}\right) + \frac{1-\alpha}{2} W_R\left(e^{-j2\pi/(N-1)} e^{j\omega T_s}\right) + \frac{1-\alpha}{2} W_R\left(e^{+j2\pi/(N-1)} e^{j\omega T_s}\right)$$

$$= \alpha W_R\left(e^{j\omega T_s}\right) + \frac{1-\alpha}{2} W_R\left[e^{-j(\omega T_s - 2\pi/(N-1))}.\right] + \frac{1-\alpha}{2} W_R\left[e^{j(\omega T_s + 2\pi/(N-1))}\right] \qquad (10.29)$$

Since the frequency response of a rectangular window, function is given by

$$W_R\left(e^{j\omega T_s}\right) = \frac{\sin\frac{wnT_s}{2}}{\sin\frac{wT_s}{2}}$$

Similarly, $$W_R\left[e^{j(\omega T_s - 2\pi/(N-1))}\right] = \frac{\sin\left[\left(\omega T_s - \frac{2\pi}{N-1}\right)\frac{N}{2}\right]}{\sin\left[\left(\omega T_s - \frac{2\pi}{N-1}\right)\frac{1}{2}\right]}$$

$$W_R\left[e^{j(\omega T_s - 2\pi/(N-1))}\right] = \frac{\sin\left[\left(\omega T_s - \frac{2\pi}{N-1}\right)\frac{N}{2}\right]}{\sin\left[\left(\omega T_s + \frac{2\pi}{N-1}\right)\frac{1}{2}\right]}$$

Substituting the above values in Eq. (10.29), we get

$$W_H^{\left(e^{j\omega T_s}\right)} = \alpha \frac{\sin\left[(\omega T_s)\frac{N}{2}\right]}{\sin\left[(\omega T_s)\frac{1}{2}\right]} + \left(\frac{1-\alpha}{2}\right)\frac{\sin\left[\left(\omega T_s - \frac{2\pi}{N-1}\right)\frac{N}{2}\right]}{\sin\left[\left(\omega T_s - \frac{2\pi}{N-1}\right)\frac{1}{2}\right]}$$

I-term II-term

$$+\left(\frac{1-\alpha}{2}\right)\frac{\sin\left[\left(\omega T_s + \frac{2\pi}{N-1}\right)\frac{N}{2}\right]}{\sin\left[\left(\omega T_s + \frac{2\pi}{N-1}\right)\frac{1}{2}\right]} \tag{10.30}$$

III-term

The spectra for the Hann and Hamming window functions can be formed by shifting frequency response of rectangular window function, $W_R(e^{j\omega T_s})$ first to the right and then to the left by 2π $(N - 1)Ts$ and then adding them to Eq. (10.30). Figure 10.4 illustrates the frequency spectra of Hann and Hamming window function. Figure 10.5 shows the equivalent of spectra of Hann and Hamming window functions.

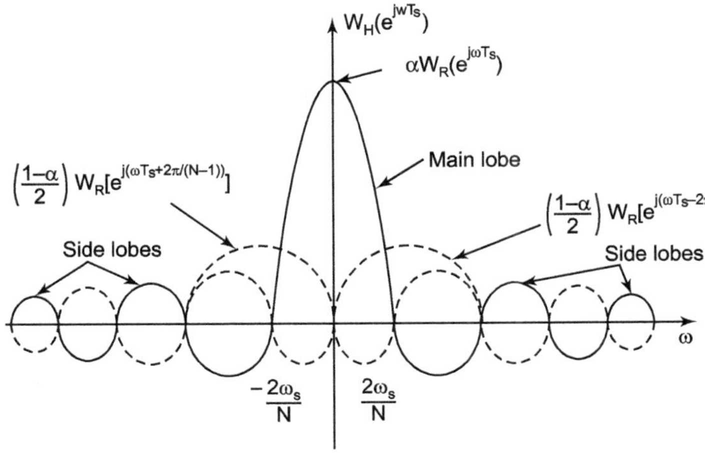

FIGURE 10.4 Spectra of Hann and Hamming window functions

II and III terms are tending to cancel the right and left sides lobes in a $\alpha W_R\left(e^{j\omega T_s}\right)$. Both Hann and Hamming window functions reduce the side lobes in comparison to rectangular window fimrimn:

$RR = 1\ 47\%$ for Hamming window function and
$RR = 2.62\%$ for Hann window function

For $N = 11$ and $\omega_s = 10$ rad/s

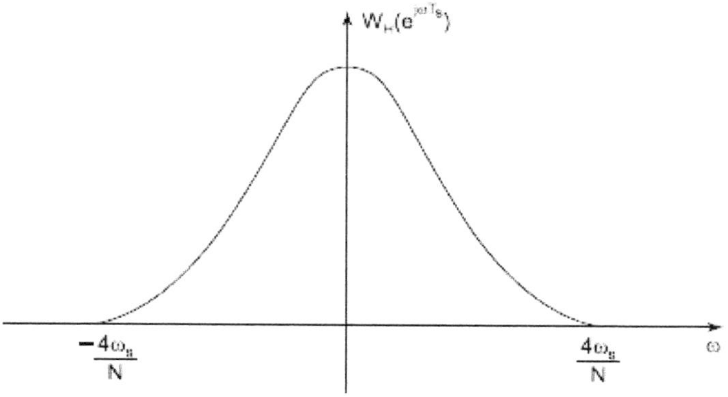

FIGURE 10.5 Equivalent spectra of Hann and Hamming window functions.

The main lobe width for these window functions is $4ws/N$ which is twice the main lobe width for rectangular window function.

10.3.2.3. Blackman Window Function

Blackman window function has one additional cosine term than Hann and Hamming window functions and is given by

$$W_B(nT_s) = \begin{cases} 0.42 + 0.50\cos\dfrac{2\pi n}{N-1} - 0.08\cos\left(\dfrac{4\pi n}{N-1}\right), & \text{for } -\left(\dfrac{N-1}{2}\right) \le n \le \left(\dfrac{N-1}{2}\right) \\ 0, & \text{otherwise} \end{cases} \tag{10.31}$$

This additional term in the Blackman window function leads to a further reduction in the amplitude of Gibb's oscillations. Ripple ratio (RR) is equal to 0.124% for $N = 11$ and $\omega_s = 10$ rad/sec. The main lobe width is increased for this window function and is equal to 6 cos/N.

EXAMPLE 10.1

Design a low-pass FIR filter using the following window functions:

i. rectangular,

ii. ann, and

iii. hamming, for $N = 7$.

Solution:

The frequency response of linear-phase FIR filters is given by

$$H\left(e^{j\omega T_s}\right) = e^{-j[(N-1)/2]\omega T_s}\left[h\left(\frac{N-1}{2}\right) + 2\sum_{n=1}^{\left(\frac{N-1}{2}\right)} h\left(\frac{N-1}{2} - n\right)\cos n\omega T_s\right] \quad (1)$$

For $N = 7$, the desired filter coefficients $h_d(n)$ for $0 \le n \le 6$ are determined.

$$\text{Delay} = \tau = \left(\frac{N-1}{2}\right)T_s = \left(\frac{7-1}{2}\right)T_s = 3T_s$$

Assume $T_s = 1$ second then Eq. (1) can be written as

$$H\left(e^{j\omega}\right) = e^{-j\omega[(N-1)/2]}\left[h\left(\frac{N-1}{2}\right) + 2\sum_{n=1}^{\left(\frac{N-1}{2}\right)} h\left(\frac{N-1}{2} - n\right)\cos n\omega\right]$$

and delay $\tau = 3$.

Also assuming $\omega_C = 1$ radian/s.

We know that the impulse response of low-pass filter is

$$h(nT_s) = \frac{\sin \omega_C n T_s}{n\pi}$$

$$h(n) = \frac{\sin n}{n\pi}$$

$$h_d(0) = h_d(6) = \frac{\sin n}{3\pi} = 0.01497$$

$$h_d(1) = h_d(5)\frac{\sin 2}{2\pi} = 0.014472$$

$$h_d(2) = h_d(4) = \frac{\sin 1}{\pi} = 0.26785$$

$$h_d(3) = \frac{1}{\pi} = 0.31831$$

Using a rectangular window function, $w_R(n) = 1$

then $h(n) = \omega_R(n)h_d(n) = 1h_d(n),\ 0 \le n \le 6$

$$H\left(e^{j\omega}\right) = e^{-j3\omega}\left[h(3) + 2h(2)\cos\omega + 2h(1)\cos 2\omega + 2h(0)\cos 3\omega\right]$$

$$= e^{-j3\omega}\left[0.31831 + 2\times 0.20089\ \cos\omega + 2\times 0.03618\ \cos 2\omega + 2\times 0\ \cos 3\omega\right]$$

$$= e^{-j3\omega}\left[0.31831 + 0.40178\ \cos\omega + 0.07236\ \cos 2\omega\right]$$

For Hamming window function

$$\omega_{Hm}(n) = 0.54 - 0.46\cos\frac{2\pi n}{7-1},\ 0 \le n \le 6$$

$$= 0.54 - 0.46\cos\frac{2\pi n}{7-1},\ 0 \le n \le 6$$

$$= 0.54 - 0.46\cos\frac{2\pi n}{6},\ 0 \le n \le 6$$

$$= 0.54 - 0.46\cos\frac{\pi n}{3},\ 0 \le n \le 6$$

$$\omega_{Hm}(0) = 0.54 - 0.46\cos 0 = 0.08 = \omega_{Hm}(6)$$

$$\omega_{Hm}(1) = 0.54 - 0.46\cos\frac{\pi}{3} = 0.54 - 0.46\times\frac{1}{2} = 0.31 = \omega_{Hm}(5)$$

$$\omega_{Hm}(2) = 0.54 - 0.46\cos\frac{2\pi}{3} = 0.54 - 0.46\times\frac{1}{2} = 0.54 + 0.23$$

$$= 0.77 = \omega_{Hm}(5)$$

$$\omega_{Hm}(3) = 0.54 - 0.46\cos\frac{3\pi}{3} = 0.54 + 0.46\times\cos\pi = 0.54 + 0.46 = 1$$

$$h(n) = \omega_{Hm}(n)h_d(n)$$

$$h(0) = \omega_{Hm}(0)h_d(0) = 0.08\times 0.01497 = 0.0011976 = h(6)$$

$$h(1) = \omega_{Hm}(1)h_d(1) = 0.31\times 0.14472 = 0.0437632 = h(5)$$

$$h(2) = \omega_{Hm}(2)h_d(2) = 0.77\times 0.26785 = 0.2172445 = h(4)$$

$$h(3) = \omega_{Hm}(3)h_d(3) = 1\times 0.31831 = 0.3183$$

$$H(e^{j\omega}) = e^{-j3\omega}[h(3) + 2h(2)\cos\omega + 2h(1)\cos 2\omega + 2h(0)\cos 3\omega]$$

$$= e^{-j3\omega}[0.3183 + 2\times 0.217445\cos\omega + 2\times 0.0437632\cos 2\omega + 2\times 0.0011976\cos 3\omega]$$

$$H\left(e^{j\omega T_s}\right) = e^{-j3\omega}[0.3183 + 0.4348890\cos\omega + 0.0875264\cos 2\omega + 0.0023952\cos 3\omega]$$

$$= e^{-j\omega[(7-1)/2]}\left[h(3) + 2\sum_{n=1}^{\left(\frac{7-1}{2}\right)} h\left(\frac{7-1}{2} - n\right)\cos n\omega\right]$$

$$= e^{-j3\omega}\left[h(3) + 2\sum_{n=1}^{3} h(3-n)\cos n\omega\right]$$

$$= e^{-j3\omega}[h(3) + 2h(2)\cos\omega] + 2h(1)\cos 2\omega + 2h(0)\cos 3\omega]$$

$$= e^{-j3\omega}[0.31831 + 2\times 0.26785\cos\omega + 2\times 0.14472\cos 2\omega + 2\times 0.01497\cos 3\omega]$$

$$= e^{-j3\omega}[0.31831 + 0.5357\cos\omega + 0.28944\cos 2\omega + 0.02994\cos 3\omega]$$

For Hann window function

$$\omega_{Hn}(n) = 0.50 - 0.50\cos\frac{2\pi n}{7-1}, 0 \le n \le 6$$

$$= \frac{1}{2}\left(1 - \cos\frac{2\pi n}{6}\right), 0 \le n \le 6 = \frac{1}{2}\left(1 - \cos\frac{\pi n}{3}\right), 0 \le n \le 6$$

$$\omega_{Hn}(0) = \frac{1}{2}(1 - \cos 0) = \frac{1}{2}(1-1) = 0 = \omega_{Hn}(6)$$

$$\omega_{Hn}(1) = \frac{1}{2}\left(1 - \cos\frac{\pi}{3}\right) = \frac{1}{2}\left(1 - \frac{1}{2}\right) = \frac{1}{4} = 0.25 = \omega_{Hn}(5)$$

$$\omega_{Hn}(2) = \frac{1}{2}\left(1 - \cos\frac{2\pi}{3}\right) = \frac{1}{2}\left(1 + \frac{1}{2}\right) = \frac{3}{4} = 0.75 = \omega_{Hn}(4)$$

$$\omega_{Hn}(3) = \frac{1}{2}\left(1 - \cos\frac{3\pi}{3}\right) = \frac{1}{2}(1+1) = \frac{2}{2} = 1 = \omega_{Hn}(3)$$

$$h(n) = \omega H_n(n)h_d(n)$$

$$h(0) = \omega H_n(0)h_d(0) = 0.01497 = 0$$

$$h(1) = \omega H_n(1)h_d(1) = 0.25\times 0.14472 = 0.03618$$

$$h(2) = \omega H_n(2)h_d(2) = 0.75\times 0.26785 = 0.20089$$

$$h(3) = \omega H_n(3)h_d(3) = 1\times 0.31831 = 0.31831$$

10.3.2.4. Kaiser Window Function

RR is decreased from rectangular to Blackman window function but the main lobe width is increased. Main lobe width can be adjusted by changing N. For achieving prescribed minimum stopband attenuation and passband ripple, we select a window function with an appropriate RR and then choose N to get the prescribed transition width.

To get the desired transition width, the order of the filter (N) is to be increased to an unnecessarily high value. A window function that easily overcomes this problem is called *Kaiser Window Function*.

$$w_K(nT_s) = \begin{cases} F_0(\beta), & \text{for} -\left(\frac{N-1}{2}\right) \le n \le \left(\frac{n-1}{2}\right) \\ 0, & \text{otherwise} \end{cases} \quad (10.32)$$

where α is an independent parameter and β is a dependent parameter, which depends upon α as

$$\beta = \alpha\sqrt{1-\left(\frac{2n}{N-1}\right)} \quad (10.33)$$

$$F_0(\beta) = 1 + \sum_{k=1}^{\infty}\left[\frac{1}{\angle k}\left(\frac{\beta}{2}\right)^k\right]^2$$

and $$F_0(\alpha) = 1 + \sum_{k=1}^{\infty}\left[\frac{1}{\angle k}\left(\frac{\alpha}{2}\right)^n\right]^2$$

where $\angle k$ is factorial of k and $F_0(\beta)$ and $F_0(\alpha)$ are zeroth-order Bessel functions of the first-kind.

Spectrum of Kaiser Window,

Now we will determine the spectrum of Kaiser window.

$$W(z) = Z\left[w(nT_s)\right] = \sum_{n=-\infty}^{\infty} w(nT_s)z^{-n}$$

Putting $z = e^{j\omega T_s}$

$$W_k\left(e^{j\omega T_s}\right) = \sum_{n=-\infty}^{\infty} w(nT_s)\left(e^{-j\omega T_s}\right)^{-n}$$

or $$W_k\left(e^{j\omega T_s}\right) = \sum_{n=-\left(\frac{N-1}{2}\right)}^{\frac{N-1}{2}} w_k(nT_s)e^{-j\omega T_s n} \text{(For Kaiser window function)}$$

$$W\left(e^{j\omega T_s}\right) = w_k(0) + 2\sum_{n=1}^{\left(\frac{N-1}{2}\right)} w_k(nT_s)\cos\omega T_s n \quad (10.34)$$

Kaiser window function in a continuous-time domain is given by

$$W_k(t) = \begin{cases} \frac{F_0(\beta)}{F_0(\alpha)}, & \text{for} -T_d \le t \le T_d \\ 0, & \text{otherwise} \end{cases} \quad (10.35)$$

where $$\beta = \alpha\sqrt{1-\left(\frac{t}{T_d}\right)^2} \quad (10.36)$$

$$t_d = \text{delay time} = \left(\frac{N-1}{2}\right)T_s$$

The spectrum of $W_k(t)$:

$$W_k(j\omega) = \frac{2}{F_0(\alpha)} \frac{\sin\left(T_d\sqrt{\omega^2 - \omega_\alpha^2}\right)}{\sqrt{\omega^2 - \omega_\alpha^2}} \tag{10.37}$$

where $\omega_\alpha = \dfrac{\alpha}{T_d}$

If $W_k(j\omega) \approx 0$ for $\omega| \geq \dfrac{\omega_s}{2}$.

The spectrum of the sampled window function $\omega_k^{'}(t)$ is equivalently the spectrum of $\omega_k(nT_s)$. It is expressed as

$$W_k(j\omega) = W_k\left(e^{j\omega T_s}\right) \approx \frac{1}{T_s} W_k(j\omega) \text{ for } 0 \leq |\omega| \leq \frac{W_s}{2} \tag{10.38}$$

From Eq. (10.37), $T_s W_k\left(e^{j\omega T_s}\right) \approx \dfrac{2}{F_0(\alpha)} \dfrac{\sin\left(T_d\sqrt{\omega^2 - \omega_\alpha^2}\right)}{\sqrt{\omega^2 - \omega_\alpha^2}}$

$$W_k\left(e^{j\omega T_s}\right) = \frac{2}{T_3 F_0(\alpha)} \frac{\sin\left(T_d\sqrt{\omega^2 - \omega_\alpha^2}\right)}{\sqrt{\omega^2 - \omega_\alpha^2}}$$

$$\approx \frac{2}{T_s\omega_\alpha F_0(\alpha)} \frac{\sin\left(T_d\omega_\alpha\sqrt{(\omega/\omega_\alpha)^2 - 1}\right)}{\sqrt{(\omega/\omega_\alpha)^2 - 1}} \tag{10.39}$$

But we know that

$$T_d = \frac{(N-1)T_s}{2} \text{ and } \omega_\alpha = \frac{\alpha}{T_d} \quad or \quad \alpha = T_d\omega_\alpha$$

$$\therefore \qquad \omega_\alpha = \frac{\alpha}{\frac{(N-1)T_s}{2}} = \frac{2\alpha}{(N-1)T_s}$$

or $$\omega_\alpha T_s = \frac{2\alpha}{(N-1)}$$

Substituting $\alpha = T_d\,\omega_\alpha$ and $\omega_\alpha T_d = \dfrac{2\alpha}{(N-1)}$ in Eq. (10.39)

$$W_k\left(e^{j\omega T_s}\right) \approx \frac{2}{\dfrac{2\alpha}{(N-1)} F_0(\alpha) \dfrac{\sin\left[\alpha\sqrt{(\omega/\omega_\alpha)^2 - 1}\right]}{\sqrt{(\omega/\omega_\alpha)^2 - 1}}}$$

$$\approx \frac{N-1}{\alpha F_0(\alpha)} \frac{\sin\left[\alpha\sqrt{(\omega/\omega_\alpha)^2 - 1}\right]}{\sqrt{(\omega/\omega_\alpha)^2 - 1}} \tag{10.40}$$

A very attractive property of the Kaiser window is that the RR can be varied continuously from the low value in the Blackman window function to the high value in the rectangular window function by simply changing the parameter a. In other window functions such as rectangular, Hann, Hamming, and Blackman window functions, the main lobe width can be adjusted by changing the order of filter (N).

For low-pass filter specifications,

The passband ripple, $R_p = 20\log\frac{1+\gamma}{1-\gamma}$, dB

and minimum stopband attenuation, $R_\alpha = -20 \log \gamma$, dB

Transition width, $B_t = \omega_\alpha - \omega_p$, radians

For a filter with passband ripple $R_p \le R_p'$,

A minimum stopband attenuation $R_a \le R_a'$

and transition width B_t can be designed by the following procedure:

Step I: Determination of $h(nT_s)$ using Fourier series method

$$H\left(e^{j\omega T_s}\right) = \begin{cases} 1, & \text{for } |\omega| \le \omega_C \\ 1, & \omega_C < |\omega| \le \frac{\omega_s}{2} \end{cases}$$

ω_C is the lower cutoff frequency

$$\omega_C = \frac{\omega_p + \omega_\alpha}{2}$$

Step II: Choice of γ

$$\gamma = \text{minimum } (\gamma_1, \gamma_2)$$

where $\gamma_1 = 10^{-0.05R'a}$

$$\gamma_2 = \left(\frac{10^{0.05R_p'} - 1}{10^{0.05R_p'} + 1}\right)$$

Step III: Calculation of R_a using Eq. $R_a = -20 \log (\gamma)$

Step IV: Choosing the parameter α:

$$\alpha = \begin{cases} 0, & \text{for } R_\alpha \le 21 \\ 0.5842(R_a - 1)^{0.4} + 0.07886(R_a - 1), & \text{for } 21 < R_a \le 50 \\ 0.1102(R_a - 8.7) & \text{for } R_\alpha > 50 \end{cases}$$

Relation for α is an empirical one and is developed by Kaiser.

Step V: Choosing parameter D:

$$D=\begin{cases} 0.9222, & \text{for } R_\alpha \le 21 \\ \dfrac{R_\alpha - 7.95}{14.36}, & \text{for } R_\alpha > 21 \end{cases}$$

It is also an empirical relation developed by Kaiser

After choosing D, we select the smallest odd value of N satisfying the following relation

$$N \ge \frac{\omega_s D}{B_t} + 1$$

Step VI: Determine $w_k(nT_s)=\begin{cases} \dfrac{F_0(\beta)}{F_0(\alpha)}, & \text{for } |n| \le \dfrac{N-1}{2} \\ 0, & \text{otherwise} \end{cases}$

Step VII: Determine $H'_w(z) z^{-[(N-1)/2]} H_w(z)$

where $H_w(z) = \mathcal{Z}\left[w_k nT_s h(nT_s)\right]$

EXAMPLE 10.2

Design a low-pass filter whose specifications are:

Frequency response $H\left(e^{j\omega T_s}\right) \approx \begin{cases} 1, & \omega_C \le \omega \le \omega_C \\ 0, & \text{for } \omega_C <|\omega| \le \dfrac{\omega_s}{2} \end{cases}$

Passband ripple in frequency range 0 to 15 rad/sec < 0.1 dB.

Minimum stopband attenuation in frequency range 2.5 to 5.0 rad/sec ≥ 40 dB

Sampling frequency ω_s = 10 rad/sec

Solution:

Step I: Impulse response of low-pass filter

$$h(nT_s) = \frac{1}{\omega_s} \int_{-\omega_s/2}^{\omega_s/2} H\left(e^{j\omega T_s}\right) e^{j\omega T_s} d\omega = \frac{1}{\omega_s} = \int_{-\omega_C}^{\omega_C} 1 e^{j\omega n T_s} d\omega$$

$$= \frac{1}{\omega_s}\left[\frac{e^{j\omega n T_s}}{jnT_s}\right]_{-\omega_C}^{\omega_C}$$

$$= \frac{1}{n\omega_s T_s}\left[\frac{e^{j\omega n T_s} - e^{j\omega_c n T_s}}{j}\right] \qquad \text{But } \omega_s = 2\pi f_s = \frac{2\pi}{T_s}$$

$$= \frac{1}{n\frac{2\pi}{T_s}T_s}\left[\frac{e^{j\omega_c nT_s} - e^{-j\omega_c nT_s}}{j}\right] = \frac{1}{n\pi}\left[\frac{e^{j\omega_c nT_s} - e^{-j\omega_c nT_s}}{2}\right]$$

$$\therefore \quad h(nT_s) = \frac{\sin \omega_c nT_s}{n\pi}$$

(It is the impulse response of the LPF filter.)

Step II: Determination of γ from γ_1 and γ_2

$$R_p' = 20\log_{10}\frac{1+\gamma_2}{1-\gamma_2}$$

$$R_\alpha' = -20\log_{10}\gamma_1$$

$$40 = -20\log_{10}\gamma_1$$

or $$\gamma_1 = 0.01$$

$$0.1 = 20\log_{10}\left(\frac{1+\gamma_2}{1-\gamma_2}\right)$$

or $$\gamma_2 = 5.7564\times 10^{-3}$$

$$\gamma = \text{Minimum}(\gamma_1, \gamma_2) = \text{Minimum}\ (0.01, 5.7564\times 10^{-3})$$

Step III: Determination of α, Kaiser's empirical relations for α is given as

$$R_a = -20\log_{10}(\gamma) = -20\log_{10}(5.7564\times 10^{-3}) = 44.797 \text{ dB}$$

Step IV: Determination of α, Kaiser's empirical relations for α is given as

$$\alpha = \begin{cases} 0, & \text{for } R_\alpha \le 21 \\ 0.5842(R_\alpha - 1)^{0.40} + 0.07886(R_\alpha - 21), & \text{for } 21 < R_\alpha \le 50 \\ 0.1102(R_\alpha - 8.7) & \text{for } R_\alpha > 50 \end{cases}$$

From step III, we have determined $R_a = 44.797$ dB

then $$\alpha = 0.5842\,(R_a - 21)^{0.40} + 0.07886\,(R_a - 21)$$

$$= 0.5842\,(44.797 - 21)^{0.40} + 0.07886\,(44.797 - 21)$$

$$= 3.9524$$

Step V: Determination of D, Kaiser's empirical relations for D is given by

$$D = \begin{cases} 0.9222, & \text{for } R_\alpha \le 21 \\ \dfrac{R_\alpha - 7.95}{14.36}, & \text{for } R_a > 21 \end{cases}$$

From Step III, we have determined $R_a = 44.797$ dB

then $$D = \frac{R_\alpha - 7.95}{14.36} = \frac{44.797 - 7.95}{14.36} = 0.5660$$

Hence the order of the filter (N) is given by

$$N \geq \frac{\omega_s D}{B_t} + 1$$

But transition width $B_t = \omega_a - \omega_p = 2.5 - 1.5 = 1.0$ rad/sec

Now $N > \dfrac{10 \times 2.5660}{1.0} + 1 \geq 26.66$

or $\quad N = 27$

Step VI: From Kaiser window

$$w_k(nT_s) = \begin{cases} \dfrac{F_0(\beta)}{F_0(\alpha)}, & \text{for } -\left(\dfrac{N-1}{2}\right) \leq n \leq \left(\dfrac{N-1}{2}\right) \\ 0, & \text{otherwise} \end{cases}$$

Step VII: $$H_w'(z) = z^{-[(N-1)/2]} \sum_{n=0}^{\frac{N-1}{2}} \frac{a_n'}{2}\left(z^n + z^{-n}\right)$$

where $$\alpha_0' = w_k(0)h(0)$$
$$\alpha_n' = 2w_k(nT_s)h(nT_s).$$

10.4 DESIGN OF FIR DIGITAL FILTER BASED ON NUMERICAL-ANALYSIS FORMULAE

A signal $s(t)$ whose sampled values are known at $t = nT_s$, where $n = 0, 1, 2, \ldots$ and T_s is called sampling period. It can be interpolated, differentiated or integrated by using Numerical-Analysis formulae such as Gregory-Newton Forward and Backward difference formulae and Bessel's, Everett's and Stirling's central difference formulae.

Gregory-Newton Forward Difference Formula

The value of signal $s(t)$ at $= nT_s + mT_s$, $0 \leq m \leq 1$ is given by Gregory-Newton forward difference formula as

$$s(t)\big|_{t=nT_s+mT_s} = s(nT_s + mT_s) = (1+\Delta)^m s(nT_s)$$

$$= \left[1 + \frac{m}{\angle 1}\Delta + \frac{m(m-1)}{\angle 2}\Delta^2 + \frac{m(m-1)(m-2)}{\angle 3}\Delta^3 + \ldots\right] \quad (10.41)$$

where $\quad \Delta s(nT_s) = s(nT_s + T_s) - s(nT_s)$

$$s(nT_s + 2T_s) - s(nT_s + T_s) \text{ and so on} \quad (10.42)$$

Δ is known as forward or ascending difference operator.

Gregory-Newton Backward Difference Formula

The value of signal $s(t)$ at $t = nT_s + mT_s$, $0 \le m < 1$ is given by Gregory-Newton backward difference formula as

$$s(t)|_{t=nT_s+mT_s} = s(nT_s + mT_s) = (1-\nabla)^{-m_s(nT_s)}$$

$$= \left[1 + \frac{m}{\angle 1}\nabla + \frac{m(m+1)}{\angle 2}\nabla^2 + \frac{m(m+1)(m+2)}{\angle 3}\nabla^3 + \ldots\right]s(nT_s) \quad (10.43)$$

∇ is known as backward or descending difference operator.

Stirling's Central Difference Formula

The value of signal $s(t)$ at $t = nT_s + mT_s$, $0 \le m < 1$ is given by Stirling's central difference formula as

$$s(t)|_{t=nT_s+mT_s} = s(nT_s + mT_s)$$

$$= \left[1 + \frac{m^2}{\angle 2}\delta^2 + \frac{m^2(m-1)}{\angle 4} + \ldots\right]s(nT_s)$$

$$+ \frac{m}{2}\left[\delta s\left(nT_s - \frac{T_s}{2}\right) + \delta s\left(nT_s - \frac{T_s}{2}\right)\right]$$

$$+ \frac{m(m^2-1)}{2\angle 3}\left[\delta^3 s\left(nT_s - \frac{T_s}{s}\right) + \delta^5 s\left(nT_s + \frac{T_s}{2}\right)\right]$$

$$+ \frac{m(m^2-1)(m^2-2^2)}{2(\angle 5)}\left[\delta^5 s\left(nT_s - \frac{T_s}{2}\right) + \delta^5 s\left(nT_s + \frac{T_s}{2}\right) + \ldots\right] \quad (10.44)$$

where $$\delta s\left(nT_s + \frac{T_s}{2}\right) = s(nT_s + T_s) - s(nT_s) \quad (10.45)$$

δ is known as central difference operator.

Differentiation of $s(t)$ at $t = nT_s + mT_s$

$$\frac{d}{dt}s(t)\bigg|_{t=nT_s+mT_s} = \frac{d}{dt}s(nT_s + mT_s)$$

$$= \frac{d}{dm}s(nT_s + mT_s)\frac{1}{T_c} \quad (10.46)$$

[But $$t = nT_s + mT_s$$

$$1 = 0 + \frac{dm}{dt}T_s$$

or $$\frac{dm}{dt} = \frac{1}{T_s}$$]

Substituting the value of $\frac{dm}{dt} = \frac{1}{T_s}$ in Eq. (10.46), we get

$$\left.\frac{d}{dt}s(t)\right|_{t=nT_s+mT_s} = \frac{d}{dm}s(nT_s + mT_s)\frac{1}{T_s}$$

$$= \frac{1}{T_s}\frac{d}{dm}s(nT_s + mT_s) \tag{10.47}$$

Integration Formula
It can be derived as

$$\int_{t=nT_s}^{q_2} s(t)dt = T_s \int_0^{m_2} s(nT_s + mT_s)dm \tag{10.48}$$

where $nT_s < q_2 < nT_s + T_s$ and $m_2 = \frac{q_2 - nT_s}{T_s}$, $0 < m^2 \leq 1$

FIR filter that perform interpolation, differentiation or integration can now be obtained. Now we assume $s(nT_s)$ and $y(nT_s)$ are the input and output of a FIR filter.

$$y(nT_s) = f[s(t)] \tag{10.49}$$

$$y(nT_s) = \left.\frac{ds(t)}{dt}\right|_{t=nT_s+mT_s} \tag{10.50}$$

$$y(nT_s) = \sum_{j=-k}^{M} A_j s(nT_s - jT_s) \tag{10.51}$$

Thus the derived transfer function is given as

$$H(z) = \sum_{j=-k}^{M} h(nT_s)z^{-n} \tag{10.52}$$

For the case of forward central difference formula. Transfer function $H(z)$ is non-causal. For converting this non-causal transfer function into causal transfer function previous one is multiplied by an appropriate negative powers of z.

EXAMPLE 10.3

A signal $s(t)$ sampled at a rate $1/T_s$ Hz. By using Stirling's central difference formula, design a sixth order differentiating filter. Its time domain response is given by

$$y(nT_s) = \left.\frac{ds(t)}{dt}\right|_{t=nT_s}$$

Solution:

From Stirling's central difference formula

$$\left.\frac{ds(t)}{dt}\right|_{t=nT_s+mT_s} = \frac{1}{T_s}\frac{ds(nT_s+mT_s)}{dm}$$

$$\left.\frac{ds(t)}{dt}\right|_{t=nT_s+mT_s} = \frac{1}{T_s}\frac{1}{dm}s(nT_s+mT_s)$$

$$= \frac{1}{T_s}\frac{1}{d_m}\left\{\left[1+\frac{m^2}{\angle 2}\delta^2+\frac{m^2(m^2-1)}{\angle 4}\delta^4+\ldots\right]s(nT_s)\right.$$

$$+\frac{m}{2}\left[\delta s\left(nT_s-\frac{T_s}{2}\right)+\delta_s\left(nT_s+\frac{T_s}{2}\right)\right]$$

$$+\frac{m(m^2-1)}{2(\angle 3)}\left[\delta^3 s\left(nT_s-\frac{T_s}{2}\right)+\delta^3 s\left(nT_s+\frac{T_s}{2}\right)\right]$$

$$\left.+\frac{m(m^2-1)(m^2-2^2)}{2(\angle 5)}\left[\delta^5\left(nT_s-\frac{T_s}{2}\right)+\delta^5 s\left(nT_s+\frac{T_s}{2}\right)\right]+\ldots\right\} \qquad (1)$$

$$= \frac{1}{T_s}\left\{\left[1+m\delta^2+\frac{(4m^2-2m)}{\angle 4}\delta^4+\ldots\right]s(nT_s)\right.$$

$$+\frac{1}{2}\left[\delta s\left(nT_s-\frac{T_s}{2}\right)+\delta_s\left(nT_s+\frac{T_s}{2}\right)\right]$$

$$+\frac{(3m^2-1)}{2(\angle 3)}\left[\delta^3 s\left(nT_s-\frac{T_s}{2}\right)+\delta^3 s\left(nT_s+\frac{T_s}{2}\right)\right]$$

$$\left.+\frac{(5m^4-15m^2+4)}{2(\angle 5)}\left[\delta^5 s\left(nT_s-\frac{T_s}{2}\right)+\delta^5 s\delta^3 s\left(nT_s+\frac{T_s}{2}\right)\right]+\ldots\right\} \qquad (2)$$

$$\left.\frac{ds(t)}{dt}\right|_{t=nT_s} = \left.\frac{ds(t)}{dt}\right|_{t=nT_s+mT_s} \text{ after putting } m=0$$

$$= \frac{1}{2T_s}\left[\delta s\left(nT_s-\frac{T_s}{2}\right)+\delta_s\left(nT_s+\frac{T_s}{2}\right)\right]$$

$$-\frac{1}{12T_s}\left[\delta^3 s\left(nT_s-\frac{T_s}{2}\right)+\delta^3 s\left(nT_s+\frac{T_s}{2}\right)\right]$$

$$+\frac{1}{60T_s}\left[\delta^5 s\left(nT_s-\frac{T_s}{2}\right)+\delta^5 s\left(nT_s+\frac{T_s}{2}\right)\right]+\ldots \qquad (3)$$

Now using the relation

$$\delta s\left(nT_s + \frac{T_s}{2}\right) = s(nT_s + T_s) - s(nT_s) \tag{4}$$

Similarly $$\delta_s\left(nT_s - \frac{T_s}{2}\right) = s(nT_s)s(nT_s - T_s) \tag{5}$$

Adding both Eqs. (4) and (5), we get

$$\delta_s\left(nT_s + \frac{T_s}{2}\right) + \delta^3 s\left(nT_s - \frac{T_s}{2}\right) = s(nT_s + T_s) - s(nT_s - T_s) \tag{6}$$

Now we will determine $\delta^3 s\left(nT_s + \frac{T_s}{2}\right) + \delta^3 s\left(nT_s - \frac{T_s}{2}\right)$

From Eq. (6)

$$\delta\left[\delta_s\left(nT_s + \frac{T_s}{2}\right) + \delta_s\left(nT_s - \frac{T_s}{2}\right)\right] = \delta\left[s(nT_s + T_s) - s(nT_s - T_s)\right]$$

or $$\delta^2 s\left(nT_s + \frac{T_s}{2}\right) + \delta^2 s\left(nT_s - \frac{T_s}{2}\right) = \delta s(nT_s + T_s) - \delta s(nT_s - T_s)$$

$$= \left[s\left(nT_s + T_s + \frac{T_s}{2}\right) - s\left(nT_s + T_s + \frac{T_s}{2}\right)\right]$$

$$-\left[s\left(nT_s - T_s + \frac{T_s}{2}\right) - s\left(nT_s - T_s - \frac{T_s}{2}\right)\right]$$

$$= \left[s\left(nT_s + \frac{3T_s}{2}\right) - s\left(nT_s + \frac{T_s}{2}\right)\right.$$

$$\left. -s\left(nT_s + \frac{T_s}{2}\right) - s\left(nT_s + \frac{3T_s}{2}\right)\right] \tag{7}$$

From Eq. (7), $$\delta\left[\delta^2 s\left(nT_s + \frac{T_s}{2}\right) - \delta^2 s\left(nT_s - \frac{T_s}{2}\right)\right]$$

$$= \delta\left[s\left(nT_s + \frac{3T_s}{2}\right) - s\left(nT_s + \frac{T_s}{2}\right)\right.$$

$$\left. -s\left(nT_s - \frac{T_s}{2}\right) + s\left(nT_s - \frac{3T_s}{2}\right)\right] \tag{8}$$

or $$\delta^3 s\left(nT_s + \frac{T_s}{2}\right) + \delta^3 s\left(nT_s - \frac{T_s}{2}\right)$$

$$= \delta s\left(nT_s + \frac{3T_s}{2}\right) - \delta s\left(nT_s + \frac{T_s}{2}\right) - \delta s\left(nT_s - \frac{T_s}{2}\right) + \delta s\left(nT_s - \frac{3T_s}{2}\right)$$

$$= \left[s\left(nT_s + \frac{3T_s}{2} + \frac{T_s}{2} \right) - s\left(nT_s + \frac{3T_s}{2} - \frac{T_s}{2} \right) \right]$$

$$- \left[s\left(nT_s + \frac{T_s}{2} + \frac{T_s}{2} \right) - ss\left(nT_s + \frac{T_s}{2} - \frac{T_s}{2} \right) \right]$$

$$- \left[s\left(nT_s - \frac{T_s}{2} + \frac{T_s}{2} \right) - s\left(nT_s - \frac{T_s}{2} - \frac{T_s}{2} \right) \right]$$

$$+ \left[s\left(nT_s - \frac{3T_s}{2} + \frac{T_s}{2} \right) - s\left(nT_s - \frac{3T_s}{2} - \frac{T_s}{2} \right) \right]$$

$$= s(nT_s + 2T_s) - s(nT_s - T_s) - s(nT_s + T_s) + s(nT_s)$$

$$-s(nT_s) + s(nT_s - T_s) + s(nT_s - T_s) - s(nT_s - T_s)$$

$$= s(nT_s + 2T_s) - 2s(nT_s + T_s) + 2s(nT_s + T_s) - s(nT_s - 2T_s)$$

or $\delta^3 s\left(nT_s + \frac{T_s}{2} \right) + \delta^3 s\left(nT_s - \frac{T_s}{2} \right) = s(nT_s + 2T_s) - 2s(nT_s + T_s) + 2s(nT_s + T_s)$

$$-s(nT_s - 2T_s) \qquad (9)$$

Similarly, we can determine

$$\delta^5 s\left(nT_s + \frac{T_s}{2} \right) + \delta^5 s\left(nT_s - \frac{T_s}{2} \right) = s(nT_s + 3T_s) - 4s(nT_s + 2T_s) + 5s(nT_s + T_s)$$

$$-5s(nT_s - T_s) + 4s(nT_s - 2T_s) - s(nT_s - 3T_s) \qquad (10)$$

Hence, from Eq. (10.44), we can get

$$y(nT_s) = \left. \frac{ds(t)}{dt} \right|_{t=nT_s} = \left. \frac{ds(t)}{dt} \right|_{t=nT_s+mT_s \text{ with } m=0}$$

$$= \frac{1}{2T_s}\left[\delta s\left(nT_s - \frac{T_s}{2} \right) + \delta s\left(nT_s + \frac{T_s}{2} \right) \right]$$

$$- \frac{1}{12T_s}\left[\delta^3 s\left(nT_s - \frac{T_s}{2} \right) + \delta^3 s\left(nT_s + \frac{T_s}{2} \right) \right]$$

$$+ \frac{1}{60T_s}\left[\delta^5 s\left(nT_s - \frac{T_s}{2} \right) + \delta^s s\left(nT_s + \frac{T_s}{2} \right) \right] + \ldots \qquad (11)$$

Putting the values of $\delta s\left(nT_s - \frac{T_s}{2} \right) + \delta_s\left(nT_s + \frac{T_s}{2} \right)$

$$\delta^3 s\left(nT_s + \frac{T_s}{2} \right) + \delta^3 s\left(nT_s + \frac{T_s}{2} \right)$$

and $\delta^5 s\left(nT_s - \frac{T_s}{2} \right) + \delta^5 s\left(nT_s + \frac{T_s}{2} \right)$

In Eq. (10), we get

$$y(nT_s)=\frac{1}{2T_s}\left[s(nT_s+2T_s)-s(nT_s-T_s)\right]$$
$$-\frac{1}{12T_s}\left[s(nT_s+2T_s)-2s(nT_s+T_s)+2s(nT_s-T_s)-s(nT_s-2T_s)\right]$$
$$+\frac{1}{60T_s}\left[s(nT_s+3T_s)-4s(nT_s+2T_s)+5s(nT_s+T_s)-5s(nT_s-T_s)\right.$$
$$\left.+4s(nT_s-2T_s)-s(nT_s+3T_s)\right]$$

or $$y(nT_s)=\frac{1}{60T_s}\left[s(nT_s+3T_s)-9s(nT_s+2T_s)+45s(nT_s+T_s)\right.$$
$$\left.-45s(nT_s-T_s)+9s(nT_s-2T_s)-s(nT_s-3T_s)\right] \quad (12)$$

This is the difference equation of the sixth order differentiator.

Taking the z-transform of Eq. (12), we get

$$y(z)=\frac{1}{60T_s}\left[z^3S(z)-9z^2S(z)+45zS(z)-45z^{-1}S(z)\right.$$
$$\left.+9z^{-2}S(z)-z^{-3}S(z)\right]$$

$$H(z)=\frac{Y(z)}{S(z)}=\frac{1}{60T_s}\left[z^3-9z^2+45z-45z^{-1}+9z^{-2}-z^{-3}\right] \quad (13)$$

$$H(z)=\frac{1}{60T_s}\left[z^3-9z^2+45z-45z^{-1}+9z^{-2}-z^{-3}\right]$$

This filter has an anti-symmetrical impulse response and is non-causal. This non-causal filter is converted into a causal filter by multiplying $H(z)$ by z^{-3}

$$H'(z)=z^{-3}H(z)=z^{-3}\left[\frac{1}{60T_s}z^3-9z^2+45z-45z^{-1}+9z^{-2}-z^{-3}\right]$$

$$\frac{1}{60T_s}\left(1-9z^2+45z^{-2}-45z^{-4}+9z^{-5}-z^{-6}\right)$$

This is the transfer function of causal sixth-order FIR digital differentiator.

10.5 DESIGN OF OPTIMAL LINEAR-PHASE FIR DIGITAL FILTERS USING M-CLELLAN–PARKS METHOD

In both frequency sampling and windowing methods of designing FIR digital filters, there was a problem with the precise control of the critical frequencies. In the optimal filter design method, we consider the Chebyshev approximation problem. It is viewed that the weighted approximation error between the actual frequency response and the desired filter response is spread across the passband and the stopband and the maximum error is minimized.

This design method results in passband and the stopband having ripples. The design procedure is explained using a low-pass filter (LPF) with passband and stopband edge frequencies w_p and w_s, respectively.

From Figure 10.6, the frequency response of the filter in the passband is given by

$$1-\delta_p \leq \left|H\left(e^{j\omega}\right)\right| \leq 1+\delta_p, |\omega| \leq \omega_p \tag{10.53}$$

The frequency response in the stopband is given by

$$-\delta_p \leq \left|H\left(e^{j\omega}\right)\right| \leq 1+\delta_s, |\omega| \leq \omega_s \tag{10.54}$$

The term δ_p represents the passband ripple, and δ_s is the maximum attenuation in the stopband.

There are four different cases that result in a linear-phase FIR digital filter, viz.,

1. Symmetric unit-sample (impulse) response and the length of the filter (M) is odd.
2. Symmetric unit-sample (impulse) response and (M) is over.
3. Anti-symmetric unit-sample (impulse) response and (M) is odd.
4. Anti-symmetric unit-sample (impulse) response and (M) is even.

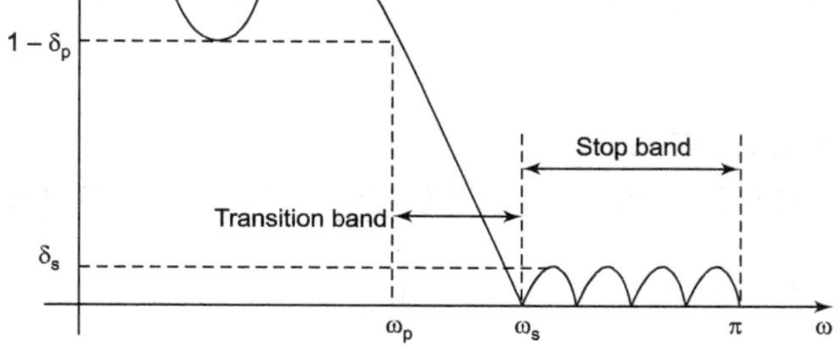

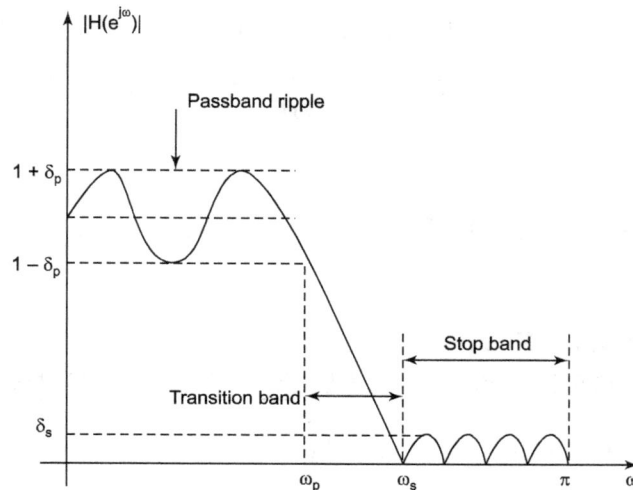

FIGURE 10.6 Frequency characteristics of physically Realizable Filters.

Here we discuss the only first case and other cases are given in Table 10.1 In the symmetric unit-sample (impulse) response case,

$$h(n) = h(M-1-n)$$

The real-valued frequency response characteristics are

$\left|H\left(e^{j\omega}\right)\right| = \left|H_r\left(e^{j\omega}\right)\right|$, given in Eq. (10.55), is

$$\left|H\left(e^{j\omega}\right)\right| = h\left[\frac{M-1}{2}\right] + 2\sum_{n=0}^{\frac{(M-3)}{2}} h(n)\cos\omega\left[\frac{M-1}{2} - n\right] \tag{10.55}$$

Let $$k = \left(\frac{M-1}{2}\right) - n$$

Eq. (10.55) can be written as

$$\left|H\left(e^{j\omega}\right)\right| = \sum_{k=0}^{\frac{(M-3)}{2}} A(k)\cos(\omega k) \tag{10.56}$$

where $$A(0) = h\left[\frac{M-1}{2}\right]$$

$$A(k) = 2h\left[\frac{M-1}{2} - k\right], \text{ for } 1 \le k \le \frac{M-1}{2} \tag{10.57}$$

The magnitude response for other there cases are similarly converted to a compact form as given in Table 10.2.

TABLE 10.2 Magnitude Response functions for Linear-Phase FIR Digital filters.

Filter Type	$Q(\omega)$	$P(\omega)$
Case I: Symmetric and M odd $h(n) = h(M-1-n)$	1	$\sum_{k=0}^{\left(\frac{M-1}{2}\right)} A(k)\cos(\omega k)$
Case II: Symmetric and M Even $h(n) = h(M-1-n)$	$\cos\frac{\omega}{2}$	$\sum_{k=0}^{\left(\frac{M}{2}-1\right)} B(k)\cos(\omega k)$.
Case III: Anti-symmetric and M odd $h(n) = -h(M-1-n)$	$\sin\omega$	$\sum_{k=0}^{\left(\frac{M-3}{2}\right)} C(k)\cos(\omega k)$
Case IV: Anti-symmetric and M even $h(n) = -h(M-1-n)$	$\sin\frac{\omega}{2}$	$\sum_{k=0}^{\frac{(M-1)}{2}} D(k)\cos(\omega k)$

From Table 10.2, it can be seen that the magnitude response function can be written as given in Eq. (10.58), for the four different cases.

$$\boxed{\left|H\left(e^{j\omega}\right)\right| = Q(\omega)P(\omega)} \tag{10.58}$$

where
$$Q(\omega) = \begin{cases} 1, & \text{for Case I} \\ \cos\frac{\omega}{2}, & \text{for Case II} \\ \sin\omega, & \text{for Case III} \\ \cos\frac{\omega}{2}, & \text{for Case IV} \end{cases} \tag{10.59}$$

and $P(\omega)$ is of the common form

$$P(\omega) = \sum_{k=0}^{L} \tilde{A}(k)\cos(\omega k) \tag{10.60}$$

$$\left\{\tilde{A}(k)\right\} = \text{Filter parameters.}$$

These filter parameters are linearly related to the unit-impulse response $h(n)$ of the filter. The upper limit L changes from case to case. In the design of optimal filters, the decreased frequency $H_d(\omega)$ and the weighting function $W(\omega)$ on the approximation error are also defined. The desired frequency response is defined to be 1 in the passband and 0 in the stopband. The weighting function $W(\omega)$ helps in selecting the relative size of the errors in the frequency bands. The weighting function is usually normalized to unity in the stopband and $W(\omega) = \delta_s/\delta_p$ in the passband:

$$W(\omega) = \begin{cases} \frac{\delta_s}{\delta_p}, & \text{passband} \\ 1, & \text{stopband} \end{cases} \tag{10.61}$$

The weighted approximation error is defined as

$$\begin{aligned} E(\omega) &= W(\omega)\left[H_d(\omega) - H(e^{j\omega})\right] \\ &= W(\omega)\left[H_d(\omega) - Q(\omega)P(\omega)\right] \\ &= W(\omega)Q(\omega)\left[\frac{H_d(\omega)}{Q(\omega)} - P(\omega)\right] \end{aligned} \tag{10.62}$$

Let us define the modified weighting function $\hat{W}(\omega)$. The modified desired frequency response $\hat{W}_d(\omega)$ is given below

$$\left.\begin{aligned} \hat{W}(\omega) &= W(\omega)Q(\omega) \\ \text{and } \hat{H}_d(\omega) &= \frac{H_d(\omega)}{Q(\omega)} \end{aligned}\right\} \tag{10.63}$$

The approximation error is then given by

$$E(\omega) = \hat{W}(\omega)\left[\hat{H}_d(\omega) - P(\omega)\right] \tag{10.64}$$

Expression for the approximation error given by Eq. (10.64) is valid for all four types of linear-phase FIR digital filters. Once the error function is given, the filter parameters $\{\tilde{A}(k)\}$ are determined such that the maximum absolute value of $E(\omega)$ is minimized. Mathematically, this is equivalent to seeking the solution to the problem

$$\begin{bmatrix} \sim & \min \\ A & (k) \end{bmatrix}\begin{bmatrix} \max & \\ \omega & \in S^{|E(\omega)|} \end{bmatrix} = \begin{bmatrix} \sim & \min \\ A & (k) \end{bmatrix}\left\{ {}_{\omega\in S}\left|\hat{W}(\omega)\left[H_d(\omega) - \sum_{k=0}^{L}\tilde{A}(k)\cos(\omega k)\right]\right|\right\} \tag{10.65}$$

where S is the set of frequency bands over which the optimization is to be performed. Parks and Mc Clellan applied the alternation theorem in the Chebyshev approximation and obtained the solution to the problem specified in Eq. (10.65). The alternation theorem is given below. Let S be a compact subset of the interval $[0, \pi]$. A necessary and sufficient condition for

$$P(\omega) = \sum_{k=0}^{L}\tilde{A}(k)\cos(\omega k) \tag{10.66}$$

to be the unique, best weighted Chebyshev approximation to $H_d(\omega)$ in S is that the error function $E(\omega)$ exhibit at least $L + 2$ external frequencies in S. That is, there must exist at least $L + 2$ frequencies $\{\omega_i\}$; in S such that

$\omega_1 < \omega_2 < \omega_3 < \ldots, \omega_{L+2}, E(\omega i) = -(\omega i + 1),$

and $$\left|E(\omega_i)\right| = \underset{\omega\in S}{\max}\,|E(\omega)|, \; i = 1,2,3,\ldots,L+2 \tag{10.67}$$

The error function $E(\omega)$ alternates in sign between two successive external frequencies. Hence, this theorem is called alternation theorem. The filter designs containing more than $L + 2$ external frequencies are called extra ripple filters. The alternation theorem guarantees a unique solution for the approximation problem and for a given set of external frequencies $\{\omega_n\}$ then the error function may be written as

$$\begin{aligned} E(\omega_n) &= (-1)^n \delta \\ &= \hat{W}(\omega_n)\left[\hat{H}(\omega_n) - P(\omega_n)\right], n = 1,2,3,\ldots,L-1 \end{aligned} \tag{10.68}$$

where δ represents the maximum value of the error function $E(\omega)$. Eq. (10.68) can be written in matrix form as

$$\begin{bmatrix} 1\cos\omega_0 & \cos 2\omega_0 & \cos 3\omega_0 & \cdots & \cos L\omega_0 & \dfrac{1}{\hat{W}(\omega_0)} \\ 1\cos\omega_1 & \cos 2\omega_1 & \cos 3\omega_1 & \cdots & \cos L\omega_1 & \dfrac{-1}{\hat{W}(\omega_1)} \\ 1\cos\omega_2 & \cos 2\omega_2 & \cos 3\omega_2 & \cdots & \cos L\omega_2 & \dfrac{1}{\hat{W}(\omega_2)} \\ . & . & . & \cdots & . & . \\ . & . & . & \cdots & . & . \\ 1\cos\omega_{L+1} & \cos 2\omega_{L+1} & \cos 3\omega_{L+1} & \cdots & \cos L\omega_{L+1} & \dfrac{(-1)^{L+1}}{\hat{W}(\omega_{L+1})} \end{bmatrix} \begin{bmatrix} \tilde{A}(0) \\ \tilde{A}(1) \\ \tilde{A}(2) \\ . \\ . \\ \tilde{A}(L) \end{bmatrix} = \begin{bmatrix} \hat{H}_d(\omega_0) \\ \hat{H}_d(\omega_1) \\ \hat{H}_d(\omega_2) \\ . \\ . \\ \hat{H}_d(\omega_{L+1}) \end{bmatrix} \quad (10.69)$$

Therefore, if the external frequencies are known as the coefficients $\{h(n)\}$, the peak error δ, and hence the frequency response of the filter, can be determined by inverting the matrix. As matrix inversion is time-consuming and insufficient, the peak error δ can be computed using the Ramez Exchange Algorithm.

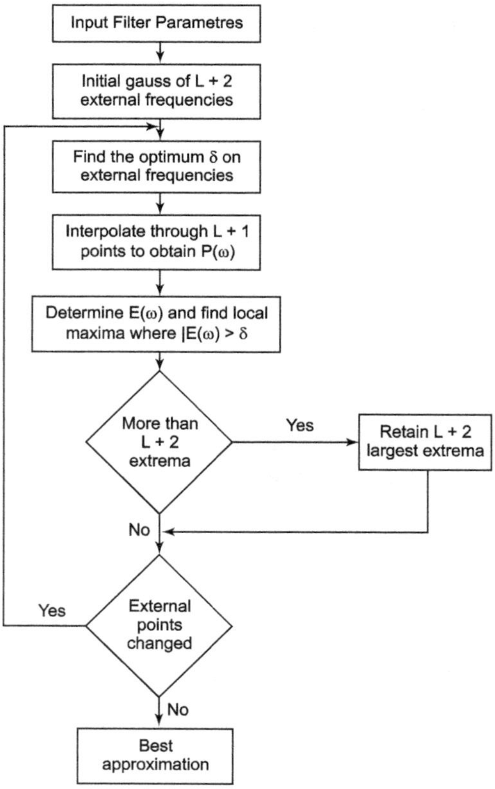

FIGURE 10.7 Flowchart of the Ramez Exchange Algorithm.

10.5.1 Ramez Exchange Algorithm

In this algorithm, a set of external frequencies is first assumed, the values of $P(\omega)$ and δ are determined and then the error function $E(\omega)$ is computed. This error function is then used to determine another set of $L + 2$ external frequencies.

This iterative process is repeated until the error function converges. to the optimal set of external frequencies. A flowchart of the Ramez Exchange Algorithm is given in Figure 10.7. A computer-aided iterative procedure for designing an optimal FIR digital filter has been developed by parks and Me Clellan.

10.6 FINITE WORD LENGTH EFFECTS IN DIGITAL FILTERS

Finite word length effects are also called finite precision effects. In digital signal processing (DSP), all the signals and systems are digital. The digital implementation has finite accuracy. When numbers are represented in digital form, errors are introduced due to their finite accuracy. These errors generate finite word length effects.

Now we consider an example of the first-order IIR digital filter to illustrate how errors are encountered in discretization process. Such an IIR digital filter can be described by following difference equation

$$y(n) = \alpha y(n-1) + s(n) \tag{10.70}$$

Taking the z-transform of both sides of Eq. (10.70)

$$\mathcal{Z}[y(n)] = \mathcal{Z}[\alpha y(n-1) + s(n)]$$

or

$$Y(z) = \alpha z^{-1} Y(z) + S(z)$$

or

$$\frac{Y(z)}{S(z)} = \frac{1}{1-\alpha z^{-1}} \tag{10.71}$$

Hence the transfer function of the system is given as

$$H(z) = \frac{Y(z)}{S(z)} = \frac{1}{1-\alpha z^{-1}}$$

or

$$H(z) = \frac{1}{1-\alpha z^{-1}} \tag{10.72}$$

Eq. (10.72) can be written as

$$H(z) = \frac{z}{z-\alpha} \tag{10.73}$$

Here we observe that α is the filter coefficient when this filter is implemented on some DSP process or software, α can have any discrete values. Let the discrete values of coefficient α be represented by $\hat{\alpha}$. The $\hat{\alpha}$ is the discrete approximation of α. Hence the actual transfer function which is implemented is given as

$$\hat{H}(z) = \frac{z}{z-\hat{\alpha}} \tag{10.75}$$

The transfer function $\hat{H}(z)$ is slightly different from $H(z)$. Hence the actual frequency response will be different from desired frequency response.

Here $e(n)$ is the error introduced during analog-to-digital (A/D) conversion process due to the finite word length of the quantizer. Similarly, the error is introduced in the multiplication of a and $y(n-1)$ in Eq. (10.70). This is because the product a $y(n-1)$ has to be quantized to one of the available discrete values. This introduces errors. Here, we introduced various sources of errors.

10.6.1 Quantization Process and Errors

The digital data can be represented by fixed-point or floating-point format. This representation as well as arithmetic operations such as addition, subtraction and multiplications give rise to various errors. Such errors are generated due to finite word length limitations of the registers. For example, the product of N-bit numbers is $2N$ bits long. This product must be quantized to N-bits, so that it will fit in the prescribed word length of the registers. Such quantization generates an error.

Similarly, in fixed-point arithmetic, the addition of the two numbers can result in a sum exceeding the word length of the register. This causes overflow and the error is generated. Again the result of addition and multiplication has to be quantized.

Let us consider that the available word length is $(N + l)$ bits and the most significant bit is representing the sign of the number. Let the data be $(N + 1)$ bit fixed-point fraction with the binary point just to the right of the sign bit. This is shown in Figure 10.8.

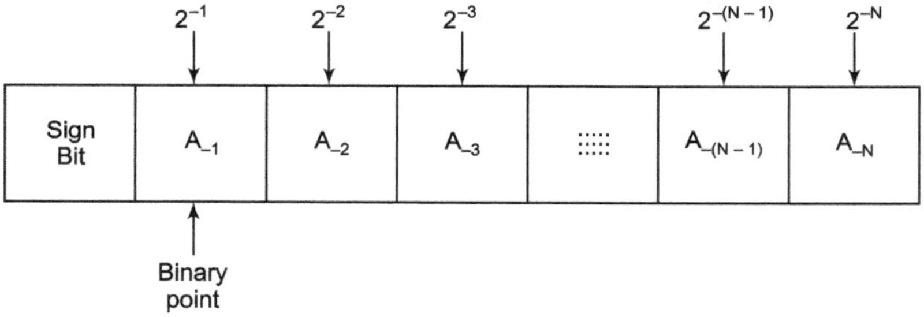

FIGURE 10.8 Illustration bit fixed-point fraction

The smallest positive number in this format will have the least significant bit of 1, will all bits zero. The decimal equivalent of such a number will be 2^{-N}. This smallest number of 2^{-N} is called the quantization step. The quantization step is denoted by δ.

Quantization step, $$\delta = 2^{-N} \qquad (10.76)$$

Let us assume that the original data x is represented by $(\delta + 1)$ bit fraction is quantized to $(N + 1)$ bits. This quantization can be performed by truncation or by rounding off. The value of b can be very large since it does not use any quantization. Ideally speaking $(\beta + 1)$ bit fraction is also called infinite precision representation. This quantization operation can be modeled as shown in Figure 10.9.

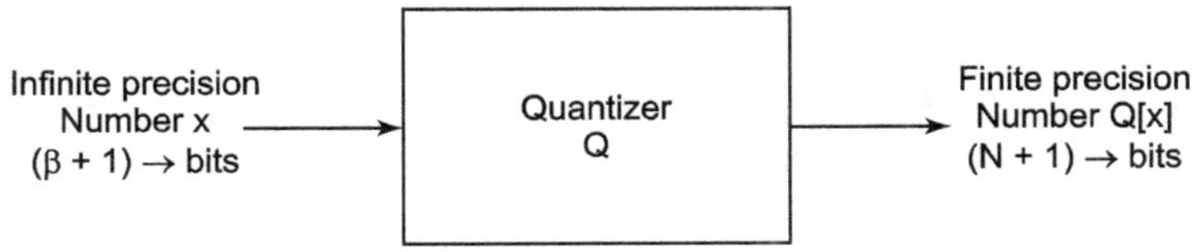

FIGURE 10.9 Quantization process model.

The range of quantization errors for fixed-point numbers is given in Table 10.3.

TABLE 10.3 Range of Quantization Errors

Type of Quantization	Number Representation	Range of Error Q[x] − x
Truncation	Positive number and 2's complement negative number	$-\delta < \varepsilon^{\tau} \leq 1$
Truncation	Sign magnitude negative number and 1's complement negative number	$0 \leq \varepsilon^{\tau} < \delta$
Rounding	All positive and negative numbers.	$-\frac{\delta}{2} < \varepsilon_r \leq \frac{\delta}{2}$
Here ε_t = Truncation error ε_r = Rounding error		

For the floating-point numbers, the quantization is carried out only on mantissa. Hence it is more relevant to consider the relative error caused by the quantization process.

Let the unquantized floating-point number is represented as $x = 2^E M$, and quantized floating-point number is represented as

$$Q[x\} = 2^E Q[M]$$

Then relative error e in this operation is given as

$$e = \frac{Q[x]-x}{x}$$

$$= \frac{Q[M]-M}{M} \tag{10.77}$$

The range of relative errors in truncation and rounding for various types of numbers is given in Table 10.4.

TABLE 10.4 Range of Relative errors

Type of quantization	Number of representation	Range of relative error
Truncation	2's complement	$-2\delta < e_t, 0, x > 0$ $0 \leq et < 2\delta, x < 0$
Truncation	sign magnitude and 1's complement	$-2\delta < e_t \leq 0$
Rounding	All numbers	$-\delta < e_r \leq \delta$
et = Relative error in truncation, *er* = Relative error in rounding		

10.6.2 Analysis of Coefficient Quantization Effects in FIR Digital Filters

In this section, we discuss the analysis of coefficient quantization effects in FIR digital filters. Consider the transfer function of the FIR digital filter of length M

$$H(z) = \sum_{n=0}^{M-1} h(n)z^{-n} \tag{10.78}$$

The quantization of impulse response $h(n)$ takes place during the implementation of filter. Let the quantized coefficient be denoted by $\hat{h}(n)$ and $e(n)$ be the error in quantization.

Then, $\hat{h}(n)$ is given by

$$\hat{h}(n) = h(n) + e(n) \tag{10.79}$$

Now, the new transfer function is given as

$$\hat{H}(z) = \sum_{n=0}^{M-1} \hat{h}(n)z^{-n} \tag{10.80}$$

Substituting Eq. (10.79) in Eq. (10.80), we get

$$\hat{H}(z) = \sum_{n=0}^{M-1} [h(n) + e(n)]z^{-n} = \sum_{n=0}^{M-1} h(n)z^{-n} + \sum_{n=0}^{M-1} e(n)z^{-n}$$

$$= H(z) + E(z) \tag{10.81}$$

where

$$E(z) = \sum_{n=0}^{M-1} e(n)z^{-n} \tag{10.82}$$

Here we observe that the FIR digital filter with quantized coefficient can be modeled as a parallel connection of two FIR digital filters $H(z)$ and $E(z)$. This is illustrated in Figure 10.10.

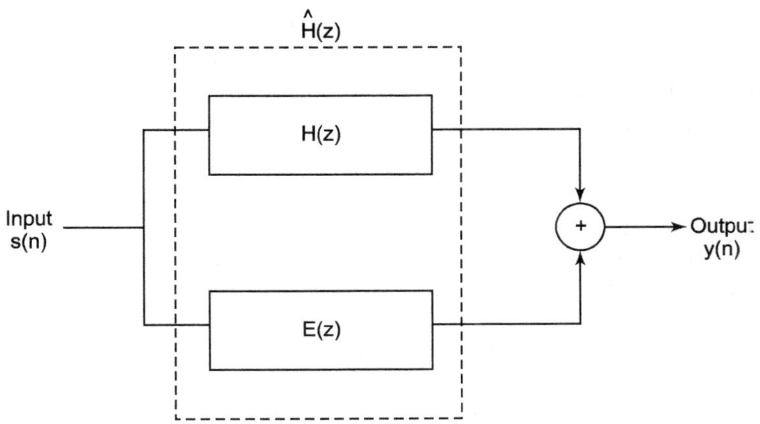

FIGURE 10.10 Model of FIR digital filter with quantized coefficient.

$H(z)$ is the transfer function of the FIR digital filter with unquantized coefficients (see Figure 10.10). Here, $E(z)$ is the transfer function of the FIR digital filter representing coefficient quantization error.

From the equation, $\hat{H}(z) - H(z) + E(z)$, we can write the frequency response of FIR digital filter with quantized coefficients as,

$$\hat{H}(\omega) = H(\omega) + E(\omega) \tag{10.83}$$

Here $E(\omega)$ is the error in the desired frequency response which is given by

$$E(\omega) = \sum_{n=0}^{M-1} e(n) z^{-j\omega n} \tag{10.84}$$

The magnitude of $E(\omega)$ is determined as

$$|E(\omega)| = \left| \sum_{n=0}^{M-1} e(n) z^{-j\omega n} \right|$$

$$\leq \sum_{n=0}^{M-1} |e(n)| \left| e^{-j\omega n} \right| \leq \sum_{n=0}^{M-1} |e(n)| \tag{10.85}$$

Since $\left| e^{-j\omega b} \right| = 1$ always.

The upper bound on the error in the frequency response is given by Eq. (10.85). From Table 10.1, it could be observed that the magnitude of error for rounding is

$$|e(n)| \leq \frac{\delta}{2} \tag{10.86}$$

Substituting Eq. (10.86) in Eq. (10.85), we get

$$|E(\omega)| \leq \sum_{n=0}^{M-1} \left(\frac{\delta}{2} \right) \leq M \left(\frac{\delta}{2} \right) \tag{10.87}$$

This upper bound is reached if all the errors have the same sign and have the maximum value in the range. If we consider $e(n)$ to be statistically independent random variables, then a more realistic bound is given by standard deviation of $E(\omega)$, that is,

$$\sigma_E(\omega) \leq \delta\sqrt{\frac{2M-1}{12}} \tag{10.88}$$

Here $\sigma_E(0)$ is the standard deviation of the error in frequency response $E(\omega)$.

10.6.3 Analog-to-Digital (A/D) Conversion Noise Analysis

Many types of continuous-time signals are processed using DSP techniques. For example, speech, music, video, environmental parameters, biomedical signals such as ECG, EEG, etc. These continuous-time signals must be converted to discrete-time (digital) sample. The conversion from continuous-time signals to digital samples is done by analog-to-digital (A/D) converters.

The A/D converters represent the digital samples by finite number of bits. This introduces errors in the A/D conversion process. In this section, we analyze the effect of such errors.

Quantization Noise Model

The analog-to-digital (A/D) converter generates a sample at the output. This sample is to be quantized to one of the finite set of discrete values. The number of these discrete values depends upon the output word length. For example, if the output word length is $(N+1)$ bits including sign bit then the number of discrete values, that is, quantization levels will be $2^{(N+1)}$. A model of a practical A/D conversion system is shown in Figure 10.11.

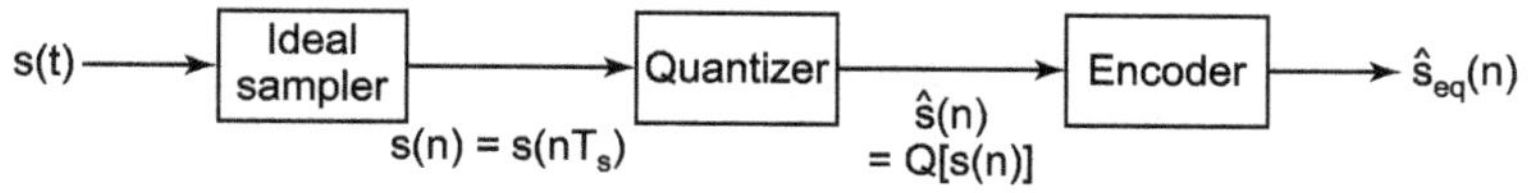

FIGURE 10.11 Model of the practical A/D conversion system.

The ideal sampler samples $s(t)$ to $s(n)$. The quantizer maps $s(n)$ to $\hat{s}(n)$. There is a limited number of values of $\hat{s}(n)$ depending upon the word length. The encoder converts $S(M)$ to its binary equivalent $\hat{s}_{eq}(n)$ depending upon the type of binary number representation. The quantizer of Figure 10.11 can use either rounding or truncation.

Let the quantization error be denoted by $e(n)$. Quantization error is the difference between quantized value $s(n)$ and input sample $\hat{s}(n)$, that is,

$$e(n) = \hat{s}(n) - s(n) \tag{10.89}$$

We consider that the quantizer has the step size of δ, which is given

$$\delta = \frac{R_{FS}}{2^{N+1}} \tag{10.90}$$

where R_{FS} is the full-scale range of A/D converter and (N + 1) output word length.

Then for rounding, the quantization error lies within,

$$-\frac{\delta}{2} < e(n) \leq \frac{\delta}{2} \tag{10.91}$$

When the sample is exactly halfway between two levels, then it round up to the nearest higher level. In such an operation the quantization error will be maximum and equal to δ/2. When the input analog sample is outside the full range of the A/D converter, then the magnitude of error $e(n)$ increases linearly with an increase in input. Such error is called saturation error or overload error. Therefore, the A/D converter output is clipped to its maximum value which is equal to $(1 - 2^{-N})$. The clipping can be avoided by scaling down the analog input, such that it remains within the full-scale range of the A/D converter.

A statistical model for the analysis of quantization error is shown in Figure 10.12.

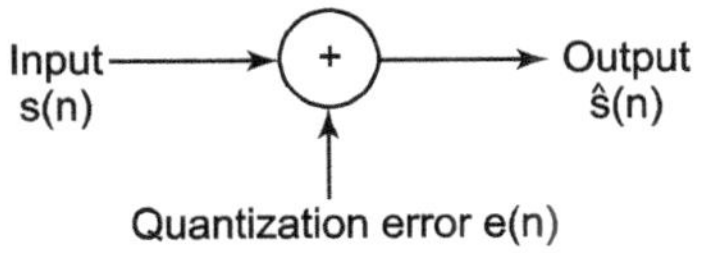

FIGURE 10.12 Statistical model of A/D converter.

It is to be assumed that quantization error $e(n)$ is a random signal. We also assume the followings:

1. The error sequence $e(n)$ is the sample sequence of a wide sense stationary process. The sequence follows uniform distribution over the range of quantization error.
2. The error sequence $e(n)$ is uncorrelated with the input sequence $s(n)$.
3. The input sequence $s(n)$ is the sample sequence of a stationary random process.

The probability density functions (pdfs) of quantization error are shown in Figure 10.13.

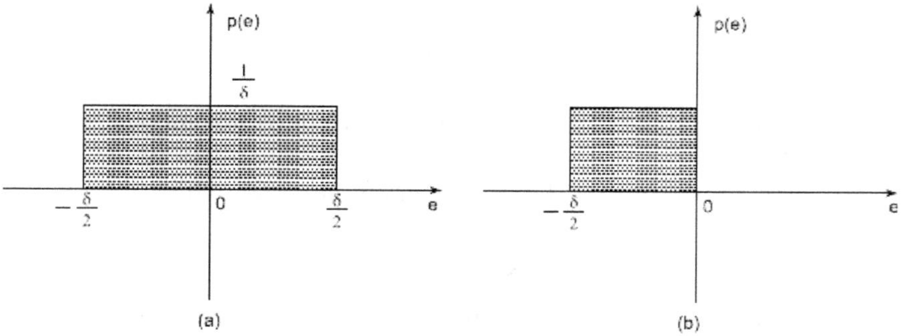

FIGURE 10.13 Probability density function (a) in rounding operation (b) in 2's complement truncation.

The practical A/D converters use either rounding or 2's complement truncation. The mean and variance of error sample from Figure 10.13(*a*), for rounding operation is given by

$$\text{Mean} = m_e(\text{rounding}) = \frac{\frac{\delta}{2} + \left(\frac{\delta}{2}\right)}{2} = 0 \tag{10.92}$$

$$\text{Variance} = \sigma_e^2(\text{rouding})$$

$$= \frac{\left[\left(\frac{\delta}{2}\right) - \left(-\frac{\delta}{2}\right)\right]^2}{2} = \frac{\delta^2}{2} \tag{10.93}$$

Similarly, the mean and variance of error sample for 2's complement truncation is given as from Figure 10.13(*b*),

$$\text{Mean} = m_e(2\text{'s comp.}) = \frac{0 + (-\delta)}{2} = -\frac{\delta}{2} \tag{10.94}$$

$$\text{Variance} = \sigma_e^2\ (\text{comp.}) = \frac{[0 - (-\delta)]^2}{12} = \frac{\delta}{2} \tag{10.95}$$

Signal-to-Quantization Noise Ratio

Now let us evaluate the signal-to-quantization noise ratio (SNR) of the quantizer. This ratio is denoted by $\text{SNR}_{\text{A/D}}$. It is given as

$$\text{SNR}_{\text{A/D}} = 10\log_{10}\left[\frac{\sigma_s^2}{\sigma_e^2}\right], \text{dB} \tag{10.96}$$

Here, σ_s^2 = Variance of the input signal, which represents input signal power

σ_e^2 = Variance of the quantization noise, which represents quantization noise power.

The variance of quantization error in rounding operation is given as

$$\sigma_e^2 = \frac{\delta^2}{12} \tag{10.97}$$

Substituting $\delta = \frac{R_{FS}}{2^{N+1}}$ in Eq. (10.97), we get

$$\sigma_e^2 = \frac{R_{FS}^2}{\left[2^{N+1}\right]^2 12} = \frac{2^{-2N} R_{FS}^2}{48} \tag{10.98}$$

Substituting Eq. (10.98) in Eq. (10.96), we get

$$\text{SNR}_{\text{A/D}} = 10\log_{10}\left[\frac{\sigma_s^2}{2^{-2N} R_{FS}^2 / 48}\right]$$

$$= 10\log_{10}\left[48 \times 2^{2N} \times \left(\frac{R_{FS}}{\sigma s}\right)^2\right]$$

or

$$\boxed{\text{SNR}_{\text{A/D}} = 16.81 + 6.02N - 20\log_{10}\left(\frac{R_{FS}}{\sigma_s}\right), \text{dB}} \tag{10.39}$$

This is the expression for SNR of the A/D converter. It shows that the signal-to-quantization noise ratio $\text{SNR}_{\text{A/D}}$ increases by 6 dB for every bit added to the word length. This equation can be used to determine the number of bits in the output word for the given power, full-scale range and acceptance SNR.

Effect of Input Scaling on SNR

Let the input is to be scaled by some constant A such that the new input will be $As(n)$. then the variance of this scaled input will be $A^2\sigma_s^2$.

Hence SNR of Eq. (10.99) becomes

$$\text{SNR}_{\text{A/D}} = 16.81 + 6.02\text{N} - 20\ \log_{10}\left(\frac{R_{FS}}{A\sigma_s}\right)$$

$$= 16.81 + 6.02N - 20\log_{10}\left(\frac{R_{FS}}{\sigma_s}\right) + 20\log_{10}(A) \tag{10.100}$$

Hence, we observe that SNR increases if A > 1. But this may exceed the full-scale range of input R_{FS}. Hence scaled input should remain within the full-scale range.

If A < 1, then SNR decreases. The last term in Eq. (10.100) becomes negative. Hence to get maximum $\text{SNR}_{\text{A/D}}$, the input should be scaled such that it uses the complete full-scale range of the A/D converter.

10.6.4 Effect of Quantization Noise on Filter Output

The output of the A/D converter is $s(n)$. This is given to the digital filter. Also, we know that $s(n) = \hat{s}(n) + e(n)$. That is quantization noise $e(n)$ is also given to the filter as input. A model of such a system is shown in Figure 10.14.

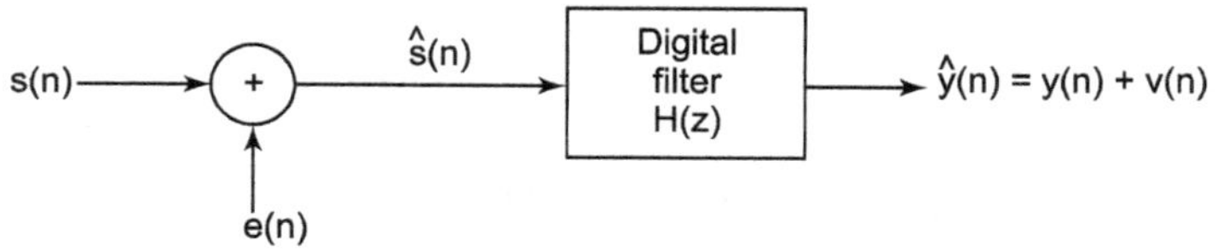

FIGURE 10.14 Model of digital filter with quantization noise input.

As given in above figure that $\hat{y}(n) = y(n) + u(n)$, that is, $\hat{y}\,(n)$ is the sum of two sequences $y(n)$ and $u(n)$. The digital filter is the discrete-time LT1 system hence we can say that $y(n)$ is the output due to input $s(n)$ and $y(n)$ being output due to error sequence $e(n)$. If $h(n)$ is the unit-sample response of the digital filter, then $y(n)$ can be expressed as,

$$u(n) = \sum_{m=-\infty}^{\infty} e(m)h(n-m) \tag{10.101}$$

The mean of output noise is given as

$$m_v = m_e H(0) \tag{10.102}$$

Here, $H(0) = H(\omega)\,|_{\omega=0}$

and m_e is the mean of output noise sequence.

Similarly, the variance of the output noise is given as

$$\sigma_v^2 = \sigma_e^2 \frac{1}{2\pi} \int_{-\pi}^{\pi} |\,H(\omega)\,|^2\, d\omega \tag{10.103}$$

The power spectrum of output noise is given as

$$P_{vv}(\omega) = \sigma_e^2\, |\,H(\omega)\,|^2 \tag{10.104}$$

The normalized output noise variance is given as

$$P_{vv}(\omega) = \frac{\sigma_v^2}{\sigma_e^2} = \frac{1}{2\pi} \int_{-\pi}^{\pi} |\,H(\omega)\,|^2\, d\omega \tag{10.105}$$

From the standard results, Eq. (10.105) can also be written as

$$\sigma_{v,n}^2 = \frac{1}{2\pi} \oint_C H(z)H\left(z^{-1}\right)z^{-1}dz \tag{10.106}$$

Here C is the counterclockwise contour in the ROC of $H(z)\ H(z^{-1})$.

10.6.5 Analysis of Arithmetic Rounding off Errors

Now, let us consider the effects of arithmetic errors particularly in multiplication and summation. The results of arithmetic operations are required to be quantized so that they can occupy one of the finite set of digital levels. Such operation can be visualized as multiplier (or other arithmetic operation) with quantizer at its output. It is shown in Figure 10.15.

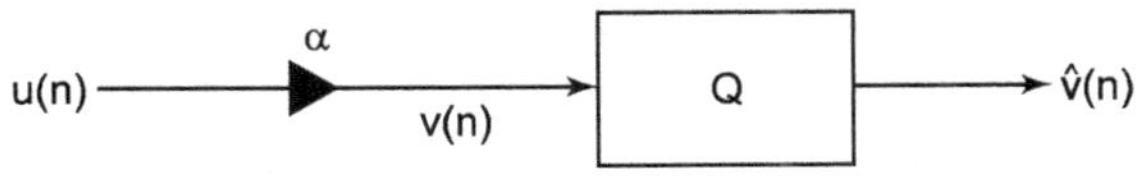

FIGURE 10.15 Quantization of multiplication or product.

The above process can be represented by a statistical model for error analysis.

The output $\hat{\upsilon}(n)$ can be considered as the sum of error-free output $\upsilon(n)$ and error $e_\alpha(n)$ in the product quantization process. It is given as

$$\hat{\upsilon}(n) = \upsilon(n) + e_\alpha(n) \tag{10.107}$$

A statistical model is shown in Figure 10.16.

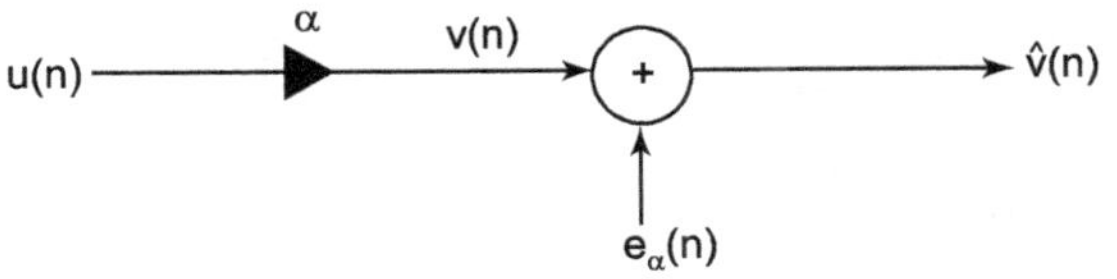

FIGURE 10.16 Statistical model for analysis of round-off error multiplication.

For the analysis purpose following assumptions are made:

1. The error sequence $\{e_\alpha(n)\}$ is the sample sequence of a stationary white noise process.
2. $e_\alpha(n)$ is having uniform distribution over the range of quantization error.
3. The sequence $\{e_\alpha(n)\}$ is uncorrelated with the sequence $\upsilon(n)$ and input sequence $s(n)$.

10.6.6 Dynamic Range Scaling

When the digital filters are implemented by using fixed-point arithmetic, the overflow can take place at some internal nodes. Such nodes can be inputs/outputs of adders or multipliers. This overflow can take place even if the inputs are scaled. Because of such overflow at intermediate points, produced totally undesired output or oscillations. The overflow can be avoided by scaling the internal signal levels with the help of scaling multipliers. These scaling multipliers are inserted at the appropriate points in the filter structure to avoid the possibilities of overflow. Sometimes these scaling multipliers are absorbed with the existing multipliers in the structure to reduce the total number and complexity.

At which node the overflow will take place is not known in advance. This is because the overflow depends upon the type of input signal samples. Hence, whenever overflow takes place at some node, the scaling should be done dynamically. Hence dynamic range scaling in the digital filter structure can avoid the effects of overflow.

Let $u_r(n)$ be the signal sample at rth node in the structure. Then the scaling should ensure that,

$$|u_r(n)| \leq 1 \text{ for all } r \text{ and } n \tag{10.108}$$

10.6.7 Low Sensitivity Digital Filters

We already know that the frequency response of the digital filter changes because of coefficient quantization. The unquantized coefficients provide perfect frequency response, but such coefficients need infinite precision. Because of the coefficient quantization, the change in frequency response may be significant and the filter may be unsuitable for a particular application. Hence it is necessary to develop the digital filter coefficient quantization. Such low sensitivity digital filters can be obtained by one of the following two methods:

1. An inherently low sensitivity analog filter is considered. This filter is converted to a digital filter such that the overall structure simulates the analog prototype. Such a digital filter is called a wave digital filter.
2. The condition for low coefficient sensitivity to be satisfied by the digital filter structure is determined. Then the realization methods are developed such that these conditions are really satisfied.

Requirements for low coefficient sensitivity

Let $H(z)$ be a causal stable real coefficient function having magnitude response $|H(\omega)|$ bounded by unity, that is,

$$|H(\omega)| \leq 1 \tag{10.109}$$

Let us assume that transfer function $H(z)$ is such that at a set of frequencies w_k, the magnitude is exactly equal to unity, that is,

$$|H(\omega_k)| = 1 \tag{10.110}$$

The typical frequency response satisfying Eq. (10.110) is shown in Figure 10.17.

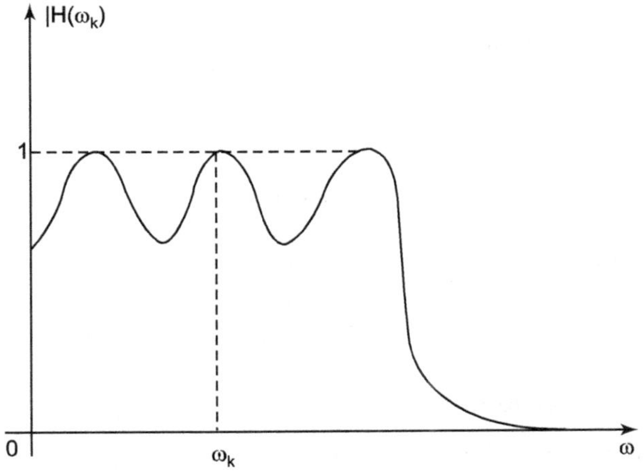

FIGURE 10.17 Magnitude response of a bounded real transfer function.

The transfer function shown in Figure 10.17 is called the bounded real transfer function.

Let the filter structure be characterized by the set of multipliers having coefficient m_i. Let the values of these coefficients with infinite precision realization be m_{i0}. Then the condition of Eq. (10.109) implies that the plot of $|H(\omega_k)|$ as a function of w, will be as shown in Figure 10.18.

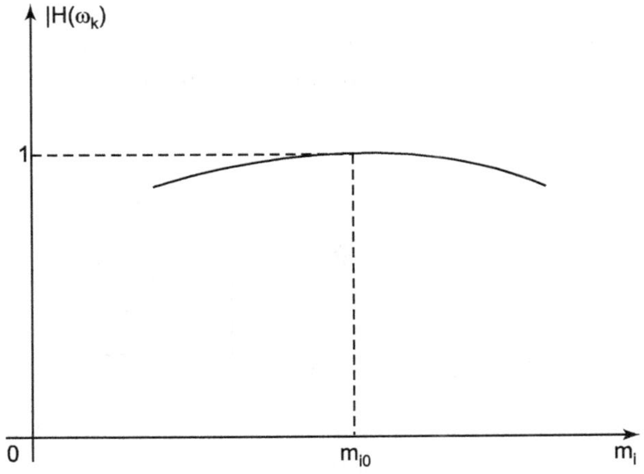

FIGURE 10.18 Illustration of zero sensitivity property.

It is observed from Figure 10.18 that at $m_i = m_{i0}$, $|H(\omega_k)| = 1$. This shows that $m_i = m_{i0}$, transfer function has zero sensitivity. But for $m_i \neq m_{i0}$, that is, quantized coefficients $|H(\omega_k)| < 1$. This condition can be represented mathematically as follows.

$$\left.\frac{\partial |H(\omega_k)|}{\partial m_k}\right|_{m_i = m_{i0}} = 0 \tag{10.11}$$

This means $|H(\omega_k)|$ has zero slope at $m_i = m_{i0}$. If the frequencies ω_k are closely spaced, then $|H(\omega)|$ will remain at unity. The sensitivity will be very low at other frequencies. If all the frequencies ω_k lie in passband, then such filters are called *low passband sensitivity filters*. Many methods are available to realize low passband sensitivity IIR and FIR digital filters.

Reduction of Product Round-off Error

When the digital filters are implemented using fixed-point arithmetic, the results of product or multiplication operations are quantized to fit into the finite word length. This quantization uses a rounding operation. Hence errors generated in such operation are called product round-off errors. The effect of product round-off errors can be analyzed using the statistical model of the quantization process. The noise due to product round-off errors reduces the signal to noise ratio at the output of the filter. Sometimes this ratio may be reduced below acceptable levels.

Hence it is necessary to reduce the effect of product round-off errors.

Now we discuss first-order error-feedback structure for reducing round-off errors.

Let the quantization error signal be given as the difference between unquantized signal $y(n)$ and quantized signal $\upsilon(n)$, that is,

$$e(n) = y(n) - \upsilon(n) \tag{10.112}$$

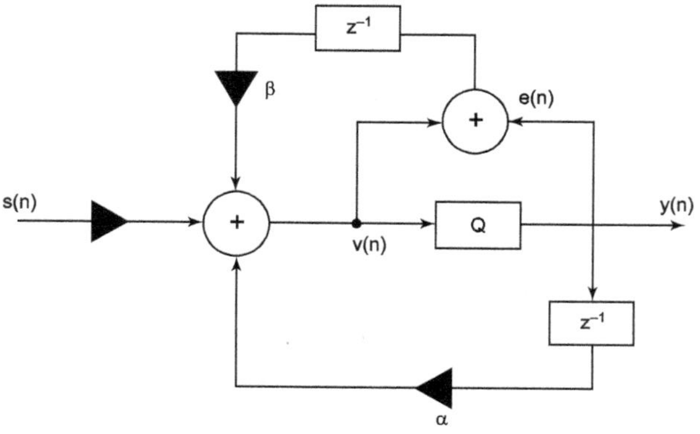

FIGURE 10.19 First-order error-feedback structure for reducing round-off errors.

The incorporation of quantization error feedback as shown in Figure 10.19 helps in reducing the noise power at the output.

10.6.8 Limit Cycles in IIR Digital Filters

The finite word length effects are analyzed using the linear model of the digital systems. But non-linearities are introduced because of the quantization of arithmetic operations. Because of these non-linearities, the stable digital filter under infinite precision may become unstable under finite precision. Because of this instability, the oscillating period output is generated. Such output is called the limit cycle. The limit cycles occur in IIR digital filters due to the feedback path. The FIR filters do not show limit cycles, since they do not have any feedback paths.

The limit cycles are of two types: (1) granular and (2) overflow.

Granular Limit Cycles

The granular limit cycles are of low amplitude. These limit cycles occur in digital filters when the input signal level is very low. The granular limit cycles are inaccessible and accessible limit cycles.

Overflow Limit Cycles

These limit cycles occur because of overflow in digital filters implemented with finite precision. The amplitudes of overflow limit cycles are very large and they can cover the complete dynamic range of the register. This further loads to overflow causing the cumulative effect. Hence overflow limit cycles are more serious than granular limit cycles.

Limit cycle free structures are normally used to avoid the effects of limit cycles.

EXERCISES

1. What is an FIR digital filter? Differentiate between FIR and IIR digital filters.
2. Discuss the properties of the FIR digital filter.
3. Describe the design of FIR digital filters using the Fourier series method.
4. What is Gibb's oscillations? Give some ways by which these oscillations can be reduced.

5. Define window function. Discuss various types of window functions used in the design of FIR digital filters.
6. Derive an expression for frequency response of a rectangular window function.
7. Derive an expression for frequency response of Hann and Hamming window functions.
8. Discuss the Blackman window function and give all the steps of design of FIR filters using Blackman window.
9. Describe FIR digital filter design based on the numerical-analysis formula.
10. What do you mean by optimal linear-phase FIR digital filter? What parameters are optimized in these filters?
11. Give the four cases that result in a linear-phase FIR digital filter.
12. State and explain the alternation theorem.
13. What are extra ripple filters?
14. What are maximal ripple filters?
15. Discuss the Ramez Exchange Algorithm used in the design of optimal filters.
16. What is the finite word length effect in Digital Filters?
17. Discuss the quantization process and error.
18. Tabulate the range of quantization errors.
19. Explain the effects of coefficient quantization in FIR Digital Filters.
20. Explain the statistical model of the A/D converter.
21. Derive the signal-to-quantization noise ratio (SKRA/D) of the analog-to-digital (A/D) converter.
22. Explain the effect of input scaling on signal-to-quantization noise ratio (SKRA/D) of A/D converter.
23. Describe the statistical model for analysis of round-off error multiplication.
24. Explain how the reduction of product round-off error is achieved in digital filters?

25. Write short notes on the following topics:

a. Dynamic range scaling
b. Low sensitivity digital filter
c. Limit cycles in IIR digital
d. Finite precision effects.

NUMERICAL EXERCISES

1. Design an FIR LPF which satisfy following specifications

$$\begin{aligned} \text{Passband ripple, } Ap &\le 0.1 \text{ dB} \\ \text{Stopband attenuation, } Aa &\le 44.0 \text{ dB} \\ \omega_p &= 20 \text{ rad/s} \\ \omega_\alpha &= 2 \text{ rad/s} \\ \text{Sampling frequency, } \omega_s &= 10 \text{ rad/s.} \end{aligned}$$

2. Design an FIR low-pass digital filter using Kaiser window. The specifications are given below:

$$\begin{aligned} \text{Passband ripple, } A_p &\le 0.1 \text{ dB}, \quad 0 \le w \le 1.5 \text{ rad/s} \\ \text{Minimum stopband attenuation, } A_\alpha &\le 40 \text{ dB}, \\ 2.5 \le \omega &\le 5.0 \text{ rad/s} \\ \text{Sampling frequency, } \omega_s &= 10 \text{ rad/s.} \end{aligned}$$

3. Design a low-pass FIR digital filter whose frequency response is given by

$$H\left(e^{j\omega T_s}\right) = \begin{cases} 1, & \omega_c \le \omega \le \omega_c \\ 0, & \omega_c < |\omega| \le \dfrac{\omega_s}{2} \end{cases}$$

where ω_s is the sampling frequency and ω_c is the cutoff frequency of the LPF.

4. Design a low-pass FIR digital filter using Hamming and Blackman windows whose frequency response is given by

$$H\left(e^{j\omega T_s}\right) = \begin{cases} 1, & \omega_c \le \omega \le \omega_c \\ 0, & \omega_c < |\omega| \le \dfrac{\omega_s}{2} \end{cases}$$

5. Derive an expression for the impulse response of constant-delay FIR digital filters given by

$$\sum_{n=0}^{N-1} h(nT_s)\sin(\omega\tau - \omega nT_s) = 0.$$

CHAPTER 11

STATISTICAL DIGITAL SIGNAL PROCESSING

11.1 INTRODUCTION TO SPECTRAL ESTIMATION

The signal processing methods which characterize the frequency content of a signal corresponds to spectral analysis. The signals which are analyzed in any communication system are either purely random or will have noisy components also. If the signal is random, then, only an estimate of the spectrum of the signal can be obtained. This is possible only if the statistical attributes of the random signal are known.

Determination of spectral estimation is useful in a variety of fields such as astronomy, seismology, communication engineering, etc. In communication engineering, spectral estimation is helpful in detecting the signal component (carrier) which has the noise component in it. In radar and sonar, spectral estimation is useful in detecting the targets.

11.2 ENERGY DENSITY SPECTRUM

In this section, we study Energy Density Spectrum in two cases:

1. Energy Density Spectrum of a Continuous-time Signal.
2. Energy Density Spectrum of a Discrete-time Signal.

Now we discuss each one in detail in subsequent subsections.

11.2.1 Energy Density Spectrum of a Continuous-time Signal

Consider a continuous-time signal $s(t)$ which is deterministic, analog, and complex-valued. Now, we sample this signal at the sampling rate f_s. Then, we obtain a discrete-time signal $s(n)$.

If $s(t)$ is a finite energy signal, then

$$\boxed{\int_{-\infty}^{\infty} |s(t)|^2\, dt < \infty} \tag{11.1}$$

Continuous-time Fourier transform (CTFT) of signal $s(t)$ is given by

$$S(f) = \text{CTFT}\,[s(t)] = \int_{-\infty}^{\infty} s(t)e^{-e^{j\omega T}}\, dt$$
$$= \int_{-\infty}^{\infty} s(t)e^{-j2\pi ft}\, dt \tag{11.2}$$

Signal energy is given by Parseval's relations

$$\boxed{\int_{-\infty}^{\infty} |s(t)|^2\, dt = \int_{-\infty}^{\infty} | s(f) |^2\, dt = \text{energy}} \tag{11.3}$$

The density of energy of signal $s(t)$ with respect to frequency is represented by $|s(f)|^2$.

This density of a signal is called the energy spectral density (*ESD*). Energy spectral density (*ESD*) is denoted by $\psi_{SS}(f)$:

$$\psi_{ss}(f) = EST[s(t)] = | S(f) |^2 \tag{11.4}$$

Let $R_{ss}(\tau)$ be the autocorrelation function of the signal $s(t)$

$$R_{ss}(\tau) = \int_{-\infty}^{\infty} s^*(t)s(t+\tau)dt \tag{11.5}$$

The CTFT of $R_{ss}(\tau)$ is given by

$$CTFT\{R_{ss}(\tau)\} = \int_{-\infty}^{\infty} R_{ss}(\tau)e^{-j\omega\tau}\, dt$$
$$= \int_{-\infty}^{\infty} R_{ss}(\tau)e^{-j2\pi f\tau}\, d\tau \tag{11.6}$$

Substituting Eq. (11.5) in Eq. (11.6), we get

$$\text{CTFT}\{R_{ss}(\tau)\} = \int_{-\infty}^{\infty}\left[\int_{-\infty}^{\infty} s^*(t)s\left(t+\tau\right)dt\right]e^{-j2\pi f\tau}\, d\tau$$
$$= \int_{-\infty}^{\infty} s^*(t)\left[\int_{-\infty}^{\infty} s(t+\tau)e^{-j2\pi f\tau}\, d\tau\right]dt$$
$$= \int_{-\infty}^{\infty} s^*(t)\left[s(f)e^{j2\pi ft}\right]dt = s(f)\int_{-\infty}^{\infty} s^*(t)e^{j2\pi ft}\, dt$$

$$= s(f)\left[\int_{-\infty}^{\infty} s(t)e^{-j2\pi ft}dt\right]$$

$$= s(f)s^*(f) = |s(f)|^2 = \Upsilon_{SS}(f) \tag{11.7}$$

Therefore, we can say that the CTFT of the autocorrelation function of a signal gives the spectral density.

11.2.2 Energy Density Spectrum of a Discrete-Time Signal

Analog signal $s(t)$ be discretized by ideally sampling $s(t)$ with the sampling period of T_s. The resultant signal is denoted by $s'(t)$ and represented by

$$s'(t) = s(t)\sum_{n=-\infty}^{\infty} \delta(t - nT_s) \tag{11.8}$$

The CTFT of the sampled signal is given by

$$s'(f) = CTFT\{s'(t)\} = FTFT\left\{s(t)\sum_{n=-\infty}^{\infty} \delta(t - nT_s)\right\}$$

$$= \int_{-\infty}^{\infty}\left\{s(t)\sum_{n=-\infty}^{\infty} \delta(t - nT_s)\right\}e^{-j2\pi ft}dt$$

$$= \sum_{n=-\infty}^{\infty}\left[\int_{-\infty}^{\infty} s(t)e^{-j2\pi ft}\left[\delta(t - nT_s)\right]dt\right]$$

$$= \sum_{n=-\infty}^{\infty} s(nT_s)e^{-j2\pi fnT_s} = f_s\sum_{k=-\infty}^{\infty} s(f - kf_s) \tag{11.9}$$

If aliasing is avoided, that is, $s(t)$ is band-limited to a frequency less than 1/2 T_s, then

$$S'(f) = f_s S(f) \tag{11.10}$$

Let $s(n)$ be a sampled version of signal $s(t)$. DTFT of $s(n)$ is given by

$$S'(f) = \text{DTFT}\{\sin(n)\} = \sum_{n=-\infty}^{\infty} s(n)e^{-j2\pi fn} \tag{11.11}$$

The autocorrelation of the sampled signal $s(n)$ is given by

$$r_{ss}(k) = \sum_{n=-\infty}^{\infty} s^*(n)s(n+k) \tag{11.12}$$

The DTFT of $r_{ss}(k)$ from the Wiener–Khintchine theorem is

$$\psi_{SS}(f) = \sum_{k=-\infty}^{\infty} r_{ss}(k)e^{-j2\pi kf} \tag{11.13}$$

The other method for computing the energy density spectrum is obtained from the DTFT of $s(n)$,

$$\boxed{\psi_{ss}(f) = |S(f)|^2 = \left| \sum_{n=-\infty}^{\infty} s(n)e^{-j2\pi fn} \right|^2}$$

Since finite energy signals possess Fourier transform spectral analysis is done with the energy spectral density (ESD) function.

11.3 ESTIMATION OF THE AUTOCORRELATION AND POWER SPECTRUM OF RANDOM SIGNALS

Here, we consider signals which do not have finite energy. For these signals, Fourier transform is not possible. But these signals possess finite average power.

For these signals, spectral analysis is done with power spectral density (psd) function.

Let $s(t)$ be a stationary random process.

The statistical autocorrelation function for this signal is given by

$$R'_{ss}(\tau) = E[s^*(t)s(t+\tau)] \tag{11.14}$$

The Fourier transform of the autocorrelation function of a stationary random process gives the power density spectrum. It is given by

$$\begin{aligned}\phi_{ss}(f) &= \text{Fourier Transform } [R_{ss}(\tau)] \\ &= \int_{-\infty}^{\infty} R'_{ss}(\tau)e^{-j2\pi f\tau}d\tau \end{aligned} \tag{11.15}$$

11.3.1 Estimate of Autocorrelation Function

Since the only single realization of the random process is considered, the true autocorrelation function is not known, and hence time-average autocorrelation function is taken. Let the observation interval be T_o.

The time-average autocorrelation function is given by

$$R_{ss}(\tau) = \frac{1}{2T_0}\int_{-T_0}^{T_0} s^*(t)s(t+\tau)dt \tag{11.16}$$

If the stationary random process is ergodic, then

$$R'_{ss}(\tau) = \lim_{T_0\to\infty} R_{ss}(\tau)$$

$$= \lim_{T_o\to\infty} \frac{1}{2T_o}\int_{-T_o}^{T_o} s^*(t)s(t+\tau)dt \qquad (11.17)$$

The time-average autocorrelation function $R_{ss}(\tau)$ is an estimate of the statistical autocorrelation function $K'_{ss}(\tau)$.

11.3.2 Estimate of Power Density Spectrum

The Fourier transform of time-average autocorrelation function is given by

$$\text{Fourier Transform} \quad \{R_{ss}(\tau)\} = P_{ss}(f) = \int_{-T_o}^{T_0} R_{ss}(\tau)e^{-j2\pi f\tau}d\tau \qquad (11.18)$$

Substituting Eq. (11.16) in Eq. (11.18), we get

$$P_{ss}(f) = \int_{-T_o}^{T_o}\left[\frac{1}{2T_o}\int_{-T_o}^{T_o} s^*(t)s(t+\tau)dt\right]e^{-j2\pi f\tau}d\tau$$

$$= \frac{1}{2T_o}\int_{-T_o}^{T_o}\left[\int_{-T_o}^{T_o} s^*(t)s(t+\tau)dt\right]$$

$$= \frac{1}{2T_o}\left|\int_{-T_o}^{T_o} s(t)e^{-j2\pi ft}dt\right|^2 \qquad (11.19)$$

The statistical power density spectrum is given by

$$\phi_{ss}(f) = \lim_{T_o\to\infty} E[P_{ss}(f)]$$

$$= \lim_{T_o\to\infty} E\left[\frac{1}{2T_o}\left|\int_{-T_o}^{T_o} s(t)e^{-j2\pi ft}dt\right|^2\right] \qquad (11.20)$$

11.3.3 Estimation from Samples

There are two estimates: (1) estimate 1 and (2) estimate 2 which are discussed below.

Estimate 1

Let $s(n)$ be an N-point sequence obtained by sampling the signal $s(t)$.

The time-average autocorrelation sequence for the sequence $s(n)$ is

$$r'_{ss}(m) = \frac{1}{N-|m|}\sum_{n=0}^{N-|m|-1} s^*(n)s(n+m), |m| = 0,1,\ldots,N-1 \qquad (11.21)$$

The Fourier transform of the autocorrelation sequence is given by

$$P'_{ss}(f) = \text{DTFT}\{r'_{ss}(m)\}$$

$$= \sum_{m=-(N-1)}^{N-1} r'_{ss}(m)e^{-j2\pi fm} \tag{11.22}$$

Mean and Variance for Estimate 1

The mean value of estimate $r'_{ss}(m)$ is given by

$$E\left[r'_{ss}(m)\right] = \frac{1}{N-|m|}\sum_{n=0}^{N-|m|-1} E\left[s^*(n)s(n+m)\right]$$

$$= \phi_{ss}(m) \tag{11.23}$$

where $N-|m|$ is the normalization factor. $r'_{ss}(m)$ is an unbiased estimate of the autocorrelation of the sequence $s(n)$.

The approximate value of variance of the estimate $r'_{ss}(m)$ is given by Jenkins and Watts as

$$Var\left[r'_{ss}(m)\right] = \frac{N}{\left(N-|m|^2\right)}\sum_{n=0}^{\infty}\left[\phi_{ss}(n)^2 + \phi^*_{ss}(n-m)\phi_{ss}(n+m)\right] \tag{11.24}$$

As N becomes infinity, the variance becomes zero, and this estimate $r'_{ss}(m)$ is consistent.

Estimate-2

If the value of m is large, then only less points are considered for the estimate.

Hence, considering a different estimate for the autocorrelation function

$$r_{ss}(m) = \frac{1}{N}\sum_{n=0}^{N-|m|-1} s^*(n)s(n+m), 0 \le m \le N-1 \tag{11.25}$$

Mean and Variance for Estimate 2

$$E\{r_{ss}(m)\} = \frac{1}{N}\sum_{n=0}^{N-|m|-1} E[s^*(n)s(n+m)]$$

$$= \left[\frac{N-|m|}{N-|m|}\right]\frac{1}{N}\sum_{n=0}^{N-|m|-1} E[s^*(n)s(n+m)]$$

$$= \left[\frac{N-|m|}{N}\right]\frac{1}{N-|m|}\sum_{n=0}^{N-|m|-1} E[s^*(n)s(n+m)] \tag{11.26}$$

Substituting Eq. (11.23) in Eq. (11.26), we get

$$E[r_{ss}(m)] = \left[1 - \frac{|m|}{N}\right]\phi_{ss}(m) \tag{11.27}$$

This is the mean of autocorrelation sequence $r_{ss}(m)$. Here, $\frac{|m|}{N}\phi_{ss}(m)$ is the bias for the estimate $r_{ss}(m)$.

The variance of the estimate is given by

$$Var[r_{ss}(m)] = \frac{1}{N}\sum_{n=-\infty}^{\infty}\left[|\phi_{ss}(n)|^2 + \phi_{ss}^*(n+m)\right] \tag{11.28}$$

As N becomes infinity, the variance value becomes zero. Hence $r_{ss}(m)$ is an asymptotically unbiased estimate.

Estimate of Power Density Spectrum

The estimate of the power density spectrum is given by

$$P_{ss}(f) = \sum_{m=-(N-1)}^{(N-1)} r_{ss}(m)e^{-j2\pi fm} \tag{11.29}$$

Here, we consider that $r_{ss}(m)$ is the estimate for the autocorrelation function. Substituting the value of $r_{ss}(m)$, the power density spectrum becomes

$$P_{ss}(f) = \frac{1}{N}\left|\sum_{n=0}^{N-1} s(n)e^{-j2\pi fn}\right|^2 \tag{11.30}$$

This estimate for the power density spectrum is called the periodogram.

Mean and Variance of Periodogram Estimate

The mean value of the periodogram estimate is given by

$$\begin{aligned} E[P_{ss}(f)] &= E\left|\sum_{m=-(N-1)}^{N-1} r_{ss}(m)e^{-j2\pi fm}\right| \\ &= \sum_{m=-(N-1)}^{(N-1)} E\{r_{ss}(m)\}e^{-j2\pi fm} \\ &= \sum_{m=-(N-1)}^{(N-1)} \left[\frac{1-|m|}{N}\right]\phi_{ss}(m)e^{-j2\pi fm} \\ &= \sum_{m=-(N-1)}^{(N-1)} \phi'_{ss}(m)e^{-j2\pi fm} \end{aligned} \tag{11.31}$$

where $$\phi'_{ss}(m) = \left[\frac{1-|m|}{n}\right]\phi_{ss}(m)$$

The mean value now given as

$$E[P_{ss}(f)] = \sum_{m=-\infty}^{\infty} \phi'_{ss}(m)e^{-j2\pi fm}$$

$$= \int_{\frac{-1}{2}}^{\frac{1}{2}} \gamma_{ss}(\alpha)W(f-\alpha)d\alpha \qquad (11.32)$$

Where $W(f)$ is the frequency domain representation of the Bartlett window. Spectral leakage problem is present in this case as only finite samples are considered. If the data sequence is a Gaussian random process, the variance is given by

$$Var[P_{ss}(f)] = \gamma_{ss}^2(f)\left[1+\left\{\frac{\sin 2\pi\ f\ N}{N\sin 2\pi f}\right\}\right] \qquad (11.33)$$

As N becomes infinity, the estimated spectrum becomes the actual spectrum. Hence, the estimate is an asymptotically unbiased estimate. But the variance does not become zero as N becomes infinity,

Hence, the periodogram is not a consistent estimate of the power density spectrum. The estimated autocorrelation is a consistent estimate, but its Fourier transform, that is, power density spectrum is not a consistent estimate.

EXAMPLE 11.1

Compute the autocorrelation function and power spectral density (psd) for the signal

$$s(t) = A_c \cos(2\pi f_{ct} + \phi_0)$$

where A_c and f_c are constants, ϕ is a random variable which is uniformly distributed over the interval $(-\pi, \pi)$.

Solution:

The probability density function (pdf) of the random variable is given by

$$\phi(f) = \begin{cases} \dfrac{1}{2\pi}, & -\pi \le \phi \le \pi \\ 0, & \text{otherwise} \end{cases}$$

The autocorrelation function for the signal $s(t)$ is given by

$$R_{ss}(\tau) = E[S(t+\tau)s(t)]$$

Putting the value of signal $s(t)$, we get

$$R_{ss}(\tau) = E\left[A_c\cos(2\pi f_c(t+\tau)+\phi)\cdot A_c\cos(2\pi f_c t+\phi)\right]$$

$$= E\left[A_c^2\cos(2\pi f_c t + 2\pi f_c\tau + \phi)\cos(2\pi f_c t+\phi)\right]$$

$$= \frac{A_c^2}{2} E\left[\cos\left(4\pi f_c t + 2\pi f_c \tau + 2\phi\right)\right] + \frac{A_c^2}{2} E\left[\cos\left(2\pi f_c \tau\right)\right]$$

$$= \frac{A_c^2}{2} \int_{-\pi}^{\pi} \frac{1}{2\pi} \left(4\pi f_c t + 2\pi f_c \tau + 2\phi\right) d\phi + \frac{A_c^2}{2} \cos\left(2\pi f_c \tau\right)$$

$$= \frac{A_c^2}{2} \cos\left(2\pi f_c \tau\right)$$

The autocorrelation function plot is shown in Figure 11.1(a)

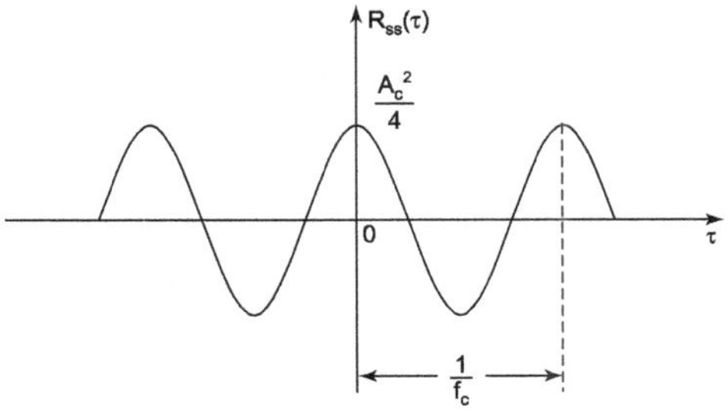

FIGURE 11.1 (a) Autocorrelation function $R_{SS}(t)$ of a sine wave with random phase.

The power spectral density (psd) function is obtained by taking the Fourier transform of the autocorrelation function and is given by

$$P_{ss}(f) = FT\left[R_{ss}(\tau)\right]$$

$$= FT\left[\frac{A_c^2}{2} \cos\left(2\pi f_c t\right)\right] = \frac{A_c^2}{2}\left[\delta\left(f - f_c\right) + \delta\left(f + f_c\right)\right]$$

The plot of psd function is shown in Figure 11.1(b)

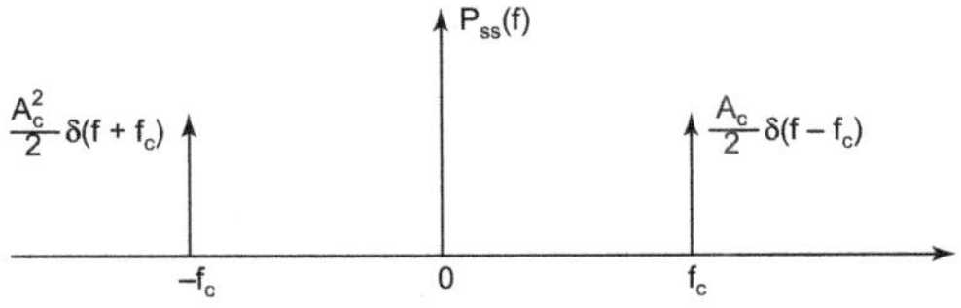

FIGURE 11.1 (b) psd of sine wave with random phase.

11.4 DFT IN SPECTRAL ESTIMATION

The periodogram is given by

$$P_{ss}(f) = \frac{1}{N}\left|\sum_{n=0}^{N-1} s(n) e^{-j2\pi f n}\right|^2 \qquad (11.34)$$

The samples of the periodogram can be obtained by using the discrete Fourier transform (DFT) algorithm.

Let $$f = \frac{k}{N}, \quad \text{where } k = 0,1,2,....,N-1$$

The periodogram can be given as

$$P_{ss}\left(\frac{k}{N}\right) = \frac{1}{N}\left|\sum_{n=0}^{N-1} s(n)e^{\frac{-j2\pi k}{N}}\right|^2 \tag{11.35}$$

where $k = 0, 1, 2, ..., N-1.$

If more samples are required in the frequency domain, the length of the sequence $s(n)$ can be increased by zero padding.

Let the new length be L, the power spectral density (psd) is given by

$$P_{ss}\left(\frac{k}{L}\right) = \frac{1}{N}\left|\sum_{n=0}^{N-1} s(n)e^{\frac{-j2\pi k}{N}}\right|^2 \tag{11.36}$$

where $k = 0,1,2,...,L-1$

This does not increase the resolution but provides the interpolated values.

EXAMPLE 11.2

We have given the discrete-time signal

$$s(n) = \cos\left(2\pi f_1 n\right) + \cos\left(2\pi f_2 n\right),\ n = 0,1,2,....7$$

Find the power spectrum for the data sequence length L = 8, 16, 32 for different values of f_1 and f_2, where $f_2 = f_1 + \Delta f$ and Δf is small deviation from f_1 or simply frequency separation.

Solution:

The power spectrum for the discrete-time signal $s(n)$ is given by

$$P_{ss}\left(\frac{k}{L}\right) = \frac{1}{N}\left|\sum_{n=0}^{N-1} s(n)e^{\frac{-j\pi nk}{N}}\right|^2$$

The power spectrum is determined for different values of L by appending zeros to the original sequence.

Figure 11.2 illustrates the plot for the power spectrum.

Let $f_1 = 0.6$ and $\Delta f = 0.05$ which gives

$$f_2 = f_1 - \Delta f = 0.6 + 0.05 = 0.65$$

$$\begin{aligned} s(n) &= \cos\left(2\pi f_1 n\right) + \cos\left(2\pi f_2 n\right) \\ &= \cos(2\pi(0.6)n) + \cos(2\pi(0.65)n) \\ &= \cos(1.2\pi n) + \cos(1.3\pi n) \end{aligned} \tag{1}$$

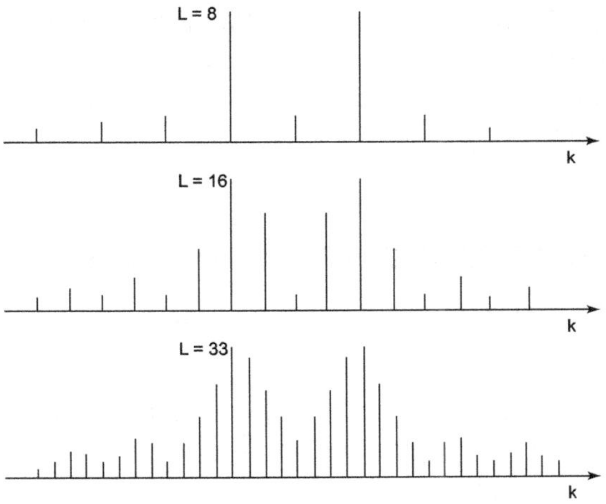

FIGURE 11.2 (*a*) Power Spectrum $1/N|S(k)|^2$ Verses *k* for $\omega_1 = 2\pi(0.6)$ and $\omega_2 = 2\pi(0.65)$

The power spectrum of $s(n)$ for various values of L is shown in Figure 11.2(a)

Let $\Delta f = 0.02$ and $f_1 = 0.60,$

Hence $f_2 = f_1 + \Delta f = 0.60 + 0.02 = 0.62$

Now, signal $s(n)$ is given by

$$s(n) = \cos(2\pi(0.6)n) + \cos(2\pi(0.62)n)$$

The power spectrum of signal $s(n)$ for various values of L is shown in Figure 11.2(b)

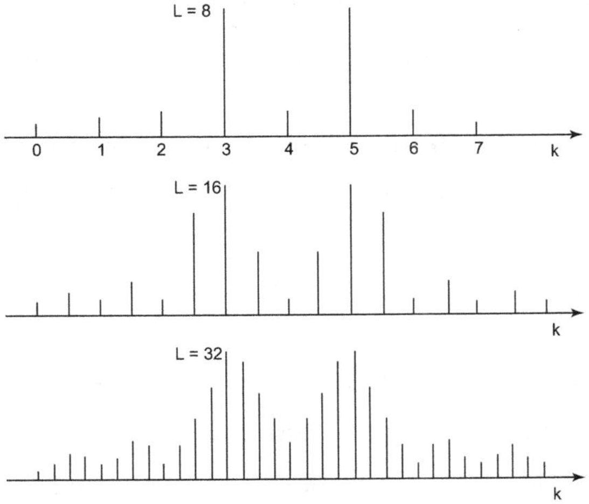

FIGURE 11.2 (*b*) Power spectrum $1/N\ |S(k)|^2$ verses *k* for $\omega_1 = 2\pi(0.6)$ and $\omega_2 = 2\pi(0.62)$

It is concluded from Figure 11.2(*a*) and (*b*) that when Δf is very small, the spectral components are not resolvable. The effect of zero padding is to provide more interpolation. It is not used to provide improvement in frequency resolution.

11.5 NON-PARAMETRIC METHODS OF POWER SPECTRUM ESTIMATION

These methods of power spectrum estimation make no assumption about how the data were generated and hence are called non-parametric methods. The estimation techniques that are discussed under non-parametric methods decrease the frequency resolution for reducing the variance of the spectral estimate.

The power spectrum estimation methods discussed here are the classical methods developed by Bartlett, Blackman and Tukey, and Welch.

Since the estimates are based entirely on a finite record of data, the frequency resolution of these methods is, at best, equal to the spectral width of the rectangular window of length N, which is approximately $1/N$ at the -3 dB points.

In this section, we discuss the following non-parametric methods of power spectrum estimation:

1. the Bartlett method (Averaging periodograms),
2. the Welch Method (Averaging modified periodograms), and
3. the Blackman and Tukey method (smoothing the periodograms).

11.5.1 The Bartlett Method (Averaging Periodograms)

The procedure for reducing the variance in the periodogram involves three steps:

1. Divide the N-point sequence $s(n)$ into k non-overlap sub-sequences of length M.
2. Find the periodogram for each sub-sequence.
3. Determine the average periodogram from k sub-sequence periodograms.

The variance of the periodogram can be reduced by averaging the periodograms and unlike either the periodogram or the modified periodogram,

Bartlett's method of periodogram averaging produces a consistent estimate of the power spectrum.

Consider the N-point sequence $s(n)$.

Divide this sequence into k non-overlapping sequences of length M.

The k-data sub-sequences are

$$s_i(n) = s(n+iM), i = 0,1,\ldots,k-1$$
$$n = 0,1,\ldots,M-1 \quad (11.37)$$

The periodogram for each sub-sequence is given by

$$P_{ss}^{(i)}(f) = \frac{1}{M}\left|\sum_{n=0}^{M-1} s_i(n)e^{-j2\pi fn}\right|^2, i = 0,1,\ldots,k-1 \quad (11.38)$$

The Bartlett power spectral estimate is obtained by averaging the periodogram of k-data sub-sequences.

$$P_{ss}^{B}(f) = \frac{1}{k}\sum_{i=0}^{k-1} P_{ss}^{(i)}(f) \quad (11.39)$$

Mean of Bartlett Power Spectral Estimate

The expected value of $P_{ss}^{B}(f)$ is given by

$$E\left[P_{ss}^{B}(f)\right] = E\left[\frac{1}{k}\sum_{i=0}^{k-1} P_{ss}^{(i)}(f)\right] = \frac{1}{k}\sum_{i=0}^{k-1} E\left[P_{ss}^{(i)}(f)\right]$$
$$= E\left[P_{ss}^{(i)}(f)\right] \quad (11.40)$$

The mean value of Bartlett power spectral estimate $P_{ss}^{B}(f)$ is identical to that of individual sub-sequence if the input is a zero-mean stationary process.

From following equations, $E[P_{ss}(f)] = \sum_{m=-(N-1)}^{N-1}\left(\frac{1-|m|}{N}\right)\phi_{ss}(m)e^{-j2\pi fm}$ and $E[P_{ss}(f)] = \int_{-1/2}^{1/2} \gamma_{ss}(\alpha)W_B(f-\alpha)d\alpha$, we have the expected value for single periodogram as

$$E\left[P_{ss}^{(i)}(f)\right] = \sum_{n=-(M-1)}^{(M-1)}\left(1-\frac{|m|}{M}\right)\phi_{ss}(m)e^{-j2\pi fm}$$
$$= \frac{1}{M}\int_{-1/2}^{1/2} \gamma_{ss}(f)\left[\frac{\sin\pi(f-\alpha)M^{\delta}}{\sin\pi(f-\alpha)}\right]^2 d\alpha \quad (11.41)$$

where $$W_B(f) = \frac{1}{M}\left(\frac{\sin\pi fM}{\sin\pi f}\right)^2 \quad (11.42)$$

$W_B(f)$ *is* the frequency characteristics of the Bartlett window function and $w_B(n)$ is given by

$$w_B(n) = \begin{cases} 1 - \dfrac{|m|}{M}, & |m| \leq M-1 \\ 0, & \text{otherwise} \end{cases} \quad (11.43)$$

Since the data sequence length is reduced from N to M, the spectral width is increased by a factor k $(= N/M)$ and the frequency resolution is reduced by a factor k.

Variance of Bartlett Power Spectral Estimate

By reducing the resolution, the variance is also reduced by a factor k. The variance of Bartlett power spectral estimate is given by

$$Var\left[P_{ss}^{B}(f)\right] = \frac{1}{k^2}\sum_{k=0}^{M-1} Var\left[P_{ss}^{(i)}(f)\right] = \frac{1}{k^2}.k\ Var\left[P_{ss}^{(i)}(f)\right]$$

$$= \frac{1}{k} Var\left[P_{ss}^{(i)}(f)\right] \quad (11.43)$$

By using equation $Var\left[P_{ss}(f)\right] = \gamma_{ss}^{2}(f)\left[1 + \left(\dfrac{\sin 2\pi fN}{N \sin 2\pi f}\right)\right]$

in Eq. (11.43), we get $Var\left[P_{ss}^{B}(f)\right] = \dfrac{1}{k}\gamma_{ss}^{2}(f)\left[1 + \left(\dfrac{\sin 2\pi fM}{M \sin 2\pi f}\right)^2\right]$ (11.44)

Therefore, the variance of the Bartlett power spectrum estimate has been, reduced by the factor k.

11.5.2 The Welch Method (Averaging Modified Periodograms)

The following two modifications were made by Welch in 1967 in the averaging periodogram or Bartlett method.

1. The Sub-sequences of $s(n)$ are allowed to overlap.
2. The data window function $w(n)$ is applied to each sub-sequence in computing the periodograms.

In the Welch method, the overlapping sequences are represented by

$$si(n) = s(n + iD),\ n = 0, 1, \ldots, M-1$$
$$i = 0, 1, \ldots, L-1 \quad (11.45)$$

Where iD is the starting point of the sub-sequence. If $D = M$, then this is same as the Bartlett method.

The modified periodogram estimate is given by

$$P_{ss}^{-(i)}(f) = \frac{1}{MU}\left|\sum_{n=0}^{M-1} s_i(n)\omega(n)e^{-j2\pi fn}dn\right|^2, i = 0,1,...L-1 \tag{11.46}$$

where U is the normalization factor for the power in the window function.
It is selected as

$$U = \frac{1}{M}\sum_{n=0}^{M-1}\omega^2(n) \tag{11.47}$$

The Welch power spectrum estimate is the average of these modified periodograms, that is,

$$P_{ss}^{\omega}(f) = \frac{1}{L}\sum_{i=0}^{L-1}\tilde{P}_{ss}^{(i)}(f) \tag{11.48}$$

Mean Value of the Welch Estimate

The expected value or mean value of the Welch estimate is given by

$$\begin{aligned} E\left[P_{ss}^{\omega}(f)\right] &= \frac{1}{L}\sum_{i=0}^{L-1}E\left[\tilde{P}_{ss}^{(i)}(f)\right] \\ &= E\left[P_{ss}^{-(i)}(f)\right] \end{aligned} \tag{11.49}$$

But the expected value of the modified periodogram given by

$$\begin{aligned} = E\left[P_{ss}^{-(i)}(f)\right] &= \frac{1}{MU}\sum_{n=0}^{M-1}\sum_{m=0}^{M-1}\omega(n)\omega(m)E\left[s_i(n)s_i^*(m)\right]e^{-j2\pi f(n-m)} \\ &= \frac{1}{MU}\sum_{n=0}^{M-1}\sum_{m=0}^{M-1}\omega(n)\omega(m)\phi_{ss}(n-m)e^{-j2\pi f(n-m)} \end{aligned} \tag{11.50}$$

Since $\phi_{ss}(n) = \int_{-1/2}^{1/2}\gamma_{ss}(\alpha)e^{-j2\pi an}d\alpha$ (11.51)

Substituting the value of $\phi_{ss}(n)$ from Eq. (11.51) in Eq. (11.50), we get

$$\begin{aligned} E\left[\tilde{P}_{ss}^{(i)}(f)\right] &= \frac{1}{MU}\int_{-1/2}^{1/2}\gamma_{ss}(\alpha)\left[\sum_{n=0}^{M-1}\sum_{m=0}^{M-1}\omega(n)\omega(m)e^{-j2\pi(n-m)(f-\alpha)}\right]d\alpha \\ &= \int_{-1/2}^{1/2}\gamma_{ss}(\alpha)W(f-\alpha)d\alpha \end{aligned} \tag{11.52}$$

where by definition

$$W(f) = \frac{1}{MU}\left|\sum_{n=0}^{M-1}\omega(n)e^{-j2\pi fn}\right|^2 \tag{11.53}$$

The normalization factor U ensures that

$$\int_{-1/2}^{1/2} W(f)df = 1 \tag{11.54}$$

Variance of the Welch Estimate

The variance of the Welch Estimate is given by

$$Var\left[P_{ss}^{w}(f)\right] = \frac{1}{L^2}\sum_{i=0}^{L-1}\sum_{j=0}^{L-1} E\left[\tilde{P}_{ss}^{(i)}(f)\tilde{P}_{ss}^{(j)}(f)\right] - \left\{E\left[P_{ss}^{w}(f)\right]\right\}^2 \tag{11.55}$$

In the case of no overlap between successive data segments ($L = k$), Welch has shown that

$$\begin{aligned} Var\left[P_{ss}^{w}(f)\right] &= \frac{1}{L} Var\left[\tilde{P}_{ss}^{(i)}(f)\right] \\ &\approx \frac{1}{L}\gamma_{ss}^{2}(f) \end{aligned} \tag{11.56}$$

In the case of 50% overlap between successive data segments ($L = 2k$), the variance of the Welch power spectrum is estimated with the Bartlett (triangular) window.

Welch also derived that

$$Var\left[P_{ss}^{w}(f)\right] = \frac{9}{8L}\gamma_{ss}^{2}(f) \tag{11.57}$$

Here, we considered only the triangular window function in the computation of the variance. Other window functions may be used.

In general, they will produce a different variance. We can improve the relevant characteristics of the estimate by varying the data segment overlapping by either more or less than 50%.

11.5.3 The Blackman and Tukey Method (Smoothing the Periodogram)

Smoothing the periodogram method was proposed by Blackman and Tukey in 1958. The autocorrelation sequence is windowed before calculating the power spectral density (psd). Windowing is used because if the value of m is large, that is, for larger lags, only less data enter in the estimation.

The Blackman–Tukey estimate is given by

$$P_{SS}^{BT}(f) = \sum_{m=-(M-1)}^{(M-1)} r_{ss}(m)\omega(m)e^{-j2\pi fm} \tag{11.58}$$

where $\omega(m)$ is the window function with length $2M - 1$. This is also called lag window. The lag window tapers away from the center.

The effect of multiplication by lag window is convolution in the frequency domain.

$$P_{SS}^{BT}(f) = \int_{-1/2}^{1/2} P_{ss}(\alpha)W(f-\alpha)d\alpha \tag{11.59}$$

where $P_{ss}(f)$ is the periodogram. The window sequence should be symmetric about $m = 0$ and the window spectrum should be non-negative.

Mean Value of Blackman–Tukey Estimate

The expected value of the Blackman–Tukey power spectrum estimate is given by

$$E\left[P_{SS}^{BT}(f)\right] = E\left[\int_{-1/2}^{1/2} P_{ss}(\alpha)W(f-\alpha)d\alpha\right]$$

$$= \int_{-1/2}^{1/2} E\left[P_{ss}(\alpha)\right]W(f-\alpha)d\alpha \tag{11.60}$$

We know that

$$E\left[P_{ss}(\alpha)\right] = \int_{-1/2}^{1/2} \gamma_{ss}(\theta)W_B(\alpha-\theta)d\theta \tag{11.61}$$

where $W_B(f)$ = Fourier transform of the Bartlett Window function or triangular window function.

Substituting Eq. (11.61) in Eq. (11.60), we get the double convolution integral.

$$E\left[P_{SS}^{BT}(f)\right] = \int_{-1/2}^{1/2}\int_{-1/2}^{1/2} \gamma_{ss}(\theta)W_B(\alpha-\theta)W(f-\alpha)d\alpha\, d\theta \tag{11.62}$$

Above Eq. (11.62) is in frequency domain.

Equivalently, in the time domain, the expected value of the Blackman–Tukey power spectrum estimate is

$$E\left[P_{SS}^{BT}(f)\right] = \sum_{m=-(M-1)}^{(M-1)} E\left[r_{ss}(m)\right]\omega(m)e^{-j2\pi fm}$$

$$= \sum_{m=-(M-1)}^{(M-1)} \gamma_{ss}(m)W_B(m)\omega(m)e^{-j2\pi fm} \tag{11.63}$$

where the Bartlett window is given by

$$\omega_B(m) = \begin{cases} \dfrac{1-|m|}{N}, & |m| < N \\ 0, & \text{otherwise} \end{cases} \tag{11.64}$$

Choose the window length such that $M << N$. Then $w(n)$ will be narrower than $\omega_B(m)$. This provides further smoothing of the periodogram. Under the conditions of $M << N$, Eq. (11.62) can be written as

$$E\left[P_{SS}^{BT}(s)\right] \approx \int_{-1/2}^{1/2} \gamma_{ss}(\theta)W(f-\theta)d\theta \tag{11.65}$$

Since

$$\int_{-1/2}^{1/2} W_B(\alpha-\theta)W(f-\alpha)d\alpha = \int_{-1/2}^{1/2} W_B(\alpha)W(f-\theta-\alpha)d\alpha$$

$$\approx W(f-\theta) \tag{11.66}$$

Variance of the Blackman–Tukey Power Spectrum Estimate

The variance of the Blackman–Tukey power spectrum estimate is given by

$$Var\left[P_{SS}^{BT}(f)\right] = E\left\{\left[P_{SS}^{BT}(f)\right]^2\right\} - \left\{E\left[P_{SS}^{BT}(f)\right]\right\}^2 \tag{11.67}$$

where the mean can be approximated as in Eq. (11.65).

The second moment in Eq. (11.67) is given by

$$E\left\{\left[P_{SS}^{BT}(f)\right]^2\right\} = \int_{-1/2}^{1/2}\int_{-1/2}^{1/2} E\left[P_{SS}(\alpha)P_{SS}(\theta)\right]W(f-\alpha)W(f-\theta)d\alpha\, d\theta \tag{11.68}$$

Now, we are assuming that process is a Gaussian random process:

$$E\left[P_{SS}(\alpha)P_{SS}(\theta)\right] = \gamma_{ss}(\alpha)\gamma_{ss}(\theta)\left\{1 + \left[\frac{\sin\pi(\theta+\alpha)}{N\sin\pi(\theta+\alpha)}\right]^2 + \left[\frac{\sin\pi(\theta-\alpha)}{N\sin\pi(\theta-\alpha)}\right]^2\right\} \tag{11.69}$$

Substituting Eq. (11.69) in Eq. (11.68), we get

$$E\left\{\left[P_{SS}^{BT}(f)\right]^2\right\} = \left[\int_{-1/2}^{1/2} \gamma_{ss}(\theta)W(f-\theta)d\theta\right]^2 + \int_{-1/2}^{1/2}\int_{-1/2}^{1/2} \gamma_{ss}(\alpha)\gamma_{ss}(\theta)W(f-\alpha)W(f-\theta)$$
$$\times\left\{\left[\frac{\sin\pi(\theta+\alpha)N}{N\sin\pi(\theta+\alpha)}\right]^2 + \left[\frac{\sin\pi(\theta-\alpha)N}{N\sin\pi(\theta-\alpha)}\right]^2\right\}d\alpha\, d\theta \tag{11.70}$$

The first term of Eq. (11.70) is simply the square of the mean of $P_{SS}^{BT}(f)$.

This term is to be subtracted out according to Eq. (11.67). This leaves the second term in Eq. (11.70), which constitutes the variance. For the case in which $N >> M$, the functions $\left[\dfrac{\sin\pi(\theta+\alpha)N}{N\sin\pi(\theta+\alpha)}\right]$ and $\left[\dfrac{\sin\pi(\theta+\alpha)N}{N(\sin(\theta+\alpha)}\right]$ are relatively narrow compared to $W(f)$ in the vicinity of $\theta = -\alpha$ and $\theta = \alpha$, respectively.

Therefore,

$$\int_{-1/2}^{1/2} \gamma_{ss}(\theta)W(f-\theta)\left\{\left[\frac{\sin\pi(\theta+\alpha)N}{N\sin\pi(\theta+\alpha)}\right]+\left[\frac{\sin\pi(\theta-\alpha)N}{N\sin\pi(\theta-\alpha)}\right]^2\right\}d\theta$$

$$\approx \frac{\gamma_{ss}(-\alpha)W(f+\alpha)+\gamma_{ss}(\alpha)W(f-\alpha)}{N} \quad (11.71)$$

Using the above approximation given in Eq. (11.71), the variance of $P_{SS}^{BT}(f)$ is given by

$$Var\left[P_{SS}^{BT}(f)\right] \approx \frac{1}{N}\int_{-1/2}^{1/2} \gamma_{SS}(\alpha)W(f-\alpha)$$
$$\times\left[\gamma_{SS}(-\alpha)W(f+\alpha)+\gamma_{SS}(\alpha)W(f-\alpha)\right]d\alpha$$
$$\approx \frac{1}{N}\int_{-1/2}^{1/2} \gamma_{SS}^2(\alpha)W^2(f-\alpha)d\alpha \quad (11.72)$$

Wherein the last step, we made the approximation

$$\int_{-1/2}^{1/2} \gamma_{SS}(\alpha)\gamma_{SS}(-\alpha)W(f-\alpha)W(f+a)d\alpha = 0 \quad (11.73)$$

When $W(f)$ *is* narrow compared to the true power spectrum $\gamma_{SS}(f)$, Eq. (11.72) in further approximated as

$$Var\left[P_{SS}^{BT}(f)\right] \approx \gamma_{SS}^2(f)\left[\frac{1}{N}\int_{-1/2}^{1/2} W^2(\theta)d\theta\right]$$
$$\approx \gamma_{SS}^2(f)\left[\frac{1}{N}\int_{m=(M-1)}^{(M-1)} \omega^2(m)\right] \quad (11.74)$$

11.5.4 Quality of Non-parametric Power Spectrum Estimators

In this section, we compare the quality or performance characteristics of the Bartlett, Welch, and Blackman and Tukey power spectrum estimates.

The quality of the estimator is given by the ratio of the square of the mean of the power spectrum estimate to its variance. It is given as

$$Q_A = \frac{\left\{E\left[P_{SS}^A(f)\right]\right\}^2}{Var\left[P_{SS}^A(f)\right]}$$

where $A = B$, W, or BT for the three power spectrum estimates.

Variability is another measure of performance. It is the reciprocal of the quality of the estimator.

Now, we shall discuss and calculate the quality of periodogram using Bartlett, Welch, and Blackman–Tukey power spectrum estimation techniques.

For reference, the periodogram has a mean and variance.

Mean of Periodogram

Mean of periodogram can be given as

$$E\left[P_{SS}(f)\right] = \int_{-1/2}^{1/2} \gamma_{ss}(\theta) W_B(f-\theta) d\theta \tag{11.75}$$

Variance of Periodogram

Variance of the periodogram is given by

$$Var\left[P_{SS}(f)\right] = \gamma_{SS}^2(f)\left[1 + \left(\frac{\sin 2\pi fN}{N \sin 2\pi f}\right)^2\right] \tag{11.76}$$

where
$$W_B(f) = \frac{1}{N}\left(\frac{\sin \pi fN}{\sin \pi f}\right)^2 \tag{11.77}$$

when $N \to \infty$,

$$E\left[P_{SS}(f)\right] \to \gamma_{ss}(f) \int_{-1/2}^{1/2} W_B(\theta) d\theta$$

$$= \omega_B(o)\gamma_{ss}(f) = \gamma_{SS}(f)$$

$$Var\left[P_{SS}(f)\right] \to \gamma_{SS}^2(f) \tag{11.78}$$

This is an asymptotically unbiased estimate, but not consistent as variance does not approach zero when $N \to \infty$.

Asymptotically, the periodogram is characterized by the quantity factor

$$Q_P = \frac{\gamma_{SS}^2(f)}{\gamma_{SS}^2(f)} = 1 \tag{11.79}$$

Here Q_P is constant and independent of N specifies the poor quantity.

Bartlett Power Spectrum Estimate

Mean of $P_{ss}^B(f)$

Mean of Bartlett power spectrum estimate is given by

$$E\left[P_{SS}^{B}(f)\right]=\int_{-1/2}^{1/2}\gamma_{SS}(\theta)W_B(f-\theta)d\theta \tag{11.80}$$

Variance of $P_{ss}^{B}(f)$

The variance of Bartlett power spectrum estimate is given by

$$Var\left[P_{SS}^{B}(f)\right]=\frac{1}{k}\gamma_{ss}^{2}(f)\left[1+\left(\frac{\sin 2\pi fM}{M\sin 2\pi f}\right)^{2}\right] \tag{11.81}$$

where

$$W_B(f)=\frac{1}{M}\left(\frac{\sin \pi fM}{\sin \pi f}\right)^{2} \tag{11.82}$$

As $N\to\infty$ and $M\to\infty$, while $k=\dfrac{N}{M}$ remains fixed, we find that

$$E\left[P_{SS}^{B}(f)\right]\to\gamma_{ss}(f)\int_{-1/2}^{1/2}W_B(f)df=\gamma_{ss}(f)w_B(0)$$

$$=\gamma_{SS}(f) \tag{11.83}$$

$$Var\left[P_{SS}^{B}(f)\right]\to\frac{1}{k^2}\gamma_{ss}^{2}(f) \tag{11.84}$$

This estimate is asymptotically unbiased and k can increase with an increase in N. The estimate is consistent. Hence, asymptotically, this estimate is characterized by the quality factor

$$Q_B=k=\frac{N}{M} \tag{11.85}$$

With 3-dB main lobe width of a rectangular window, the frequency resolution is given by

$$\Delta f=\frac{0.9}{M} \tag{11.86}$$

Hence,

$$M=\frac{0.9}{\Delta f} \tag{11.87}$$

Substituting Eq. (11.87) in Eq. (11.85), we get the quality factor

$$Q_B=\frac{N}{M}=\frac{N}{\left(\dfrac{0.9}{\Delta f}\right)}$$

$$\text{or} \quad Q_B = \frac{N}{\left(\dfrac{0.9}{\Delta f}\right)} = 1.1N.\Delta f \tag{11.88}$$

Welch Power Spectrum Estimate

Mean of $P_{ss}^{B}(f)$

The mean of Welch power spectrum estimate is given by

$$E\left[P_{SS}^{B}(f)\right] = \int_{-1/2}^{1/2} \gamma_{ss}(\theta)W(f-\theta)d\theta \tag{11.89}$$

$$\text{where} \quad W(f) = \frac{1}{MU}\left|\sum_{n=0}^{M-1} \omega(n)e^{-j2\pi fn}\right|^2 \tag{11.90}$$

Variance of $P_{ss}^{B}(f)$

The variance of Welch power spectrum estimate is given by

$$Var\left[P_{SS}^{B}(f)\right] = \begin{cases} \dfrac{1}{L}\gamma_{SS}^{2}(f), & \text{for no overlap} \\ \dfrac{9}{8L}\gamma_{SS}^{2}(f), & \text{for 50\% overlap and triangular window} \end{cases} \tag{11.91}$$

As $N \to \infty$ and $M \to \infty$, the mean of $P_{SS}^{W}(f)$ converges to

$$E\left[P_{SS}^{W}(f)\right] \to \gamma_{ss}(f) \tag{11.92}$$

As $N \to \infty$ and $M \to \infty$, the variance of $P_{SS}^{W}(f)$ converges to zero. Therefore the estimate is consistent.

The quantity factor for two conditions is given in Eq. (11.91) are

$$Q_W = \begin{cases} L = \dfrac{N}{M}, & \text{for no overlap} \\ \dfrac{8L}{9} = \dfrac{16M}{9M}, & \text{for 50\% overlap and triangular window} \end{cases} \tag{11.93}$$

With spectral width of the triangular window at 3dB points,

$$\Delta f = \frac{1.28}{M} \tag{11.94}$$

Consequently, the quality factor expressed in terms of Δf and N is

$$Q_W = \begin{cases} 0.78\, N\Delta f, & \text{for no overlap} \\ 1.39\Delta f, & \text{for 50\% overlap and triangular window} \end{cases} \tag{11.95}$$

Blackman–Tukey Power Spectrum Estimate

Mean of $P_{SS}^{BT}(f)$

The mean of the Blackman–Tukey power spectrum estimate is approximated as

$$E\left[P_{SS}^{BT}(f)\right] \cong \int_{-1/2}^{1/2} \gamma_{SS}(\theta)W(f-\theta)d\theta \qquad (11.96)$$

Variance of $P_{SS}^{BT}(f)$

The variance of the Blackman–Tukey power spectrum estimate is approximated as

$$Var\left[P_{SS}^{BT}(f)\right] = \gamma_{SS}^{2}(f)\left[\frac{1}{N}\sum_{n=-(M-1)}^{(M-1)} \omega^2(m)\right] \qquad (11.97)$$

where $\omega(m)$ is the window sequence used to taper the estimated autocorrelation sequence.

For the rectangular window, we have

$$\frac{1}{N}\sum_{m=-(M-1)}^{(M-1)} \omega^2(m) = \frac{2M}{N} \qquad (11.98)$$

For the triangular window, we have

$$\frac{1}{N}\sum_{n=-(M-1)}^{(M-1)} \omega^2(n) = \frac{2M}{3N} \qquad (11.99)$$

The mean value of this estimate is asymptotically unbiased.

The quality factor of this estimate for the triangular window is given by

$$Q_{BT} = 1.5\frac{N}{M} \qquad (11.100)$$

Since the window length is $2M - 1$, the frequency resolution measured at the 3-dB points is

$$\Delta f = \frac{1.28}{2M} = \frac{0.64}{M} \qquad (11.101)$$

Eq. (11.101) can be written as

$$M = \frac{0.64}{\Delta f} \qquad (11.102)$$

Substituting Eq. (11.102) in Eq. (11.100), we get

$$Q_{BT} = \frac{1.5}{0.64}N.\Delta f$$

$$= 234N\Delta f \qquad (11.103)$$

Quality factor increases when N is increased. For a desired quality level, decrease Δf by increasing N.

Results of the Quality factor for Bartlett, Welch, and Blackman–Tukey power spectrum estimates are summarized in Table 11.1.

TABLE 11.1 Quality of Power Spectrum Estimates

S. No.	Estimate	Quality factor
1.	Bartlett	$1.11\, N\Delta f$
2.	Welch (50% overlap)	$1.39\, N\Delta f$
3.	Blackman–Tukey	$2.34\, N\Delta f$

11.5.5 Computational Requirements of Non-parametric Power Spectrum Estimates

The other important aspect of the non-parametric power spectrum estimates is their computational requirements. Here, we consider the following assumptions:

Fixed data length = N
Frequency resolution = Δf
Radix-2 FFT algorithm for computation.

Bartlett Power Spectrum Estimate

$$\text{FFT length} = M = \frac{0.9}{\Delta f}$$

$$\text{Number of FFTs} = \frac{N}{M} = 1.11\, N\, \Delta f$$

$$\text{Number of computations} = \frac{N}{M}\left(\frac{M}{2}\log_2 M\right) = \frac{N}{2}\log_2\left(\frac{0.9}{\Delta f}\right)$$

Welch Power Spectrum Estimate (50% Overlap)

$$\text{FFT length} = M = \frac{1.28}{\Delta f}$$

$$\text{Number of FFTs} = \frac{2N}{M} = 1.56\, N\, \Delta f$$

$$\text{Number of computations} = \frac{2N}{M}\left(\frac{M}{2}\log_2 M\right) = \frac{N}{2}\log_2\left(\frac{1.28}{\Delta f}\right)$$

For windowing $\dfrac{2N}{M} \times M$ multiplications are required.

$$\text{Total computations} = 2N + N\log_2\left(\frac{128}{\Delta f}\right) = N\log_2\left(\frac{5.12}{\Delta f}\right)$$

Blackman–Tukey Power Spectrum Estimate

In the Blackman–Tukey method, the autocorrelation $r_{ss}(m)$ can be computed efficiently *via* the FFT algorithm. For large data points, FFT can be done by segmenting the data into $k = N/2M$ (windowing to $2m - 1$ points or samples)

By using this approach,

$$\text{FFT length} = 2M = \frac{1.28}{\Delta f}$$

$$\text{Number of FFTs} = 2k + 1$$

$$= 2\left(\frac{N}{2N}\right) + 1 = \frac{N}{M} + 1 \cong \frac{N}{M}$$

$$\text{Number of computations} = \frac{N}{M}(M\log_2 2M) = N\log_2\left(\frac{128}{\Delta f}\right)$$

11.5.6 Limitations of Non-parametric Methods for Power Spectrum Estimation

These methods have the following limitations

1. It requires long data sequences to obtain the necessary frequency resolution.
2. Spectra] leakage effects because of windowing.
3. The assumption of the autocorrelation estimate $rss(m)$ to be zero for $m \geq N$. This assumption limits the frequency resolution and quality of the power spectrum estimate.
4. Assumption that the data are periodic with period N. These assumptions may not be realistic.

11.6 PARAMETRIC METHODS OF POWER SPECTRUM ESTIMATION

The non-parametric methods of power spectrum estimation which we studied in the previous section are relatively simple, well understood, and easy to compute using the FFT algorithms. However, these methods require the availability of long data records in order to obtain the necessary frequency resolution required in many applications. Furthermore, non-parametric methods of power spectrum estimation suffer from spectral leakage, due to windowing that is inherent in finite-length data records. Often, the spectral leakage masks weak signals that are present in the data.

Parametric methods provide better frequency resolution since this modeling does not require window function and the assumptions that autocorrelation sequence to be zero for $|m| \geq N$ is not required.

It extrapolates the values for $|m| \geq N$. But it requires prior information about the generation of the data sequence. A model for the signal generation can be obtained from the observed data. These methods are useful for data sequences that are short.

The parametric spectral estimation has three steps:

1. First of all select the model.
2. Estimate the model parameters from the observed measured data or the correlation sequence which is estimated from the data.
3. Obtain the power spectral estimate with the help of the estimated model parameters.

11.6.1 Basics of Auto-Regressive (AR), Moving Average (MA) and Auto-Regressive Moving Average (ARMA) Models

Let $s(n)$ be the observed signal which is modeled as the output of a linear system represented by

$$H(z) = \frac{D(z)}{C(z)} = \frac{\sum_{k=0}^{q} B_k z^{-k}}{1 + \sum_{k=1}^{p} A_k z^{-k}} = \frac{Y(z)}{W(z)} \tag{11.104}$$

The difference equation of this system is given by

$$y(n) = \sum_{k=1}^{p} A_k y(n-k) + \sum_{k=0}^{q} B_k \omega(n-k) \tag{11.105}$$

where $\omega(n)$ is the input sequence to the linear system shown in Figure 11.3 and $y(n)$ is the output sequence from the linear system.

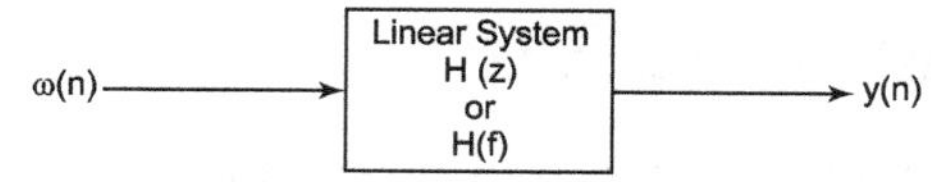

FIGURE 11.3 Linear system for analyzing parametric models.

If the observed data is considered as a stationary random process, then the input can also be assumed as a stationary random process Let $\gamma_{ww}(f)$ be the power spectral density of the input sequence $\omega(n)$, $\gamma_{yy}(f)$ be the power spectral density of the output sequence $y(n)$ and $H(f)$ be the frequency response of the linear system then

$$\gamma_{yy}(f) = |H(f)|^2 \gamma_{ww}(f) \tag{11.106}$$

If the sequence $w(n)$ is a zero-mean white noise of the process of variance σ_w^2, the autocorrelation sequence is given by

$$\gamma_{ww}(m) = \sigma_w^2 \delta(m) \tag{11.107}$$

The power spectral density of the input sequence is given by

$$\gamma_{ww}(m) = \sigma\omega^2 \tag{11.108}$$

Hence, the power spectral density of the output sequence $y(n)$ is given as

$$\boxed{\gamma_{ff}(f) = |H(f)|^2 \gamma_{ww}(f)}$$
$$= \sigma_w^2 \, |H(f)|^2$$
$$= \sigma_w^2 \left| \frac{D(f)}{C(f)} \right|^2 \tag{11.109}$$

The process modeled by such a linear system is the ARMA process of order (p, q). It is represented as ARMA (p, q).

If $q = 0$, $B_o = 1$ and $B_k = 0$ for $1 \le k \le q$, then

$$H(z) = \frac{1}{1 + \sum_{k=1}^{p} A_k z^{-k}} \tag{11.110}$$

The model for this system is called the AR process of order p. It is represented as AR (p)

If $A_k = 0$ for $1 \le k \le p$, then $\quad H(z) = \dfrac{\sum_{k=0}^{q} B_k Z^{-k}}{}$ (11.111)

The model for this system is called moving average (MA) process of order q. It is represented as MA (q). Of these three linear models, the AR model is by far the most widely used. There are two reasons.

1. The AR model is suitable for representing spectra with narrow peaks (resonances).
2. The AR model results in very simple linear equations for the AR parameters.

On the other hand, the MA model, as a general rule, requires many more coefficients to represent a narrow spectrum. Consequently, it is rarely used by itself as a model for spectrum estimation. By combining poles and zeros, the ARMA model provides a more efficient representation, from the viewpoint of the number of model parameters, of the spectrum of a random process.

The decomposition theorem due to Wold (1938) assets that any AR MA or MA process can be represented uniquely by an AR model of possibly infinite order, and any ARMA or AR process can be represented by a MA model of possibly infinite order. In view of this theorem, the issue of model selection reduces to selecting the model that requires the smallest number of parameters that are also easy to compute. Usually, the choice in practice is the AR model.

The ARMA model is used to a lesser extent.

ARMA Process

The output power spectral density for ARMA (p, q) process is given by

$$\gamma_{yy}(f) = |H(f)|^2 \, \gamma_{\omega\omega}(f) = \sigma_\omega^2 \frac{|D(f)|^2}{|C(f)|^2}$$

$$= \frac{\sigma_\omega^2 \left|1 + \sum_{k=1}^{q} B_k e^{-j2\pi fk}\right|^2}{\left|1 + \sum_{k=1}^{p} A_k e^{-j2\pi fk}\right|^2} \qquad (11.112)$$

The power spectral density depends only on the white noise variance σ_w^2 and filter coefficients A_k and B_k. The filter coefficients are to be estimated.

If the observed N data points are greater than $p + q + 1$, then a good estimate for the unknown parameters can be obtained.

Consider the general ARMA(p, q) model represented by the difference equation,

$$y(n) = -\sum_{k=1}^{p} A_k y(n-k) + \sum_{k=0}^{q} B_k \omega(n-k) \qquad (11.113)$$

Let $B_o = 1$ and $w(n)$ be a zero-mean white noise process with variance.

Multiplying both sides of Eq. (11.113) by $y^*(n - m)$ and taking the expectation, we get

$$E\left[y(n)y^*(n-m) = -\sum_{k=1}^{p} A_k E[y(n-k)y^*(n-k)] + \sum_{k=0}^{1} B_k E[\omega(n-k)y^*(n-k)] \right]$$

or $$r_{xy}(m) = -\sum_{k=1}^{p} A_k r_{yy}(m-k) + \sum_{k=0}^{q} B_k r_{\omega y}(m-k) \qquad (11.114)$$

where $$r_{yy}(m) = E[y(n)y^*(n-m)$$

$$r_{yy}(m-k) = E[y(n-k)y^*(n-k)$$

$$r_{wy}(m-k) = E[\omega(n-k)y^*(n-k)$$

The cross-correlation function $r_{\omega y}(m)$ is given by

$$r_{\omega y}(m) = E[y^*(n)(n+m)$$

$$= E\left[\sum_{k=0}^{\infty} h(k)\omega^*(n-k)\omega(n+m)\right]$$

$$= \sum_{k=0}^{\infty} h(k)E\left[\omega^*(n-k)\omega(n+m) = \sigma_w^2 h(-m)\right. \qquad (11.115)$$

Since, $$E[\omega^*(n)\omega(n+m)] = \sigma_w^2 \delta(m)$$

$$r_{\omega y}(m)\begin{cases} 0, & m \geq 0 \\ \sigma_\omega^2 h(-m), & m \leq 0 \end{cases} \qquad (11.116)$$

Therefore,

$$r_{yy}(m) = \begin{cases} -\sum_{k=1}^{p} A_k r_{yy}(m-k), m > q \\ -\sum_{k=1}^{p} A_k r_{yy}(m-k) + \sigma_\omega^2 \sum_{k=0}^{q-m} h(k)B_{k+m}, 0 \leq m \leq q \\ r_{yy}^*(-m), m < 0 \end{cases} \qquad (11.117)$$

The relationships in Eq. (11.117) provide a formula for determining the model parameters $\{A_k\}$ by restricting our attention to the case $m > q$.

The set of linear equations is used in solving for $\{A_k\}$ right with the autocorrelation estimates.

$$\begin{bmatrix} r_{yy}(q) & r_{yy}(q-1) & \cdots\cdots\cdots r_{yy} & (q-p+1) \\ r_{yy}(q+1) & r_{yy}(q) & \cdots\cdots\cdots r_{yy} & (q-p+2) \\ r_{yy}(q+2) & r_{yy}(q+1) & \cdots\cdots\cdots r_{yy} & (q-p+3) \\ \vdots & \vdots & \vdots & \vdots \\ \vdots & \vdots & \vdots & \vdots \\ r_{yy}(q+p-1) & r_{yy}(q+p-2) & \cdots\cdots r_{yy}(q) & \end{bmatrix} \begin{bmatrix} A_1 \\ A_2 \\ A_3 \\ \\ \\ A_p \end{bmatrix} = -\begin{bmatrix} r_{yy}(q+1) \\ r_{yy}(q+2) \\ r_{yy}(q+3) \\ \vdots \\ \vdots \\ r_{yy}(q+p) \end{bmatrix} \qquad (11.118)$$

AR Process

If $q = 0$, then AR model parameters can be obtained

$$r_{yy}(m) = \begin{cases} -\sum_{k=1}^{p} A_k r_{yy}(m-k), \; m > 0 \\ -\sum_{k=1}^{p} A_k r_{yy}(m-k) + \sigma_\omega^2, \; m = 0 \\ r_{yy}^*(-m), \; m < 0 \end{cases} \tag{11.119}$$

In this case, the AR parameters $\{A_k\}$ are obtained from the solution of the Yule–Walker or normal equations

$$\begin{bmatrix} r_{yy}(0) & r_{yy}(-1) & \vdots & r_{yy}(-p+1) \\ r_{yy}(1) & r_{yy}(0) & \vdots & r_{yy}(-p+2) \\ r_{yy}(2) & r_{yy}(1) & \vdots & r_{yy}(-p+3) \\ \vdots & \vdots & \vdots & \\ r_{yy}(p-1) & r_{yy}(p-2) & \vdots & r_{yy}(0) \end{bmatrix} \begin{bmatrix} A_1 \\ A_2 \\ A_3 \\ \\ A_p \end{bmatrix} = -\begin{bmatrix} r_{yy}(q+1) \\ r_{yy}(q+2) \\ r_{yy}(q+3) \\ \vdots \\ r_{yy}(q+p) \end{bmatrix} \tag{11.120}$$

Using a set of Eq. (11.120), coefficients $\{A_k\}$ can be obtained and the variance is determined as

$$\sigma_\omega^2 = r_{yy}(0) + \sum_{k=1}^{p} A_k r_{yy}(-k) \tag{11.121}$$

Eqs. (11.120) and (11.121) are usually combined into a single matrix equation of the form

$$\begin{bmatrix} r_{yy}(0) & r_{yy}(-1) & \cdots\cdots\cdots r_{yy} \\ r_{yy}(1) & r_{yy}(0) & \cdots\cdots\cdots r_{yy}(-p+1) \\ r_{yy}(2) & r_{yy}(1) & \cdots\cdots\cdots r_{yy}(-p+2) \\ \vdots & \vdots & \vdots \\ \vdots & \vdots & \vdots \\ \vdots & \vdots & \vdots \\ r_{yy}(p) & r_{yy}(p-1) & \cdots\cdots r_{yy}(0) \end{bmatrix} \begin{bmatrix} 1 \\ A_1 \\ A_2 \\ \vdots \\ \vdots \\ \vdots \\ A_p \end{bmatrix} \begin{bmatrix} \sigma_\omega^2 \\ 0 \\ 0 \\ \vdots \\ \vdots \\ \vdots \\ 0 \end{bmatrix} \tag{11.122}$$

The correlation matrix is a Toeplitz and non-singular and can be solved with Levinson – Durbin Algorithm for obtaining the inverse matrix.

Thus all the system parameters in the AR (p) model are easily determined from knowledge of the autocorrelation sequence $r_{yy}(m)$ for $0 \le m \le p$.

MA Process

If $A_k = 0$ for $1 \le k \le p$ and $h(k) = B_k$ for $0 \le k \le q$, then $MA(q)$ model is obtained

$$r_{yy}(m) = \begin{bmatrix} \sigma_\omega^2 \sum_{k=0}^{q} B_k B_{k+m}, & 0 \le m \le q \\ 0, & m > q \\ r_{yy}^*(-m), & m < 0 \end{bmatrix} \tag{11.123}$$

11.6.2 Power Spectrum Estimation Using *AR(p)*, *ARMA(p, q)*, and *MA(q)* Models

With the background that is discussed in the previous section, we now describe the power spectrum estimation methods for the $AR(p)$, $ARMA(p, q)$, and MA (q) models.

The Yule–Walker Method for the AR Model Parameters

In this method, we simply estimate the autocorrelation from the data and use the estimates given by Eq. (11.120) to solve for the AR model parameters

Let the autocorrelation estimate (based) be,

$$r_{yy}(m) = \frac{1}{N} \sum_{n=0}^{N-|m|} y^*(n-m)y(n+m).m \ge 0 \tag{11.124}$$

The autocorrelation matrix should be positive semidefinite to yield a stable AR model. The power spectral estimate is given by

$$p_{yy}^{yw}(f) = \frac{\sigma_{\omega p}^{'2}}{\left|1 + \sum_{k=1}^{p} A_p^{'}(k) e^{-j2\pi fk}\right|} \tag{11.125}$$

where,

$A_p^{'}(k)$ are the estimates of the AR parameters obtained from the Levinson-Derbin recursions or algorithm, $\sigma_{\omega p}^{'2}$ is the estimated minimum mean square value for the pth order predictor

$$\sigma_{\omega p}^{'2} = r_{yy}(0) \pi_{k=1}^{p} \left[1 - \left|A_p^{'}(k)\right|^2\right] \tag{11.126}$$

In estimating the power spectrum of sinusoidal signals *via* AR models, Lacos showed in 1971 that spectral peaks in an AR spectrum estimate are proportional to the square of the power of the sinusoidal signal. On the other hand, the area under the peak in the power density spectrum is linearly proportional to the power of the sinusoid. This characteristic behavior holds for all AR model-based estimation methods.

MA Model

The *MA* (q) model for the observed data is represented by

$$r_{yy}(m) = \begin{cases} \sigma_\omega^2 \sum_{k=0}^{q} B_k B_{k+m}, & 0 \le m \le q \\ 0, & m > q \\ r_{yy}^*(-m), & m < 0 \end{cases} \tag{11.127}$$

The power spectral estimate is given by

$$P_{yy}^{MA}(f) = \sum_{m=-q}^{q} r_{yy}(m) e^{-j2\pi fm} \tag{11.128}$$

$$= |D(z)|^2$$

Hence, *MA* parameters $\{B_k\}$ need not be calculated, but an estimate of autocorrelation for $|m| \le q$ will suffice.

ARMA Model

For certain type of processes, the *ARMA* model is used to estimate the spectrum with fewer parameters. This model is mostly used when the data is corrupted by noise.

We construct a set of linear equations for $m > q$ and use the method of least squares on the set of equations. Here, we use linear prediction,

$$r_{yy}(m) = \sum_{k=1}^{q} A_k r_{yy}(m-k), \quad m = q+1, q+1, \ldots, M < N \tag{11.129}$$

where $r_{yy}(m)$ is the estimated autocorrelation sequence. Select the parameters $\{A_k\}$ such that the squared error is minimized.

Squared error is given by

$$\varepsilon = \sum_{k=q+1}^{m} |e(n)|^2 \tag{11.130}$$

Eq. (11.130) can be written as

$$\varepsilon = \sum_{k=q+1}^{m} \left| r_{yy}(m) + \sum_{k=1}^{p} A_k r_{yy}(m-k) \right|^2 \tag{11.131}$$

Thus the *AR* parameters can be obtained and the procedure is known as the least-squares modified Yule–Waker method. Suitable weightage can also be applied to the autocorrelation sequence. After finding the *AR* model parameters, the system obtained is represented by

$$C'(z) = \sum_{k-1}^{p} A_k' z^{-k} \tag{11.132}$$

Consider the sequence $s(n)$ to be filtered by *FIR* filter, that is, *AR* model is $C'(z)$. Let the output be $z(n)$.

$$z(n) = y(n) + \sum_{k=1}^{p} A_k' y(n-k),\ n = 0,1,\ldots N-1 \tag{11.133}$$

MA model $D(z)$ can be obtained by cascading $ARMA(p, q)$ model with $C(z)$.

The sequence $z(n)$ for $p \le n \le N - 1$ is used to obtain the correlation sequence $r_{zz}(m)$. The *MA* power spectral estimate is given by

$$p_{zz}^{MA}(f) = \sum_{m=-q}^{q} r_{zz}(m) e^{-e^{j2\pi fm}} \tag{11.134}$$

The ARMA power spectral estimate is given by

$$p_{yy}^{ARMA}(f) = \frac{p_{zz}^{MA}(f)}{\left|1 + \sum_{k=1}^{p} A_k' e^{-j2\pi fk}\right|^2} \tag{11.135}$$

Among the three models, the AR model is widely used for power spectrum estimation due to

i. *AR* model is suitable for representing power spectra with narrow peaks.

ii. *AR* model is represented by a simple linear equation.

11.6.3 The Burg Method for the AR Model Parameters

The method devised by Burg for estimating the AR parameters can be viewed as an order recursive least-squares lattice method, based on the minimization of the forward and backward errors in linear predictors, with the constraint that the AR parameters satisfy the Levinson–Durban recursion or algorithm.

For a given data sequence $s(n)$, $n = 0, 1, \ldots, N - 1$ the mth order forward and backward linear prediction estimates are

$$\hat{s}(n) = -\sum_{k=1}^{m} A_m(k) s(n-k)$$

$$\hat{s}(n-m) = -\sum_{k=1}^{m} A_m^*(k)s(n+k-m) \tag{11.136}$$

The corresponding forward and backward errors are $f_m(n)$ and $g_m(n)$. These errors are defined as

$$f_m(n) = s(n) - \hat{s}(n)$$

and $\quad g_m(n) = s(n-m) - \hat{s}(n-m)$

where $A_m(k)$, $0 \le k \le m-1$, $m = 1, 2, \ldots, p$ are the prediction coefficients. Total least squared error is given by

$$\varepsilon_m \sum_{n=m}^{N-1} |f_m(n)|^2 + |g_m(n)|^2 \tag{11.137}$$

This error is to be minimized by selecting the prediction coefficients, subject to the constraint that they satisfy the Levinson–Durbin recursion given by

$$A_m(k) = A_{m-1}(k) + K_m A_{m-1}^*(m\ -\ k),\ 1 \le k \le m, (1 \le m \le p) \tag{11.138}$$

where $K_m = A_m(m)$ is the mth reflection coefficient in the lattice filter realization of the predictor.

When Eq. (11.138) is substituted into expressions for $f_m(n)$ and $g_m(n)$, the result is the pair of order recursive equations for the forward and backward prediction errors given by equation:

$$f_0(n) = g_0(n) = s(n)$$

$$\left.\begin{aligned} f_m(n) &= f_{m-1}(n) + K_m g_{m-1}(n-1), m = 1, 2\ \ldots, p \\ g_m(n) &= K_m^* f_{m-1}(n) + g_{m-1}(n-1), m = 1, 2\ \ldots, p \end{aligned}\right\} \tag{11.139}$$

Now, if we substitute Eq. (11.139) into Eq. (11.138) and perform the minimization of ε_m with respect to the complex-valued reflection coefficient K_m, we obtain the result

$$\hat{K}_m = \frac{-\sum_{n=m}^{N-1} f_{m-1}(n) g_{m-1}^*(n-1)}{\frac{1}{2}\sum_{n=m}^{N-1}\left[|f_{m-1}(n)|^2 + |g_{m-1}(n-1)|^2\right]},\ m = 1, 2, \ldots p \tag{11.140}$$

The term in the numerator of Eq. (11.140) is an estimate of the cross-correlation between the forward and backward prediction errors. With the normalization factors in the denominator of Eq. (11. 140), it is apparent that $|K_m| < 1$, so that the all pole model obtained from the data is stable.

The denominator of $\hat{K}_m$ is the least-squares estimate of the forward and backward errors. Hence,

$$\hat{K}_m = \frac{\sum_{n=m}^{N-1} f_{m-1}(n) g^*_{m-1}(n-1)}{\frac{1}{2}\left[\hat{E}^f_{m-1} + \hat{E}^b_{m-1}\right]} \qquad (11.141)$$

$$m = 1, 2, \ldots, p$$

where $\hat{E}^f_{m-1} + \hat{E}^b_{m-1}$ is an estimate of the total squared error E_m.

$\hat{E}^f_{m-1}$ Least-squares estimate of the forward errors

$\hat{E}^b_{m-1}$ Least-squares estimate of the backward errors.

An estimate of total squared error (E_m) is given by

$$\hat{E}^f_{m-1} + \hat{E}^b_{m-1} = \hat{E}_m$$

$$\hat{E}_m = \left[1 - \left|\hat{K} - 1\right|^2\right]\hat{E}_{m-1} - \left|f_{m-1}(m-1)\right|^2 - \left|g_{m-1}(m-2)\right|^2 \qquad (11.142)$$

This result is due to Anderson.

In Burg's algorithm, with the estimates of AR parameters, the power spectrum estimate becomes

$$P_{ss}^{BU}(f) = \frac{\hat{E}_p}{\left|1 + \sum_{k=1}^{p} \hat{A}_p(k) e^{-j2\pi fk}\right|^2}$$

Advantages of Burg's Method

Burg's method has the following advantages:

1. high-frequency solution,
2. stable AR model, and
3. computationally efficient.

Disadvantages of Burg's Method

This method has the following disadvantages:

1. exhibits spectral line splitting at high SNR_s,
2. sensitive to the initial phase of a sinusoid, and
3. produces spurious peaks in high-order models.

Several modifications have been proposed to overcome some of the more important limitations of the Burg method: namely the line splitting, spurious peaks, and frequency bias. Basically, the modifications involve the introduction of a weighing (window) sequence on the squared forward and backward errors. That is, the least-squares optimization is performed on the weighted squared errors:

$$\varepsilon_m^{WB} = \sum_{n=m}^{N-1} \omega_m(n)\left[\left|f_m(n)\right|^2 + \left|g_m(n)\right|^2\right] \tag{11.143}$$

upon minimization, we obtain the reflection coefficient estimates

$$\hat{K}_m = \frac{\sum_{n=m}^{N-1} \omega_{m-1}(n)\left[f_{m-1}(n)g_{m-1}^*(n-1)\right]}{\frac{1}{2}\sum_{n=m}^{N-1} w_{m-1}(n)\left[\left|f_{m-1}(n)^2\right| + \left|g_{m-1}(n)^2\right|\right]} \tag{11.144}$$

These windowing and energy weighing methods have proved effective in reducing the occurrence of line splitting and spurious peaks, and are also effective in reducing frequency bias.

11.6.4 Unconstrained Least-Squares Method for the AR Model Parameters

Burg's method for determining the parameters of the AR model is basically a least-squares lattice algorithm with the added constraint that the predictor coefficients satisfy the Levinson recursion.

As a result of this constraint, an increase in the order of the AR model requires only a single parameter optimization at each stage. In contrast to this approach, we may use an unconstrained least-squares algorithm to determine the AR parameters.

Minimize the sum of forward and backward errors as

$$\begin{aligned}\varepsilon_p &= \sum_{n=p}^{N-1}\left[\left|f_p(n)\right|^2 + \left|g_p(n)\right|^2\right] \\ &= \sum_{n=p}^{N-1}\left[\left|s(n) + \sum_{k=1}^{p} A_p(k)s(n-k)\right|^2 \left|s(n-p) + \sum_{k=1}^{p} A_p^*(k)s(n+k-p)\right|^2\right]\end{aligned} \tag{11.145}$$

Without using the Levinson–Durbin constraint for AR parameters, the minimization of ε_p with respect to the prediction coefficients results in the set of linear equations

$$\sum_{k=1}^{p} A_p(k) r_{ss}(l,k) = r_{ss}(l,0), l = 1,2,\ldots,p \tag{11.146}$$

where, by definition, the autocorrelation $r_{ss}(l, k)$ is given by

$$r_{ss}(l,k) = \sum_{n=p}^{N-1}\left[s(n-k)s^*(n-l) + s(n-p+l)s^*(n-p+k)\right] \quad (11.147)$$

The resulting residual least-squares error is given by

$$\varepsilon_P^{LS} = r_{ss}(0,0) + \sum_{k=1}^{p} \hat{A}_p(k) r_{ss}(0,k) \quad (11.148)$$

Hence, the unconstrained least-squares power spectrum estimate is

$$P_{SS}^{LS}(f) = \frac{\varepsilon_P^{LS}}{\left|1 + \sum_{k=1}^{p} \hat{A}_p(k) e^{-j2\pi fk}\right|^2} \quad (11.149)$$

This method can also be called as unwanted data least-squares method. The limitations of Burg's method are not present in this method. Also, computational efficiency is comparable with the Levinson–Durbin algorithm. But stability is not guaranteed.

In the above methods, AR parameter estimates are obtained from a block of data $s(n)$, $n = 0, 1, \ldots, N-1$. If continuous data is not available then segments of data with N-point block is taken for estimation. Another way is using sequential estimation methods using adaptive filtering.

11.6.5 Maximum Likelihood Method

This method is also known as Capon method. The Capon method was initially used for large seismic array of frequency wavenumber estimation.

Later on, this was extended to single time series spectrum estimation by Lucas.

Consider an *FIR* filter, where the filter coefficients to be determined are A_k, where $0 \le k \le p$. Hence, there is no restriction that A_o should be unity.

Let $s(n)$, $0 \le n \le N-1$ be the data sequence that is passed through the FIR filter.

The output filter response is given by

$$y(n) = \sum_{k=0}^{p} A_k s(n-k) = S^t(n) A \quad (11.150)$$

where $\quad S^t(n) = \left[s(n)s(n-1), s(n-2) \ldots s(n-p)\right]$

$\Rightarrow$ Data Vector

$A =$ Filter Coefficient Vector

Let us assume that the mean of the data sequence is zero, that is, $E[s(n)] = 0$.

The variance of $y(n)$ is given by

$$\sigma y^2 = Var[y(n)] = E[|y(n)|^2]$$

$$= E\left[A^{*t}S^*(n)S^t(n)A\right] = A^*\gamma_{ss}A \qquad (11.151)$$

where r_{ss} = Autocorrelation matrix of $s(n)$ with elements $y_{ss}(m)$.

Choose the filter coefficients such that the frequency response is normalized to unity at a particular selected frequency f_l.

The constraint is given by

$$\sum_{k=0}^{p} A_k e^{-j2\pi k f_1} = 1$$

In matrix form $E^*t(f_l)\ A = 1$

where $$E^t(f_l) = \left[2e^{j2\pi f_l} \ldots .e^{j2\pi f_l}\right]$$

Minimize the variance σy^2 subjected to the constraint specified, yields an FIR filter which allows the f_l frequency components undistorted. Other frequency components are alternated.

This yields,

$$\tilde{A} = \frac{\gamma_{ss}^1 E^*(f_l)}{E^t(f_l)\gamma_{ss}^{-1}E^*(f_l)}$$

The variance becomes

$$\sigma_{\min}^2 = \frac{1}{E^t(f_l)\gamma_{ss}^{-1}E^*(f_l)} \qquad (11.152)$$

The minimum variance power spectrum estimate at frequency f_l is represented in the above equation. By varying the frequency f_l from 0 to 0.5, the power spectrum estimate can be obtained. Even if f_l changes, γ_{ss}^{-1} is computed only once. The denominator of $\sigma_{\min}^2$ can be computed using single DFT. If R_{ss} is the estimate of γ_{ss}, R_{ss} can replace γ_{ss} and the minimum variance power spectrum estimate of capon's method is given by

$$P_{ss}^{mv}(f) = \frac{1}{E^t(f)R_{ss}^{-1}E^*(f)} \qquad (11.153)$$

This estimate results in spectrum peaks estimate proportional to the power at that frequency.

11.6.6 The Pisarenko Harmonic Decomposition Method

This method provides the estimate for the signal components which are sinusoids corrupted by additive white noise.

A real sinusoid signal can be obtained from the difference equation

$$s(n) = -A_1 s(n-1) - A_2 s(n-2) \tag{11.154}$$

where
$$A_1 = 2\cos 2\pi f_k$$
$$A_2 = 1$$

Initial conditions,

$$s(-1) = -1$$
$$s(-2) = 0$$

This system has complex-conjugate poles at $f = f_k$ and $f = -f_k$, obtaining the sinusoid

$$s(n) = \cos 2\pi f_k n,\ n \geq 0$$

Consider p sinusoid components available in the signal,

$$s(n) = -\sum_{m=1}^{2p} A_m s(n-m) \tag{11.155}$$

The system function is given by

$$H(z) = \frac{1}{1 + \sum_{m=1}^{2p} A_m z^{-m}} \tag{11.156}$$

The denominator of $H(z)$ has $2p$ roots on the unit-circle. They correspond to the sinusoid frequencies. Assume that the sinusoids are corrupted by an additive white noise $\omega(n)$.

$$y(n) = s(n) + \omega(n) \tag{11.157}$$

and
$$E\left[|\omega(n)|^2\right] = \sigma\omega^2$$

From Eq. (11.157), we get

$$s(n) = y(n) - \omega(n) \tag{11.158}$$

Substituting the values of $s(n)$ in the difference equation provides,

$$y(n) - \omega(n) = -\sum_{m=1}^{2p} A_m [y(n-m) - \omega(n-m)]$$

or
$$\sum_{m=0}^{2p} A_m y(n-m) = \sum_{m=0}^{2p} A_m \omega(n-m) \tag{11.159}$$

Let $A_o = 1$

The difference equation for *ARMA* (*p*, *p*) process has both *AR* and *MA* parameters identical.

In matrix form,

$$Y^t A = W^t A$$

where $Y^t = [y(n)y(n-1)y(n-2)....y(n-2p)]$ - Observed data vector

$$W^t = [\omega(n)\omega(n-1)\omega(n-2)....\omega(n-2p)] \text{ - Noise vector}$$

$$A = [1A_1A_2A_3....A_{2p}] \text{ - Coefficient vector}$$

Premultiply by Y and take expected value,

$$E\left[Y\,Y^t\right]A = E\left[Y\,W^t\right]A = E\left[(X+W)W^t\right]A$$

$$\gamma_{yy}A = \sigma_\omega^2 A \tag{11.160}$$

This expression is obtained with the following assumptions:

1. $\omega(n)$ is zero mean white noise
2. X is a deterministic signal.

In Eigen equation form

$$\left(\gamma_{yy} - \sigma_\omega^2 I\right)A = 0$$

Where σ_ω^2 is the eign value of the autocorrelation matrix γ_{yy}.
This forms the basis for the decomposition method.
The autocorrelation values are given by

$$\gamma_{yy}(o) = \sigma_\omega^2 + \sum_{i=1}^{p} P_i$$

$$\gamma_{yy} - (k) = \sum_{i=1}^{p} P_i \cos 2\pi f_i k,\; k \neq 0 \tag{11.161}$$

where $P_i \dfrac{A_i^2}{2} \rightarrow$ Average power in the ith sinusoid.

$$\begin{bmatrix} \cos 2\pi f_1 & \cos 2\pi f_2 & \cos 2\pi f_3 & \cos 2\pi f_p \\ \cos 4\pi f_1 & \cos 4\pi f_2 & \cos 4\pi f_3 & \cos 4\pi f_p \\ \cos 6\pi f_1 & \cos 6\pi f_2 & \cos 6\pi f_3 & \cos 6\pi f_p \\ \vdots & \vdots & \vdots & \vdots \\ \vdots & \vdots & \vdots & \vdots \\ \vdots & \vdots & \vdots & \vdots \\ \cos 2\pi p f_1 & \cos 2\pi p f_2 & \cos 2\pi p f_3 & \cos 2\pi p f_p \end{bmatrix} \begin{bmatrix} P_1 \\ P_2 \\ P_4 \\ \vdots \\ \vdots \\ \vdots \\ P_P \end{bmatrix} \begin{bmatrix} \gamma_{yy}(1) \\ \gamma_{yy}(2) \\ \gamma_{yy}(2) \\ \vdots \\ \vdots \\ \vdots \\ \gamma_{yy}(p) \end{bmatrix} \tag{11.162}$$

The powers of sinusoids can be obtained if the frequencies are known.
$\gamma_{ss}(m)$ can be replaced by its estimate $r_{ss}(m)$.
The noise variance is obtained from,

$$\sigma_\omega^2 = r_{yy}(o) - \sum_{i=1}^{p} P_i \tag{11.163}$$

EXAMPLE 11.3

Show that

a. $$E\left[P_{ss}(f_1)P_{ss}(f_2)\right] = \sigma_s^2\left\{1 + \left[\frac{\sin\pi(f_1+f_2)N}{N\sin\pi(f_1+f_2)}\right]^2 + \left[\frac{\sin\pi(f_1-f_2)N}{N\sin\pi(f_1-f_2)}\right]\right\}$$

b. $$Var\left[P_{SS}(f)\right] = \sigma_s^2\left\{1 + \left[\frac{\sin 2\pi\, f\, N}{N\sin 2\pi f}\right]^2\right\}$$

Use the expression for the 4th joint moment for Gaussian random variable.

$E[S_1\, S_2\, S_3\, S_4] = E\,[S_1\, S_2]\, E\,[S_3\, S_4] + E\,[S_1\, S_3]\, E\,[S_2\, S_4] + E\,[S_1\, S_4]\, E\,[S_2\, S_3]$.

Solution:

a. $$E\left[P_{SS}(f_1)P_{SS}(f_2)\right] = E\left[\frac{1}{N^2}\left|\sum_{n_1=0}^{N-1} s(n_1)e^{-j2\pi f_1 n_1}\right|^2 \cdot \left|\sum_{n_2=0}^{N-1} s(n_2)e^{-j2\pi f_2 n_2}\right|^2\right]$$

$$= \frac{1}{N^2}\sum_{n_1=0}^{N-1}\sum_{n_2=0}^{N-1}\sum_{n_3=0}^{N-1}\sum_{n_4=0}^{N-1} E\left[s(n_1)s(n_2)s(n_3)s(n_4)e^{-j2\pi f_1(n_1-n_3)}e^{-j2\pi f_2(n_2-n_4)}\right] \tag{1}$$

But, $$E\left[S(n_1)S(n_2)S(n_3)S(n_4)\right] = E\left[S(n_1)S(n_2)\right]E\left[S(n_3)S(n_4)\right] + E\left[s(n_1)s(n_3)\right]E\left[s(n_2)s(n_4)\right] + E\left[s(n_1)s(n_4)\right]E\left[s(n_2)s(n_3)\right]$$

$$= \sigma_S^4, \text{ for } n_1 = n_2 \text{ and } n_3 = n_4;$$
$$n_1 = n_3 \text{ and } n_2 = n_4;$$
$$n_1 = n_4 \text{ and } n_2 = n_3, \tag{2}$$

For $n_1 = n_4$ and $n_2 = n_3$,

$$e^{-j2\pi f_1(n_1-n_3)}e^{-j2\pi f_2(n_2-n_4)} = e^{-j2\pi f_1(n_1-n_3)}e^{-j2\pi f_2(n_3-n_1)}$$
$$= e^{-j2\pi f_1(n_1-n_3)}e^{+j2\pi f_2(n_1-n_3)}$$
$$= e^{-j2\pi(f_1-f_2)(n_1-n_3)} \tag{3}$$

For $n_1 = n_2$ and $n_3 = n_4$

$$e^{-j2\pi f_1(n_1-n_3)}e^{-j2\pi f_2(n_2-n_4)} = e^{-j2\pi f_1(n_1-n_3)}e^{-j2\pi f_2(n_1-n_3)}$$
$$= e^{-j2\pi(f_1-f_2)(n_1+n_3)} \tag{4}$$

Substituting Eq. (2) in Eq. (1), we get

$E[P_{SS}(f_1)\, P_{SS}(f_2)]$

$$=\frac{1}{N^2}\sum_{n_1=0}^{N-1}\sum_{n_2=0}^{N-1}\sum_{n_3=0}^{N-1}\sum_{n_4=0}^{N-1}\{E[s(n_1)s(n_2)]E[s(n_3)s(n_4)]$$

$$+E[s(n_1)s(n_3)E[s(n_2)s(n_4)]+E(s(n_1)s(n_4)]E[s(n_2)s(n_3)]\}$$

$$\times e^{j2\pi f_1(n_1-n_3)}e^{j2\pi f_2(n_2-n_4)}\Big]$$

$$=\frac{1}{N^2}\left\{\sum_{n_1=0}^{N-1}\sum_{n_2=0}^{N-1}\sigma_s^4+\sum_{n_1=0}^{N-1}\sum_{n_3=0}^{N-1}\sigma_s^4 e^{-j2\pi(f_1-f_2)(n_1-n_3)}+\sum_{n_1=0}^{N-1}\sum_{n_2=0}^{N-1}\sigma_s^4 e^{-j2\pi(f_1+f_2)(n_1-n_3)}\right\}$$

$$=\frac{\sigma_S^4}{N^2}\left\{N^2+\sum_{n_1=0}^{N-1}e^{-j2\pi(f_1-f_2)n_1}\sum_{n_3=0}^{N-1}e^{j2\pi(f_1-f_2)n_3}+\sum_{n_1=0}^{N-1}e^{-j2\pi(f_1+f_2)n_1}\sum_{n_3=0}^{N-1}e^{j2\pi(f_1+f_2)n_3}\right\}$$

$$=\sigma_S^r\left\{1+\frac{1-e^{-j2\pi(f_1-f_2)N}}{1-e^{-j2\pi(f_1-f_2)N}}\times\frac{1-e^{j2\pi(f_1-f_2)N}}{1-e^{j2\pi(f_1-f_2)N}}\times\frac{1}{N^2}+\frac{1-e^{-j2\pi(f_1+f_2)N}}{1-e^{-j2\pi(f_1+f_2)N}}\times\frac{1-e^{j2\pi(f_1+f_2)N}}{1-e^{j2\pi(f_1+f_2)}}\times\frac{1}{N^2}\right\}$$

$$=\sigma_S^4\left\{1+\frac{2-e^{-j2\pi(f_1-f_2)N}-e^{j2\pi(f_1-f_2)N}}{2-e^{-j2\pi(f_1-f_2)}-e^{j2\pi(f_1-f_2)}}\times\frac{1}{N^2}+\frac{2-e^{-j2\pi(f_1-f_2)N}-e^{j2\pi(f_1-f_2)N}}{2-e^{-j2\pi(f_1-f_2)N}-e^{j2\pi(f_1-f_2)}}\times\frac{1}{N^2}\right\}$$

$$=\sigma_S^4\left\{1+\frac{1}{N^2}\left[\frac{2-2\cos 2\pi(f_1-f_2)N}{2-2\cos 2\pi(f_1-f_2)}\right]+\frac{1}{N^2}\left[\frac{2-2\cos 2\pi(f_1-f_2)N}{2-2\cos 2\pi(f_1-f_2)}\right]\right\}$$

$$=\sigma_S^4\left\{1+\left[\frac{\sin\pi(f_1-f_2)N}{N\sin\pi(f_1-f_2)}\right]^2\left[\frac{\sin\pi(f_1-f_2)N}{N\sin\pi(f_1-f_2)}\right]^2\right.\quad\textbf{Proved.}$$

b. $Var[P_{SS}(f)]=E[P_{SS}^2(f)]-\{E[P_{SS}^2(f)]\}^2$

But, $E[P_{ss}(f)]=\sigma_s^2$

$$E[P_{SS}^2(f)]=\frac{1}{N^2}\sum_{n_1=0}^{N-1}\sum_{n_2=0}^{N-1}\sum_{n_3=0}^{N-1}\sum_{n_4=0}^{N-1}E[s(n_1)s(n_2)s(n_3)s(n_4)]e^{-j2\pi f(n_1-n_2+n_3-n_4)}$$

and $E[s(n_1)s(n_2)s(n_3)s(n_4)]=E[s(n_1)s(n_2)]E[s(n_3)s(n_4)]$

$+E[s(n_1)s(n_3)E[s(n_2)s(n_4)]]$

$+E[s(n_1)s(n_4)]E[s(n_2)s(n_3)]$

$=\sigma_S^4,\ n_1=n_2$ and $n_3=n_4$;

$n_1=n_3$ and $n_2=n_4$;

$n_1=n_4$ and $n_2=n_3$

Similar to part (a), the summation can be expanded and the value is

$$E\left[P_{SS}^2(f)\right] = \frac{\sigma_S^4}{N^2}\left\{2N^2 + \left[\frac{\sin 2\pi fN}{N\sin 2\pi f}\right]^2\right\}$$

$$Var\left[P_{SS}(f)\right] = \frac{\sigma_S^4}{N^2}\left\{2N^2 + \left[\frac{\sin 2\pi fN}{N\sin 2\pi f}\right]^2\right\} - \sigma S_s^4$$

$$= \sigma_s^4\left\{1 + \left[\frac{\sin 2\pi fN}{N\sin 2\pi f}\right]^2\right\}.$$

EXAMPLE 11.4

Find the frequency resolution of the Bartlett, Welch, and Blackman–Tukey methods of power spectrum estimation for quality factor $Q = 10$. Assume that overlap in Welch's method is 50%. Given the length of the sample, the sequence is 1000.

Solution:

Data given:

Quality factor, $Q = 10$
Length of the Sample Sequence, $N = 1000$
Overlap in Welch's method, $= 50\%$

Bartlett method

$$\text{Quality factor } Q_B = 1.11\, N\, \Delta f$$

$$\Rightarrow \qquad \Delta f = \text{frequency resolution} = \frac{Q_B}{1.11\, N}$$

$$= \frac{10}{1.11 \times 1000} = 0.009$$

Welch Method

$$\text{Quality factor } Q_w = 1.39\, N\, \Delta f$$

$$\Rightarrow \qquad \Delta f = \text{frequency resolution} = \frac{Q_W}{1.39\, N}$$

$$= \frac{10}{1.39 \times 1000} = 0.0072$$

Blackman–Tukey Method

$$Q_{BT} = \text{Quality factor } Q_w = 2.34\, N\, \Delta f$$

$$\Rightarrow \qquad \Delta f = \text{frequency resolution} = \frac{Q_{BT}}{2.34\, N}$$

$$= \frac{10}{2.34 \times 1000} = 0.0042$$

EXAMPLE 11.5

We have given an auto-regressive (AR) process of order 1. This process is described by

$$s(n) = A\, s(n-1) + \omega(n)$$

where A is a constant and $\omega(n)$ is a white noise process of zero mean and variance σ^2. Find the mean and autocorrelation function of the process $\{S(n)\}$.

Solution:

Given,
$$s(n) = A\, s(n-1) + \omega(n)$$

or
$$s(n) - A\, s(n-1) = \omega(n) \tag{1}$$

This is a difference equation with constant coefficients. The complementary solution and particular solution can be obtained.

Complementary solution of Eq. (1)

Consider the homogeneous equation

$$s(n) - A\, s(n-1) = 0 \tag{2}$$

The solution of Eq. (2) is of the form $C(A)n$, where C is a constant.

A particular solution of Eq. (1)

Substituting $s(n-1) = Z^{-1}\, s(n)$ in Eq. (1), we get

$$s(n) - A.Z^{-1}s(n) = \omega(n)$$

or
$$s(n) = \frac{1}{1 - AZ^{-1}}\omega(n)$$

or
$$s(n) = \left[\sum_{k=0}^{\infty} A^k Z^{-k}\right]\omega(n) \tag{3}$$

Since
$$\frac{1}{1 - AZ^{-1}} = 1 + AZ^{-1} + A^2Z^{-2} + A^3Z^{-3} + \ldots$$

$$= \sum_{k=0}^{\infty} A^k Z^{-k} \tag{4}$$

From Eq. (3),

$$s(n) = \sum_{k=0}^{\infty} A^k Z^{-k} \omega(n)$$

$$= \sum_{k=0}^{\infty} A^k \omega(n-k) \tag{5}$$

Since $\qquad Z^{-k}\omega(n) = \omega(n-k)$

Complete solution of Eq. (1) can be obtained by adding both solutions

$$s(n) = CA^n + \sum_{k=0}^{\infty} A^k \omega(n-k) \tag{6}$$

Assuming $\qquad s(o) = 0.$

$$C = -\sum_{k=0}^{\infty} A^k \omega(-k) \tag{7}$$

Substituting the value of C from Eq. (7) in Eq. (6), we get

$$s(n) = A^n \sum_{k=0}^{\infty} A^k \omega(-k) + \sum_{k=0}^{\infty} A^k \omega(n-k)$$

$$= -\sum_{k=n}^{\infty} A^k \omega(n-k) + \sum_{k=0}^{\infty} A^k \omega(n-k)$$

or

$$s(n) = A^n \sum_{k=0}^{\infty} A^k \omega(-k) + \sum_{k=0}^{\infty} A^k \omega(n-k) \tag{8}$$

Mean:

$$E[\omega(n)] = 0$$

$$E[s(n)] = E\left[\sum_{k=0}^{n-1} A^k \omega(n-k)\right]$$

$$= \sum_{k=0}^{n-1} A^k E[\omega(n-k)] = 0 \text{ for all } n.$$

Autocorrelation:

$$r_{ss}(l) = E[s(n)s(n-l)]$$

$$= E\left[\sum_{k=0}^{n-1} A^k \omega(n-k) \sum_{i=0}^{n-1} A^i \omega(n-l-i)\right]$$

$$= E\left[\sum_{k=0}^{n-1} \sum_{i=0}^{n-1} A^{k+i} \omega(n-k)\omega(n-l-i)\right]$$

$$= \sum_{k=0}^{n-1} \sum_{i=0}^{n-1} A^{k+i} E[\omega(n-k)\omega(n-l-i) \text{ correct it} \tag{9}$$

$$E[\omega(n-k)\omega(n-l-i)] = \begin{cases} \sigma^2, & k = l+i \\ O, & k \neq l+i \end{cases} \quad (10)$$

Since $\{\omega(n)\}$ is a white noise process.

Substituting Eq. (10) in Eq. (9), we get

$$r_{ss}(l) = \sum_{k=0}^{n-1}\sum_{i=0}^{n-1} A^{k+i}\sigma^2 \text{ for } k = l+i$$

$$r_{ss}(l) = \sigma^2 A^{-l}\left\{\frac{1-A^{2n}}{1-A^2}\right\}$$

EXAMPLE 11.6

A second-order *AR* process is described by

$$s(n) + A_1 s(n-1) + A_2 s(n-2) = \omega(n)$$

where $\{\omega(n)\}$ is a white noise process of zero mean and variance σ^2. Find the conditions required for this *AR* process to be asymptotically stationary up to order 2.

Solution:

Given $\quad s(n) + A_1 s(n-1) + A_2 s(n-2) = \omega(n) \quad (1)$

Substituting $s(n-1) = Z^{-1}s(n)$ and $s(n-2) = Z^{-2}s(n)$ in Eq. (1), we get

$$s(n) + A_1 Z^{-1}s(n) + A_2 Z^{-2}s(n) = \omega(n)$$

or

$$s(n)\left[1 + A_1 Z^{-1} + A_2 Z^{-2}\right] = \omega(n) \quad (2)$$

Let

$$A(z) = 1 + Z_1 Z^{-1} + A_2 Z^{-1}$$

$$= \left(1 - r_1 Z^{-1}\right)\left(1 - r_2 Z^{-1}\right)$$

where r_1 and r_2 are roots of $A(z)$.

A particular solution of Eq. (1):

$$s(n) = \frac{1}{\left(1 - r_1 z^{-1}\right)\left(1 - r_2 Z^{-1}\right)}\omega(n)$$

$$= \frac{1}{r_1 - r_2}\left[\frac{r_1}{1 - r_z z^{-1}} - \frac{r_2}{1 - r_2 z^{-1}}\right]\omega(n)$$

$$= \frac{1}{r_1 - r_2}\left[\sum_{k=0}^{\infty} r_1^{k+1} z^{-k} - \sum_{k=0}^{\infty} r_2^{k+1} z^{-k}\right]\omega(n)$$

$$= \frac{1}{r_1 - r_2}\left[\sum_{k=0}^{\infty} r_1^{k+1} z^{-k}\omega(n) - \sum_{k=0}^{\infty} r_2^{k+1} z^{-k}\omega(n)\right]$$

$$= \frac{1}{r_1 - r_2}\left[\sum_{k=0}^{\infty} r_1^{k+1}\omega(n-k) - \sum_{k=0}^{\infty} r_2^{k+1}\omega(n-k)\right]$$

$$= \sum_{k=0}^{\infty}\left[\frac{r_1^{k+1} - r_2^{k+1}}{r_1 - r_2}\right]\omega(n-k) \qquad (3)$$

Complementary solution of Eq. (1):
The solution of the homogeneous equation

$$s(n) + A_1 s(n-1) + A_2 s(n-2) = 0 \qquad (4)$$

is of the form

$$C_1 r_1^n + C_2 r_2^n$$

where C_1 and C_2 are constants.

The general solution of Eq. (1) is the sum of complementary solution and particular solution.

$$s(n) = C_1 r_1^n + C_2 r_2^n + \sum_{k=0}^{\infty}\left[\frac{r_1^{k+1} - r_2^{k+1}}{r_1 - r_2}\right]\omega(n-k) \qquad (5)$$

For asymptotically stationary process $\left(C_1 r_1^n + C_2 r_2^n\right)$ must decay to zero as $n \to \infty$.

Hence, $|r_1| < 1$ and $|r_2| < 1$.

EXERCISES

1. What is the need for spectral estimation?
2. How can the energy density spectrum be determined?
3. Define autocorrelation function.
4. Give the relationship between autocorrelation function and spectral density.
5. Give the estimate of autocorrelation function and power density for random signals.
6. Find the expression for mean and variance for the autocorrelation function of random signal.
7. Give the time and frequency domain representation for Bartlett Window.
8. What is periodogram?
9. Explain how DFT and FFT algorithms are useful in power spectral estimation.

10. Discuss power spectrum estimation using the Bartlett method.
11. Obtain the mean and variance for the Welch method of power spectrum estimation.
12. How is the Blackman and Tukey method used in smoothing the periodogram?
13. Derive the mean and variance of the power spectral estimate of the Blackman and Tukey method.
14. Give the limitations of non-parametric methods of power spectrum estimation.
15. How do the parametric methods of power spectrum estimation overcome the limitations of the non-parametric methods?
16. Give the various steps involved in the parametric estimation process.
17. Describe the following models:
    a. AR model
    b. MA model
    c. ARMA model.
18. Why is the AR model widely used?
19. Give the relationship between input and output power spectral density of a linear system.
20. Give the expression for cross-correlation.
21. Give the expression for power spectrum estimates of AR, MA, and ARMA models.
22. Derive the power spectrum estimates using the Burg method.

NUMERICAL EXERCISES

1. The discrete-time sequence is given by

   $s(n) = \sin 2\pi(0.12)n + \cos 2\pi(0.122)n,\ n = 0, 1, \ldots, 15$

   Find the power spectrum

   $P(f) = \frac{1}{N}|S(f)|^2$ at the frequencies $f_k = \frac{k}{L}$, 0, 1, 2, ..., $L - 1$, for $L = 16$, 32, 64, and 128.

2. In the Welch method, find the variance of the Welch power spectrum estimate with the Bartlett window if there is a 50% overlap between successive sequences.

3. Using the 4th joint moment for Gaussian random variables, show that

$$\text{Covar}\left[P_{SS}(f_1)P_{SS}(f_2)\right] = \sigma_S^4\left\{\left[\frac{\sin\pi(f_1+f_2)N}{N\sin\pi(f_1+f_2)}\right]+\left[\frac{\sin\pi(f_1-f_2)N}{N\sin\pi(f_2-f_2)}\right]\right\}$$

Under the condition that the sequence $s(n)$ is a zero-mean white Gaussian noise sequence with variance $\left(\sigma_s^2\right)$.

4. For the *AR* process of order 2,

$$s(n) = A_1 s(n-1) + A_2 s(n-2) + \omega(n)$$

where A_1 and A_2 are constants. $\omega(n)$ is a white process of zero mean and variance σ^2. Determine the mean and autocorrelation of $\{s(n)\}$.

5. Find the mean and the autocorrelation of the sequence $s(n)$ generated by the *MA* process described by the difference equation.

$$s(n) = \omega(n) - A\omega(n-1) + B\omega(n-2)$$

where $w(n) \Rightarrow$ White noise process with variance σ^2.

6. Find the power spectra for the random process generated by

$$s(n) = \omega(n) - s(n-2)$$

where $w(n) \Rightarrow$ White noise process with variance σ^2.

7. Suppose we have $N = 500$ samples from a sample sequence of a random process

a. Determine the frequency resolution Δf of the following:

i. Bartlett method,

ii. Welch method,

iii. Blackman–Tukey method.

Overlapping is 50%.

Quality factor $Q = 12$.

b. Find the record length (M) for the Bartlett, Welch (50% overlap), and Blackman–Tukey methods.

CHAPTER 12

MULTIRATE DIGITAL SIGNAL PROCESSING (MDSP)

12.1 INTRODUCTION

There are many applications where the signal of a given sampling rate (sampling frequency) requires to be converted into an equivalent signal with a different sampling rate. For example, there are three sampling rates that are used in digital audio, 32 kHz in broadcasting, 44.1 kHz in digital compact disk, and 48 kHz in digital audio tape. Conversion of sampling rates of audio signals between these three different sampling rates is often necessary under many circumstances.

In telecommunication systems that transmit and receive different types of signals such as speech, video, teletype, facsimile, etc., there is a requirement to process the various signals at different sampling rates commensurate with the corresponding bandwidths of signals.

In digital video, the sampling rates for composite video signals are 14.3181818 MHz and 17.734475 MHz for NTSC and PAL TV systems, respectively. But the sampling rates for the digital component of video signals are 13.5 MHz and 6.75 MHz for luminance and color difference signals, respectively.

The process of converting or alternating a signal from a given sampling rate to a different sampling rate is called sampling rate conversion. The systems which employ multisampling rates in the processing of digital signals are called multirate digital signal processing (MDSP) systems.

Different sampling rates can be obtained using an upsampler and downsampler. The basic operations in MDSP to achieve this are decimation and interpolation.

Decimation is for reducing the sampling rate and interpolation is for increasing the sampling rate. In digital transmission systems like teletype, facsimile, low-bit-rate speech where data has to be handled at different rates, MDSP is used. MDSP finds its applications in (*i*) subband coding for speech and image, (*ii*) voice privacy using analog phone lines, (*iii*) signal compression by subsampling, (*iv*) A/D and D/A converters, etc.

There are various areas in which MDSP is used. Some of few are given as under:

1. radar systems,
2. antenna systems,
3. speech and audio processing systems, and
4. communication systems.

Advantages of using MDSP

1. computational requirements are less,
2. storage for filter coefficients is less,
3. finite arithmetic effects are less,
4. filter order required in multirate application is low, and
5. sensitivity to filter coefficient lengths is less.

12.1.1 Sampling Rate Conversion Methods

There are two sampling rate conversion methods that are used in MDSP:

First Method: In this method, digital signal is passed through D/A converter then it is filtered if necessary, and then it is resampled at the desired sampling rate. The resampling of analog signal is performed by using an A/D converter.

Figure 12.1 shows the block diagram of sampling rate conversion using D/A converter and A/D converter. This block diagram comprises three blocks. These blocks are:

D/A Converter, Linear Filter, and A/D Converter.

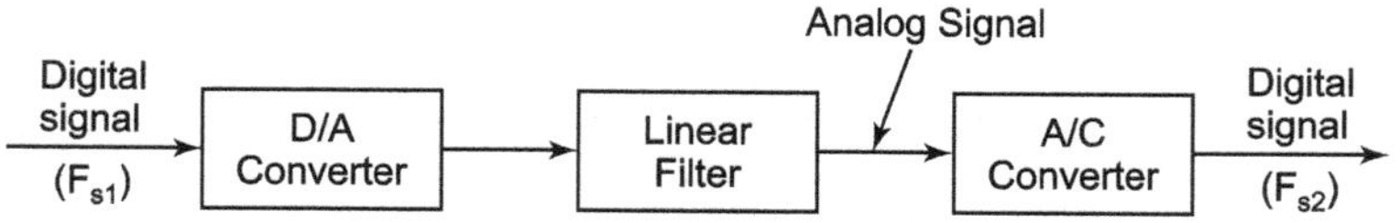

FIGURE 12.1 Sampling rate conversion using D/A converter and A/D converter.

Second Method: In this method, sampling rate conversion is performed entirely in the digital-domain. This method does not require any D/A and A/D converters. This method uses interpolator, or decimator or both depending upon the sampling rate conversion factor. Interpolator and decimator will be discussed in the next section.

The advantages of the first method are that the new sampling rate can be arbitrarily selected and this new sampling rate has no special relationship with the old sampling rate.

Major disadvantages of the first method are that there is a signal distortion introduced by the D/A converter in the signal reconstruction and by the quantization noise in the A/D conversion.

Sampling rate conversion performed in the digital domain avoids the major disadvantages of the first method.

12.2 SAMPLING RATE CONVERSION

There are two cases of sampling rate conversion.

Decimation: The process of reducing the sampling rate by an integer factor (D) is called the decimation of the sampling rate. It is also called downsampling by factor (D). Figure 12.2 shows the block diagram of a decimator.

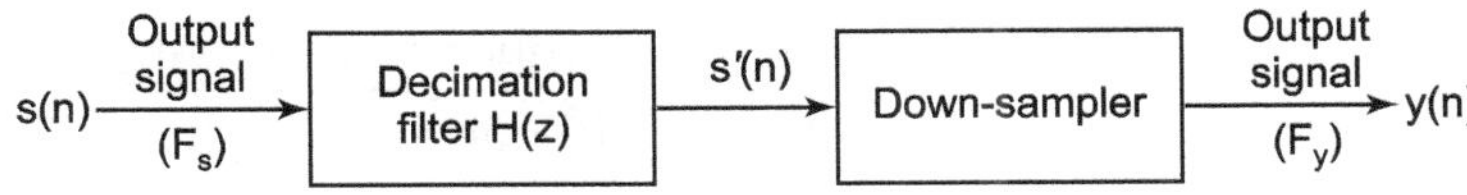

FIGURE 12.2 Block diagram of a decimator.

This decimator comprises two blocks such as decimation filter and downsampler. Decimation filter is used to band limit the signal before decimation operation. Downsampler decreases the sampling rate by an integer factor (D).

Decimation filter is used to avoid aliasing caused by downsampling the signal $s(n)$.

Prior to downsampling the signal $s'(n)$ should be band limited to $|\omega| < \dfrac{\pi}{M}$ by means of a low pass filter (LPF), $H(z)$, called the decimation filter.

Interpolation: The process of increasing the sampling rate by an integer factor (I) is called interpolation of the sampling rate. It is also called upsampling by factor (I).

Figure 12.3 shows the block diagram of an interpolator. The interpolator comprises two blocks such as upsampler and interpolation filter. Here upsampler is used to increase the sampling rate by an integer (I) and interpolation filter removes the unwanted images that are yielded by upsampling.

The unwanted images are removed by using a LPF, $H(z)$ called the interpolation filter.

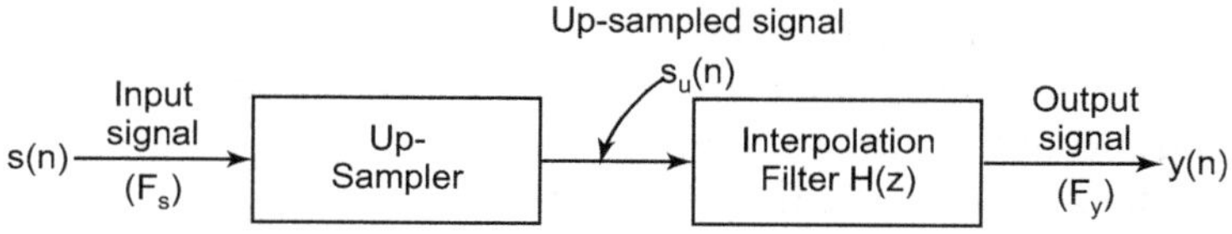

FIGURE 12.3 Block diagram of an interpolator.

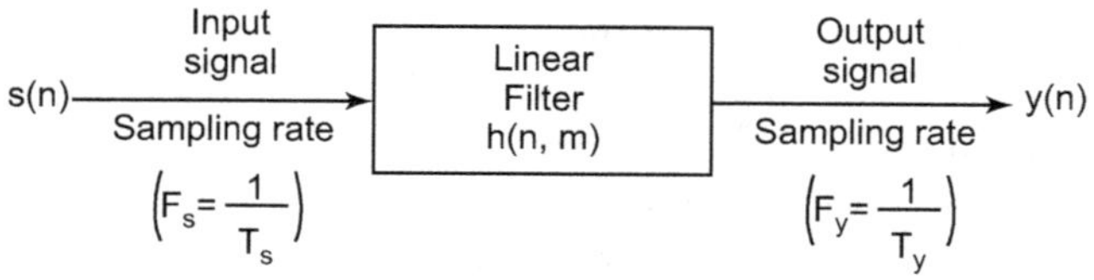

FIGURE 12.4 Linear filter.

The process of sampling rate conversion in the digital domain can be viewed as a linear filtering operation. It is shown in Figure 12.4

The input signal $s(n)$ is characterized by the sampling rate $F_s = \dfrac{1}{T_s}$ and the output signal $y(m)$ is characterized by the sampling rate $F_y = \dfrac{1}{T_y}$. Here T_s and T_y are corresponding sampling intervals:

$$\frac{F_y}{F_s} = \frac{\text{Sampling frequency of output signal}}{\text{Sampling frequency of input signal}}$$

$$= \frac{I}{D} = \frac{\text{Prime Integer } (I)}{\text{Prime Integer } (D)}$$

where I is the integer factor by which interpolation of sampling rate is performed and the D is the integer factor by which decimation of the sampling rate is performed.

But for the case of ratio $\dfrac{I}{D}$, both I and D should be prime integer. Linear filter is characterized by a time-varying impulse response, $h(n, m)$.

Hence the input $s(n)$ and output $y(n)$ are related by convolution sum for the time-varying system.

The sampling rate conversion process can also be performed by digital resampling of the same analog signal. Let us have an analog signal $s(t)$ which is sampled at first rate F_s to generate $s(n)$. The purpose of sampling rate conversion is to obtain another sequence $y(m)$ directly from $s(n)$, which is equal to the sampled values of $s(t)$ at second sampling rate F_y, $y(m)$ *is* shifted version of $s(n)$.

Time shift in sampling rate conversion can be realized by using a linear filter which has a flat magnitude response and a linear phase response. If both the sampling rates are not equal, then the required amount of time-shifting will vary from sample to sample as shown in Figure 12.5.

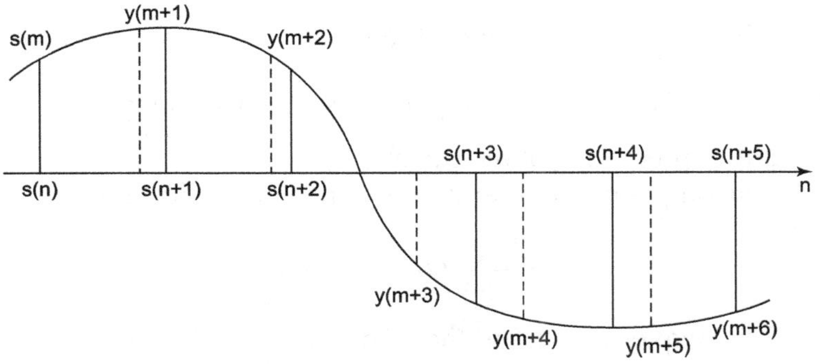

FIGURE 12.5 Sampling Rate conversion viewed as linear filtering process.

12.2.1 Decimation of Sampling Rate by an Integer Factor (D)

The process of reducing the sampling rate of a signal is called decimation. Let us assume that the discrete-time signal $s(n)$ with spectrum $S(\omega)$ is to be downsampled by an integer factor D. The spectrum $s(\omega)$ is assumed to be non-zero in the frequency interval $0 \le |\omega| \le \pi$ it or equivalently, $F''\frac{F_s}{2}$. We know that if we reduce the sampling rate simply by selecting every Dth value of signal $s(n)$, the resulting signal will be an aliased version of $s(n)$ with folding frequency of $\frac{F_s}{2D}$. For avoiding aliasing, we must first reduce the bandwidth of $s(n)$ to $F_{max} = \frac{F_s}{2D}$ or $\omega_{max} = \frac{\pi}{D}$. Then we may down-sample by D and thus avoid aliasing.

The block diagram of the decimation process is given in Figure 12.6.

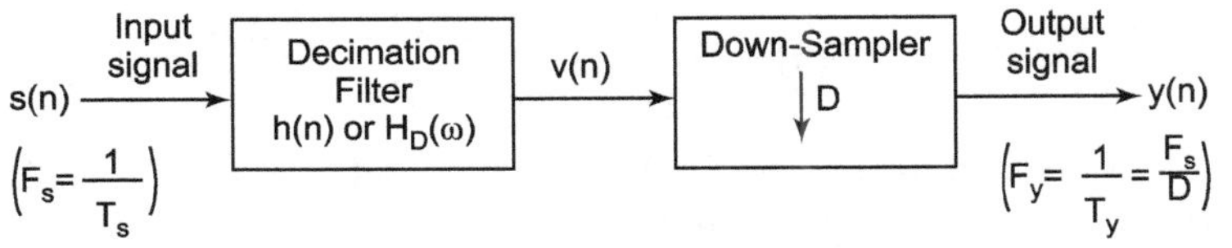

FIGURE 12.6 Block Diagram of decimation process.

The input signal or sequence $s(n)$ is passed through a decimation filter, that is, LPF. This LPF is characterized by the impulse response $h(n)$ and frequency response $HD(\omega)$.

The LPF ideally satisfies the condition

$$H_D(\omega) = \begin{cases} 1, & |\omega| \le \dfrac{\pi}{D} \\ 0, & \text{Otherwise} \end{cases} \tag{12.1}$$

Thus the filter eliminates the spectrum of $S(\omega)$ in the range $\frac{\pi}{D} < \omega < \pi$. Here, only the frequency components of $s(n)$ in the range $\omega| \leq \frac{\pi}{D}$ are of interest in further processing of the signal.

The output of the filter is a sequence $\upsilon(n)$ which is given as

$$\upsilon(n) = \sum_{k=0}^{\infty} h(k)s(n-k) \tag{12.2}$$

The sequence $\upsilon(n)$ is then downsampled by the factor D to produce another sequence $y(m)$. Thus

$$y(m) = \upsilon(mD) \tag{12.3}$$

Substituting Eq. (12.2) in Eq. (12.3), we get

$$y(m) = \sum_{k=0}^{\infty} h(k)s(mD-k) \tag{12.4}$$

Although the filtering operation on $s(n)$ is linear and time-invariant, the downsampling operation in combination with the filtering results in a time-varying or time-variant system. Now we can easily verify it. Given the fact that $s(n)$ produces $y(m)$, we note that $s(n - n_0)$ does not imply $y(n - n_0)$ unless n_o is a multiple of D. Consequently, we can say that the overall linear operation (linear operation followed by downsampling) on sequence $s(n)$ *is* not time-invariant.

The frequency-domain characteristics of the output sequence $y(m)$ can be obtained by relating the spectrum of $y(m)$ to the spectrum of the input sequence $s(n)$.

For convenience, we define a sequence $\tilde{\upsilon}(n)$ as

$$\tilde{\upsilon}(n) = \begin{cases} \upsilon(n), & n-0, \pm D, \pm 2D, \pm 3D, \ldots. \\ 0, & \text{otherwise} \end{cases} \tag{12.5}$$

Clearly, the sequence $\tilde{\upsilon}(n)$ can be viewed as a sequence obtained by multiplying $\upsilon(n)$ with a periodic train of impulses $p(n)$, with period D. It is illustrated in Figure 12.7.

The discrete Fourier series representation of $p(n)$ is given by

$$p(n) = \frac{1}{D}\sum_{k=0}^{D-1} e^{\frac{j2\pi kn}{D}} \tag{12.6}$$

Hence,

$$\tilde{\upsilon}(n) = \upsilon(n)p(n) \tag{12.7}$$

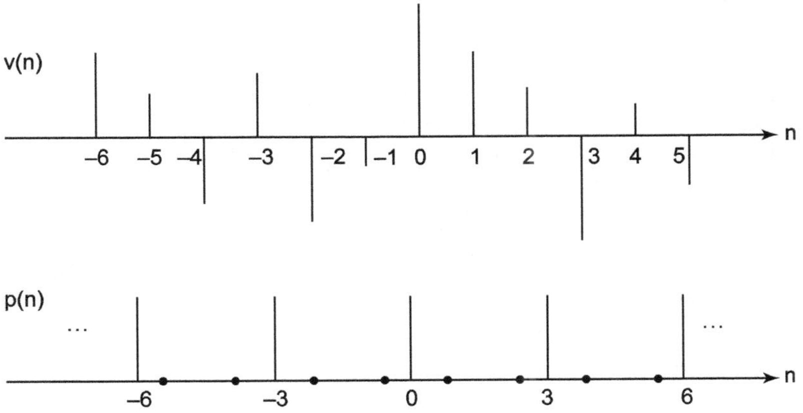

FIGURE 12.7 Multiplication of $v(n)$ and $p(n)$ with period $D = 3$.

and
$$y(m) = \tilde{\upsilon}(mD) = \upsilon(mD)p(mD) = \upsilon(mD) \tag{12.8}$$

Now, taking z-transform of the output sequence $y(m)$.

$$Y(z) = \mathcal{Z}[y(m)] = \sum_{m=-\infty}^{\infty} y(m)z^{-m}$$

$$= \sum_{m=-\infty}^{\infty} \tilde{\upsilon}(mD)z^{-m} \tag{12.9}$$

Eq. (12.9) can be written as

$$Y(z) = \sum_{m=-\infty}^{\infty} \tilde{\upsilon}(m)z^{-\left(\frac{m}{D}\right)} \tag{12.10}$$

Eq. (12.10) follows from the fact that $\tilde{\upsilon}(m) = 0$, except at multiples of D. Using Eqs. (12.6) and (12.7) in Eq. (12.9), we get

$$Y(z) = \sum_{m=-\infty}^{\infty} \upsilon(m)\left[\frac{1}{D}\sum_{k=0}^{D-1} e^{\frac{j2\pi mk}{D}}\right] z^{-\left(\frac{m}{D}\right)}$$

$$= \frac{1}{D}\sum_{k=0}^{D-1}\sum_{m=-\infty}^{\infty} \upsilon(m)\left[e^{\frac{-j2\pi k}{D}} z^{\frac{1}{D}}\right]^{-m} \tag{12.11}$$

Eq. (12.11) can be written as

$$Y(z) = \frac{1}{D}\sum_{k=0}^{D-1} V\left[e^{\frac{-j2\pi k}{D}} z^{\frac{1}{D}}\right] \tag{12.12}$$

where
$$V\left[e^{\frac{-j2\pi k}{D}} z^{\frac{1}{D}}\right] = \sum_{m=-\infty}^{\infty} \upsilon(m)\left[e^{\frac{-j2\pi k}{D}} z^{\frac{1}{D}}\right]^{-m}$$

We know that $V(z) = H_D(z)\, S(z)$

or $$V\left[e^{\frac{-j2\pi k}{D}} z^{\frac{1}{D}}\right] = H_D\left[e^{\frac{-j2\pi k}{D}} z^{\frac{1}{D}}\right] S\left(e^{\frac{-j2\pi k}{D}} z^{\frac{1}{D}}\right) \tag{12.13}$$

Substituting Eq. (12.13) in Eq. (12.12), we get

$$Y(z) = \frac{1}{D}\sum_{D}^{D-1} H_D\left(e^{\frac{-j2\pi k}{D}} z^{\frac{1}{D}}\right) S\left(e^{\frac{-j2\pi k}{D}} z^{\frac{1}{D}}\right) \tag{12.14}$$

By evaluating $Y(z)$ in the unit circle, that is, $|z| = 1$, we obtain the spectrum of the output signal $y(m)$. Since the sampling rate is $F_y = \frac{1}{T_y}$, the frequency variable, which we denote as ω_y, is in radians and relative to the sampling rate F_y,

$$\omega_y = \frac{2\pi F}{F_y} = 2\pi F T_y \tag{12.15}$$

Since, $$F_y = \frac{F_s}{D} \tag{12.16}$$

$$\omega_s = \frac{2\pi F}{F_s} = 2\pi F T_s \tag{12.17}$$

But, ω_y and ω_s are related by

$$w_y = D\omega_s \tag{12.18}$$

Thus, as expected, the frequency range $0 \le |\omega_s| \le \frac{\pi}{D}$ is stretched into the corresponding frequency range $0 \le |\omega_y| \le p$ by the downsampling process.

Now, we conclude that the spectrum $Y(\omega_y)$, which is obtained by evaluating Eq. (12.11) on the unit circle, can be expressed as

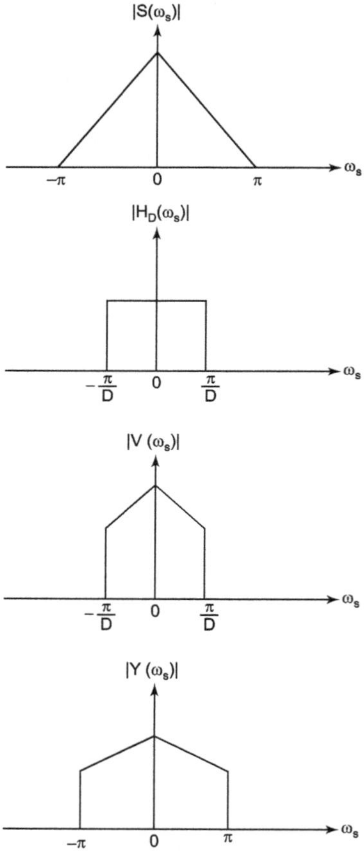

FIGURE 12.8 Spectra of Signals in the decimation of $s(n)$ by a factor D.

$$Y\left(\omega_y\right) = \frac{1}{D}\sum_{k=0}^{D-1} H_D\left[\frac{\omega_y - 2\pi k}{D}\right] S\left[\frac{\omega_y - 2\pi k}{D}\right] \tag{12.19}$$

With a properly designed filter $H_D(\omega)$, the aliasing is eliminated and consequently, all but the first term in Eq. (12.19) vanish.

Hence, $$Y(\omega_y) = \frac{1}{D} H_D\left(\frac{\omega_y}{D}\right) S\left(\frac{\omega_y}{D}\right)$$

$$= \frac{1}{D} S\left(\frac{\omega_y}{D}\right) \quad (12.20)$$

for $0 \le |\omega_y| \le \pi$. The spectra for the sequences (n), $\upsilon(n)$ and $y(m)$ are shown in Figure 12.8.

12.3 INTERPOLATION OF SAMPLING RATE BY A INTEGER FACTOR (*I*)

Increasing the sampling rate of a signal is called interpolation. An increase in the sampling rate by an integer factor I can be accomplished by interpolating $(I - 1)$ new samples between successive values of the signals. The interpolation process can be accomplished by various types of methods. Here we shall describe a process that preserves the spectral shape of the signal sequence $s(n)$.

Let $\upsilon(m)$ is a signal sequence with a sampling rate $F_y = IF_s$. This signal sequence $\upsilon(m)$ is obtained from $s(n)$ by adding $(I - 1)$ zeros between successive values of $s(n)$.

Thus, $$\upsilon(m) = \begin{cases} s\left(\frac{m}{I}\right) m = 0, & \mp I, \pm 2I, \ldots \\ 0, & \text{otherwise} \end{cases} \quad (12.21)$$

The sampling rate of $\upsilon(m)$ is identical to the sampling rate of $y(m)$.

Taking z-transform of Eq. (12.21)

$$V(z) = \mathcal{Z}[\upsilon(m)]$$

or $$V(z) = \sum_{m=-\infty}^{\infty} \upsilon(m) z^{-m}$$

$$= \sum_{m=-\infty}^{\infty} S\left(\frac{m}{u}\right) z^{-m}$$

$$= \sum_{m=-\infty}^{\infty} s(m) z^{-mI} \quad (12.22)$$

or

$$V(z) = S(z^I) \tag{12.23}$$

The corresponding spectrum of $\upsilon(m)$ is obtained by evaluating Eq. (12.22) on the unit circle. Thus

$$V(\omega_y) = S(\omega_y I) \tag{12.24}$$

where ω_y denotes the frequency variable relative to the new sampling rate F_y.

Here $\omega_y = \frac{2\pi F}{F_y}$. Now the relationship between sampling rates is $F_y = IF_s$ and hence, the frequency variables ω_s and ω_y are related according to the formula

$$\omega_y = \frac{\omega_s}{I} \tag{12.25}$$

Figure 12.9 illustrates the spectra $S(\omega_s)$ and $V(\omega_y)$. We observe that the sampling rate increases (obtained by the addition of I-1 zero samples between successive values of $s(n)$) result in a signal whose spectrum $V(\omega_y)$ is an I-fold periodic repetition of the input-signal spectrum $S(\omega_s)$.

Since only the frequency components of input sequence $s(n)$ in the range $0 \le \omega_y \le \frac{\pi}{I}$ are unique, the images of $s(\omega)$ above $\omega_y = \frac{\pi}{I}$ should be rejected by passing the sequence $\upsilon(m)$ through an LPF. This LPF has a frequency response of $H_I(\omega_y)$. Ideal frequency response of the above LPF is given as

$$H_I(\omega_y) = \begin{cases} C, & 0 \le |\omega_y| \le \frac{\pi}{I} \\ 0, & \text{otherwise} \end{cases} \tag{12.26}$$

where C is the scale factor.

The scale factor C requires proper normalization of the output sequence $y(m)$. Consequently, the output spectrum is given by

$$Y(\omega_y) = \begin{cases} CS(\omega_y I), & 0 \le |\omega_y| \le \frac{\pi}{I} \\ 0, & \text{otherwise} \end{cases} \tag{12.27}$$

The scale factor C is selected so that the output $y(m) = s\left(\frac{m}{I}\right)$ for $m = 0$, $\pm I$, $\pm 2I$, ...

For mathematical convenience, we select the point $m = 0$. Thus

$$y(0) = y(m)/m = 0 = \frac{1}{2\pi}\int_{-\pi}^{\pi} Y(\omega_y)\, d\omega_y \tag{12.28}$$

Substituting Eq. (12.27) in Eq. (12.28), we get

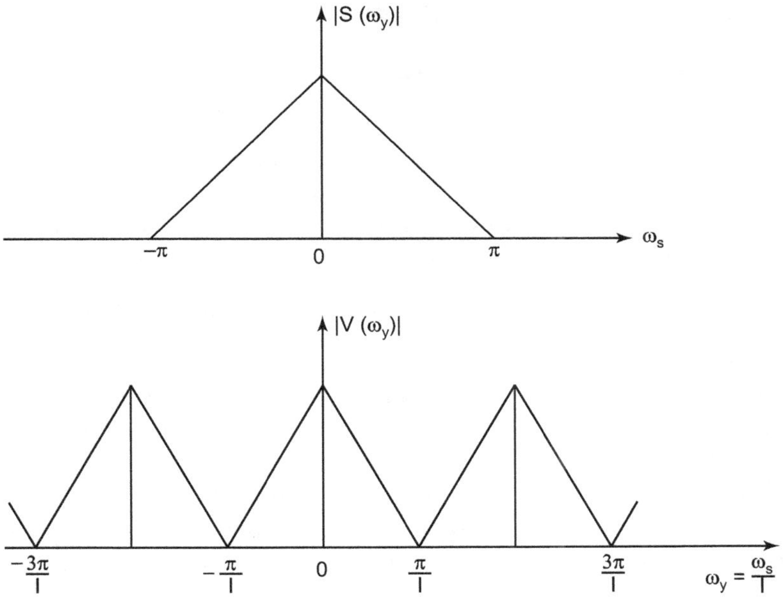

FIGURE 12.9 Illustration of spectra of $s(n)$ and $v(n)$ where $V(\omega_y) = S(\omega_y I)$.

$$y(0) = \frac{1}{2\pi}\int_{\pi}^{\pi} CS\left(\omega_y I\right) d\omega_y$$

$$= \frac{C}{2\pi}\int_{\frac{-\pi}{I}}^{\frac{\pi}{I}} S\left(\omega_y I\right) d\omega_y \qquad \left(\text{Since } \omega_y = \frac{\omega_s}{I}\right)$$

or

$$y(0) = \frac{C}{I}\frac{1}{2\pi}\int_{-\pi}^{\pi} S\left(\omega_s\right) d\omega_s$$

$$= \frac{C}{I} s(0) \qquad (12.29)$$

Therefore, $C = I$ is the desired normalization factor.

Finally, output sequence $y(m)$ can be expressed as a convolution of the sequence $\upsilon(n)$ with the unit-sample response $h(n)$ of LPF.

Thus,

$$y(m) = \sum_{k=-\infty}^{\infty} h(m-k)\upsilon(k) = h(m) * \upsilon(m) \qquad (12.30)$$

Since $\upsilon(k) = 0$ except at multiples of I, where $\upsilon(kI) = s(k)$, above Eq. (12.30) can be written as

$$y(m) = \sum_{k=-\infty}^{\infty} h(m-kI)s(k) \qquad (12.31)$$

12.4 SAMPLING RATE ALTERNATION OR CONVERSION BY A RATIONAL FACTOR $\left(\frac{I}{D}\right)$

We have already discussed two special cases of decimation and interpolation. We now consider the general case of sampling rate conversion by first performing interpolation by the factor I and decimating the output of the interpolator by the factor D. In other words, a sampling rate conversion by the rational factor $\frac{I}{D}$ is accomplished by cascading an interpolator with a decimator. It is illustrated in Figure 12.10.

The importance of performing the interpolation first and the decimation second is to preserve the desired spectral characteristics of $s(n)$. Furthermore, the two filters shown in Figure 12.10 with impulse response $\{h_u(l)\}$ are operated at the same sampling rate, namely IF_s, and hence can be combined into a single LPF with impulse response $[h(l)]$. It is shown in Figure 12.11, The frequency response $H(\omega_v)$ of the combined filter must incorporate the filtering operations for both interpolation and decimation, and hence this LPF ideally possess the frequency response characteristic.

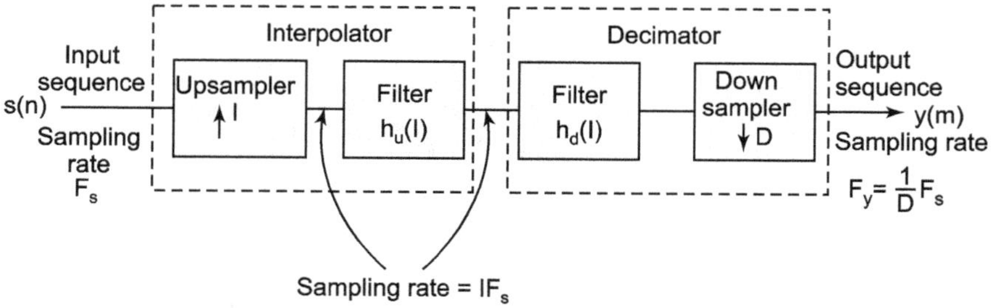

FIGURE 12.10 Block diagram of a method for sampling rate conversion by a factor $\left(\frac{I}{D}\right)$.

$$H(\omega_v) = \begin{cases} I, & 0 \le |w_v| \text{min. of} \left(\frac{I}{D}, \frac{\pi}{I}\right) \\ o, & \text{otherwise} \end{cases} \tag{12.32}$$

where $$\omega_v = \frac{2\pi F}{F_v} = \frac{2\pi F}{IF} = \frac{\omega_s}{I}$$

In the time domain, the output of the upsampler is given as

$$v(l) = \begin{cases} s\left(\frac{l}{I}\right), & l = 0, \pm I, \pm 2I, \ldots \\ 0, & \text{otherwise} \end{cases} \tag{12.33}$$

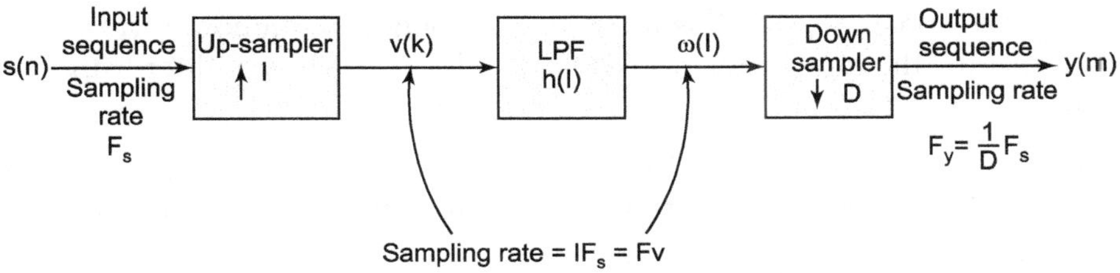

FIGURE 12.11 Block diagram of a method for sampling rate conversion by a factor $\left(\frac{I}{D}\right)$. Here two filters $h_u(l)$ and $h_d(l)$ are combined in a single LPF $h(l)$.

The output of the linear time-invariant filter is given as

$$\omega(l) = \sum_{k=-\infty}^{\infty} h(l-k)\upsilon(k) \tag{12.34}$$

or

$$\omega(l) = \sum_{k=-\infty}^{\infty} h(l-kI)s(k) \tag{12.35}$$

Finally, the output of the sampling rate converter $\{y_{(m)}\}$ is obtained by downsampling the sequence $\{\omega\ (l)\}$ by a factor D. Thus

$$y(m) = \omega(mD) \tag{12.36}$$

Substituting Eq. (12.35) in Eq. (12.36), we get

$$y(m) = \sum_{k=-\infty}^{\infty} h(mD-kI)s(k) \tag{12.37}$$

We can express Eq. (12.37) in different form by making a change in variable.

Let
$$k = \left\lfloor \frac{md}{I} \right\rfloor - n \tag{12.38}$$

The notation $\lfloor A \rfloor$ denotes the largest integer contained in A.
Substituting Eq. (12.38) in Eq. (12.37), we get

$$y(m) = \sum_{n=-\infty}^{\infty} h\left[mD - \left\lfloor \frac{mD}{I} \right\rfloor I + nI\right] s\left[\left\lfloor \frac{mD}{I} \right\rfloor\right] - n \tag{12.39}$$

Here, we note that

$$mD - \left\lfloor \frac{mD}{I} \right\rfloor I = mD \quad \text{modulo } I$$
$$= (mD)_I$$

Consequently, Eq. (12.39) can be expressed as

$$y(m) = \sum_{n=-\infty}^{\infty} h[nI + (mD)_I s\left[\left\lfloor \frac{mD}{I} \right\rfloor - n\right] \tag{12.40}$$

It is apparent from this form that the output sequence $y(m)$ is obtained by passing the input sequence $s(n)$ through a time-varying filter with impulse response

$$g(n,\ m) = h\left[nI\ + (mD)I\right], -\infty < m, n < \infty \tag{12.41}$$

where $h(k)$ is the impulse response of time-invariant *LPF* operating at the sampling rate *IFs*.

For any integer k, impulse response is given by

$$\begin{aligned} g(n, m + kI) &= h\left[nI + (mD + \ kDI)_I\right] \\ &= h\left[nI + (mD)_I\right] \\ &= g(n, m) \end{aligned} \tag{12.42}$$

Hence, $g(n,\ m)$ is periodic in the variable m with period I.

The frequency-domain relationships can be obtained by combining the results of the interpolation and decimation processes. Thus the spectrum at the output of the linear filter with impulse response $h(l)$ is given by

$$\begin{aligned} V(\omega_\upsilon) &= H(\omega_\upsilon)S(\omega_\upsilon I) \\ &= \begin{cases} IS(\omega_\upsilon I), & 0 \le |\omega_u| \le \min\left(\dfrac{\pi}{D}, \dfrac{\pi}{I}\right) \\ 0, & \text{otherwise} \end{cases} \end{aligned} \tag{12.43}$$

The spectrum of the output sequence $y(m)$ [Obtained by decimating the sequence $\upsilon(n)$ by a factor D] is given as

$$Y(\omega_y) = \frac{1}{D}\sum_{k=0}^{D-1} V\left[\frac{\omega_y - 2\pi k}{D}\right] \tag{12.41}$$

where $\qquad \omega_y = D\omega_\upsilon.$

Since the linear filter prevents aliasing as implied by Eq. (12.43), the spectrum of the $y(m)$ given by Eq. (12.44) reduces to

$$Y(\omega_y) = \begin{cases} \dfrac{I}{D} S\left(\dfrac{\omega_y}{D}\right), & 0 \le |\omega_y| \le \min.\left(\pi, \dfrac{\pi D}{I}\right) \\ 0, & \text{otherwise} \end{cases} \tag{12.45}$$

12.5 FILTER DESIGN AND IMPLEMENTATION FOR SAMPLING RATE ALTERNATION OR CONVERSION

Sampling rate alternation by a factor $\left(\dfrac{I}{D}\right)$ can be achieved by first increasing the sampling rate by integer factor I then downsampling the filtered signal by the integer factor D. Interpolation is accomplished by inserting I-1 zeros between successive values of the input signal $S(n)$. Before downsampling, interpolated signal is linearly filtered to eliminate the unwanted images of $S(\omega)$. Here, we discuss the design and implementation of the linear filter. We discuss the following types of linear filters:

1. Direct-form FIR digital filter structures.
2. Polyphase digital filter structures.
3. Time-varying digital filter structures.

12.5.1 Direct-Form FIR Digital Filter Structures

In principle, the simplest realization of the digital filter is the direct-form FIR digital filter structure. Its system function is given as

$$H(z) = \sum_{k=0}^{M-1} h(k)z^{-k} \tag{12.46}$$

where $\{h(k)\}$ is the unit-sample response of the FIR digital filter. The LPF can be designed to have linear phase. It also has a specified passband ripple and stop band attenuation. Any other standards are the same as for FIR digital filter techniques, that is, window method, frequency sampling method.

Thus we will have the filter parameters $\{h(k)\}$. These filter parameters allow us to implement the FIR digital filter directly. It is shown in Figure 12.12.

Although this realization is simple, it is also very inefficient. The inefficiency results from the fact that the upsampling process introduces $(I - 1)$ zeros between successive points of the input signal $s(n)$.

If I is large, most of the signal components in the FIR digital filter are zero. Consequently, most of the multiplications and additions result in zeros. Furthermore, the downsampling process at the output of the filter implies that only one out of every D output sample is required at the output of the filter. Consequently, only one out of every D possible value at the output of the filter should be computed.

12.5.1.1 Efficient Implementation of a Decimator

For developing a more efficient filter structure we consider a decimator that reduces the sampling rate by an integer factor D.

A decimator is obtained by passing the input sequence $s(n)$ through an FIR digital filter and then downsampling the filter output by a factor D. It is shown in Figure 12.13(a). In this configuration, the filter is operating at a high sampling rate F_s, while only one out of every D output sample is actually needed.

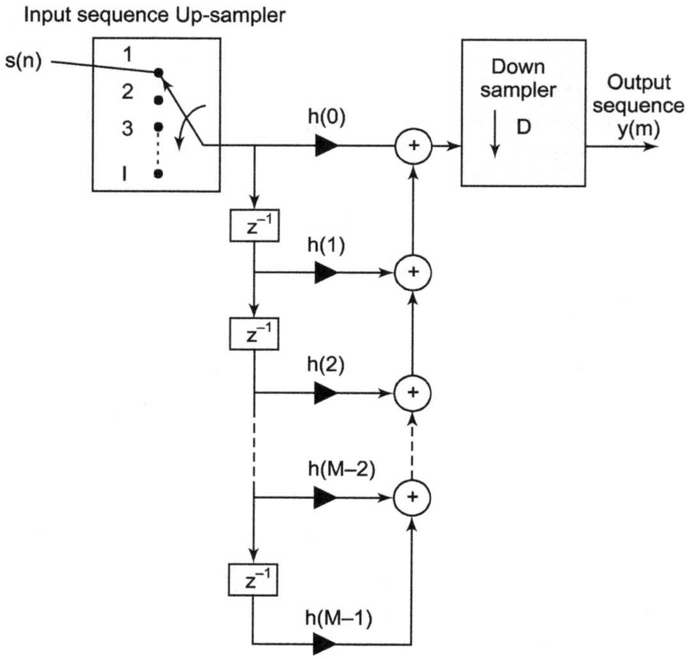

FIGURE 12.12 Direct-form realization of FIR digital filter in sampling rate conversion by factor $\left(\frac{I}{D}\right)$.

The logical solution to this inefficiency problem is to imbed the downsampling operation within the filter. Such a filter realization is shown in Figure 12.13(b). In this filter structure, all the multiplications and the additions are performed at the lower sampling rate $\frac{F_s}{D}$. Thus, we have achieved the deserved efficiency.

Additional reduction in computation can be achieved by exploiting the symmetry property of $\{h(k)\}$. Efficient realization of the decimator in which the FIR digital filter has linear phase is shown in Figure 12.14 Hence $\{h(k)\}$ is symmetric.

Efficient Implementation of an Interpolator

Here we consider the efficient implementation of an interpolator. This interpolator is realized by first inserting I-1 zeros between samples of sequence $s(n)$ and then filtering the resulting sequence. The direct-form realization of an FIR digital filter in interpolation by an integer factor I is shown in Figure 12.15. The major problem with this structure is that the filter computations are performed at the high sampling rate IF_s.

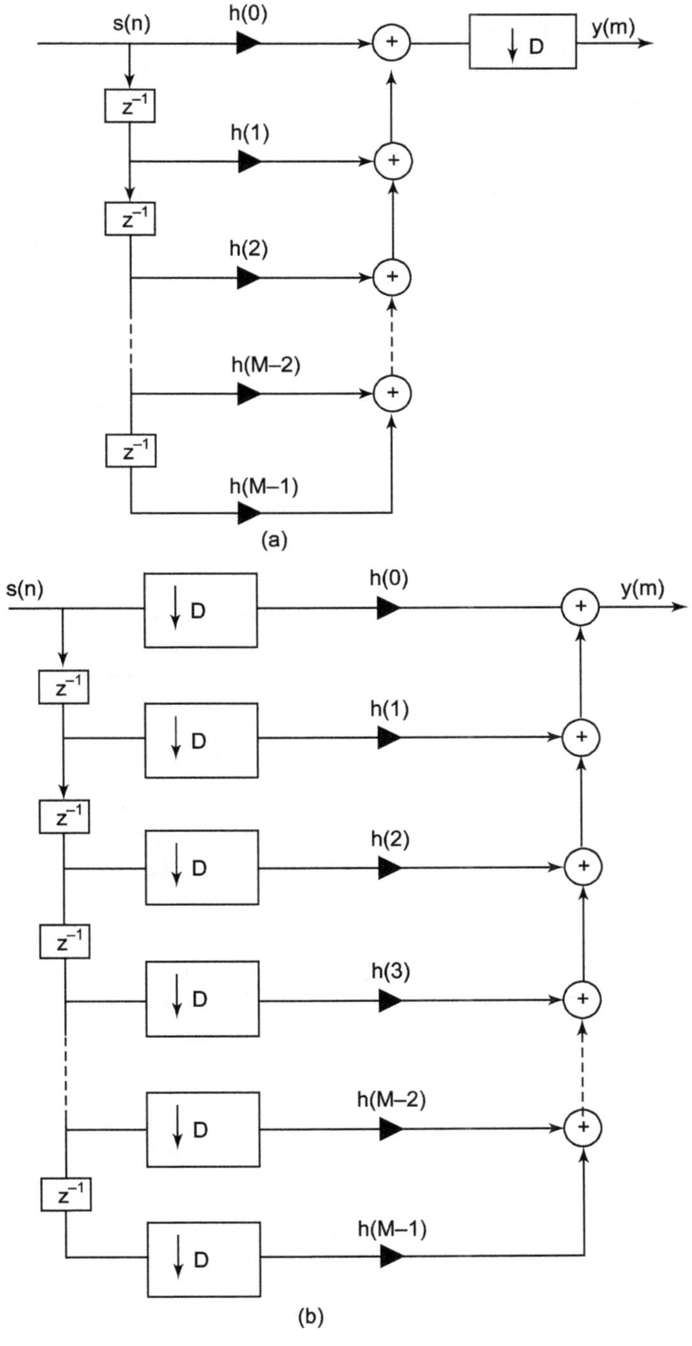

FIGURE 12.13 Decimation of a signal by a factor *D*.

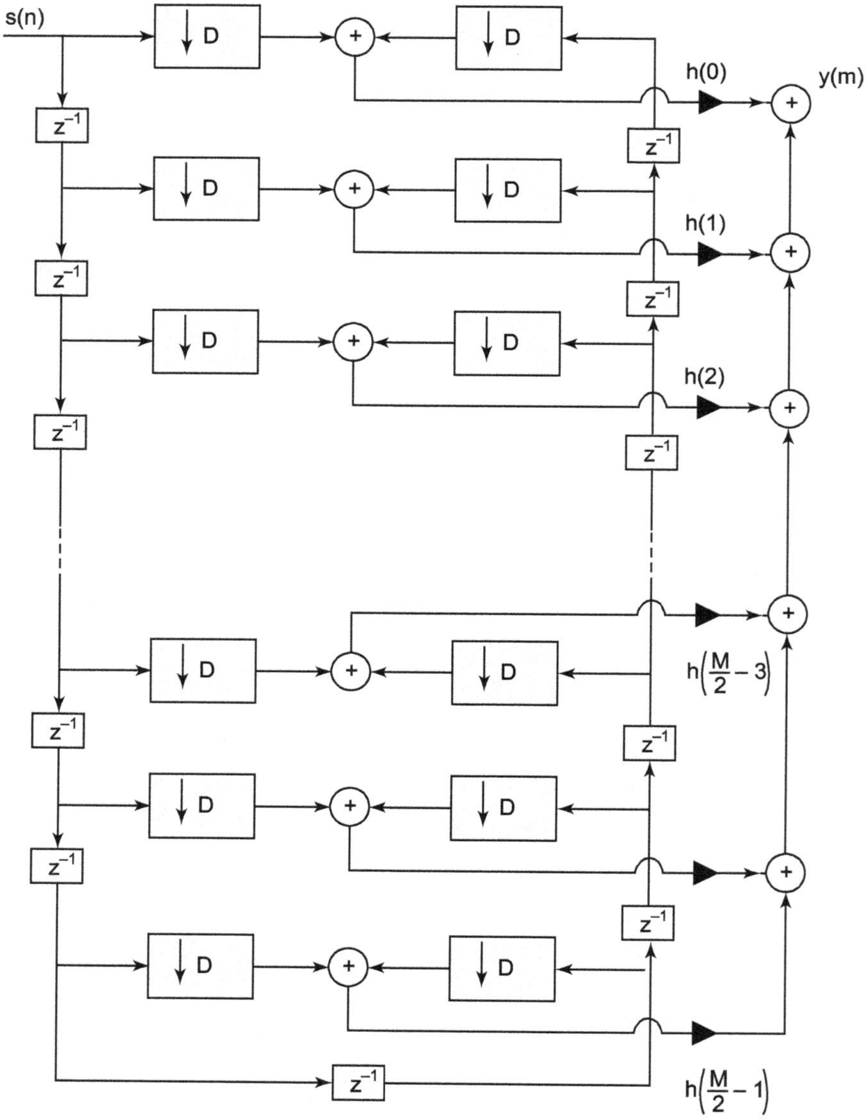

FIGURE 12.14 Efficient realization of a decimator. This decimator exploits the property of symmetry in the FIR digital filter.

The desired simplification is achieved by first using the transposed form of FIR digital filter [it is shown in Figure 12.16(*a*)] and then embedding the upsampler within the filter.

It is shown in Figure 12.16(*b*). Thus, all the filter multiplications are performed at the low sampling rate F_s, while the upsampling process introduces

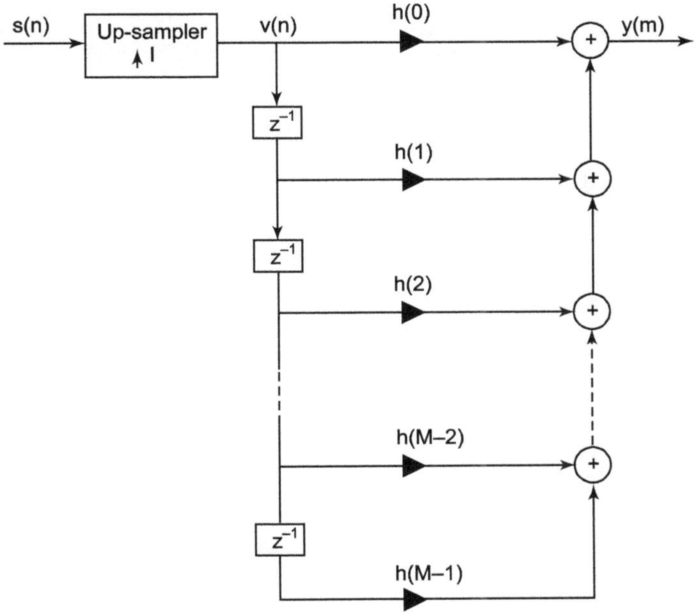

FIGURE 12.15 Direct-form realization of FIR filter in interpolation by a factor *I*.

I-1 zeros in each of the filter branches of the structure of Figure 12.16(*b*). Filter structures depicted in Figure 12.16(*a*) and (*b*) are equivalents.

Here we note that the structure of the interpolator [shown in Figure 12.16(*b*)] can be obtained by transposing the structure of the decimator [Figure 12.13]. We observe that the transpose of a decimator is an interpolator, and vice-versa. These relationships are shown in Figure 12.17. Figure 12.17(*b*) is obtained by transposing Figure 12.17(*a*) and (*d*) is obtained by transposing Figure 12.17(*c*). Consequently, a decimator is the dual of an interpolator, and vice-versa. From the above relationships, it follows that there is an interpolator whose structure is the dual of the decimator shown in Figure 12.14. Figure 12.14 exploits the symmetry in $h(n)$.

12.5.2 Polyphase Filter Structure

The computational efficiency of the filter structure given in Figure 12.15 can also be achieved by reducing the large *FIR* digital filter of length M into a set of smaller filters of length $K = \frac{M}{I}$, where M is selected to be a multiple of I.

For demonstrating how the computational efficiency of the filter is achieved, we consider the interpolator given in Figure 12.15. Since the upsampling process inserts I–1 zeros between successive values of $s(n)$, only K out of

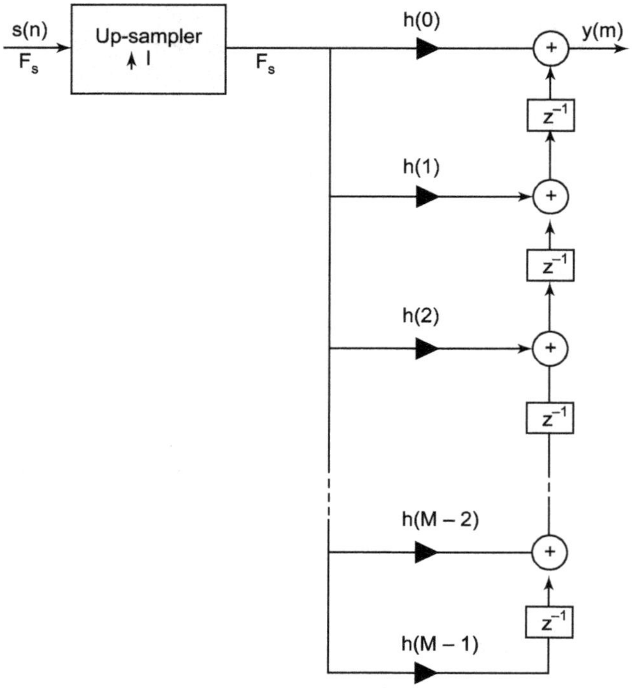

(a)

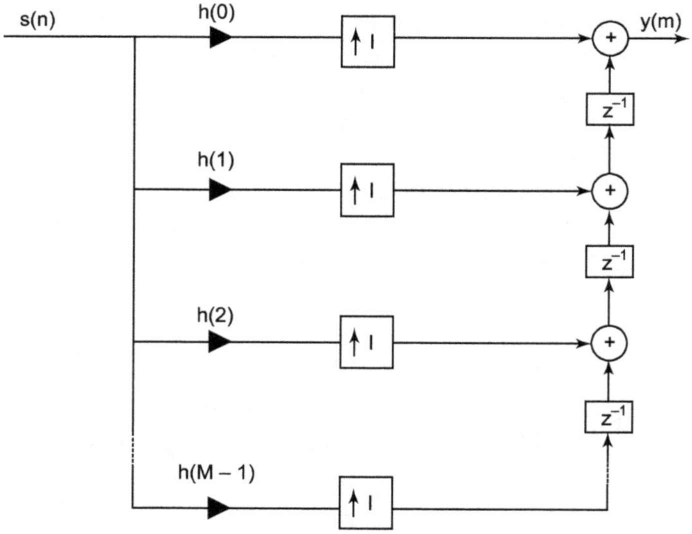

(b)

FIGURE 12.16 Efficient realization of interpolator.

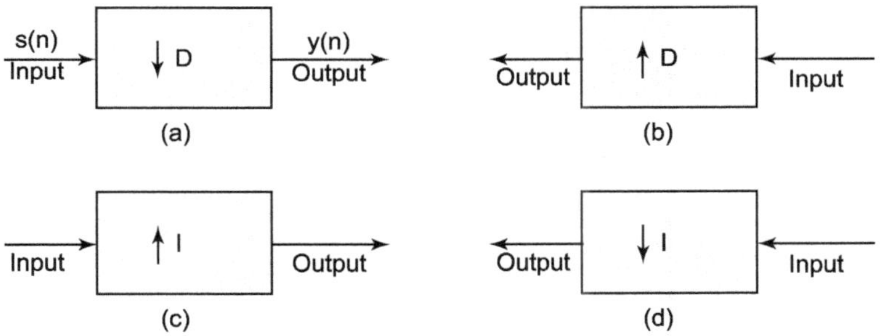

FIGURE 12.17 Duality relationships are obtained through transposition.

the M input values stored in the FIR digital filter at any one time is non-zero. At one time-instant, these non-zero values coincide and are multiplied by the filter coefficients $h(o), h(I), h(2I), ..., h(M-1)$.

In the following time-instant, the non-zero values of the input-sequence coincide and are multiplied by the filter coefficients $h(1)$, $h(I+1)$, $h(I+2)$, ..., $h(M-I+1)$, and so on. This observation leads us to define a set of smaller filters. These smaller filters are called polyphase filters. The unit-sample responses of polyphase filters are given by

$$p_k(n) = h(k+nI),\ k = 0, 1, 2, ..., I-1$$

$$n = 0, 1, 2, ...,\ K-1 \tag{12.47}$$

where $\quad K = \dfrac{M}{I}$ is in integer.

Now we conclude that the set of I polyphase filters can be arranged as a parallel realization and the output of each filter can be selected by the commutator. It is shown in Figure 12.18(a). The rotation of the commutator is in the counterclockwise direction beginning with the point at $m = 0$. Thus, the polyphase filters perform the commutations at the low sampling rate F_s, and the sampling rate conversion results from the fact that I output samplers are generated, one from each of the filters, for each input sample.

The decomposition of $[h(k)]$ into the set of I subfilters with impulse response $p_k(n)$, $k = 0, 1, ..., I-1$,. The input signal was being filtered by a periodically time-varying linear filter with impulse response

$$g(n,m) = h\left[nI + (mD)_I\right] \tag{12.48}$$

where $D = 1$ in the case of the interpolator. We already know that $g(n, m)$ varies periodically with period I. Consequently, a different set of coefficients are used to generate the set of I output samples $y(m)$, $m = 0, 1, ... I-1$.

We can gain additional insight about the characteristics of the set of polyphase subfilters by noting that $p_k(n)$ is obtained from $h(n)$ by decimation

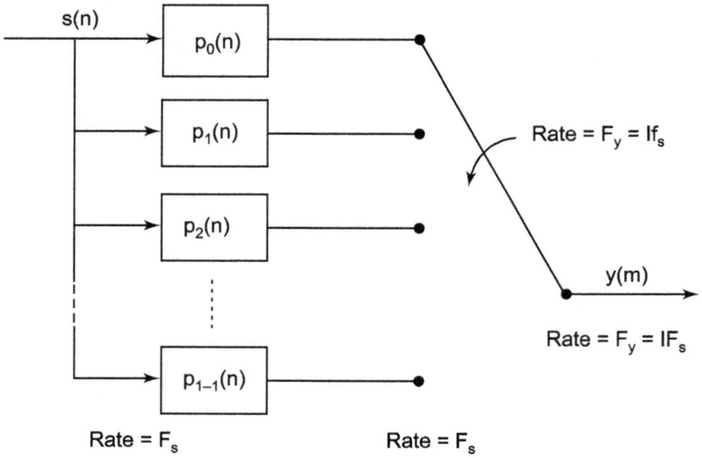

FIGURE 12.18 (a) Interpolation by use of polyphase filters.

with a factor I. Consequently, if the original filter frequency response $H(\omega)$ is flat over the range $O \leq |\omega| \leq \frac{\omega}{I}$, each of the polyphase subfilter possesses a relatively flat response over the range $o \leq |\omega| \leq \pi$ (i.e., the polyphase subfilters are basically all passfilters and differ primarily in their phase characteristics).

The polyphase filter can also be viewed as a set of I subfilters connected to a common delay line. Ideally, the kth subfilter will generate a forward time shift of $\left(\frac{k}{I}\right) T_s$, for $k = 0, 1, 2, \ldots, I-1$, relative to the zeroth subfilter. Therefore, if the zeroth filter generates zero delays, the frequency response of the kth subfilter is given as

$$p_k(\omega) = e^{\frac{j\omega k}{I}} \tag{12.49}$$

A time shift of integer number of input sampling intervals (e.g., lT_s) can be generated by shifting the input data in the delay line by l samples and using the same subfilters. By combining these two methods, we can generate an output that is shifted forward by an amount $\left(l + \frac{i}{I}\right) T_s$ relative to the previous output.

By transposing the interpolator structure given in Figure 12.18(a), we obtain a commutator structure for the decimator based on the parallel bank of polyphase filters. It is shown in Figure 12.18(b). The until sample responses of the polyphase filters are now defined as

$$pk(n) = h(k + nD), k = 0, 1, 2, \ldots. D - 1$$

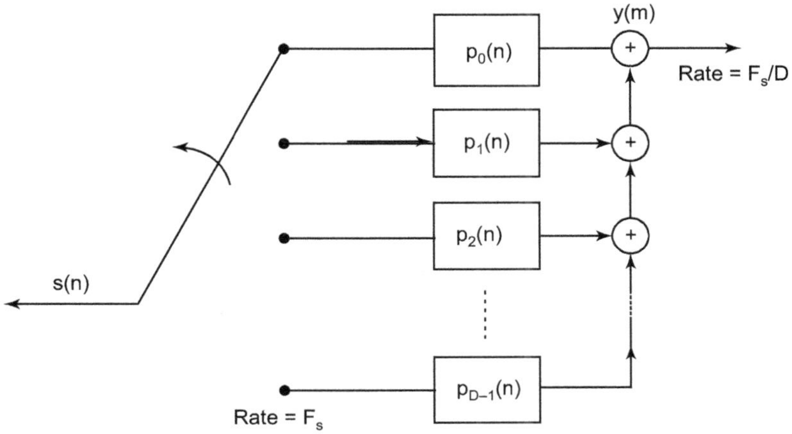

FIGURE 12.18 (b) Decimation process by use of polyphase filters.

$$n = 0, 1, 2, \ldots, K-1 \tag{12.50}$$

where $K = \dfrac{M}{D}$ is an integer when M is selected to be a multiple of D. The commutator rotates in counterclockwise duration starting with the filter $p_o(n)$ at $m = 0$.

Two commutator structures for the interpolator and the decimator rotate in a counter or counterclockwise direction.

It is also possible to derive an equivalent pair of commutator structures having a clockwise direction. In this alternative formulation, the sets of polyphase filters are defined to have impulse responses:

$$pk(n) = h(nI - k),\, k = 0, 1, 2, \ldots, I-1 \tag{12.51}$$

$$pk(n) = h(nD - k),\, k = 0, 1, 2, \ldots, D-1 \tag{12.52}$$

for the interpolator and decimator, respectively.

12.5.3 Time-Varying Digital Filter Structures

To date, we have described the filter implementation for a decimator and an interpolator. Let us now consider the general problem of sampling rate conversion by a rational factor $\left(\dfrac{I}{D}\right)$. In the general case of sampling rate conversion by a factor $\left(\dfrac{I}{D}\right)$, the filtering can be accomplished by means of the linear time-varying filter. This filter is described by the response function.

$$g(n, m) = h\left[nI - (mD)_I\right] \tag{12.53}$$

where $h(n)$ is the impulse response of the LPF, For convenience, we select the length of the *FIR* digital filter $\{h(n)\}$ to a multiple *of* I (i.e., $M = KI$). Consequently, the set of coefficients $\{g(n, m)\}$ for each $m = 0, 1, 2, ... I - 1$, contains K elements. Since sequence $g(n, m)$ is also periodic with the period I. The output $y(m)$ can be expressed as

$$y(m) = \sum_{n=0}^{K-1} g\left[n, m - \left\lfloor \frac{m}{I} \right\rfloor I\right] S\left[\left\lfloor \frac{mD}{I} \right\rfloor - n\right] \tag{12.54}$$

Conceptually, we can think of performing the computations specified by Eq. (12.54) by processing blocks of data of length K by a set of K filter coefficients

$$g\left[n, m - \left\lfloor \frac{m}{I} \right\rfloor I\right], n = 0, 1, 2, ..., K - 1.$$

There are I such sets of coefficients, one set for each block of I output points of $y(m)$. For each block of I output points, there is a corresponding block of D input points of $s(n)$ that enter the computation.

For computing Eq. (12.54), we can visualize a block processing algorithm. This algorithm is shown in Figure 12.19. A block of D input samples is buffered and shifted into a second buffer occurs at a sampling rate of one sample each time the quantity $\left\lfloor \frac{mD}{I} \right\rfloor$ increases by one.

For each output sample $y(l)$, the samples from the second buffer are multiplied by the corresponding set of filter coefficients $g(n, I)$ for $n = 0, 1, 2, ...,$

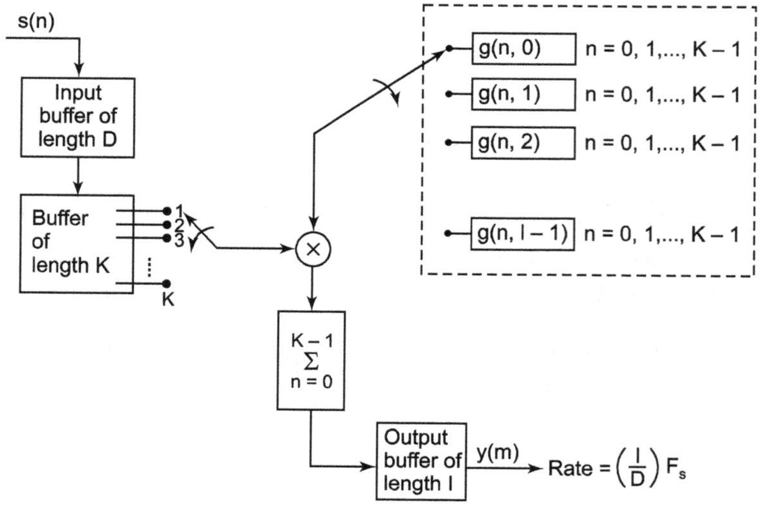

FIGURE 12.19 Efficient implementation of sampling rate conversion by block processing.

$K - 1$, and the K products are accumulated to give $y(l)$, for $l = 0, 1, \ldots, I - 1$. Thus this computation produces I output. It is then repeated for a new set of D input samples, and so on.

An alternative method for computing the output of the sample rate converter [Eq. (12.54)] is by means of an *FIR* digital filter structure with periodically varying filter coefficients. Such a structure is shown in Figure 12.20. The input samples $s(n)$ are passed into a shift register that operates at the sampling rate F_s and is of length $K = \frac{M}{I}$, where M is the length of the time-varying *FIR* filter.

The frequency response of the above filter is given by

$$H(\omega_v) = \begin{cases} I, & 0 \le \omega_v \,|\le \min\left(\frac{\pi}{D}, \frac{\pi}{I}\right) \\ 0, & \text{otherwise} \end{cases} \tag{12.55}$$

Each stage of the register is connected to a hold-and-sample device that serves to couple the input sampling rate F_s to the output sampling rate $F_y = \left(\frac{I}{D}\right)F_s$. The sample at the input to each hold-and-sample device is held until the next input sample arrives and then is discarded. The output samples of the hold-and-sample devices are taken at times $\frac{mD}{I}$, $m = 0, 1, 2, \ldots$ When both the input and output sampling times varying coefficients coincide (i.e., when $\frac{mD}{I}$ is an integer), the input to the hold-and-sample is changed first and then the output samples the new input. The K outputs from the K hold-and-sample devices are multiplied by the periodically time-varying coefficients $g\left[n, m - \left\lfloor \frac{m}{I} \right\rfloor I\right]$, for $n = 0, 1, 2, \ldots, K - 1$, and the resulting products are summed to yield $y(m)$. The computations at the output of the hold-and-sample devices are repeated at the output sampling rate of

$$F_y = \left(\frac{I}{D}\right)F_s$$

Finally, sampling rate conversion by a rational factor $\frac{I}{D}$ can be performed by the use of polyphase filters having I subfilters. If we assume that the mth sample $y(m)$ is computed by taking the output of the i_mth subfilter with input data $s(n), s(n - 1), s(n - 2), \ldots, s(n - k - 1)$, in the delay line. The next sample $y(m + 1)$ is taken from the (i_{m+1})th subfilter after shifting l_{m+1} new samples in the delay lines where $i_{m+1} = (i_m + D)_{\text{mod } I}$ and $l_m + 1$ is the integer part of $\left(\frac{i_m + D}{I}\right)$. The integer i_{m+1} should be saved to be used in determining the subfilter from which the next sample is taken.

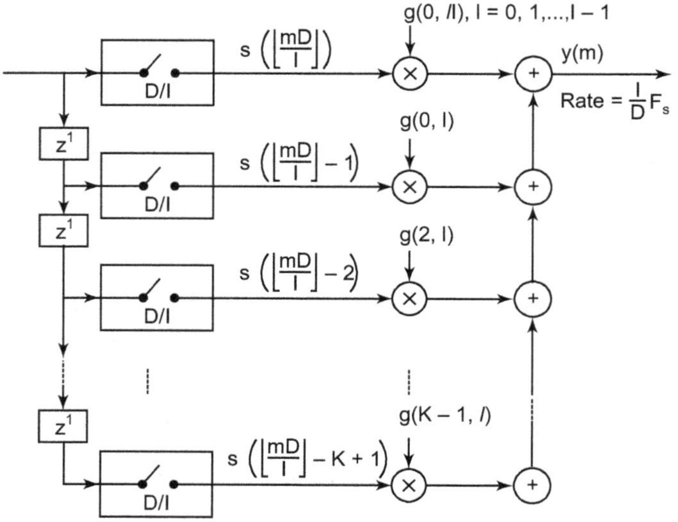

FIGURE 12.20 Efficient realization of sampling rate conversion by a factor $\left(\frac{I}{D}\right)$

12.6 SAMPLING RATE CONVERSION BY AN ARBITRARY FACTOR

In some previous sections, we have shown how to perform sampling rate conversion exactly by a rational number $\left(\frac{I}{D}\right)$. But in some applications, it is inefficient or impossible to use such an exact sampling rate conversion scheme.

Here we consider the following two cases:

Case I: We need to perform sampling rate conversionby a rational number $\left(\frac{I}{D}\right)$, where I is a large integer $\left[\text{For example, } I = 1023 \text{ and } D = 511, \text{ i.e.,} \frac{I}{D} = \frac{1023}{511}\right]$. Although we can achieve exact sampling rate conversion by this number, we would use a polyphase filter with $I = 1023$ subfilters, such an exact implementation is obviously inefficient in memory usage because we need to store a large number of filter coefficients.

Case II: In some applications, the exact conversion sampling rate is not known when we design the sampling rate converter, or the sampling rate is continuously changing during the conversion process. For example, we may counter the situation where the input and output samples are controlled by two independent clocks. It is possible to define a nominal conversion rate that is a rational number, the actual sampling rate would be slightly difficult. Sampling rates depend on the frequency difference between the two clocks obviously, it is not possible to design an exact sampling rate converter in this case.

Non-exact sampling rate conversion scheme will introduce some distortion in the converted output signal. It is also possible that an exact rational sampling rate converter introduces some distortion because the polyphase filter is never ideal. Such a converter will be adequate, as long as the total distortion does not exceed the specification required in the application.

We can use first-order, second-order, and higher-order approximations for non-exact sampling rate conversion. It depends on the application requirements and implementation constraints.

Here, we discuss only first-order approximation and second-order approximation methods.

12.6.1 First-Order Approximation Method

Let us denote the arbitrary sampling rate by R_a. Suppose that the input to the sampling rate converter is the sequence $\{s(n)\}$. We need to generate a sequence of output samples separated in time by $\frac{T_s}{R_a}$, where T_s is the sampling interval for $\{s(n)\}$. By constructing a polyphase filter with large number of subfilters, we can approximate such a sequence with a non-uniformly spaced sequence.

In the general case, we can express $\frac{1}{R_a}$ as

$$\frac{1}{R_a} = \frac{k}{I} + \beta$$

where k and I are positive integers and b is a number in the range.

$$0 < \beta < \frac{1}{I}$$

Now, the boundaries of $\frac{1}{R_a}$ is given as

$$\frac{k}{I} < \frac{1}{R_a} < \frac{k+1}{I}$$

where I corresponds to interpolation factor. Interpolation factor I will be determined to satisfy the specification on the amount of tolerable distortion introduced by the sampling rate converter. Also, I is equal to the number of polyphase filters.

For example

Suppose that $R_a = 2.2$ and that we have determined, as we will demonstrate, that $I = 6$ polyphase filters are required to meet the distortion specification. Then

$$\frac{k}{I} \equiv \frac{2}{6} < \frac{1}{R_a} < \frac{3}{6} \equiv \frac{k+1}{I}$$

So, that $k = 2$.

The time spacing between samples of the interpolated sequence is $\frac{T_s}{I}$. However, the desired conversion rate $R_s = 2.2$ for $I = 6$ corresponds to decimation factor of 2,727, which falls between $k = 2$ and $k = 3$.

In the first-order approximation method, we achieve the desired decimation rate by selecting the output sample from the polyphase filter closest in time to the desired sampling time. It is illustrated in Figure 12.21 for $I = 6$.

In general, to perform sampling rate conversion by factor R_a, we employ a polyphase filter to perform interpolation and therefore to increase the frequency of the original sequence of a factor of I. The time spacing between the samples of the interpolated sequence is equal to $\frac{T_s}{I}$. If the ideal sampling time of the mth sample, $y(m)$, of the desired output sequence in between the sampling times of two samples of the interpolated sequence, we select the sample closer to $y(m)$ as its approximation.

We assume that the mth selected sample is generated by the (i_m)th subfilter using the input samples $s(n)$, $s(n-1)$,..., $s(n-k+1)$ in the delay line. The normalized sampling time error is denoted by t_m. The normalized sampling time error is the time difference between the selected sampling time and desired sampling time normalized by T_s. The sign of t_m is positive if the desired sampling time leads to the selected sampling time, and negative otherwise. The value of time t_m will be $|t_m|" \frac{0.5}{I}$. The normalized time advance from the mth output $y(m)$ to the $(m+1)$th output $y(m+1)$ is equal to $\left(\frac{1}{R_a}\right) + t_m$.

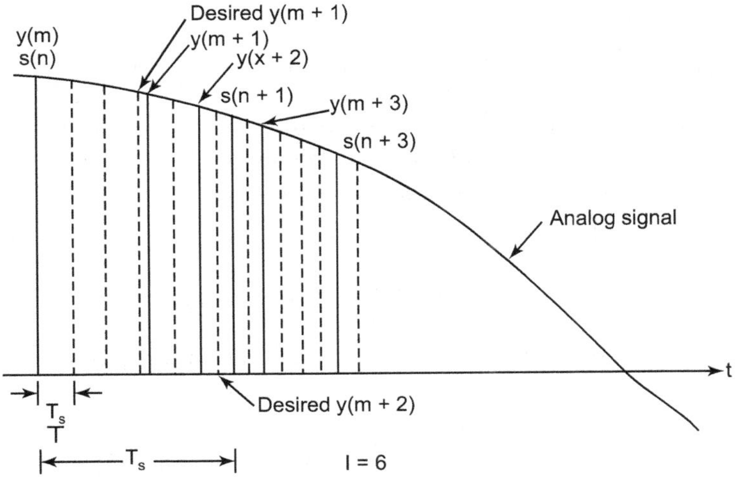

FIGURE 12.21 Sampling rate conversion by use of first-order approximations.

For computing the next output, we first determine a number closest to $\left[\frac{i_m}{I}+\frac{1}{R_a}+t_m+\frac{k_m}{I}\right]$ that is of the form $\left(l_{m-1}+\frac{i_{m+1}}{I}\right)$ where both l_{m+1} and i_{m+1} are integers and $i_{m+1} < I$. Then, the $(m+1)$th output $y(m+1)$ is computed using the (i_{m+1})th subfilter after shifting the signal in the delay lines by l_{m+1} input samples. The normalized timing error for the $(m+1)$th sample is $t_{m+1}=\left(\frac{i_m}{I}+\frac{1}{R_a}+t_m\right)-\left(l_{m+1}+\frac{i_{m+1}}{I}\right)$. It is saved for the computation of the next output sampler.

By increasing the number of subfilters used, we can arbitrarily increase the conversion accuracy. An increased number of subfilters require more memory to store a large number of filter coefficients. Hence it is describable to use as few subfilters as possible while keeping the distortion in the converted signal below the specification. The distortion introduced due to the sampling time approximation is most conveniently evaluated in the frequency domain.

We suppose that the input data sequence $\{s(n)\}$ has a flat spectrum from $-\omega_s$ to ω_s, where $\omega_s < \pi$, with a magnitude A. Its total power can be computed using Parseval's theorem.

It is given as
$$P=\frac{1}{2\pi}\int_{-\omega_s}^{\omega_s}|S(\omega)|^2\,d\omega=\frac{A^2\omega_s}{\pi} \tag{12.56}$$

We know that for each output, $y(m)$, the time difference between the desired filter and the filter actually used to t_m, where $t_m \mid'' \frac{0.5}{I}$. Hence the frequency response of these filters can be written as $e^j\omega^t$ and $e^{j\omega t(\tau - t_m)}$, respectively. When I is large, ωt_m is small. We ignore higher-order errors and we can express the difference between the frequency responses as

$$e^{j\omega t} - e^{j\omega(\tau - t_m)} = e^{j\omega\tau} - e^{j\omega(\tau - t_m)}$$

$$= e^{jj\omega t}\left[1 - \cos\omega t_m + j\sin\omega t_m\right] \approx e^{j\omega\tau}\,\omega t_m \qquad (12.57)$$

Using the bound $t_m \mid'' \frac{0.5}{I}$, we obtain an upper bound for the total error power as

$$P_e = \frac{1}{2\pi}\int_{-\omega_s}^{\omega_s} \left| s(\omega)e^{j\omega\tau} - S(\omega)e^{j\omega(\tau - t_m)} \right|^2 d\omega$$

$$\approx \frac{1}{2\pi}\int_{-\omega_s}^{\omega_s} \left| s(\omega)\, j e^{j\omega\tau}\,\omega t_m \right|^2 d\omega$$

$$\leq \frac{1}{2\pi}\int_{-\omega_s}^{\omega_s} A^2 \left(\frac{0.5}{I}\right)^2 \omega^2 d\omega$$

$$= \frac{A^2\omega_s^2}{12\pi I^2} \qquad (12.58)$$

This bound shows that the error power is inversely proportional to the square of the number of subfilters I. Therefore, the magnitude of error is inversely proportional to I. Hence we call the approximation of sampling rate conversion method described above a first-order approximation. The ratio of the signal-to-distortion due to a sampling time error for the first-order approximation is determined by using Eqs. (12.56) and (12.58) as

$$SDR_1 = \frac{P}{P_e} \geq \frac{12I^2}{\omega_s^2} \qquad (12.59)$$

We can conclude from Eq. (12.59) that the signal-to-distortion ratio is proportional to the square of the number of subfilters.

12.6.2 Second-Order Approximation Method

The disadvantage of the first-order approximation method is the large number of subfilters needed to achieve a specified distortion requirement.

Now we describe a method that uses linear interpolation (second-order approximation) to achieve the same performance with a reduced number of subfilters.

The implementation of the linear interpolation method is very similar to the first-order approximation. In linear interpolation, we compute two adjacent samples with the desired sampling time falling between their sampling times (see Figure 12.22). But in first-order approximation, we use the sample from the interpolating filter closest to the desired conversion output as the approximation. The normalized time spacing between these two samples is $\frac{1}{I}$.

We assume that the sampling time of the first sample lags the desired sampling time by t_m, the sampling time of the second sample is then leading the desired sampling time by $\left(\frac{1}{I}\right)-t_m$. If the two samples are denoted by $y_1(m)$ and $y_2(m)$ then by using linear interpolation, we can compute the approximation to the desired output as

$$y(m)=(1-\alpha_m)y_1(m)+\alpha_m y_2(m) \tag{12.60}$$

where $\alpha_m = It_m$.

Note that $0 \le \alpha_m \le 1$

The implementation of linear interpolation is similar to that for the first-order approximation. Normally, both $y_1(m)$ and $y_2(m)$ are determined using ith and $(i + 1)$th subfilters, respectively, with the same set of input data

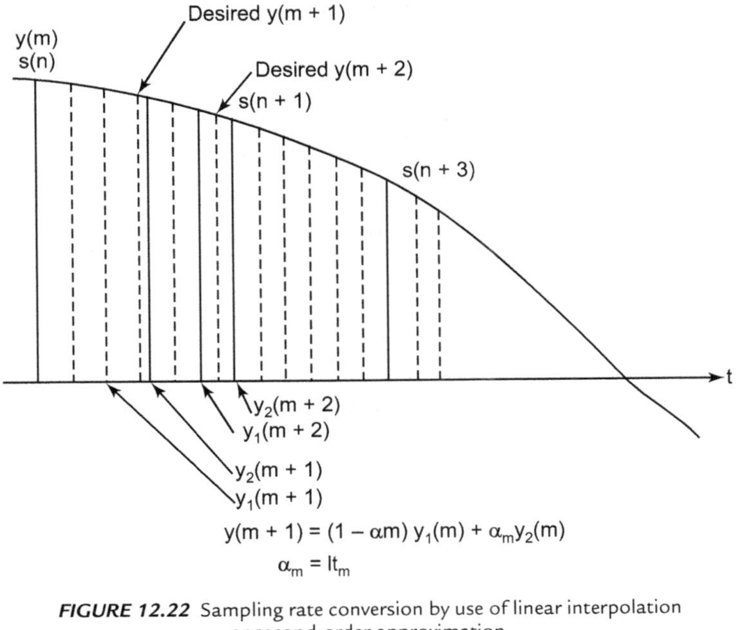

FIGURE 12.22 Sampling rate conversion by use of linear interpolation or second-order approximation.

samples in the delay line. The only exception, in the boundary case, where $i = I - 1$. In this case, we use the $(I - 1)$th subfilter to compute $y_1(m)$, but the second sample $y_2(m)$ is determined using the zeroth subfilter after new input data are shifted into the delay line.

Analysis of error introduced by the second-order approximation is done by first writing the frequency responses of the desired filter and the two subfilters used to compute $y_1(m)$ and $y_2(m)$.

The frequency responses of the desired filter, first-subfilter, and second subfilter are $e^{j\omega\tau}$, $e^{j\omega(\tau - t_m)}$, and $e^{j\omega\left(\tau - t_m + \frac{1}{I}\right)}$, respectively. Because linear interpolation is a linear operation, we can also use linear interpolation to compute the frequency response of the filter that generates $y(m)$ as

$$\begin{aligned}
&(1 - It_m)e^{j\omega(\tau - t_m)} + It_m e^{j\omega\left(\tau - t_m + \frac{1}{I}\right)} \\
&= e^{j\omega\tau}\left[(1 - \alpha_m)e^{-j\omega t_m} + \alpha_m e^{j\omega\left(-t_m + \frac{1}{I}\right)}\right] \\
&= e^{j\omega\tau}(1 - \alpha_m)[\cos\omega t_m - 1\sin\omega t_m] \\
&\quad + e^{j\omega\tau}\alpha_m\left[\cos\omega\left(\frac{1}{I} - t_m\right) + j\sin\omega\left(\frac{1}{I} - t_m\right)\right]
\end{aligned} \tag{12.61}$$

By ignoring higher-order errors, we can write Eq. (12.61) as

$$\begin{aligned}
&e^{j\omega\tau} - (1 - \alpha_m)e^{j\omega(\tau - t_m)} - \alpha_m e^{j\omega\left(t - t_m + \frac{1}{i}\right)} \\
&= e^{j\omega\tau}\left\{\left[1 - (1 - \alpha_m)\cos\omega t_m - \alpha_m\cos\omega\left(\frac{1}{I} - t_m\right)\right]\right. \\
&\quad \left. + j\left[(1 - \alpha_m)\sin\omega t_m - \omega_m\sin\omega t_m - \alpha_m\sin\omega\left(\frac{1}{I} - t_m\right)\right]\right\} \\
&\approx e^{j\omega\tau}\left[\omega^2(1 - \alpha_m)\frac{\alpha_m}{I^2}\right]
\end{aligned} \tag{12.62}$$

Using $(1 - \alpha_m)\,\alpha_m \leq \frac{1}{4}$, we obtain an upper bound for the total error power as

$$\begin{aligned}
P_e &= \frac{1}{2\pi}\int_{-\omega_s}^{\omega_s} /S(\omega)\left[e^{j\omega\tau} - (1 - \alpha_m)e^{j\omega(\tau - t_m)} - \alpha_m e^{j\omega\left(\tau - t_m + \frac{1}{t}\right)}\right]/^2\, d\omega \\
&\approx \frac{1}{2\pi}\int_{-\omega_s}^{\omega_s} /S(\omega)e^{j\omega\tau}\left[\omega^2(1 - \alpha_m)\frac{\alpha_m}{I_2}\right]/^2\, d\omega
\end{aligned}$$

$$\leq \frac{1}{2\pi}\int_{-\omega_s}^{\omega_s} A^2\left(\frac{0.25}{I^2}\right)^2 \omega^4 d\omega = \frac{A^2(\omega_s)^5}{80\pi I^4} \tag{12.63}$$

This result shows that the magnitude of errors is inversely proportional to I^2.

Hence, we call the approximation using linear interpolation a second-order approximation. The ratio of signal-to-distortion due to sampling time error for second-order approximation is determined using Eqs. (12.56) and (12.63) as

$$SDR_2 = \frac{P}{P_e} \geq \frac{80I^4}{\omega_s^4} \tag{12.64}$$

Therefore, the signal-to-distortion ratio is proportional to the fourth power of the number of subfilters (I).

12.7 APPLICATION OF MULTIRATE DIGITAL SIGNAL PROCESSING

Multirate digital signal processing has the following applications:

1. Design of phase shifters.
2. Interfacing of digital systems with different sampling rates.
3. Implementation of narrow band low pass filters (NB-LPF).
4. Implementation of digital filter banks.
5. Subband coding of speech signals.
6. Quadrature mirror filters.
7. Transmultiplexers.
8. Oversampling A/D and D/A conversion.

SOLVED EXAMPLES

EXAMPLE 12.1

Obtain the decimated signal $y(n)$ by a factor 3 from the input signal $s(n)$ shown in Figure 12.23(a).

Solution:

The decimated signal is given by

$y(n) = s(Dn)$, where D is the decimation factor and equal to 3.

The decimated signal $y(n)$ is shown in Figure 12.23(b)

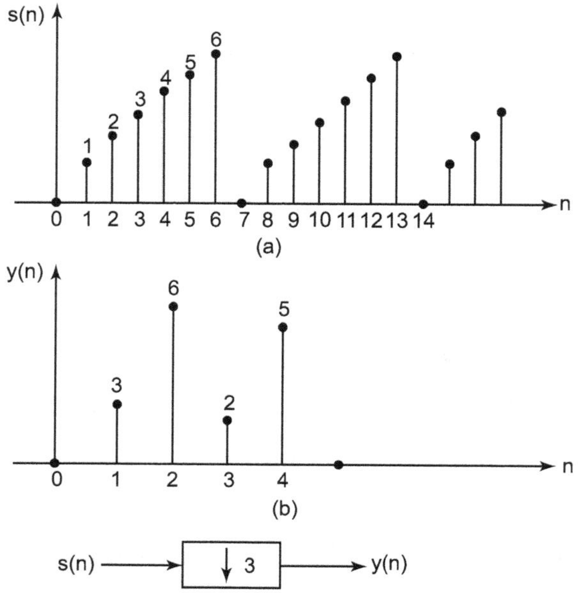

FIGURE 12.23 Illustration of decimation process with factor 3.

EXAMPLE 12.2

Obtain the two-fold expanded signal $y(n)$ of the input signal $s(n)$.

$$s(n) = \begin{cases} n, & n > 0 \\ 0, & \text{otherwise} \end{cases}$$

Solution:

The output signal $y(n)$ is given by

$$y(n) = \begin{cases} s\left(\dfrac{n}{I}\right), & n = \text{multiple of } 1 \\ 0, & \text{otherwise} \end{cases}$$

where $I = 2$

$$s(n) = 0, 1, 2, 3, 4, 5, 6, \; \ldots\ldots .$$
$$y(n) = 0, 0, 1, 0, 2, 0, 3, 0, 4, 0, 5, 0, 6.0, \ldots\ldots .$$

In general, to obtain the expanded signal $y(n)$ by a factor I, $(I - 1)$ zeros are inserted between the samples of the original signal $s(n)$.

The z-transform of the expanded signal is

$$Y(z) = s\left(z^{I}\right), I = 2.$$

The input and output signals are shown in Figure 12.24.

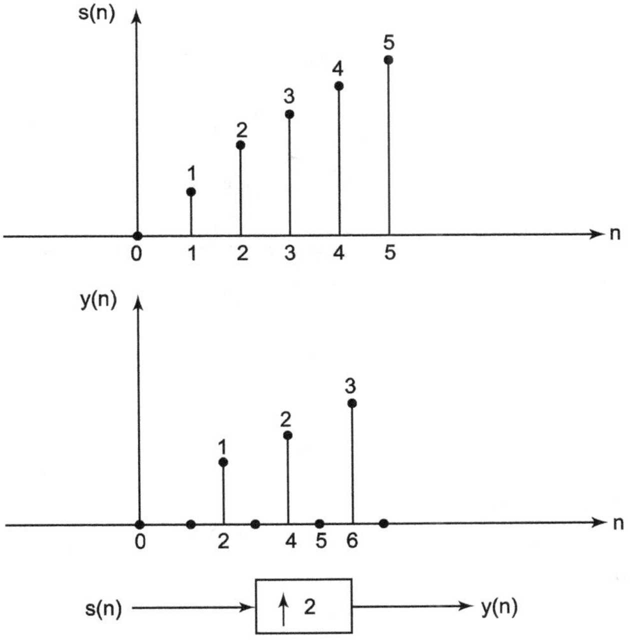

FIGURE 12.24.

EXAMPLE 12.3

Find the expression for the output $y(n)$ in terms of input $s(n)$ for the multi-sampling rate system given as follows:

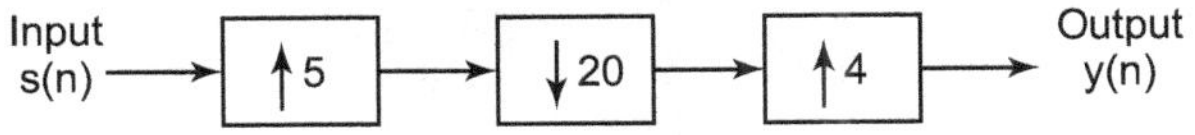

FIGURE 12.25.

Solution:

The decimation with factor 20 can be represented as a cascade of two decimators with factors 5 and 4. The resultant system is given as

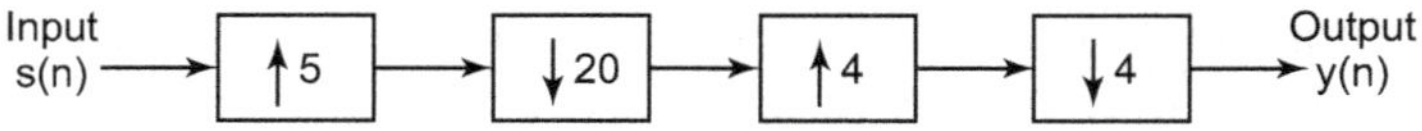

FIGURE 12.26.

Systems 1 and 2 can be combined. The upsampler operation of system 1 is canceled by the downsampler operation the system 2.

Therefore, $s_2(n) = s(n)$

Now, Figure 12.26 reduces to Figure 12.27

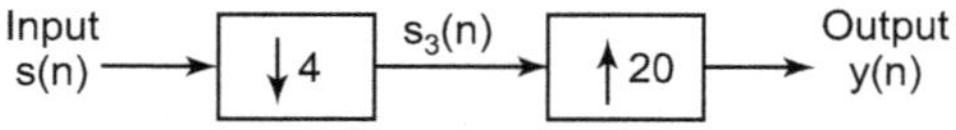

FIGURE 12.27.

Combining systems 3 and 4. The downsampler operation of system 3 is canceled by the upsampler operation of system 4.

It means that

$$s(n) = y(n)$$

where
$$s(n) = \begin{cases} 1\; n = 0, & \pm, 4, \pm 8, \pm 12, \ldots.. \\ 0, & \text{otherwise} \end{cases}$$

EXAMPLE 12.4

Find the polyphase decomposition of the *IIR* Digital System with transfer function.

$$H(z) = \frac{1 - 4z^{-1}}{1 + 5z^{-1}}.$$

Solution:

$$H(z) = H_o(z) + z^{-1}H_1(z) \tag{1}$$

where $H_o(z)$ and $H_1(z)$ are polyphase components of the *IIR* digital system $H(z)$:

$$H(z) = \frac{(1 - 4z^{-1})}{(1 + 5z^{-1})} = \frac{(1 - 4z^{-1})(1 - 5z^{-1})}{(1 + 5z^{-1})(1 - 5z^{-1})}$$

$$= \frac{1 - 9z^{-2} + 20z^{-2}}{1 - 25z^{-2}}$$

$$= \left(\frac{1 + 20z^{-2}}{1 - 25z^{-2}}\right) + z^{-1}\left(\frac{-9}{1 - 25z^{-2}}\right) \tag{2}$$

By comparing Eqs. (1) and (2), we get polyphase components of $H(z)$:

$$H_0(z) = \frac{1+20z^{-2}}{1-25z^{-2}}$$

$$H_1(z) = \frac{-9}{1-25z^{-2}}.$$

EXAMPLE 12.5

Implement a two-stage decimator for the following specifications:

Sampling rate of the Input signal $s(n)$

$$Fs = 20{,}000 \text{ Hz}$$

Decimation Factor, $D = 100$

Pass Band = 0 to 40 Hz

Transition Band = 40 to 50 Hz

Passband ripple = 0.01

Stopband ripple = 0.002.

Solution:

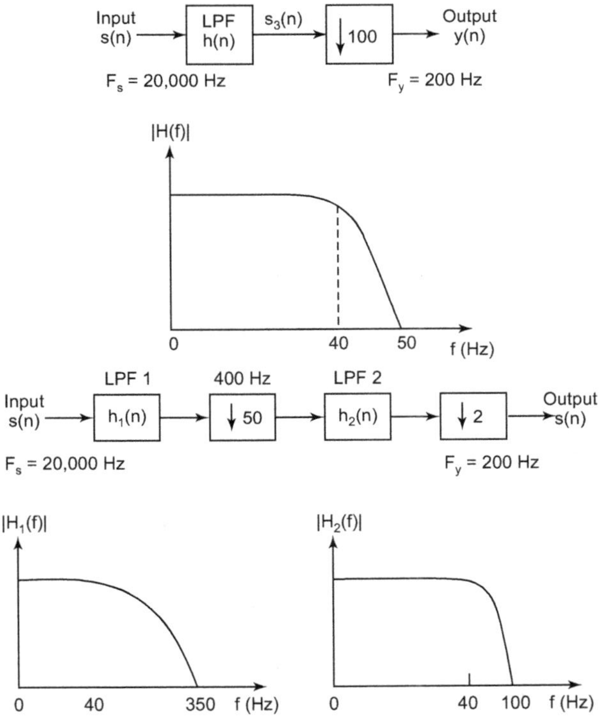

FIGURE 12.28 Illustration of single-stage and two-stage network for decimator.

The implementation of the system in shown in Fig. (12.28)

$$\text{Upper limit of passband } F_p = 40 \text{ Hz}$$
$$\text{limit of stopband } F_s = 50 \text{ Hz}$$
$$\text{Pass band ripple } \delta_p = 0.01$$
$$\text{Stopband ripple } \delta = 0.002$$
$$D = \text{Decimation factor} = 100$$

Sampling rate of the input signal $s(n)$

$$= F_s^1 = 20{,}000 \text{ Hz}$$
$$= 20 \text{ kHz}$$

Normalized transition bandwidth

$$\Delta f = \frac{F_s - F_p}{F_s'}$$

$$= \frac{50-40}{20{,}000} = \frac{10}{20{,}000} = 5\times 10^{-4}$$

For an equiripple linear phase *FIR* digital filter, the length N is given by

$$N = \frac{-20\log_{10}\sqrt{\delta_p\delta_s - 13}}{14.6\Delta f}$$

$$= \frac{-20\log_{10}\sqrt{(0.01)(0.02) - 13}}{14.6\left(5\times 10^{-4}\right)} = 4656$$

In the single-stage implementation, the number of a multiplication per second is,

$$N_{M,H} = N\frac{F_s'}{D} = \frac{4656\times 20{,}000}{100}$$
$$= 931200.$$

Two-Stage Realization

$H(z)$ can be implemented as a cascade realization in the form $G(z^{50})F(z)$. The steps in the two-stage realization of the decimator structure are shown in Figure 12.29 and the magnitude response is shown in Figure 12.30.

For a cascade realization, the overall ripple is the sum of the passband ripples of $F(z)$ and $G(z^{50})$. To maintain the stopband ripple at least as good as $F(z)$ or $G(z^{50})$, ds for both can be 0.002. The specification for the interpolated *FIR* digital filters is given by

For $\quad G(z).\ \delta_p = 0.005.\ \delta_s = 0.002$

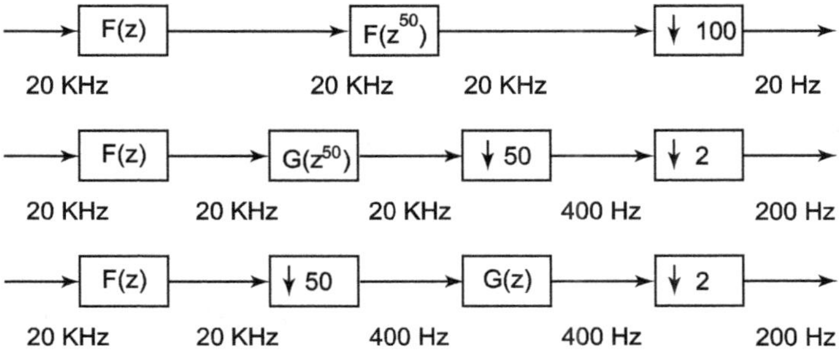

FIGURE 12.29 Two-stage realization of the decimators structure.

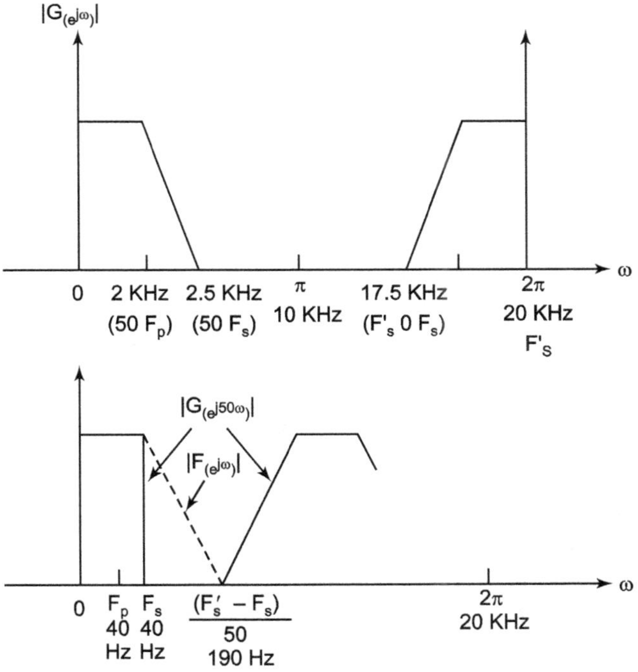

FIGURE 12.30 Magnitude response for a two-stage decimeter.

$$\Delta f = \frac{500}{20{,}000} = 2.5 \times 10^{-2}$$

For $$F(z), \delta_p = 0.005, \delta_s = 0.002$$

$$\Delta f = \frac{150}{20{,}000} = 7.5 \times 10^{-3}$$

The filter lengths are calculated as follows:

For $G(z)$,

$$N = \frac{-20\log_{10}\sqrt{(0.005)(0.002)} - 13}{14.6\left(\dfrac{2.5\times 10^3 - 2\times 10^3}{20\times 10^3}\right)}$$

$$= 101$$

For $F(z)$,

$$N = \frac{-20\log_{10}\sqrt{(0.005)(0.002)} - 13}{14.6\left(\dfrac{190 - 40}{20\times 10^3}\right)}$$

$$= 337$$

The length of the overall filter in cascade is given by

$$337 + (50 + 101) + 2 = 5389$$

The filter length in cascade realization has increased but the number of a multiplication per second can be reduced,

$$N_{M,G}\frac{101\times 400}{2} = 20{,}200$$

$$N_{M,F}\frac{337\times 20000}{50} = 134800$$

The total number of a multiplication per second is given by

$$N_{M,G} + N_{M,F} = 20{,}200 + 1{,}34{,}800$$

$$= 1{,}55{,}000.$$

EXAMPLE 12.6

Compare the single-stage, two-stage, three-stage, and multistage realization of the decimator with the following specifications.

A sampling rate of a signal has to be reduced from 10 kHz to 500 Hz. The decimation filter $H(z)$ has the passband edge (F_p) to be 150 Hz, stopband edge (F_s) to be 180 Hz.

Passband ripple $\delta_p = 0.002$

Stopband ripple $\delta_s = 0.001$

Solution:

The length N of an equiripple linear phase *FIR* digital filter is given by

$$N = \frac{-20\log_{10}\sqrt{\delta_p\delta_s} - 13}{14.6\Delta f}$$

where $\quad \Delta f$ = Normalized transition band-width

$$= \frac{F_s - F_p}{F_s'}$$

Given $$F_s' = 10\text{ kHz}$$

$$N = \frac{-20\log_{20}\sqrt{(0.002)(0.001)} - 13}{14.6\left(\dfrac{180-150}{10{,}000}\right)}$$

$$\approx 1004$$

For the single-stage implementation of the decimator with a decimation factor of 20, the number of multiplications per second is given by

$$N_{M,H} = \frac{N F_s'}{D}$$

$$= \frac{1004 \times 10{,}000}{20} = 50{,}2000$$

Two-stage realization

$H(z)$ can be implemented as a cascade realization in the form of $G(z^{10})\,F(z)$. The steps in the two-stage realization of the decimator structure are shown in Figure 12.31(a) and the response is shown in Figure 12.31(b)

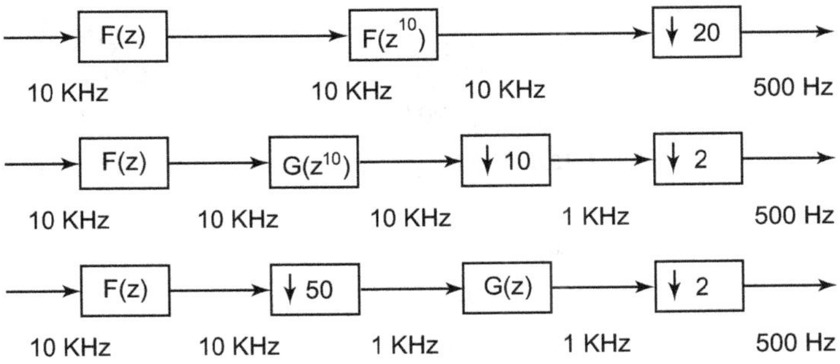

FIGURE 12.31 (a) Two-stage realization of the decimators structure.

For the cascade realization, the overall ripple is the sum of the passband ripples of $F(z)$ and $G(z^{10})$. To maintain the stopband ripple at least as good as $F(z)$ or $G(z^{10})$, δs for both can be 0.001. The specification for the interpolated *FIR* digital filters is given by

For $$G(z), \delta p = 0.001$$

$$\delta_s = 0.001$$

$$\Delta_f = \frac{300}{10,000}$$

For $F(z), \delta_p = 0.001$

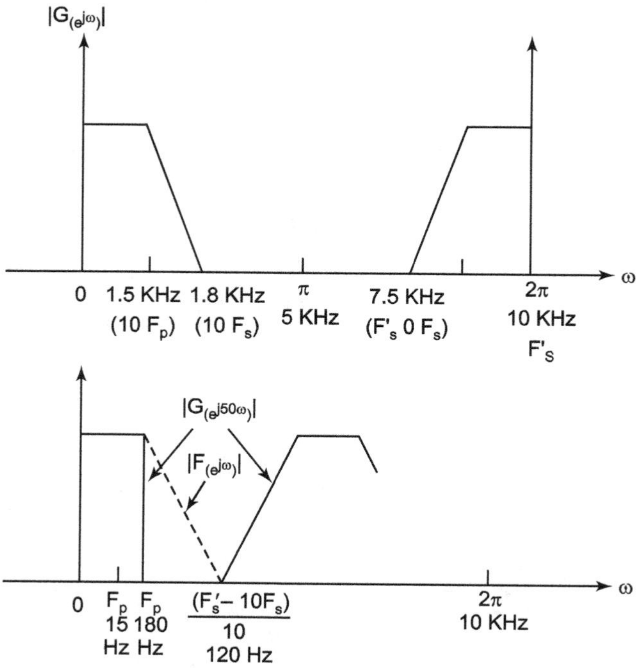

FIGURE 12.31 (b) Magnitude response for a two-stage decimator.

$$\Delta_s = 0.001$$

$$\Delta f = \frac{570}{10,000}$$

The filter length N can be determined as follows:
For $G(z)$,

$$N = \frac{-20\log_{10}\sqrt{\delta_p\delta_s} - 13}{14.6(\Delta f)} = \frac{-20\log_{10}\sqrt{\delta_p\delta_s} - 13}{14.6\left[\dfrac{F_s - F_p}{F_s^{'}}\right]}$$

For $F(z)$,

$$= \frac{-20\log_{10}\sqrt{(0.001)(0.001)} - 13}{14.6\left[\dfrac{18\times10^3 - 15\times10^3}{10\times10^3}\right]} = 107$$

The length of the overall filter in cascade is given by

$$56 + (10 \times 107) + 2 = 1128$$

The filter length in cascade realization has increased but the number of multiplications per second can be reduced.

$$N_{M,G} = 107\frac{1000}{2} = 53,500$$

$$N_{M,F} = 56\frac{10,000}{10} = 56,000$$

The total number of a multiplication per second is

$$= N_{M,G} + N_{M,F}$$
$$= 53,500 + 56,000$$
$$= 1,09,500$$

Three-Stage Realization

The decimation filter $F(z)$ can be realized in the cascade form $P(z)$ $Q(Z^5)$.

The specifications are given as follows:

For $G(z)$, $\delta p = 0.005$

$$\delta_s = 0.001$$

$$\delta f = \frac{570}{10,000} \times 5 = 0.285$$

$$N = 12$$

For $P(z)$, $\delta_p = 0.0005$

$$\delta_s = 0.001$$

$$\delta f = \frac{1130}{10,000} = 0.113$$

$$N = 30$$

The three-stage realization is shown in Figure 12.32.

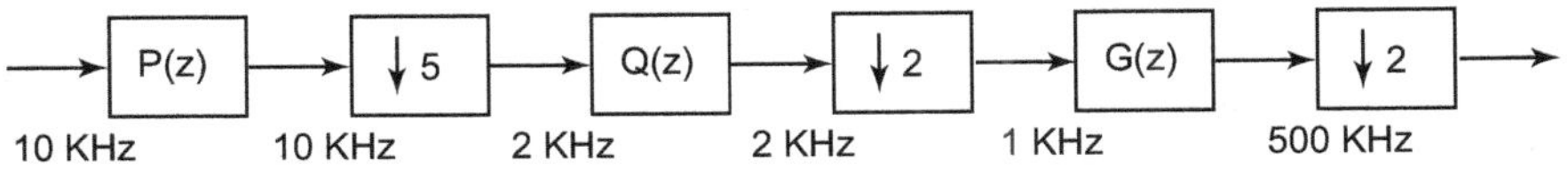

FIGURE 12.32 Frequency response for a three-stage decimation.

The number of multiplications per second is given by

$$N_{M,Q} = \frac{12 \times 2000}{2} = 12,000$$

$$N_{M,P} = \frac{30 \times 1000}{2} = 60{,}000$$

The overall number of multiplications per second for a three-stage realization is given by

$$N_{M,G} + N_{M,Q} + N_{M,P} = 53{,}500 + 12{,}000 + 60{,}000$$
$$= 1{,}25{,}500$$

The number of multiplications per second for a three-stage realization is more than that of a two-stage realization. Hence higher than two-stage realization may not lead to an efficient realization.

EXAMPLE 12.7

We have given a multisampling rate system shown in Figure 12.33. Find $y(n)$ as a function of $s(n)$.

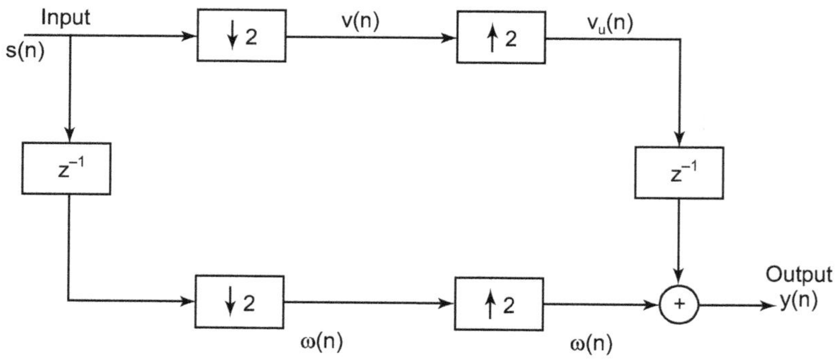

FIGURE 12.33 Multisampling rate system.

Solution:

From Figure 12.33, the outputs of the downsampler are given as

$$V(z) = \frac{1}{2}S\left(z^{\frac{1}{2}}\right) + \frac{1}{2}S\left(-z^{\frac{1}{2}}\right) \tag{1}$$

$$W(z) = \frac{z^{\frac{-1}{2}}}{2}S\left(z^{\frac{1}{2}}\right) - \frac{z^{\frac{-1}{2}}}{2}S\left(-z^{\frac{-1}{2}}\right) \tag{2}$$

The outputs of the upsampler are

$$V_u(z) = \frac{1}{2}S(z) + \frac{1}{2}S(-z) \tag{3}$$

$$W_u(z) = \frac{Z^{-1}}{2}S(z)\frac{-Z^{-1}}{2}S(-z) \tag{4}$$

$Y(z)$ is given by

$$Y(z) = z^{-1}V_u(z) + W_u(z) \tag{5}$$

Substituting Eqs. (3) and (4) in Eq. (5), we get

$$Y(z) = Z^{-1}\left[\frac{1}{2}S(z) + \frac{1}{2}S(-z)\right] + \left[\frac{z^{-1}}{z}S(z) - \frac{z^{-1}}{2}S(-z)\right]$$

or

$$Y(z) = Z^{-1}S(z) \tag{6}$$

Taking the inverse z-transform of both sides of Eq. (6), we get

$$y(n) = s(n-1).$$

EXERCISES

1. What is multirate digital signal processing (MDSP)?

2. What is the need for multirate digital signal processing?

3. Give some examples of multirate digital systems.

4. Write short notes on the following topics:

a. MDSP, **b.** Decimator, **c.** Decimation filter, **d.** Interpolator, and **e.** Interpolation filter.

5. Explain the interpolation process for an integer factor I with an example.

6. Explain the decimation process for an integer factor D with an example.

7. The signal $s(n)$ is defined by

$$s(n) = \begin{cases} A^n, & n > 0 \\ 0, & \text{otherwise} \end{cases}$$

a. Obtain the decimated signal with a factor of 3
b. Obtain the interpolated signal with a factor of 3.

8. Explain polyphase decomposition process.

9. Describe the sampling rate conversion by a rational factor $\left(\frac{I}{D}\right)$.

10. Obtain the polyphase structure of the filter with the transfer function.

$$H(z) = \frac{1-3z^{-1}}{1+4z^{-1}}.$$

11. Give the name of some areas where *MDSP* systems are used.
12. Give the advantages of using *MDSP* systems.
13. Discuss filter design and implementation for sampling rate conversion.
14. Describe and draw a direct-form *FIR* digital filter structure.
15. Write short notes on the following:
    a. Polyphase digital filter structure
    b. Time-varying digital filter structure
16. Describe the sampling rate conversion by an arbitrary factor.
17. Write short notes on the following:
    a. Sampling rate conversion by use of the first-order approximation method.
    b. Sampling rate conversion by use of the second-order (Linear) approximation method.
18. List some applications *of MDSP.*

CHAPTER 13

Applications of Digital Signal Processing to Speech

13.1 INTRODUCTION

Some of the most important applications of digital signal processing techniques have been in the area of speech processing. In fact, a large percentage of the theoretical background of digital signal processing (DSP) has been derived from speech studies. A speech signal is a one-dimensional signal. DSP techniques can be applied to a wide range of problems in speech such as spectrum analysis, channel vocoders, homomorphic processing systems, speech synthesizers, linear prediction systems, and computer voice response systems. All modern speech processing systems are dependent on DSP algorithms.

Speech processing problems are divided into three categories:

1. Speech Analysis.
2. Speech Synthesis.
3. Speech Compression.

We now discuss them below.

Speech Analysis: In this category, speech analysis is performed to extract some desirable information of speech signal. An example of speech analysis is in the automatic speech recognition system. This system starts with an analysis of speech waveform and the desired results are used for speech recognition and speaker identification.

Speech Synthesis: In this type of problem, input is in written text form and the output is a speech signal. An example of speech synthesis is automatic reading machines. It is used to retrieve data from a computer via telephone line in the form of speech at a remote place.

Speech Compression: In this type of problem speech signal is coded in such a manner that there will be some compression in a speech signal. Speech compression is done in order to have smaller bandwidth for its transmission. Compression means reducing the number of bits required for representing its samples. Purpose of speech compression to preserve bandwidth for its transmission. Another purpose of speech compression is to reduce to the number of bits required per sample. By doing so memory requirement (Number of bytes) is reduced.

13.2 MODEL OF SPEECH PRODUCTION

13.2.1 Model of Speech Waveform

Basic techniques of speech analysis and synthesis, use the knowledge of a model of the speech waveform. Speech is considered as the response of a slowly time-varying system. There are two types of excitations used. One is periodic and the other is noise-like. The speech production mechanism consists of an acoustic tube, a vocal tract that is excited by an appropriate source to generate the desired sound.

Figures 13.1 and 13.2 illustrate the cross-sectional view of the speech production system of a human (vocal tract system) and schematic diagram of human speech production system (vocal tract system) respectively.

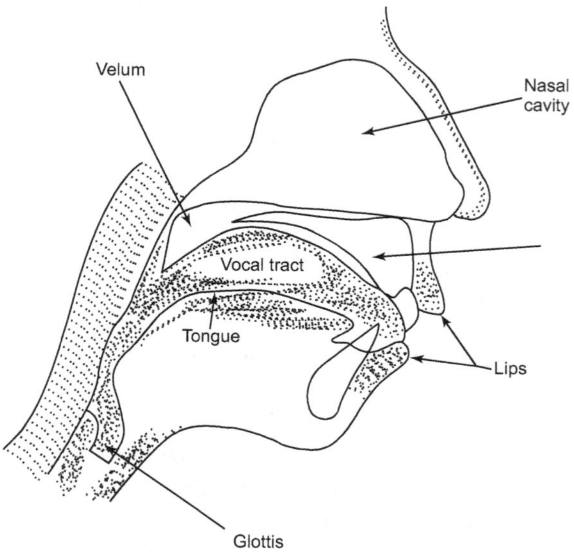

FIGURE 13.1 Cross-sectional view of the speech production system (vocal tract).

Figure 13.2 illustrates a schematic diagram of

the human speech production mechanism. In a normal speech production mechanism, the chest cavity expands and contracts to force air from the lungs out through the trachea past the glottis.

If the vocal cords which are available in the trachea are tensed then voiced sounds like vowels are produced. These vocal cords vibrate in the mode of a relaxation oscillator and modulate the air into discrete pulses. The air stream passes through the pharynx cavity, then passes through a tongue, and at the last it passes through either nasal cavity or mouth cavity depending on the position of the trap door velum. The air stream is expelled at either the nose or mouth or both and it is perceived as speech.

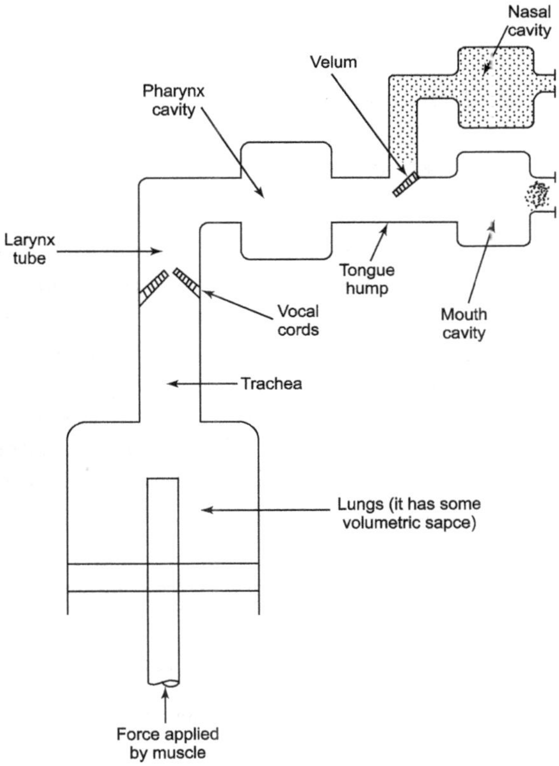

FIGURE 13.2 Schematic diagram of the human speech production system (vocal tract).

In the case of unvoiced sounds, the vocal cords are spread apart and one or two conditions are possible. Either a turbulent flow is produced as the air passes through the narrow constriction in the vocal tract or a brief transient excitation occurs following a build-up of pressure behind a point of total closure along the vocal tract.

Shapes of various cavities can be changed drastically by changing the position of various articulators such as lips, tongue, jaw, and velum during continuous speech.

As we know now that speech sounds are of two types:

1. **Voiced Sounds:** Voiced sounds are produced by forcing air through the glottis with the tension of vocal cords so adjusted that they vibrate in the mode of a relaxation oscillator and produce quasi-periodic pulses of air that excite the vocal tract. Voiced speech is repetitive in nature and its

waveform contains high energy. Rich melodious sounds are generally voiced sounds.

2. **Unvoiced Sounds:** It is also called fricative sounds. These sounds are generated by forcing air through a constriction in the vocal tract at high enough velocity to produce turbulence. Unvoiced speech is non-repetitive and random in nature. Hissing sounds are unvoiced.

Speech waveform can be modeled as the response of a linear time-varying system (Vocal tract) with appropriate excitation. If the vocal tract (acoustic tube) has a fixed shape, the output of the system is determined by the convolution of excitation and vocal tract impulse response. Various types of sounds are produced by changing the shape of the vocal tract. If the vocal tract changes slowly, the output is still approximated on a short-time basis and determined by the convolution of excitation and vocal tract impulse response.

Waveforms of voiced and unvoiced speech for a long duration (600 ms) are shown in Figure 13.3.

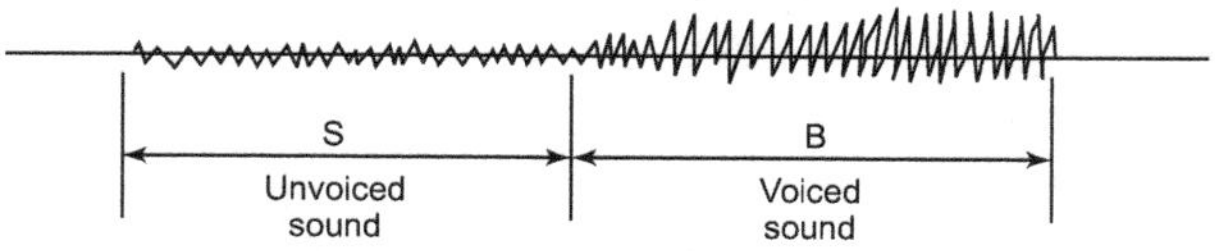

FIGURE 13.3 Waveform for voiced and unvoiced speech (long segment ≠ 600 ms).

A waveform of voiced speech for a short duration (25 ms) is shown in Figure 13.4.

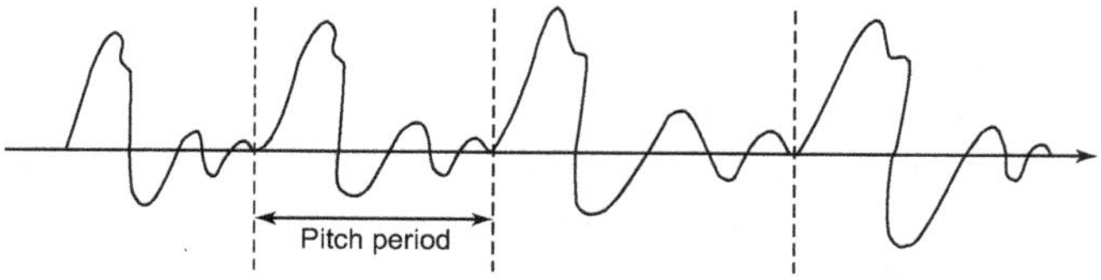

FIGURE 13.4 Waveform for voiced speech (short segment ≠ 25 ms).

A waveform of an unvoiced signal for a short duration (25 ms) is shown in Figure 13.5.

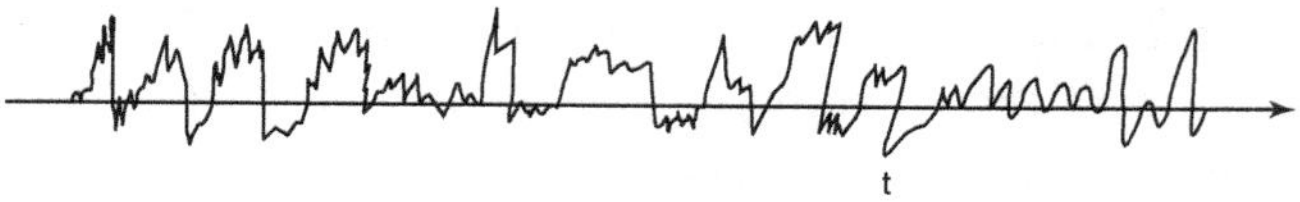

FIGURE 13.5 Waveform for unvoiced speech (short segment ≠ 25 ms).

If the input is periodic on a short-time basis for voiced speech, the output to the corresponding fundamental frequency is also periodic.

Now if we want to see the system in the frequency domain. The Fourier transform of speech waveform is the product of Fourier transforms of excitation function and impulse response of the vocal tract system.

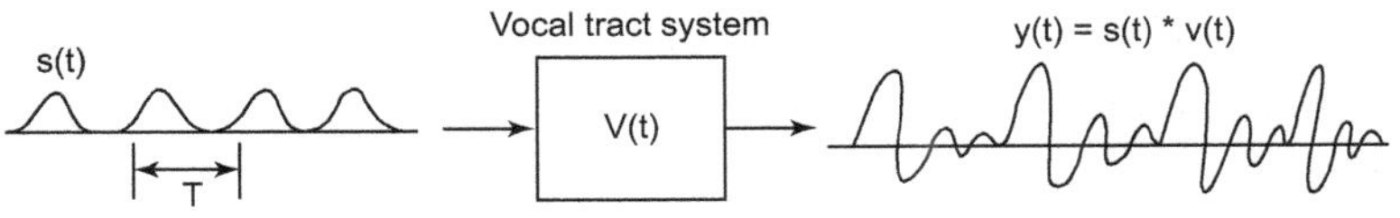

FIGURE 13.6 Time-domain characterization of a vocal tract system.

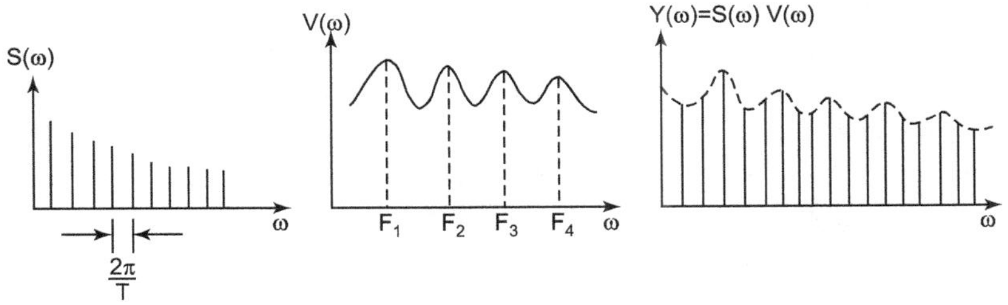

FIGURE 13.7 Frequency domain, characterization of vocal tract system.

The resonant frequency of the vocal tract is called formants, it is denoted by F_1, F_2, F_3, and so on. The human vocal tract system extends from the glottis to the lips. Its length is approximately 17 cm. The first formant will be at:

$$F_1 = \frac{\lambda_1}{4} = \frac{1}{4}\left(\frac{v}{L}\right) = \frac{1}{4}\left(\frac{340}{0.17}\right) = 500 \text{ Hz}$$

where v' = velocity of sound = 340 m/s and L = vocal tract length = 17 cm.

The non-uniform cross-sectional area of the vocal tract depends strongly on the position of the articulators and varies from 0 cm^2 at closure to about 20 cm^2. The vocal tract has certain normal resonant modes of operation. These resonant modes of operation are called formants. These formants depend heavily on the exact position of the articulators.

There are three primary modes for exciting the vocal tract system. For voiced sounds, the source is at the glottis and consists of broadband quasi-periodic puffs of air produced by the vibrating vocal cords.

For unvoiced sounds like *s*, the source is at the point constriction and consists of turbulent quasi-random airflow.

Finally, for unvoiced sounds like *p* as in pop, the source *is* at the point of closure and consists of a rapid release of the air pressure built up behind the total constriction.

The basic assumption of almost all speech processing systems is that the source of excitation and the vocal tract system are independent. Source-system independence allows us to discuss the transmission function of the vocal tract and to let it be excited by any of the possible sources. The validity of the assumption above is quite good for the majority of cases of interest. There are some cases, however, when the assumption is invalid and the basic model breaks down such as during transient sounds like *p* in a pot.

Based on the ideas above, a simple digital model of speech production is shown in Figure 13.8.

The sources of excitation are an impulse generator and a random number generator. The impulse generator is controlled from the outside world by the pitch-period signal. The impulse generator produces an impulse (corresponding to the initiation of a puff of air) once every N_0 sample. This duration is referred to as the pitch period and its reciprocal is the pitch frequency or rate of oscillation of the vocal cords. The random number generator output simulates both the quasi-random turbulence and the pressure build-up waveform for unvoiced sounds.

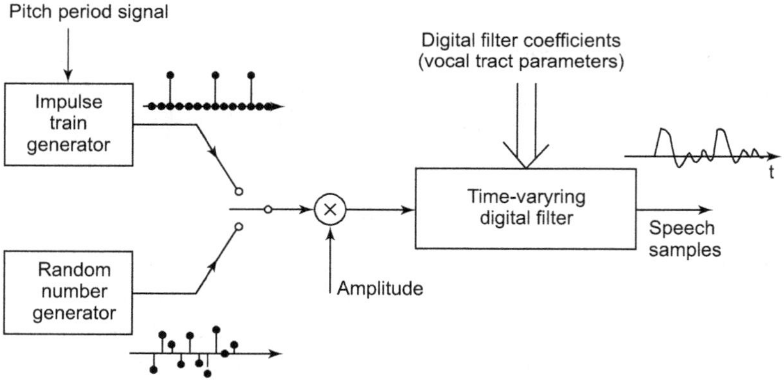

FIGURE 13.8 Digital model of speech production.

Either or both of these sources may be applied as input to a linear, time-varying digital filter. This filter simulates the vocal tract system and thus the filter coefficients specify, the vocal tract as a function of time during continuous speech. Once every 10 ms, on average, the filter coefficients are varied, indicating a new vocal track configuration.

Finally, gaining control between the source and system allows certain flexibility in the acoustic level of the output. The digital waveform at the output of the filter corresponds to the final speech output, sampled at the appropriate rate.

To control the model above requires a knowledge of the appropriate parameters as a function of time. Various parameters are pitch period, switch position, amplitude, and filter coefficients. This is the goal of almost all speech analysis systems. The purpose of speech analysis systems is to estimate the appropriate model parameters from real speech.

The goal of most speech synthesis systems is to use these parameters, obtained in any reasonable manner, to derive a synthetic speech signal that is indistinguishable perceptually from the original signal. Speech analysis-synthesis systems combine the two problems with the twin goals of efficiency and flexibility. Efficiency means to lower the bit rate of the synthesis system below that required for conventional waveform representations. Flexibility is the ability to modify and alter the speech in some desired manner through manipulation of the model parameters.

13.3 SHORT-TIME FOURIER TRANSFORM (STFT)

13.3.1 Short-Time Fourier Transform (STFT)

Fourier transform can be used for time-varying signals such as speech signals. In the frequency domain, speech can be represented on short-time basis in terms of the product of the Fourier transform of a transfer function of the vocal tract and Fourier transform of the excitation. Consequently, the spectrum of speech must be based on a STFT which of course changes with time.

STFT of sampled speech sequence is defined as

$$STFT\left[s(n) = S(\omega, n)\right] = \sum_{k=-\infty}^{\infty} s(k)h(n-k)e^{-j\omega k} \tag{13.1}$$

It represents the Fourier transform of a windowed segment of speech signal where window $h(n)$ slides with time.

13.3.1.1 Implementation of STFT

Now let us suppose that $h(n)$ is the impulse response of the vocal tract and $s(n)$ is the input speech signal. Convolution between $h(n)$ and $s(n)$ multiplied by $e^{-j\omega k}$ summed over $-\infty$ to ∞ is termed as $STFT$.

$$S(w,n) = \sum_{k=-\infty}^{\infty} s(k)h(n-k)e^{-j\omega k}$$

$$= e^{-j\omega k} \sum_{k=-\infty}^{\infty} s(k)h(n-k).e^{-j\omega k}e^{-j\omega k}$$

or $$S(\omega,n) = e^{-j\omega k} \sum_{k=-\infty}^{\infty} s(k)h(n-k)e^{+j\omega k(n-k)} \qquad (13.2)$$

Eq. (13.2) can also be written

$$S(\omega,n) = e^{-j\omega k}\{s(n) * [h(n)]e^{j\omega n}\} \qquad (13.3)$$

Figure 13.9 shows the block diagram for the determination of STFT. Here original speech signal $s(n)$ is first-multiplied by $e^{-j\omega k}$. The output of multiplier, that is, $h(n)\ e^{-j\omega k}$ is linearly convoluted with $h(n)$ to get STFT of the speech signal. It is given by the following equation.

$$\text{STFT of } s(n), S(\omega,n) = \left[s(n)e^{-j\omega k}\right] * h(n)$$

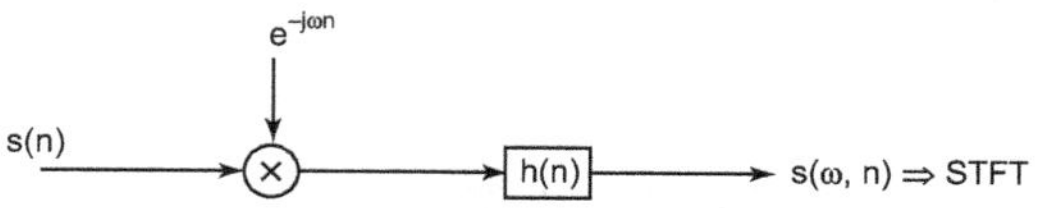

FIGURE 13.9 Determination of STFT.

Figure 13.10 illustrates the block diagram of an alternative method of determination of STFT. In this method, speech signal $s(n)$ is linearly convoluted with the product of $h(n)$ and $e^{-j\omega n}$ to get $S(\omega, n)\ e^{-j\omega n}$. It is given by the following equation

$$DTFT \text{ of } s(n), S(\omega,n) = e^{-j\omega k}\left\{s(n) * \left[h(n)e^{-j\omega k}\right]\right\}$$

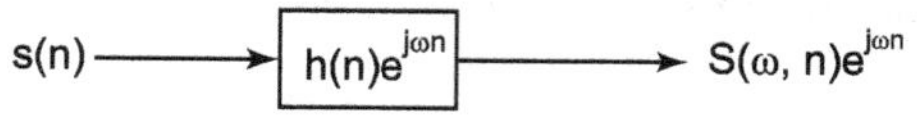

FIGURE 13.10 An alternative method of determination of STFT.

There are two common ways in which we can implement the STFT given by Eq. (13.1).

1. The first method uses a filter bank when a spectral analysis is done with an analog system.

2. The other method uses a digital circuitry (FFT algorithm) for the implementation of STFT.

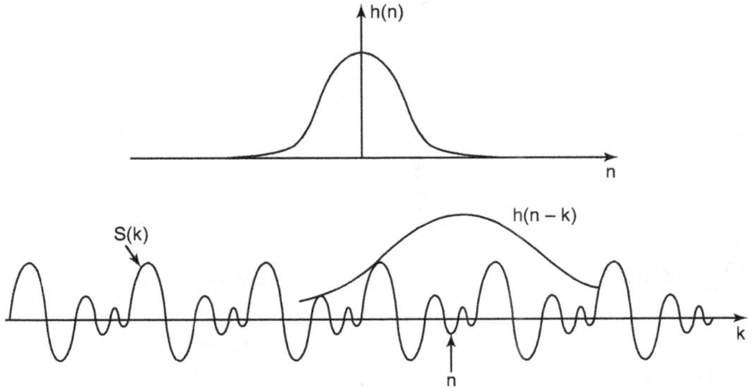

FIGURE 13.11 Short-time Fourier transform (STFT) analysis explanation.

A more useful measure of the energy content of a speech waveform *is* the STFT. STFT of sampled speech sequence $s(n)$ is defined as

$$STFT\ [s(n) = S(\omega, n) = \sum_{k=-\infty}^{\infty} s(k)h(n-k)e^{-j\omega k} \tag{13.1}$$

where $e^{-j\omega k} = \cos\omega k - j\sin\omega k$.

This can also be written as

$$S(\omega, n) = A(\omega, n) + j\,B(\omega, n) \tag{13.2}$$

where $A(\omega, n)$ and $B(\omega, n)$ are the real and imaginary parts of the STFT, $S(\omega, n)$ of sequence $s(n)$.

From Eqs. (13.1) and (13.2), we get

$$A(\omega, n) = \sum_{k=-\infty}^{\infty} s(k)h(n-k)\cos(\omega k) \tag{13.3}$$

and

$$B(\omega, n) = -\sum_{k=-\infty}^{\infty} s(k)\,h(n-k)\sin(\omega k) \tag{13.4}$$

The equations given above suggest a simple technique for measuring STFT that is shown in Figure 13.12.

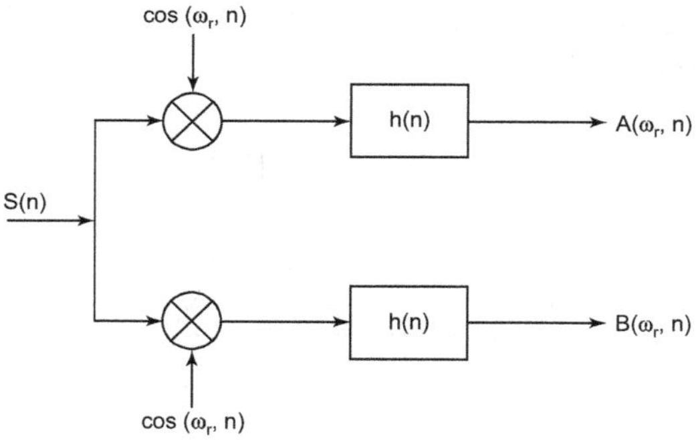

FIGURE 13.12 Simple techniques for analyzing speech based on STFT Analysis.

Generally, $H(e^{j\omega})$ is chosen to approximate the ideal low pass filter with cutoff frequency ω_c, as shown in Figure 13.13, where $H(e^{j\omega})$ *is* the Fourier transform of $h(n)$. Thus $S(\omega, n)$ is the energy of the speech waveform at time n and at frequency w. The energy measurement reflects the speech energy in the band from $(\omega - \omega_c)$ to $(\omega + \omega_c)$.

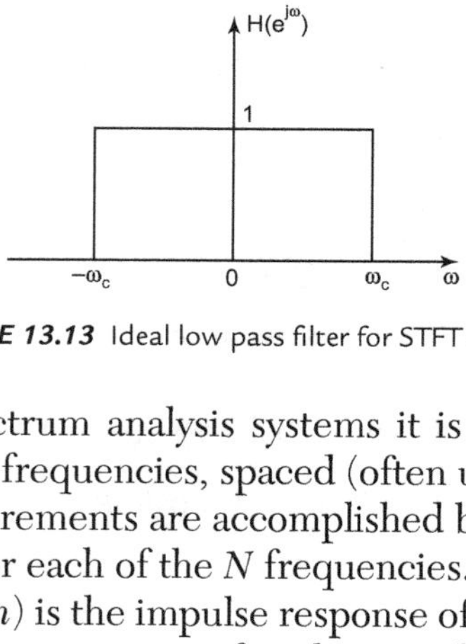

FIGURE 13.13 Ideal low pass filter for STFT analysis.

In most speech spectrum analysis systems it is desired to measure the STFT at a finite set of N frequencies, spaced (often uniformly) over the band $0 \leq \omega \leq 2\pi$. These measurements are accomplished by iterating the measurement technique above for each of the N frequencies.

In the case where $h(n)$ is the impulse response of an FIR digital filter and where the analysis frequencies are uniformly spaced, the FFT algorithm can be used to simultaneously make all the desired measurements in an extremely efficient manner. To see this, let $h(n)$ be non-zero for $0 \leq n \leq M - 1$ and let the center frequencies for analysis ω_r be chosen as

$$\omega_r = \frac{2\pi}{N} r, r = 0, 1, \ldots, N-1 \tag{13.5}$$

Eq. (13.1) can be written as

$$S(\omega_r,n)=\sum_{k=n-m+1}^{n} s(k)h(n-k)e^{-j\omega_r k} \tag{13.6}$$

$$=\sum_{m=0}^{\left[\frac{M}{n}\right]+1}\ \sum_{k=n-(m+1)N+1}^{n-mN} s(k)h(n-k)e^{-j\omega_r k} \tag{13.7}$$

where $\left[\dfrac{M}{N}\right]$ stands for the greatest integer less than or equal to *M*/*N*.

If we let $I=n-mN-k$,

$$S(\omega_r,n)=\sum_{m=0}^{\left[\frac{M}{N}\right]+1}\sum_{l=0}^{N-1} s(n-k-mN)h(l+mN)e^{-j\omega_r(l-n+mN)}$$

Substituting Eq. (13.5) in Eq. (13.8), we get

$$S(\omega_r,n)=e^{-j\left(\frac{2\pi}{N}\right)}\sum_{l=0}^{N-1}\left[\sum_{m=0}^{\left[\frac{M}{N}\right]+1} s(n-l-mN)h(l+mN)e^{j\left(\frac{2\pi}{N}\right)rl}\right] \tag{13.9}$$

where we have replaced $e^{j2\pi m}$ by 1.

Eq. (13.9) can now be written as

$$S(\omega_r,n)=e^{-j\left(\frac{2\pi}{N}\right)rn}\underbrace{\sum_{l=0}^{N-1} g(l,n)e^{j\left(\frac{2\pi}{N}\right)lr}}_{DFT} \tag{13.10}$$

where

$$g(l,n)=\sum_{m=0}^{\left[\frac{M}{N}\right]+1} s(n-l-mN)h(l+mN) \tag{13.11}$$

Eq. (13.11) shows that $S(\omega_r, n)$ may be obtained as the product of the sequence $e^{-j\left(\frac{2\pi}{N}\right)rn}$ and the DFT of the sequence $g(l, n)$.

Figure 13.14 illustrates how the sequence $g(k, n)$ is obtained term by term from the individual sequences $s(k)$ and $h(k)$.

Thus STFT analysis of speech is readily performed either directly using a bank of digital filters and modulators or indirectly using the FFT algorithms.

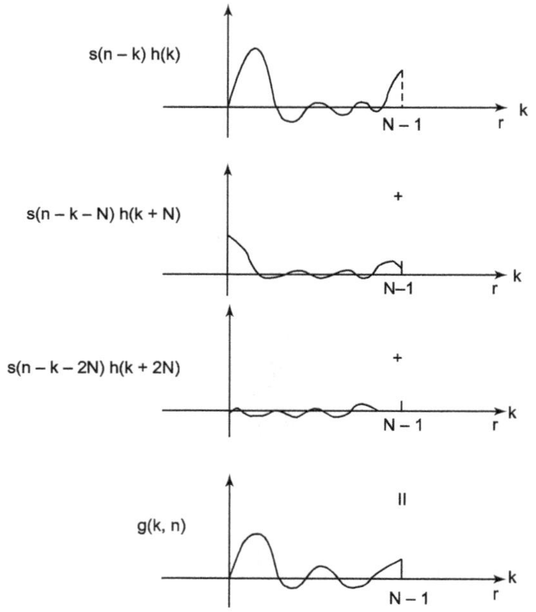

FIGURE 13.14 The construction of $g(r, n)$ from $x(nT)$ and $h(nT)$.

13.4 SPEECH ANALYSIS-SYNTHESIS USING STFT

The principles of measuring the short-time spectrum of speech may be applied to an entire analysis-synthesis system. The basic idea is to measure the outputs of a bank of M bandpass filters (BPFs) and reconstruct the speech from these M signals. A simplified schematic diagram of this system is shown in Figure 13.15.

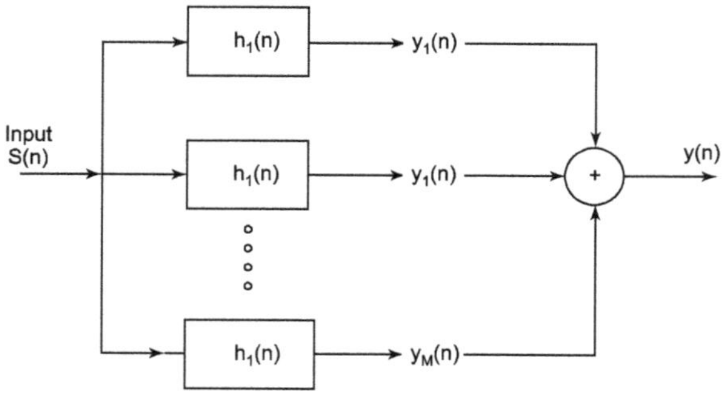

FIGURE 13.15 Schematic diagram of an analysis-synthesis system based on STFT or short-time spectrum analysis.

The input speech is $s(n)$ and the reconstructed synthetic waveform is $y(n)$. The M individual BPFs have impulse responses $h_r(n)$, $r = 1, 2, ..., M$. The bandpass outputs are labeled $y_r(n)$, $r = 1, 2 ..., M$. If the bandpass filter impulse responses are restricted to be of the form

$$h_r(n) = h(n)\ \cos(\omega_r,\ n) \tag{13.12}$$

where $h(n)$ is the impulse response of a low pass filter (LPF) (i.e., the bandpass impulse response is a modulated low pass impulse response), then the bandpass outputs $y_r(n)$ can be written as

$$y_r(n) \sum_{k=-\infty}^{\infty} s(k)h(n-k)\cos[\omega_r(n-k)] \tag{13.13}$$

or

$$y_r(n) = \text{Real}\left[e^{j\omega_r n} S(\omega_r, n)\right] \tag{13.14}$$

where $S(\omega_r, n)$ is the STFT. Thus each channel of the system can be obtained in the manner shown in Figure 13.16.

Since $S(\omega_r, n)$ can be written in terms of real and imaginary components as given in Eq. (13.14). Now, Eq. (13.14) can be put in the form.

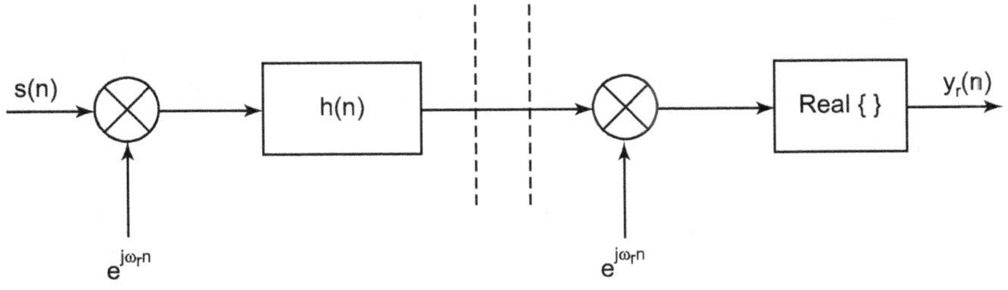

FIGURE 13.16 The operations for the *r*th channel.

$$y_r(n) = A(\omega_r, n)\cos(\omega_r, n) + B(\omega_r, n)\sin(\omega_r, n) \tag{13.15}$$

Realization of Eq. (13.15) is shown in Figure 13.17. The dashed lines in Figures 13.16 and. 13.17 indicate points of transmission and reception when the system is implemented as a speech bandwidth compression system. The straight path between the dotted lines represents the communication channel (assumed error-free here).

The transmitted parameters $A(\omega_r, n)$ and $B(\omega_r, n)$ would have to be sampled to a lower rate than the speech transmission rate and quantized to achieve any significant bandwidth reduction.

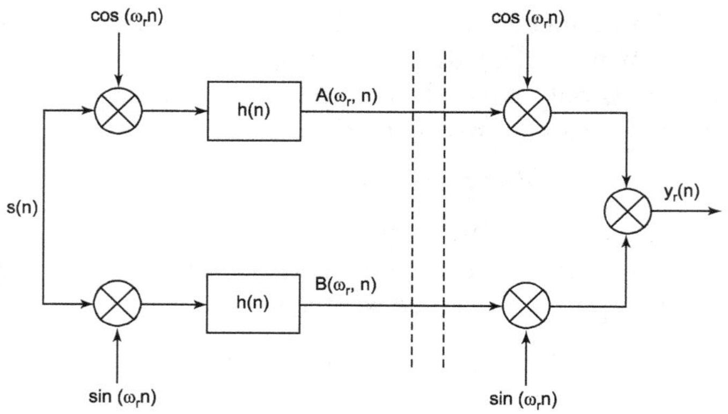

FIGURE 13.17 The operations for the *r*th channel using real processing.

13.5 ANALYSIS CONSIDERATIONS

The quality with which this system can represent speech is dependent on the extent to which the bank of M filters adequately represents the speech spectrum. One simple way of measuring this is to determine the overall impulse response of the system and examine its Fourier transform. If we denote the impulse response of the composite bank as $\hat{h}(n)$, then

$$\hat{h}(n) = \sum_{r}^{M} h_r(n) = \sum_{r}^{M} h(n)\cos(\omega_r n)$$
$$= -h(n)\sum_{r}^{M}\cos(\omega_r n) \tag{13.16}$$

If we denote $\sum_{r}^{M}\cos(\omega_r n)$ by $d(n)$. Then we can write as

$$d(n) = \sum_{r=1}^{M}\cos(\omega_r n) \tag{13.17}$$

Now Eq. (13.17) can be written as

$$\hat{h}(n) = h(n)d(n) \tag{13.18}$$

It means that the composite filter bank impulse response is the product of the prototype LPF impulse response and response depends entirely on the number of filters M and their center frequencies ω_r.

To see how well $\hat{h}(n)$ approximates a digital impulse (with some delay), we can examine the behavior of either $\hat{h}(n)$ or its Fourier transform.

In the special case of a uniform filter bank with

$$\omega_r = \Delta\omega r \text{ (for uniform spacing)} \tag{13.19}$$

If Δw is a constant, then $d(n)$ can be solved by using Eq. (13.17). It gives

$$d(n) = \sum_{r=-M}^{M} e^{jr\Delta\omega n} - 1 \tag{13.20}$$

Eq. (13.20) can be simplified as follows:

$$d(n) = \frac{\sin\left[\left(M + \frac{}{2}\right)\Delta\omega n\right]}{\sin\left[(\Delta\omega / z)n\right]} - \tag{13.21}$$

If $\Delta\omega = \frac{2\pi}{N}$, with N as integer, then the sequence $d(n)$ *is* periodic with period N samples.

If $\Delta\omega = \frac{2\pi}{N}$ is not an integer, then the sequence $d(n)$ is not periodic but still has peaks at intervals of N seconds.

Another interesting case is that we choose parameters as follows.

Let N be an odd integer and $M = \left(\frac{N-1}{2}\right)$. For $\Delta\omega = \frac{2\pi}{N}$, it can easily be seen that this corresponds to evaluating the STFT at equally spaced frequencies in the range $0 < \omega < \pi$. If, in addition, we include a channel centered on zero frequency. It can be shown that

$$d(n) = \frac{\sin(\pi n)}{\sin\left(\frac{\pi n}{N}\right)} = \begin{cases} N, & n = 0, \pm N, \pm 2N, \ldots \\ 0, & \text{elsewhere} \end{cases} \tag{13.22}$$

Thus, for these conditions $d(n)$ is a periodic train of impulses, with a period N that is inversely proportional to the frequency spacing between channels. Since $\hat{h}(n) = h(n)\, d(n)$, it is clear that the composite impulse response will also be an impulse train. Since the ideal composite impulse response is a delayed impulse, we must choose the prototype low pass impulse response $h(n)$ so as to eliminate all but one of the impulses in $d(n)$. Suppose we fix N, corresponding to fixed frequency spacing $\Delta\omega$.

Then if we choose a very narrow impulse response (e.g., of duration less than $2N$) the composite impulse response will appear as shown in Figure 13.17(*a*). Here we have shown the prototype low pass response or data window as a dotted curve superimposed on the impulse train that represents the composite response. Clearly, there is only one impulse. However, such a narrow impulse response $h(n)$ corresponds to a rather wide band low pass filter (LPF) that would not give satisfactory frequency resolution. If we use a narrower bandwidth filter, the impulse response will become proportionately greater in duration as shown in Figure 13.17(*b*). In this case, the composite impulse response consists of several impulses that would give rise to a reverberant quality in the output speech. Thus, we see that good frequency resolution (i.e., narrowband channels) seems to be at odds with low reverberation.

Figure 13.17(*c*) suggests one way in which, at least theoretically, the output can match the input exactly. Here we have used a wider filter but have constrained the values of $h(n)$ to be zero at integer multiples of the period N. In this case the composite response is a single impulse delayed by $2N$. Thus the output is a delayed and scaled replica of the input. Such a data window can be designed. Therefore, the STFT can theoretically represent the speech signal exactly.

13.6 OVERALL ANALYSIS-SYNTHESIS SYSTEM

Figure 13.18 shows the block diagram of an overall analysis-synthesis system. This figure shows all the processing required for the kth channel. Each of the M channels requires similar processing. This figure is conveniently segmented into three parts. These parts are an analysis section, a bit-rate reduction section, and a synthesis section.

The analysis section works as described in the previous section, computing $A(\omega_r, n)$ and $B(\omega_r, n)$ for each channel.

To achieve any bit-rate reduction (bandwidth compression), these signals must be sampled at a lower rate (i.e., once every T_1 seconds) and quantized to a smaller number of bits. These are the functions of the sampler and quantizer of the bit-rate reduction section. Appropriate values of T_1 and the number of bits per sample must be obtained from speech perception experiments.

The synthesis stage is used to interpolate the received values from quantizers of the bit-rate reduction section. Interpolating low pass filters are used to interpolate the received values of $A(\omega_r, n)$ and $B(\omega_r, n)$ to the appropriate synthesis rate, T_2 seconds, which need not be identical to the analysis rate.

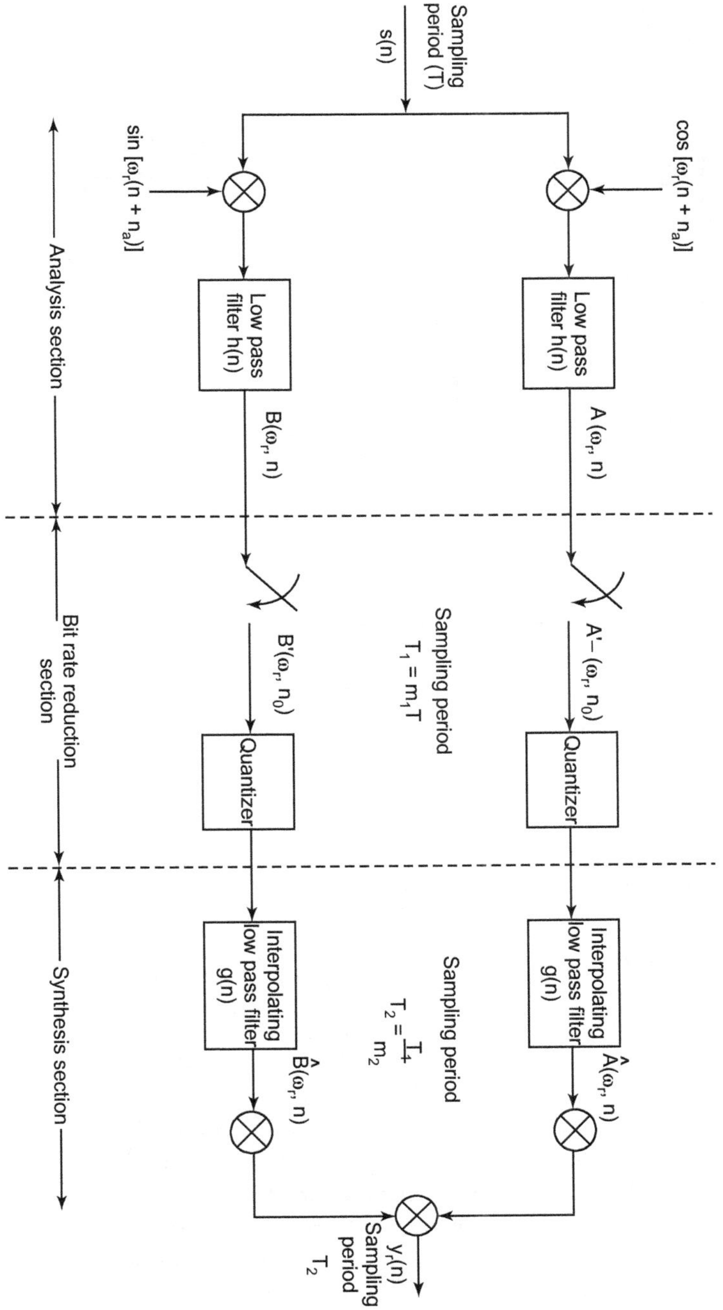

FIGURE 13.18 Block diagram of overall analysis-synthesis system.

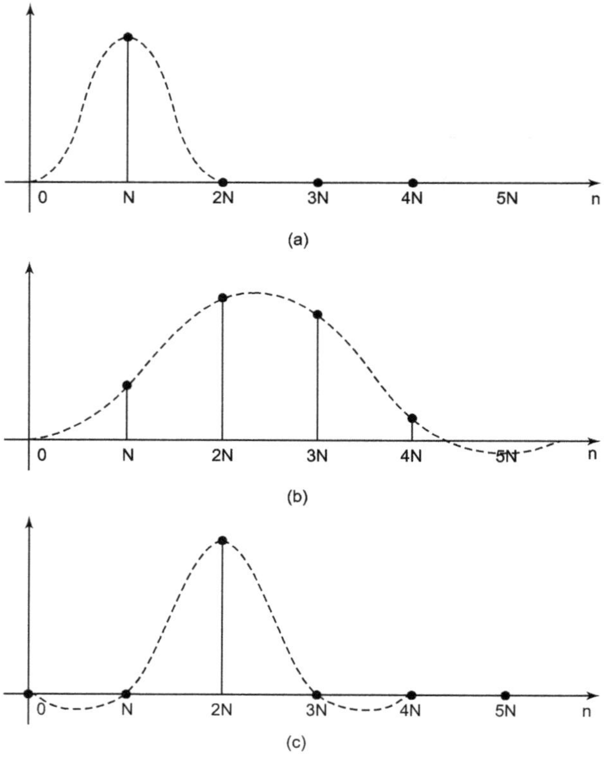

FIGURE 13.19 Time versus frequency resolution trade-offs.

13.7 CHANNEL VOCODER

It is commonly used for bandwidth compression of analog speech. It is also known as a channel voice coder. Its digital implementation can be carried out by designing the digital bandpass and lowpass filters. Spectral analysis of these filter hanks is performed by using discrete Fourier transform (DFT). Figure 13.20 illustrates the block diagram of a channel vocoder (voice coder).

Channel vocoder comprises of two units out of which one is the channel vocoder analyzer and the other is the channel vocoder synthesizer.

***(a)* Channel Vocoder Analyzer:** The input speech signal is fed into an analysis filter bank consisting of M BPFs which are contiguous in frequency. The output of each hand-pass filter is fed to a rectifier and LPF for low pass filtering. These LPF signals are the output of the channel vocoder analyzer. Here, for speech analysis, we use STFT. We should note here that DFT and FFT cannot be used directly as speech signal is time-varying signal.

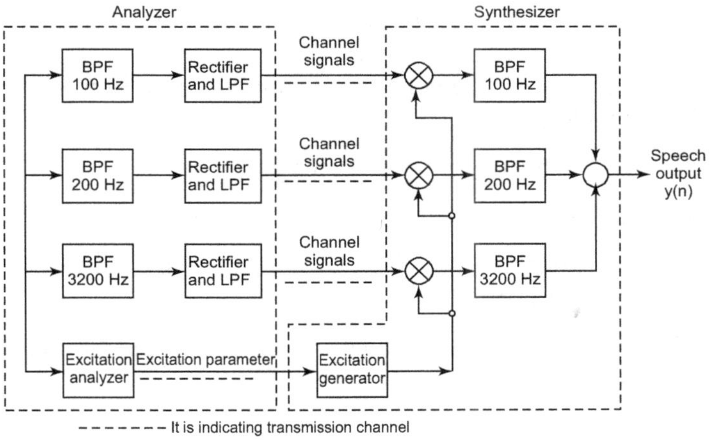

FIGURE 13.20 Block diagram of a channel vocoder.

(*b*) Channel Vocoder Synthesizer: In the synthesizer, the channel signals obtained from analyses are used as amplitudes. These amplitudes are multiplied by the excitation parameter generated by the excitation generator. Each multiplier output is bandpass filtered using a bank of BPFs. This bank of BPFs is the same as used in the channel vocoder analyzer.

The excitation parameters obtained from the channel vocoder analyzer are used to generate an excitation consisting of pulses spaced by the pitch period for voiced speech and a noise-like sequence for unvoiced speech. The pitch period is the time period after which the voiced speech repeated its waveform.

13.8 PITCH DETECTION AND VOICED–UNVOICED DECISIONS

To complete these sections on the channel vocoder, it is appropriate to mention the problems of pitch detection and voiced–unvoiced decision making in the analyzer. There exists a wide variety of algorithms for estimating the pitch period. For the sake of illustration, we shall discuss a particularly efficient algorithm that works in the time domain and uses parallel processing techniques to make its final decisions. The problems of pitch detection and voiced–unvoiced decisions are really a combination of signal processing and feature extraction. Since pitch detectors are embedded in a large number of speech processing systems, however, it is worthwhile discussing them here.

A block diagram of a pitch-period estimation algorithm is shown in Figure 13.21.

The algorithm is conveniently segmented into four distinct processing or decision-making parts.

1. Low pass filtering of the speech signal.
2. Generation of six functions of the peaks of the filtered speech signal.
3. Six identical simple pitch-period estimators, each working on one of the six functions above.
4. Final pitch-period computation, based on examination of the results from each simple pitch-period estimator.

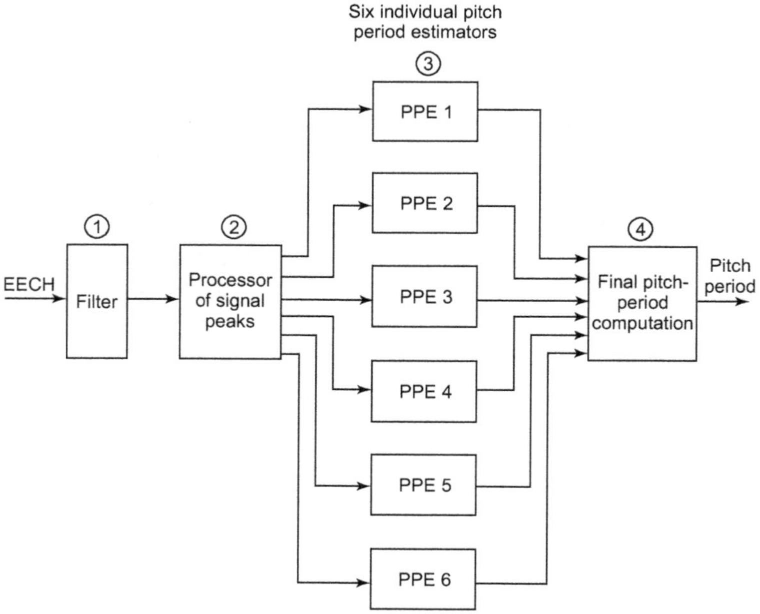

FIGURE 13.21 Block diagram of the pitch-period estimation algorithm.

The primary purpose of the lowpass filter is to filter out higher harmonics of the speech waveform. A lowpass filter with a cutoff of about 600 Hz works well.

The second part of the algorithm generates pulses at various peaks in the lowpass filtered waveform as illustrated in Figure 13.22. Pulses of height m_1, m_2, and m_3 are generated at every positive peak, while pulses of height m_4, m_5, and m_6 are generated at each negative peak. Measurements m_1 and m_4 are simple peaks (positive and negative) measurements, whereas m_2 and m_5

are peak-to-valley and valley-to-peak measurements, respectively, and m_3 and m_6 are peak-to-previous-peak and valley-to-previous-valley measurements, respectively. All the *m's* are converted into positive pulse trains. Thus if a current peak (valley) is not so large as the previous peak (valley), measurement $m_3(m_6)$ is set to zero.

The choice of this particular set of measurements was based on consideration of two extreme cases as shown in Figure 13.23. For the case when only the fundamental is present (as on the left), measurements m_3 and m_6 fail but measurements m_1, m_2, m_4, and m_5 provide strong indications of the period.

For the case when a very strong second harmonic and some fundamental are present (as on the right) measurements m_3 and m_6 will be correct and all others will fail. In this case, although four of the six measurements may fail, it will be shown below how the final computation has a high probability of being correct.

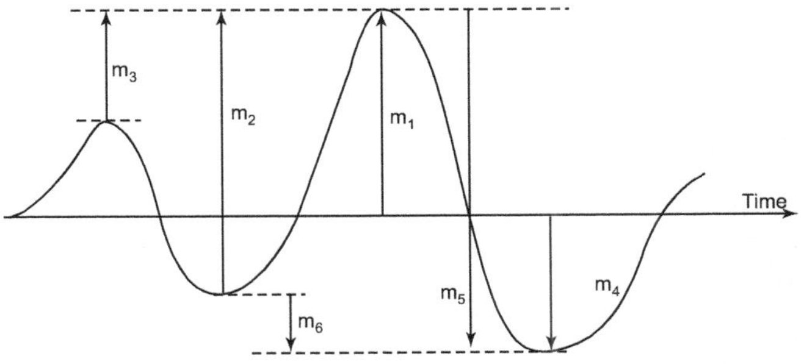

FIGURE 13.22 Measurements for determining pitch period.

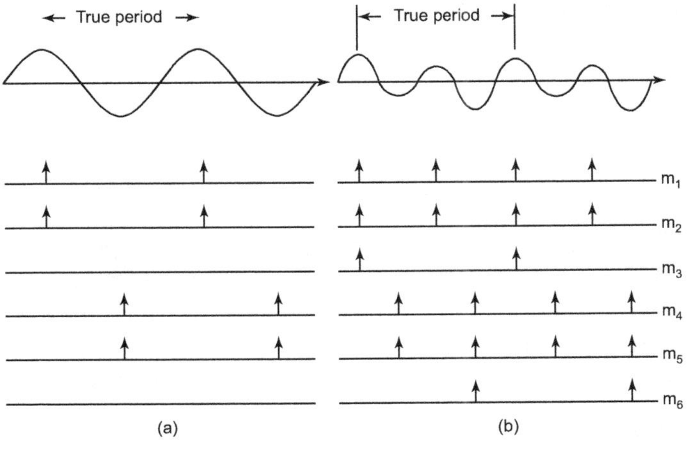

FIGURE 13.23 Pitch measurements for two extreme conditions.

The six sets of pulse trains are applied to six identical pitch detectors, each of which operates as shown in Figure 13.24. In essence, each pitch-period estimator is a peak-detecting run-down circuit. Following each detected pulse there is a blanking interval (during which no pulses can be detected) followed by a simple exponential decay. Whenever a pulse exceeds the level of the run-down circuit (during the decay), it is detected and an exponential run-down circuit is reset. The run-down time constant and the blanking time of each detector are made to be functions of the smoothed estimate of pitch period P_{av}, of that detector P_{av} is derived from the iteration.

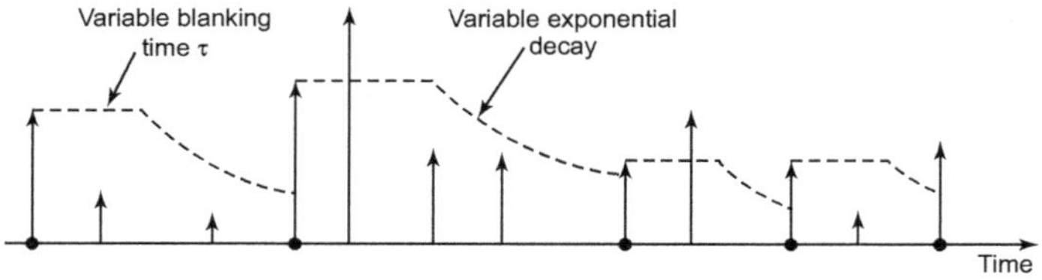

FIGURE 13.24 Operations of each pitch detector.

$$P_{av}(n) = \frac{P_{av}(n-1) + P_{new}}{2} \tag{13.23}$$

where P_{new} *is* the most recent estimate of pitch period, $P_{av}(n)$ is the current smoothed estimate of pitch period, and $P_{av}(n-1)$ is the previous smoothed estimate of the pitch period.

Each time a new peak is detected, P_{av} is updated according to the iteration. To prevent extremes of values of blanking time or run-down time constant, P_{av} is limited to be greater than 4 ms and less than 10 ms. Within these limits, the dependence of blanking time $T(\tau)$ and run-down time constant-β on P_{av} is given by

$$\left.\begin{aligned} \tau &= 0.4P_{av}, \\ B &= \frac{P_{av}}{0.695} \end{aligned}\right\} \tag{13.24}$$

The final computation of the pitch period is performed by block 4 (as shown in Figure 13.21), which may be thought of as a special purpose computer with a memory, an arithmetic algorithm, and control hardware to steer all the incoming signals.

At any time an estimate of the pitch period is made by

1. Forming a 6 × 6 order matrix of estimates of the pitch period. The columns of the matrix represent the individual detectors and the rows are

estimates of the period. The first three rows are the three most recent estimates of the period. The fourth row is a sum of the first and second rows; the fifth row is the sum of the second and third rows, and the sixth row is a sum of the first three rows. The technique for forming the matrix is shown in Figure 13.25.

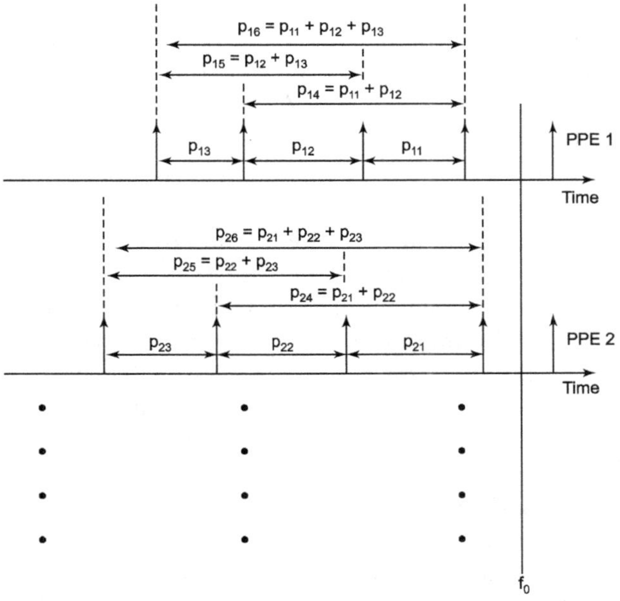

PPE NO

Eliminates of period

1	2	3	4	5	6
P_{11}	P_{21}	P_{31}	P_{41}	P_{51}	P_{61}
P_{12}	P_{22}	P_{32}	P_{42}	P_{52}	P_{62}
P_{13}	P_{23}	P_{33}	P_{43}	P_{53}	P_{63}
P_{14}	P_{24}	P_{34}	P_{44}	P_{54}	P_{64}
P_{15}	P_{25}	P_{35}	P_{45}	P_{55}	P_{65}
P_{16}	P_{26}	P_{36}	P_{46}	P_{56}	P_{66}

FIGURE 13.25 Technique for forming the matrix of estimates of pitch periods.

The reason for the last three rows of the matrix is that sometimes the individual detectors will indicate second or third harmonic rather than fundamental and it will be entries in the last three rows that are correct rather than the three most recent estimates of the pitch period.

2. Comparing each of the entries in the first row of the matrix to the other 35 entries of the matrix and counting the number of coincidences. That particular P_{i1} ($i = 1, 2, 3, 4, 5, 6$) that is the most popular (greatest number of coincidences) is used as the final estimate of the pitch period.

13.8.1 Coincidence

First, to determine whether two pitch-period estimates "coincide." It seems more appropriate to observe their ratios rather than their differences. However, the ratio measurement can be very approximate to avoid the need for a divide computation.

Second, because during many parts of the speech there are sizable variations of successive pitch-period measurements, it is useful to include several threshold values to define coincidence and then to try to select, for each overall pitch-period computation, the threshold that yields the most consistent answer. With this explanation, we now define the computation of block-4.

Figure 13.26 shows a table of 16 coincidence window widths. As indicated in Figure 13.25, only the most recent estimated pitch period from a given detector is a candidate for a final choice. This candidate is thus one of six possible choices for the correct pitch period. To determine the "winner," each candidate is numerically compared with all the remaining 35 pitch numbers.

This comparison is repeated four times, corresponding to each column in the table of Figure 13.26. From each column, the appropriate window width is chosen as a function of the estimate associated with the candidate. Thus, if this estimate, for example, were 4 μs, the coincidence between the candidate and any compared interval would mean that their difference was less than or equal to ±200 ms at a sampling rate of 10 kHz. After the number of coincidences is tabulated, a bias of 1 is subtracted from that number. The measurement is then repeated for the second column; this time the windows are wider and this increases the probability of coincidence but, in compensation, a bias of 2 is subtracted from the computation. After the computation has been repeated in this way for all four columns, the largest biased number is used as the number of coincidences that represent that particular pitch-period estimate. The entire procedure is now repeated for the

Pitch-period range (m sec) \ Bias	1	2	5	7
1.6 – 3.1	1	2	3	4
3.1 – 6.3	2	4	6	8
6.3 – 12.7	4	8	12	16
12.7 – 25.5	8	16	24	32

Coincidence window width in hundreds of microseconds

FIGURE 13.26 Table of coincidence window widths.

remaining five candidates and the winner is chosen to be that number with the greatest number of biased coincidences.

In the course of this computation, a total of $6 \times 4 \times 35 = 840$ coincidences measurements (comparison of the magnitude of a difference with a fixed number) have to be made. Repetition of the complete computation every 5 ms suffices to follow even rapid pitch period variations.

Demonstration of typical results obtained with this algorithm is shown in Figure 13.27. This figure shows a comparison between fundamental frequency estimates obtained by the method above and the true values as used in generating the synthetic utterance used in the test. The algorithm clearly works very well in this case.

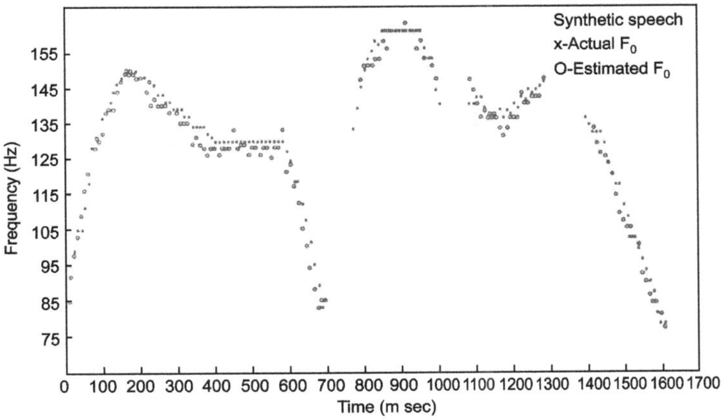

FIGURE 13.27 Comparison between synthetic pitch and that generated by the algorithm.

13.9 VOICED–UNVOICED (BUZZ-HISS) DETECTION

The pitch-period estimation algorithm described above can readily be converted to give voiced–unvoiced estimates. Whenever the speech is unvoiced, the number of coincidences observed by the individual detectors will be small. Quantitative measurements can be made to set thresholds for the appropriate decisions. In addition, if the energy measurements out of the pitch detection LPF are below a fixed threshold, this strongly indicates either silence or unvoiced speech. By combining this with functions of the pitch detector output, as indicated above, a voice-unvoiced algorithm can be implemented as discussed by Gold.

13.10 VOICED–UNVOICED (BUZZ-HISS) DETECTION

The term "homomorphic processing" is generally applied to a class of systems that obey a generalized principle of superposition, and this generalized superposition can be stated as follows:

If $s_1(n)$ and $s_2(n)$ are input to a homomorphic system, $y_1(n)$ and $y_2(n)$ are the respective outputs, and C is any scalar, then if

$$y_1(n) = \phi[s_1(n)] \tag{13.25}$$

$$y_2(n) = \phi[s_2(n)] \tag{13.26}$$

$$\phi[s_1(n)\,\Delta\, s_2(n)] = \phi[s_1(n)] \boxtimes \phi[s_1(n)] \tag{13.27}$$

and

$$\phi[C\Diamond s_1(n)] = Cy_1(n) \tag{13.28}$$

where Δ, $\boxtimes$, $\Diamond$, and $\bigcirc$ correspond to unspecified mathematical operations such as multiplication, addition, and convolution.

The importance of this type of processing lies in the fact that the operation of the homomorphic system can be decomposed into a cascade of operations as shown in Figure 13.28.

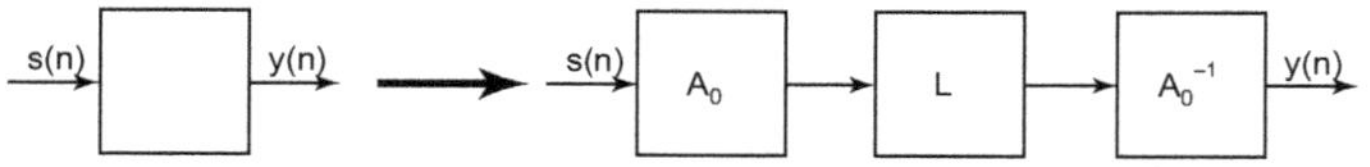

FIGURE 13.28 A homomorphic processing system.

The systems A_0 and A_0^{-1} are the inverse systems. The system L is a linear time-invariant (LTI) system, that is, a simple filter. Thus, with the decomposition as shown in Figure 13.28, one can process the output of system A_0 using standard techniques and perform processing in a relatively straightforward manner. Systems A_0 and A_0^{-1} are readily determined from ϕ.

Speech waveform is modeled as the convolution of three components. These components are:

1. a train of impulses representing the pitch,
2. the excitation pulse, and
3. the vocal tract impulse response.

The effects of radiation from the mouth also enter into the model but these effects are generally combined with excitation pulse. Now we are using the following notation:

$p(n)$ to represent the train of pitch impulses,

$e(n)$ to represent the excitation pulse,

$u(n)$ to represent the vocal tract impulse response, and

and $w(n)$ to represent a time-limited window through which the speech waveform $s(n)$ is viewed.

We find that

$$s(n) = [p(n)\ \ e(n)\ \ u(n)]w(n) \tag{13.29}$$

Since $w(n)$ is generally a smooth sequence Eq. (13.29) can be simplified to the approximate form

$$s(n) \approx [p(n).\ w(n)\ e(n)\ u(n)] \tag{13.30}$$

$$s(n) \approx p(n)\ \ e(n)\ \ u(n) \tag{13.31}$$

Eq. (13.31) shows $s(n)$ to be triple convolution. This convolution can readily be converted to a summation by Fourier transforming Eq. (13.2) and then taking the logarithm of the result. The resulting waveform may then be processed by a linear time-invariant (LTI) system to process each of the components of $s(n)$ in some desired manner. To recover a processed waveform, the inverse system A_0^{-1} consists of an exponentiator and an inverse Fourier transformation. Thus the homomorphic system for processing speech is shown in Figure 13.29.

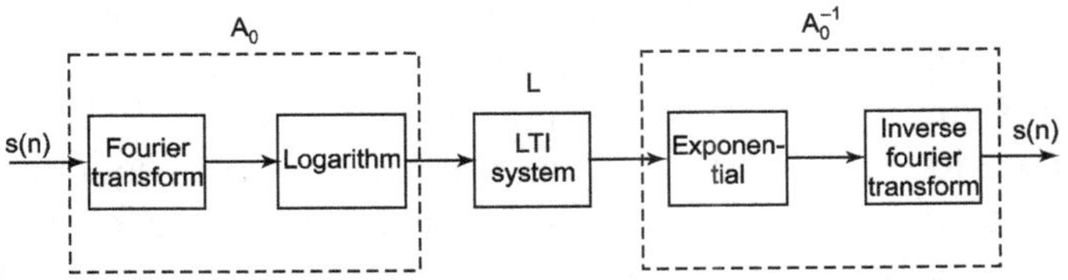

FIGURE 13.29 A homomorphic system for processing speech.

Depending on the specific application, several variations on the system above have been used to process speech.

A system is shown in Figure 13.30 has been used to estimate parameters of both the vocal tract transmission function and the excitation function. In this case, the excitation is considered to be $p(n)\ e(n)$ and the vocal tract impulse response is $u(n)$. Thus $s(n)$ is a simple discrete convolution.

$$s(n) = u(n)g(n) \tag{13.32}$$

where $g(n)$ is the excitation signal

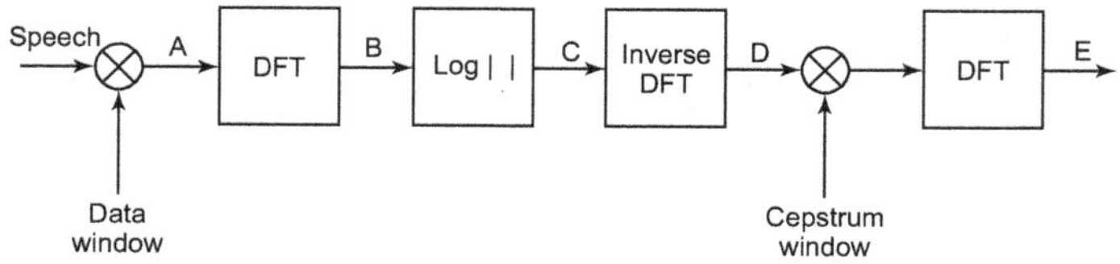

FIGURE 13.30 Homomorphic processing for speech.

In this case, s(n) is the signal at point A in Figure 13.30. The application of DFT gives a signal at point B that is the product of the DFTs of $u(n)$ and g(i). The next block takes the log magnitude of the signal at point B giving a signal at point C. Signal C is the sum of the log magnitudes of the DFTs of g(n) and $u(n)$. The sequence of blocks following point C (an inverse DFT, a windowing, and a DFT) is readily seen to be linear filtering of the signal at point C. The filtering is carried out in the transform domain (as a multiplicative operation) for the reasons discussed below. Since the inverse DFT is linear, the signal at point D (called the cepstrum of the signal at point A) is the sum of the cepstra of the excitation and the vocal tract impulse response.

It can be argued that the cepstrum at point D serves to separate the excitation from the vocal tract impulse response in the following manner. The excitation signal can be viewed as a sequence of quasi-periodic pulses whose Fourier transform consists approximately of a line spectrum where the lines are spaced at harmonics of fundamental frequency. The process of taking the log magnitude does not affect the general characteristics of the excitation spectrum. The IDPT operation yields another quasi-periodic waveform with pulses spaced at the fundamental period. Thus, the cepstrum of the excitation signal should consist of pulses around $n = 0, T, 2T, ...,$ where T is the pitch period. The vocal tract impulse response is a sequence that generally is non-negligible for about 20–30 ms. Its Fourier transform is a slowly varying function of frequency. The process of taking the log magnitude and IDFT yields a sequence that is non-negligible for only a small number of samples (generally less than the number of samples in a pitch period).

It can be shown that for a sequence that decays as $\frac{1}{n}$, its cepstrum decays as $\frac{1}{n^2}$. Thus the cepstrum serves to differentiate the excitation information from the vocal tract impulse response information.

Homomorphic processing on both voiced and unvoiced speech is shown in Figure 13.31. Figure 13.31(a) shows typical waveforms obtained at points A to E for voiced sequence. Figure 13.31(b) shows typical waveforms obtained at points A to E for an unvoiced sequence. Here we have included three distinct pitch periods in this analysis. The wiggly curve in Figure 13.31(a)(iii) shows the log magnitude of the transform of this sequence, which consists of a rapidly varying periodic component (due to the excitation) and a slowly varying component due to the vocal tract transmission. Figure 13.31(a)(ii) shows the resulting cepstrum. The strong peak at about 9 ms shows the pitch period.

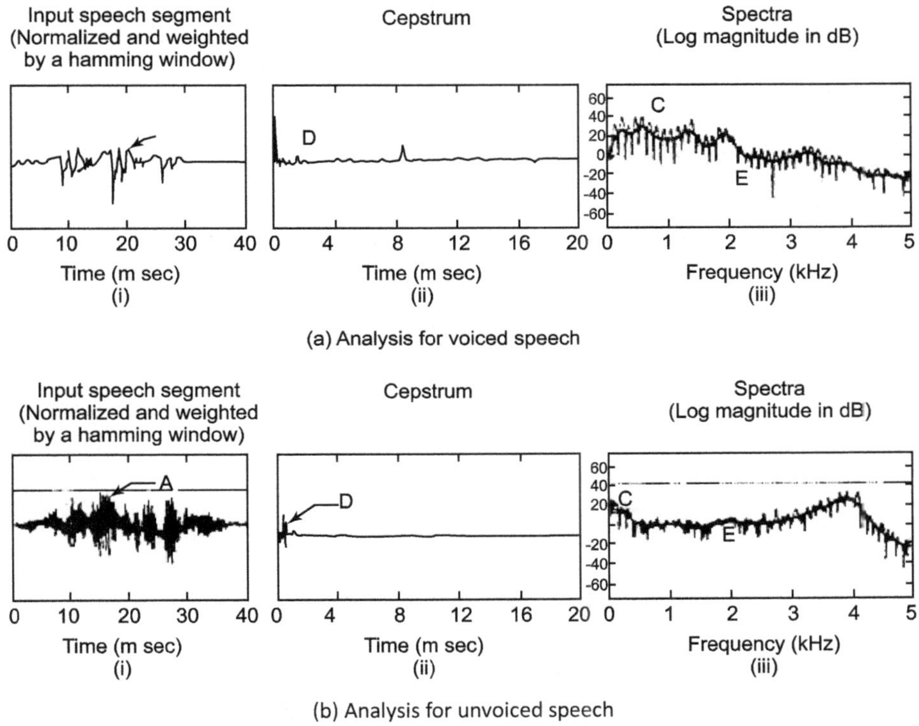

FIGURE 13.31 Homomorphic analysis for voiced and unvoiced speech.

The low-time portion corresponds to the cepstrum of the vocal tract impulse response. Application of a low-time cepstral window (to eliminate excitation information) the DFT transformation yields the slowly varying curve in Figure 13.31(a)(iii). Based on the peaks in the resulting spectrum at point E, efficient algorithms exist for estimating formant resonances corresponding to the particular vocal tract transmission function. Figure 13.31(a)(i) shows a Hamming window-weighted voiced sequence.

For unvoiced speech, the excitation is a random input rather than a quasi-periodic pulse train. In this case, the waveform at point A is as shown in Figure 13.31(b)(i). The random nature of the input is evident from this plot. The log magnitude of the DFT is as shown in the rapidly varying curve in Figure 13.31(b)(iii). The curve may again be modeled as the linear combination of a random component due to the source and a slowly varying component due to the vocal tract. The resulting cepstrum as shown in Figure 13.31(b)(iii) displays no strong peak indicating the absence of voicing. The result of using a cepstrum window and the DFT is shown as the slowly varying curve in Figure 13.31(b)(iii).

This curve represents the transmission of the vocal tract. Generally, both poles and zeros are used to represent the shape of the unvoiced spectrum.

Homomorphic Vocoder

The analysis scheme above may be readily combined with a synthesizer of the type shown in Figure 13.32 to compromise an entire vocoder system.

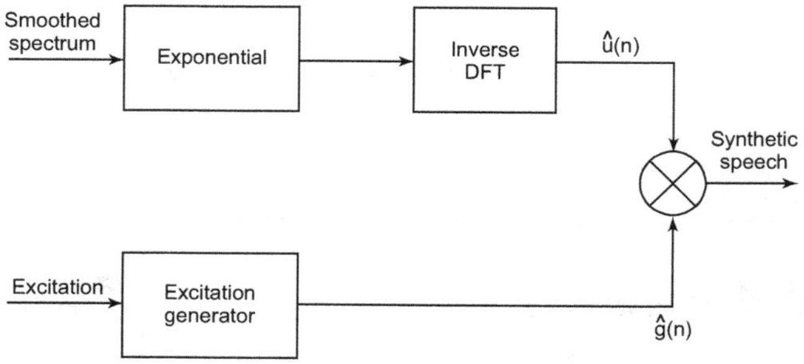

FIGURE 13.32 Block diagram of a homomorphic vocoder.

Instead of coding the vocal tract impulse response spectrum into either formants or a pole-zero representation, it is preserved and put through an inverse system to the original non-linear processing. This inverse system consists of an exponentiator (to undo the logarithm) and an inverse DFT transformation to give $\hat{u}(n)$, an estimate of the vocal tract impulse response.

The excitation period (as obtained from a cepstral measurement) is used to create either a quasi-periodic pulse train or a random train of impulses to act as an estimate of the true excitation. These two sequences are convolved to give the synthetic speech.

Formant Synthesis

One of the most important speech research problems concerns techniques for synthesizing speech from appropriate excitation parameters. Speech synthesis applications include several types of computer voice response systems and provide important insight into the basic mechanism of speech production and perception. One of the most basic sets of such parameters is the set of formant frequencies as a function of time. Formant synthesis lends a considerable degree of flexibility and efficiency to the various applications of synthetic speech. Here we present some of the signal processing problems associated with synthesizing speech from formant data. It is assumed that an analysis system is available for deriving the formant data from natural speech.

A schematic block diagram of a general-purpose formant synthesizer of the type used in several computer voice response studies is shown in Figure 13.33. There are two excitation sources. An externally controllable impulse generator (the source for voiced sounds), whose output consists of a unit pulse once every pitch period (*P* samples), and a pseudorandom uniform number generator (the source for unvoiced sounds), whose output approximates a white noise generator.

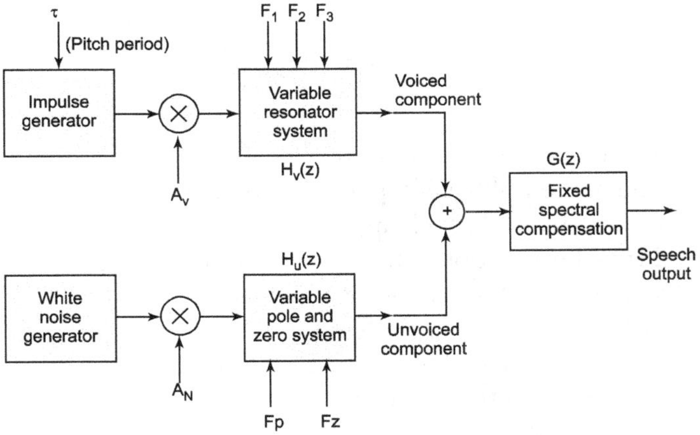

FIGURE 13.33 Schematic block diagram of a formant synthesizer.

There are two basic signal processing paths in the synthesizer. The upper path consists of an intensity modulator (A_v) and a time-varying digital filter consisting of a cascade of L variable resonators (poles). The transfer function of this filter (under steady-state conditions) is

$$H_v(z) = \prod_{k=1}^{L} \left[\frac{1 - e^{(a_k T)} 2\cos(b_k T) + e^{(-2a_k T)}}{1 - e^{(-a_k T)} 2\cos(b_k T)_Z^{-1} + e^{(-2a_k T)_z^{-2}}} \right] \tag{13.33}$$

where α_k is the radian bandwidth of the kth pole, b_k is the radian center frequency of the kth pole, and T is the sampling period.

A typical z-plane plot of the pole locations for a vowel ($L = 5$) is shown in Figure 13.34. Although all the pole center frequencies and bandwidths can be controlled, generally only the lowest three center frequencies are varied as shown by the control signal inputs (F_1, F_2, and F_3) to the variable resonator system in Figure 13.33. The variable resonator system accounts for the effects of the time-varying shape of the vocal tract on the speech spectrum.

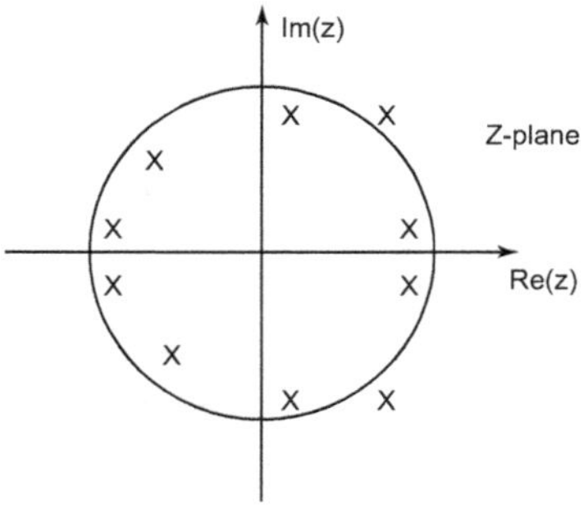

FIGURE 13.34 Pole locations for a typical vowel.

The effects of radiation of sound from the mouth (or nose) into air and glottal excitation pulse shape must be accounted for. This is the function of the fixed spectral compensation network whose transfer function is of the form

$$G(z) = \frac{\left[1 - e^{(-\alpha T)}\right]\left[1 + e^{(-\beta T)}\right]}{\left[1 - e^{(-\alpha T)_z^{-1}}\right]\left[1 + e^{(-\beta T)_z^{-1}}\right]} \tag{13.34}$$

This network consists of two real axis poles (one in the right-half z-plane and one in the left-half z-plane), which approximates the desired transfer function. The z-plane plot of pole locations for this network is as shown in Figure 13.35.

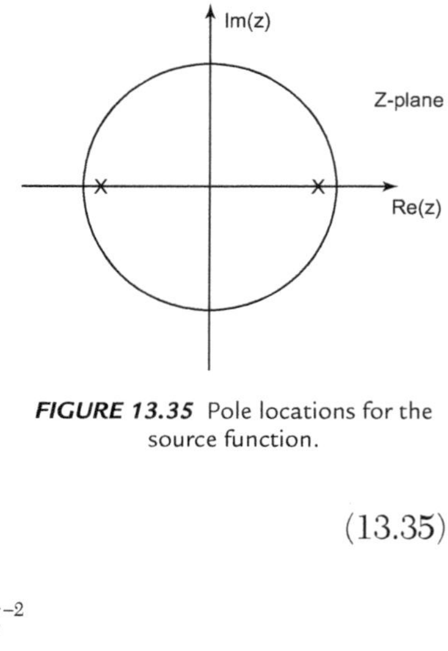

FIGURE 13.35 Pole locations for the source function.

The lower path in Figure 13.33 consists of a modulator (A_N) that controls the variance of the noise generator output and another time-varying digital filter consisting of a cascade of a pole and zero. Its transfer function is of the form

$$H_u(z) = \frac{H_1(1)H_2(z)}{H_1(z)H_2(1)} \tag{13.35}$$

where $\quad H_1(z) = 1 - 2e^{-aT}\cos(bT)Z^{-1} + e^{-2aT}Z^{-2}$

$$H_2(z) = 1 - 2e^{-cT}\cos(dt)Z^{-1} + e^{-2c}Z^{-1}$$

where a, b, c, and d are the radian bandwidths and center frequencies of the time-varying pole and zero. Generally, the bandwidths of the pole and zero are fixed and only the center frequencies vary as shown by the control signal inputs F_p and F_z to the variable pole and zero systems in Figure 13.33. The z-plane pole-zero plot for a typical unvoiced sound is given in Figure 13.36. The output of this system is passed to the fixed spectral compensation system to provide the final unvoiced speech output.

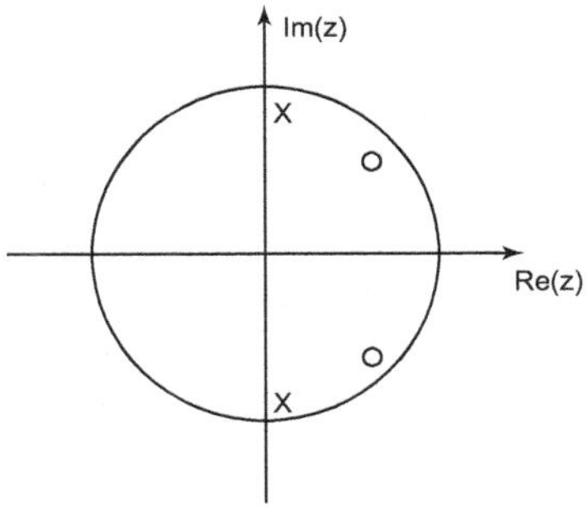

FIGURE 13.36 Pole-zero locations of a typical unvoiced sound.

It should be noted that each of the transfer functions [Eqs. (13.33) to (13.35)] of the synthesizer has the property that at zero frequency the transfer function is unity independent of the center frequencies and bandwidths of any pole or zero. This property is essential to account for the unity transmission of a vocal tract at zero frequency and is achieved by using resonators that are individually normalized to have this property.

The synthesizer configuration above is incomplete in its ability to synthesize the sounds of speech in several aspects that are desirable in a general-purpose synthesizer. For example, there is no provision for a network to produce the nasal consonants n and m or a network to produce the voiced fricatives z (as in zoo), v (as in very), and th (as in there). To synthesize nasal consonants, a network consisting of a time-varying pole and zero must be placed in a cascade with the variable resonator system of Figure 13.33.

To synthesize voice fricatives adequately, a network that modulates the noise generator output by the voiced path output is necessary. Also, for additional flexibility in the synthesizer, provision should be made to allow the noise generator output to excite the voiced processing path in order to produce whispered speech.

A more versatile synthesizer is shown in Figure 13.37. In this synthesizer, remedies for the aforementioned problems have been obtained. This synthesizer possesses both simulated and built-in digital hardware. The synthesizer derives its time-varying control parameters synchronously that is, it changes all parameters once per pitch period, at the beginning of the period. At this time, each of the filters has minimum energy and the adverse effects of any large change in any control parameter are minimized. The control parameters are supplied to the hardware from a control computer (indicated as honey well DDP-5/6 computer in Figure 13.37).

This synthesizer is similar to that discussed in Figure 13.27. Specifically, the upper signal processing path consists of six two-pole digital filters [$L = 6$ in Eq. (13.33)] and one two-zero filter, where the bandwidth and center frequency of each filter is controllable. The six two-pole filters and the two-zero filter account for a nasal pole and zero and cancel each other during non-nasal sounds.

The exact cancellation of a pole by zero is easily accomplished in a digital system. Four of the two-pole filters (or possibly five during non-nasal sounds) are used to represent the time-varying vocal tract transfer function $H_v(z)$ and the last two-pole filter provides the desired spectral compensation $G(z)$.

The unvoiced signal processing path consists of two two-pole filters and one two-zero filter. Again the bandwidths and center frequencies of each of the filters can be varied externally. One two-pole and one two-zero filter are used to represent $H_u(z)$ and the remaining two-pole filter is used to provide the necessary spectral compensation $G(z)$. In this synthesizer, for added flexibility, the voiced and unvoiced spectral compensation networks may be different since they are included separately in each path of the synthesizer.

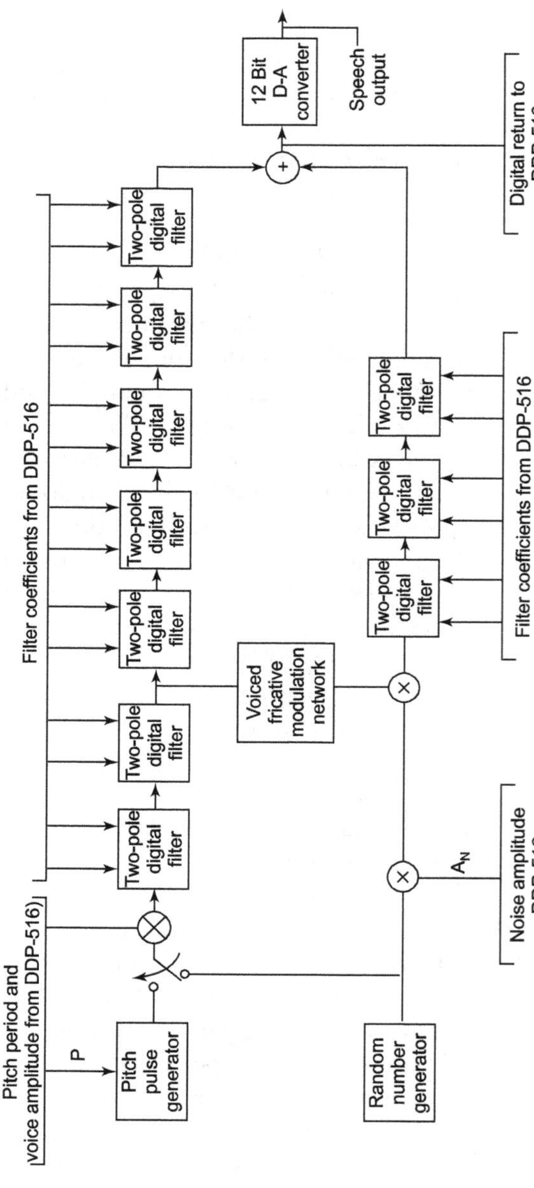

FIGURE 13.37 Block diagram of a hardware synthesizer.

13.11 VOICED FRICATIVE EXCITATION NETWORK

The voiced fricative excitation network connects the output at one point in the voicing path to the unvoiced path. It is used to model the production of the unvoiced component of voiced fricatives. The relevant networks used to synthesize the entire voiced fricative are shown in Figure 13.38.

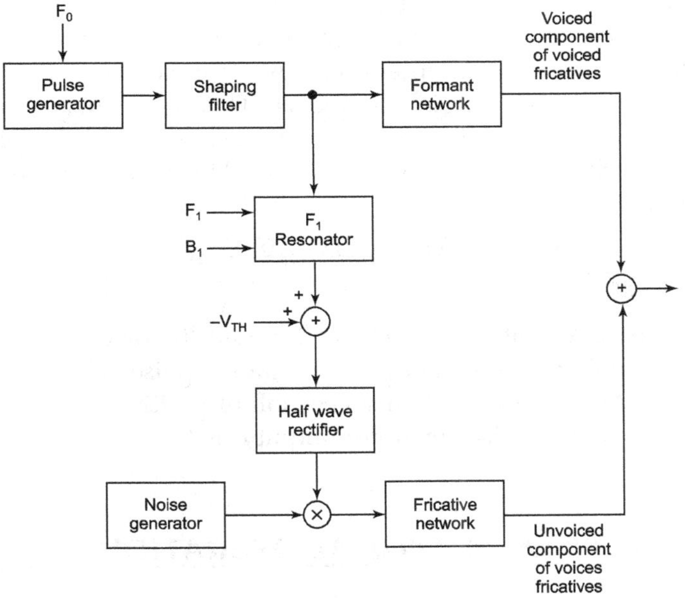

FIGURE 13.38 Excitation network for voiced fricatives.

The unvoiced excitation is produced as follows:

The pitch pulses excite a resonator tuned to the first formant of the voiced component of the fricative. This resonator is the first-order approximation to the transfer function of volume velocity from the glottis through the point of constriction of the vocal tract. A threshold level (V_{Th}) is subtracted from the output of the resonator and the result is half-wave rectified.

These operations model the physical observation that turbulence is not produced until the volume velocity of the airflow exceeds a threshold value.

The output of the half-wave rectifier modulates the output of the noise generator, producing a pitch. Synthesis excitation for the unvoiced component of the fricative. The final unvoiced component is produced by feeding this excitation into a fricative network (i.e., the lower branch of the synthesizer).

The voiced component is produced by exciting the formant networks in the usual manner.

13.12 RANDOM NUMBER GENERATOR

Generation of pseudorandom numbers needed as the source for unvoiced sounds uses any of a large number of available algorithms. For the hardware realization, the specific pseudorandom number generator used is a 16-bit maximal length shift register sequence. This algorithm generates a random bit from mod-2 sums of the previous 16 bits, shifts out the bit generated 16 clock pulses earlier, and shifts in the new bit. The algorithm used to generate the current bit is

$$X_n = X_{n-1} \boxtimes X_{n-12} \boxtimes X_{n-14} \boxtimes X_{n-15}$$
$$n = 1,2,3,..... \tag{13.36}$$

where each X is either 1 or 0 and a 1 physically corresponds to a positive excitation pulse and a 0 to a negative excitation pulse. Thus, the noise generator output consists of a random succession of positive and negative pulses. The spectrum of the noise generator output is flat.

13.13 PRINCIPLES OF DIGITAL OPERATION

The basic principle behind the digital hardware is the multiplexing of a single arithmetic unit among all the two-pole filters and the two-zero filters. The arithmetic operations required to realize a two-pole filter (for example) are two additions, two subtractions, and two multiplications for each output. Sample high-speed integrated circuits are currently capable of doing about 25 times this number of arithmetic operations in the time between output samples (100 μs at a 10 kHz sampling rate). Thus, the notation of sharing a single arithmetic unit among many filters attains practical significance in the synthesizer. By providing storage for the filter coefficients and the delayed outputs of the filters and by dynamically controlling which inputs go into the arithmetic unit and where the outputs go, a single arithmetic unit can service the entire synthesizer.

A schematic block diagram of the digital logic used to realize the synthesizer is shown in Figure 13.39. The arithmetic unit consists of a three-input adder, a shift register delay, a subtracter, and a multiplex shift register delay that holds the delayed filter variables. The length of the shift register delay is 480 bits (20 delayed variables × 24 bits per variable).

Another shift register memory of 320 bits (20 filter coefficients times 16 bit per coefficient) holds the multiplexers for each of the filter sections. This arithmetic unit can perform a multiplication, an addition, and a subtraction simultaneously in about 3.9 µs; therefore, each filter section requires about 7.8 µs per iteration. In this manner, the 10 filter sections of the synthesizer require about 78 µs. Thus, the synthesizer can operate at sampling frequencies up to 12.8 kHz.

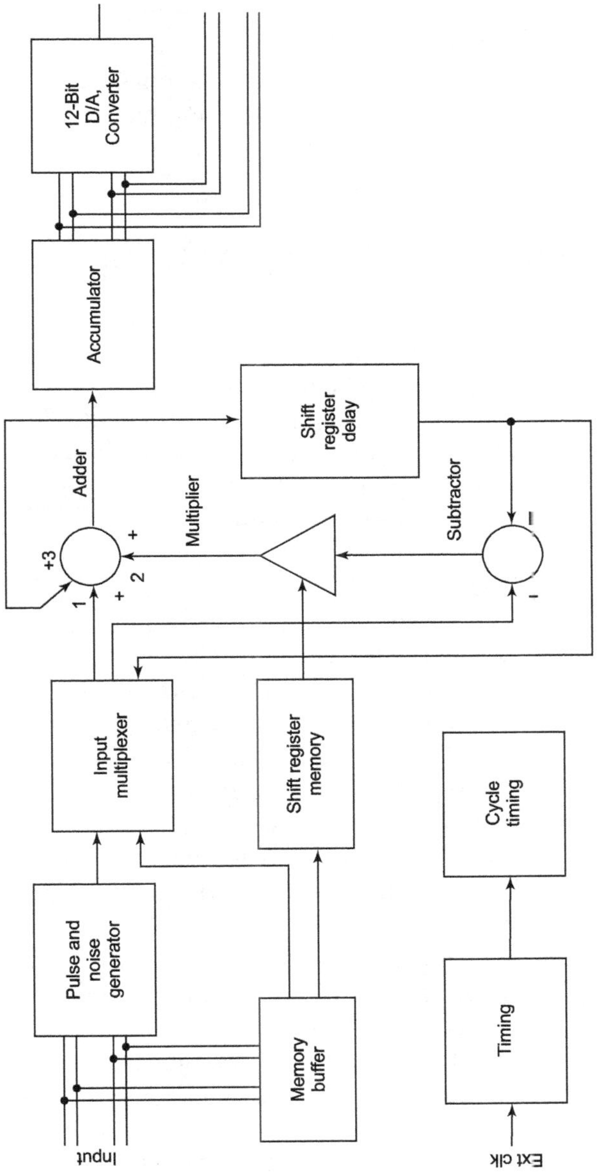

FIGURE 13.39 Block diagram of logic in hardware synthesizer.

The synthesizer control signals come from the computer output line to the input of the synthesizer. A memory buffer transfers the gain coefficients and pitch period to the pulse and noise generator and to the input multiplexer. The memory buffer shifts the filter coefficients to the shift register memory. The pulse and noise generator provides excitation to the arithmetic unit via the input multiplexer. An accumulator sums the voiced and unvoiced outputs and sends the 16 most significant bits (MSB_s) back to the computer, simultaneously converting the 12 MSB_s to analog form. The switching and timing logic is determined from timing logic, which uses an externally supplied clock to determine the basic synthesizer sampling rate. The sampling rate is thus easily changed without any interval modifications to the synthesizer.

13.14 LINEAR PREDICTION OF SPEECH

The basic idea of formant analysis and synthesis is that speech production is well modeled by exciting a cascade of linear time-varying second-order section digital filters (formant resonators) with either quasi-periodic pulses or noise. The major difficulty with this idea lies in assigning computed formants to specific second-order sections. Formants seem to disappear during certain sounds and additional formants are seen to be present during other sounds. A large number of errors of either of these types can quickly render the synthetic output unintelligible or at best make its quality unacceptable. Such errors are generally not uncommon across sentence length utterances.

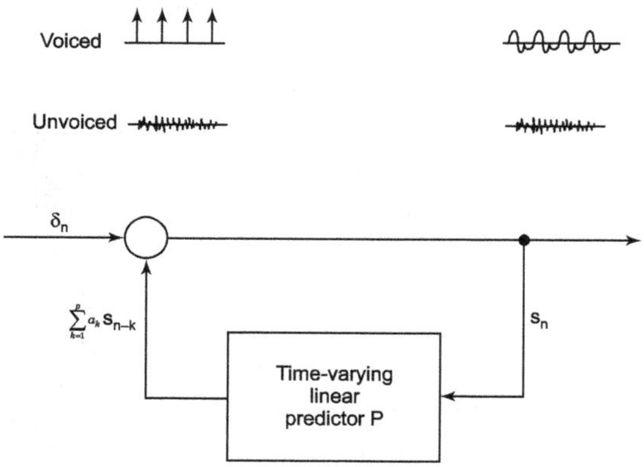

FIGURE 13.40 Linear prediction model of speech production.

The aforementioned problems can be overcome by modifying the basic speech synthesis (production) model to the form as shown in Figure 13.40.

The L individual second-order systems of the formant model are combined to give one pth-order linear system (where $P \geq 2L$). This system accounts for the vocal tract transmission, the source pulse shape, and the radiation characteristics.

The input $s(n)$ is either a stream of digital impulses or a quasi-random input.

The transfer function $H(z)$ of the filter is of the form

$$H(z) = \frac{S(z)}{\delta(z)} = \frac{1}{1 - \sum_{k=1}^{p} a_k z^{-k}} \tag{13.37}$$

The analysis of speech to determine pitch and the voiced–unvoiced decision is performed as for any other system using a pitch detector of the type shown earlier or any other algorithm that is desired. The predictor coefficients $\{a_k, k = 1, 2, ..., p\}$ are determined from a minimum mean square error analysis.

The difference equation of the system shown in Figure 13.40 is written as

$$s(n)\sum_{k=1}^{p} a_k s(n-k) + \delta(n) \tag{13.38}$$

For voiced speech, $\delta(n) = 0$ except for one sample at the beginning of every pitch period.

Thus, except in this sample when $\delta(n)$ is non-zero.

Now, Eq. (13.38) can be written as

$$s(n) = \sum_{k=1}^{p} a_k s(n-k) \tag{13.39}$$

Theoretically, we can say that if the linear prediction model of speech production is perfect, the speech samples $s(n)$ are completely predictable from Eq. (13.39). Since the speech waveform does not fit the model perfectly, it is possible to define an error between $s(n)$, the true value at the sample n, and $\hat{s}(n)$, the value predicted by Eq. (13.39). Let $E(n)$ be the prediction error, that is,

$$E(n) = s(n) - \hat{s}(n) \tag{13.40}$$

or

$$E(n) = s(n) - \sum_{k=1}^{p} a_k s(n-k) \tag{13.41}$$

The predictor coefficients are chosen so as to minimize the mean square prediction error $<E^2(n)>$, averaged over all n.

The expression for the mean square error can be put into the form

$$\langle E^2(n)\rangle = \sum_{n=1}^{\infty}\left[s(n) - \sum_{k=1}^{p} a_k s(n-k)\right]^2 \tag{13.42}$$

We can solve Eq. (13.42) for the predictor coefficients. Now, Eq. (13.42) is differentiated with respect to a_j, $j = 1, 2, ..., p$, and the result is set to zero giving the set of equations

$$\sum_{k=1}^{p} a_k \sum_{n=1}^{\infty} s(n-k)s(n-j) = \sum_{n=1}^{\infty} s(n)s(n-j)$$

$$h = 1, 2, 3, ..., p \tag{13.43}$$

In matrix formulation, the set of equations given by Eq. (13.43) can be written as

$$\phi_a = \psi \tag{13.44}$$

where $$\phi_{ij} = \sum_{n=1}^{\infty} s(n-i)s(n-j) \tag{13.45}$$

and $$\psi_j = \phi_{oj} \tag{13.46}$$

Thus ϕ is a matrix of autocorrelations and ψ is a vector of autocorrelations. Since ϕ is symmetric and positive definite, there exist several efficient methods of solving the set of equations implied by Eq. (13.43). Thus, the analysis required by linear prediction is relatively straightforward.

For synthesis, the system shown in Figure 13.41 is used to give a high-quality representation of the natural signal. The distinctions between linear prediction synthesizer and formant synthesizer are worth noting.

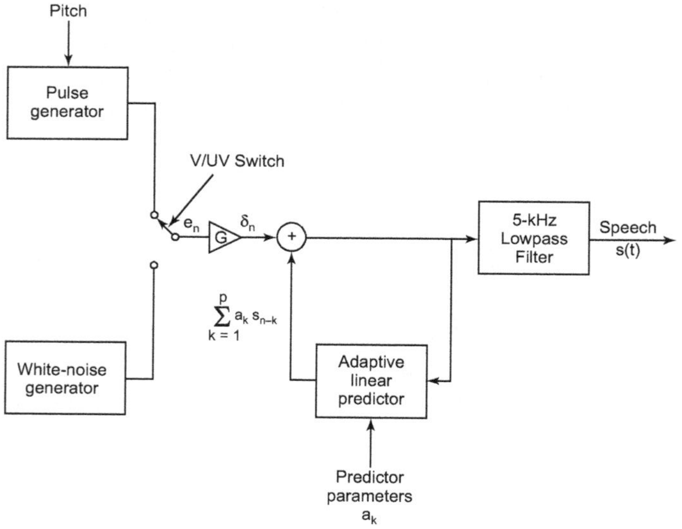

FIGURE 13.41 Linear prediction synthesizer.

The most important difference is the use of a single pth-order recursive filter in place of the cascade of second-order filters. In the time-invariant case, for example, a steady vowel, these two models are exactly equivalent. In the time-varying case, for example, most of the time during a speech, these two configurations are not equivalent. In the formant synthesizer, it is essential that each resonance be assigned to the proper formant or improper operation results.

This is not necessary for the linear prediction case as all formants are synthesized by one recursive filter. Another important difference is that the amplitude of the pitch pulses, as well as the white noise, is adjusted by the gain network G to provide the correct *rms* value of the synthetic speech samples. No such adjustment is generally made for synthetic speech.

13.15 A COMPUTER VOICE RESPONSE SYSTEM

Representation of speech in parametric form (e.g., in terms of pitch and formants) has two important advantages in terms of its utility in computer voice response systems.

1. Since the formants change at rates comparable to the motions of the vocal tract, they can be sampled and quantized to low bit rates. Hence representation of speech by formant parameters constitutes as an economical form for digital storage of speech information.
2. The advantage of the formant representation of speech is its inherent flexibility. Since contextual information is contained in the formant data, and prosodic data (e.g., inflection, rate of speaking, etc.) is contained in the pitch data and the timing information, the formant representation enables you to separate "What is sad," from "the manner in which it is sad." This flexibility and economy form the basis for a simple computer voice response system where isolated vocabulary elements are smoothly assembled into connected speech.

A block diagram of the system used for the synthesis of connected speech from a vocabulary of formant coded words is shown in Figure 13.42. Naturally spoken, isolated words (or phrases) are analyzed by a formant analyzer to give three formants (F_1, F_2, and F_3), voiced and unvoiced amplitude (A_V and A_N), pitch period (P), and unvoiced pole, and zero (F_p and F_Z) once every 10 ms. These control parameters are smoothed by programmed digital filters, sampled at their Nyquist rates, quantized, and stored in the word catalog as the reference library. The typical bit rate used for storage of these data is 700 bps when the pitch signal is saved. When the pitch is not saved, the bit rate for the stored data is 533 bps.

A breakdown of how these bit rates are achieved is given in Table.

The data in this table are derived from experimentation and investigation of the effects of smoothing and quantization on the perception of the synthetic output.

As shown in Table at every 10 ms interval the speech is classified as voiced on unvoiced (V/U) by a 1-bit signal. Thus, for each frame, storage is required for either voiced parameters or unvoiced parameters but not for both. It should be noted that the control parameter frame rate ($33\frac{1}{3}$ per second) is one-third the rate of the V/U signal.

Once input words and phrases are coded in terms of the formant representation, they can easily be modified for use with the synthesis program. Words can be lengthened or shortened, formants can be changed easily, and a pitch contour, different from the one originally spoken, can be superimposed on the data. Thus, the vocal resonance data is available to the synthesis program in a form flexible enough to confirm the timing and pitch generated by the concatenation program.

The lower portion of Figure 13.42 shows how the system assembles a synthetic message composed of words and phrases from the reference library.

First, the answer-back program requests the word sequence for a specific message. The word concatenation program first determines timing data for the message from an auxiliary program. The timing data is in the form of a word duration for each word in the output message. The concatenation program then accesses, in a sequence, the control parameters for each of the words in the string. A duration modification adjustment on each word is first made so the word duration in context matches the duration specified by the timing rules. Next, the concatenation program smoothly interpolates the formant control parameters when the final part of any word and the initial part of the following word are both voiced. An interpolation algorithm designed to produce realistic formant transitions is used. Finally, a continuous function for pitch variation is produced for the whole message. All computed control parameters are outputted to a hardware digital speech synthesizer. Digital-to-analog (D/A) conversion produces a continuous synthetic speech output.

The computer voice response system has been applied to several specific problems including automatic generation of telephone numbers and computer-aided voice wiring.

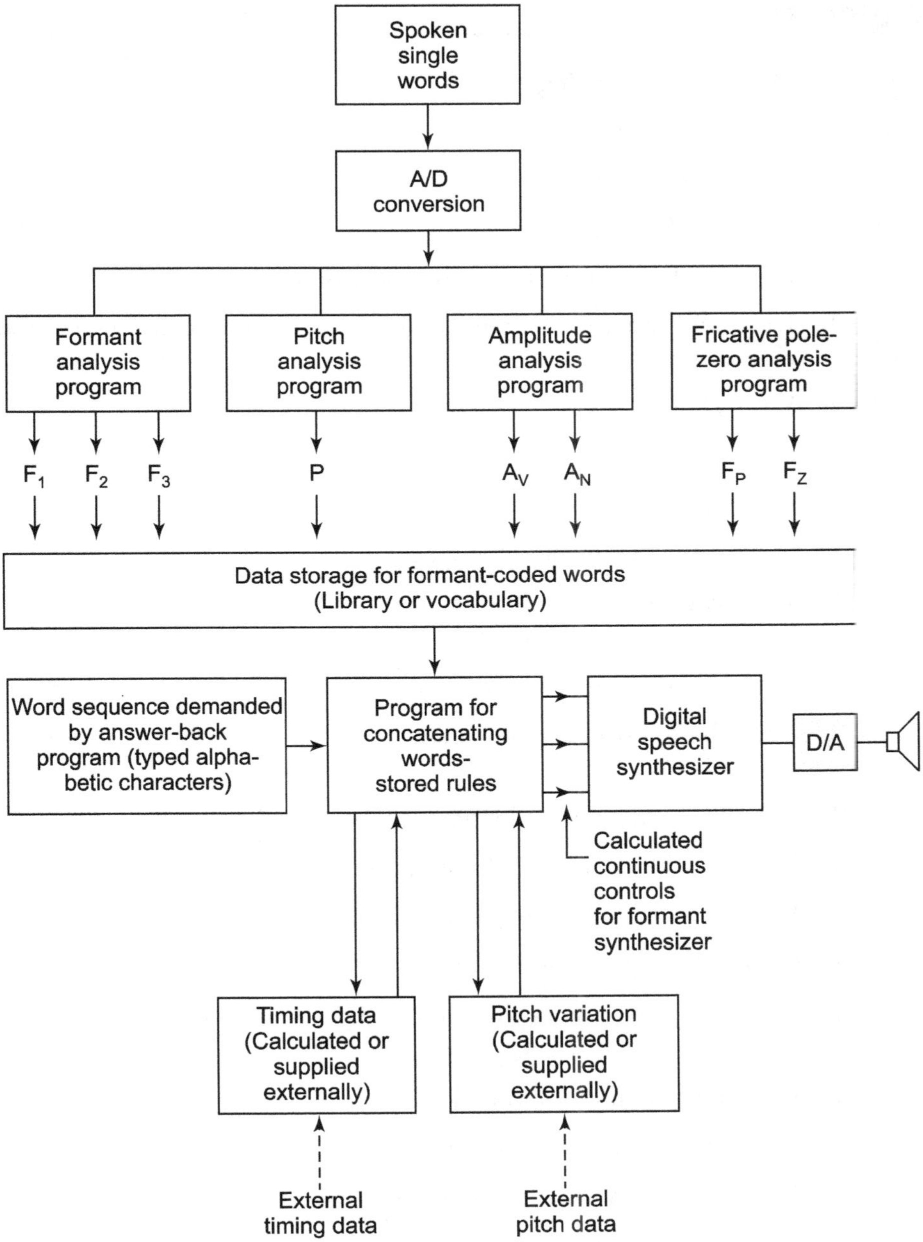

FIGURE 13.42 Block diagram of concatenation voice response system.

EXERCISES

1. Discuss the following:
   - i. Speech analysis,
   - ii. Speech synthesis,
   - iii. Speech compression.
2. Describe the model of speech production.
3. Draw and discuss a schematic diagram of the human speech production mechanism.
4. Write short notes on:
   - i. Voiced sounds.
   - ii. Unvoiced sounds.
5. Explain the digital model of speech production with the help of a diagram.
6. What is short-time Fourier transform (STFT)? Discuss the advantage of STFT in speech analysis and synthesis.
7. Discuss speech analysis-synthesis using STFT.
8. Describe the overall speech analysis-synthesis system with the help of a block diagram.
9. Describe channel vocoder with the help of block diagram.
10. Describe the pitch-period estimation algorithm.
11. Discuss voiced–unvoiced (Buzz-Hiss) detection.
12. What is the homomorphic processing of speech?
13. Discuss the homomorphic processing system with the help of a block diagram.
14. Explain homomorphic vocoder.
15. Draw and discuss the schematic block diagram of a formant synthesizer.
16. Describe voiced fricative excitation network.
17. Discuss random number generators.
18. Explain principles of digital operation.

19. What do you mean by linear prediction of speech?
20. Describe a linear prediction synthesizer with the help of a block diagram.
21. Describe a computer voice response system with the help of a diagram.
22. What do you mean by acoustic characteristics of speech signals? Draw the block diagram of the speech analysis procedure with parameters. Why is the short-term or short-time spectrum of speech preferred?
23. What is meant by cepstrum? Draw the block diagram of cepstrum analysis for extracting spectral envelope and fundamental period. Explain each block of your diagram with suitable mathematical expressions.

CHAPTER 14

Applications of Digital Signal Processing to Radar

14.1 INTRODUCTION

"Radar" is an acronym for radio detection and ranging. It was named by U.S. Navy in 1942 (during World War II). It is basically a means of gathering information about distant objects or targets by sending electromagnetic (EM) waves to them and thereafter analyzing reflected waves or echo signals. Radar was actually developed for the first time a few years before World War II. It was radar that gave birth to microwave technology. Radar can detect static or mobile objects or targets and is the most effective method for guiding a pilot with regards to his location in space as well as for warning the approach of an enemy plane for similar purposes (early warning radars). Electromagnetic waves have some special properties that they encounter a sudden change in conductivity, permittivity, or permeability in the medium, a part of the electromagnetic energy gets absorbed by the second medium and is re-radiated. This sudden change in the electrical property of the medium constitutes the target.

The re-radiated energy on being received back at the radar station gives information about the location of the target. The location of the target includes range, angle, and velocity parameters. The range is the distance of the target from the radar station. The angle could be azimuth or elevation angle for static targets and velocity for moving or mobile targets.

To find the location of the target satisfactorily, the echo power (power of a received signal) must be appreciable. Accordingly, the amount of power (energy) required to be radiated by the radar transmitter must be tremendous, typically few KW to MW. Such high power at high frequencies can be generated using magnetrons.

We know that the development of digital computer technology has led to great sophistication of radar tracking algorithms. In addition, computers in conjunction with electronically steerable phased array antennas have led to refined methods of scheduling of the radar's repertoire of transmitted signals. We anticipate that future radars will incorporate high-speed digital hardware to perform the desired filtering and thresholding algorithms. Several radars are incorporated with digital signal processing (DSP) hardware. DSP processors are more flexible than their analog counterpart, that is, analog signal processing processors.

14.2 APPLICATIONS AND ADVANTAGES OF RADARS

Radars can be used in civilian applications as well as in military applications. Some of the applications are given below:

1. Detection and range of enemy targets even at night.
2. Early warning regarding approaching aircraft or ships.
3. For directing guided missiles.
4. For aiming guns at aircraft and ships.
5. For bombing ships, aircraft, or cities even during overcast or at night.
6. For searching submarines, landmasses, and buoys.
7. Navigational aids on ground and sea. But navigation is not affected by poor visibility or darkness.
8. Police radars for directing and detecting speeding vehicles.
9. Airborne radar for satellite surveillance.
10. Radar altimeters for determining the height of the plane above ground.
11. Radars for determining the speed of moving targets such as automobiles, shells, guided missiles, etc.
12. Radar blind lander for aiding aircraft to land under poor visibility, at night, under adverse weather conditions, etc.

Advantages of Using Radar

Radar has the following advantages:

1. Radar can see through darkness, haze, fog, rain, and snow.
2. They can determine the range, angle, that is, the location of the target very accurately. It can also determine the speed of moving objects.

14.3 LIMITATIONS OF USING RADAR

Radar has the following limitations:

1. Radar cannot resolve in detail like the human eye, especially at short distances.
2. They cannot recognize the color of the target.

Before embarking on our study of applications of DSP to radar, let us first discuss the various radar system radar parameters. We will also study the chirp z-transform which is used extensively in the digital processing of radar signals.

14.4 CHIRP z-TRANSFORM (CZT) ALGORITHM

We have already studied the computation of DFT in a very efficient manner, that is, fast Fourier transform (FFT) algorithms.

CZT algorithms are equivalent to efficient computation of samples of the z-transform of a finite length sequence taken at equally spaced points around the unit circle. In order to achieve this efficiency in evaluating the z-transform, N is required to be a highly composite number. Also, we will have an interest either in sampling the z-transform on some other contour or we may not require samples of the z-transform over the entire unit circle. Thus, the scheme for increasing the flexibility of DFT computations is of considerable interest.

Now, we are interested in obtaining samples of the z-transform of a finite length sequence on a circle that is concentric with the unit circle, and the samples are to be equally spaced in angle around this circle.

In such a case, we can use an FFT algorithm after a minor modification. Specifically, if we have a finite duration sequence $s(n)$ of length N, then the DFT of the sequence $s(n)\alpha^{-n}$ will provide N samples equally spaced in angle around a circle of radius α in the z-plane.

If we are interested in obtaining frequency samples equally spaced over a small portion of the unit circle, the most efficient approach is often to use an FFT algorithm to computer frequency samples with the desired spacing, but obtaining samples outside the frequency range of interest. For example, we have a sequence $s(n)$ with $N = 128$. Now we are interested in obtaining 128 samples of z-transform on the unit circle between $\omega = \frac{-\pi}{8}$ and $\omega = \frac{\pi}{8}$. The most efficient procedure is to compute a 1024 point DFT sequence by augmenting the original sequence with zeroes and retain only the 128 spectral points desired.

An alternative procedure is the use of the CZT algorithm. In many situations, CZT algorithm is the most efficient method. The algorithm is directed toward the computation of samples of the z-transform on a spiral contour equally spaced in angle over some portion of the spiral.

Specifically, let $s(n)$ is an N-point sequence and $s(n)$ is the z-transform of $s(n)$. Using the CZT algorithm, $s(n)$ can be computed at the points z_k given by

$$z_k = AW^{-k}, k = 0, 1, 2, \ ..., M^{-1} \tag{14.1}$$

where,

$$W = W_0 e^{-\phi_0} \tag{14.2}$$

$$A = A_0 e^{j\theta_0} \tag{14.3}$$

A and W are positive real numbers.

Consequently, the contour along which the samples are obtained is shown in Figure 14.1.

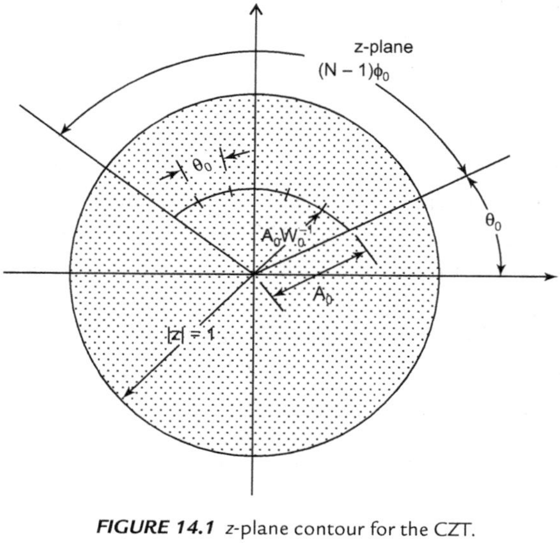

FIGURE 14.1 *z*-plane contour for the CZT.

This contour is a spiral in z-plane. The parameter W_0 controls the rate at which the contour spirals.

1. If W_0 is greater than unity, the contour spirals toward the origin as k increases.
2. If W_0 is less than unity, the contour spirals outward as k increases. The parameters A_0 and ϕ_0 are the location in radius and angle, respectively, of the first sample, that is, $k = 0$. The remaining samples are located along the spiral contour with an angular spacing of ϕ_0.
3. If $W_0 = 1$, the spiral is, in fact, a circular arc, and if $A_0 = 1$, this circular arc is a part of the unit circle, that is, $|z| = 1$.
   z-transform of sequence $s(n)$ is given be

$$S(z) = Z[s(n)] = \sum_{n=-\infty}^{\infty} s(n)^{z^{-n}} \tag{14.4}$$

where

$$z = \text{Complex variable.}$$

Substituting $z = z_k$ and $s(n)$ is finite length sequence, that is, $s(n)$ = Non-zero, $0 \le n \le N - 1$ = Zero, elsewhere in Eq. (14.4), we get

$$S(z_k) = \sum_{n=0}^{\infty} s(n) z_k^{-n} \tag{14.5}$$

Substituting Eq. (14.1) in Eq. (14.5), we get

$$S(z_k) = \sum_{n=0}^{N-1} s(n)\left[AW^{-k}\right]^{-n}$$

$$= \sum_{n=0}^{N-1} s(n) A^{-n} W^{nk}, k = 0,1,2,\ldots,M-1 \tag{14.6}$$

where N is the length of the sequence $s(n)$.

Now we have an identity

$$nk = \frac{1}{2}\left[n^2 + k^2 - (k-n)^2\right] \tag{14.7}$$

Substituting identity given by Eq. (14.7) in Eq. (14.6), we get

$$S(z_k) = \sum_{n=0}^{N-1} s(n) A^{-n} W^{nk}$$

$$= \sum_{n=0}^{N-1} s(n) A^{-n} W^{\frac{1}{2}\left[n^2 + k^2 - (k-n)^2\right]}$$

$$= \sum_{n=0}^{N-1} s(n) A^{-n} W^{\frac{n^2}{2}} W^{\frac{k^2}{2}} W^{\frac{-(k-n)^2}{2}}$$

or
$$S(z_k) = W^{\frac{k^2}{2}} \sum_{n=0}^{N-1} s(n) A^{-n} W^{\frac{n^2}{2}} W^{\frac{-(k-n)^2}{2}} \tag{14.8}$$

Let us assume that

$$g(n) = S(n) A^{-n} W^{\frac{a^2}{2}} \tag{14.9}$$

Substituting Eq. (14.9) in Eq. (14.8), we get

$$S(z_k) = W^{\frac{k^2}{2}} \sum_{n=0}^{N-1} g(n) W^{\frac{-(k-n)^2}{2}}, k = 0, 1, 2, \ldots, M-1. \tag{14.10}$$

Eq. (14.16) expresses $S(z_k)$ in terms of the new sequence $g(n)$. We recognize the summation as corresponding to the convolution of the sequence $g(n)$ with the sequence $W^{-\frac{n^2}{2}}$.

Therefore Eq. (14.10) can be written as

$$S(z_k) = W^{\frac{k^2}{2}} \left[g(n) * W^{-\frac{n^2}{2}} \right] \tag{14.11}$$

Sign * shows the convolution sum between two sequences.

Let us assume that $W^{-\frac{n^2}{2}}$ is another sequence and given as

$$h(n) = W^{-\frac{n^2}{2}} \tag{14.12}$$

Interpretation of Eq. (14.11) in terms of a linear system with impulse response

$h(n) = W^{-\frac{n^2}{2}}$ is shown in Figure 14.2.

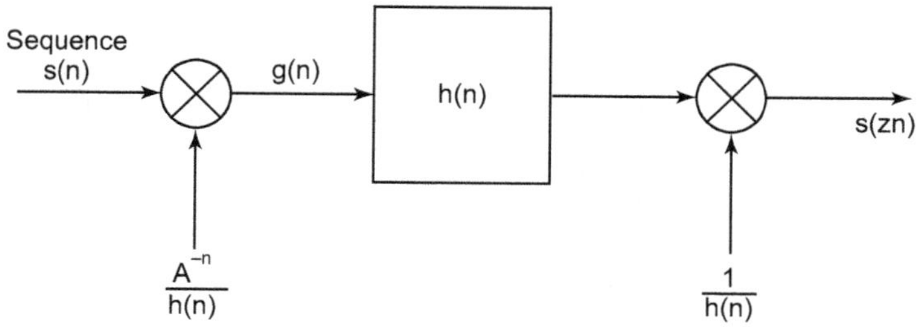

FIGURE 14.2 Interpretation of Eq. (14.11) in terms of a linear system with impulse response $h(n)$.

When A and W_0 are unity, the sequence $h(n)$ can be thought of as a complex exponential sequence with linearly increasing frequency. In radar systems such signals are chirp signals; hence the name CZT. A system similar to shown in Figure 14.2 is commonly used for spectrum, analysis in radar problems.

Since the sequence $g(n)$ is of finite duration, the convolution in Eq. (14.6) can be carried out by means of the DFT (of course computed by using FFT algorithms).

Whereas the sequence $g(n)$ is of finite duration, the sequence $W^{-\frac{n^2}{2}}$ is of infinite duration. Consequently, if the convolution is to be implemented using the DFT, then it is necessary to section the sequence $W^{-\frac{n^2}{2}}$.

We also note that, while the result of the convolution is of indefinite length, we are only interested in the result of the convolution for $k = 0, 1, 2, \ldots, M - 1$. Consequently, in sectioning the sequence $W^{-\frac{n^2}{2}}$, it would be advantageous to choose the sections in such a way that the result of the computation of one section results in the M desired output points.

Various sequences involved in this process for the case $N = 10$ and $M = 6$ are shown in Figure 14.3. The sequence $g(n)$ and $W^{-\frac{n^2}{2}}$ are depicted in Figure 14.3(a) and (b), respectively.

In implementing the convolution of sequences $s(n)$ and $W^{-\frac{n^2}{2}}$, the only part of $W^{-\frac{n^2}{2}}$ that is required to compute the result of the convolution in the interval 0 to $M - 1$ is that part from $-N + 1$ to $M - I$, including both of these end points. That part of the sequence $W^{-\frac{n^2}{2}}$ is between the dashed lines labeled A and B in Figure 14.3(b).

Consequently, the convolution can be implemented by computing the ($M + N - 1$) point DFT of $g(n)$ (augmented of course with $M - 1$ zeroes) and the ($M + N - 1$)-point DFT of the part of the sequence $W^{-\frac{n^2}{2}}$ in the region from A and B in Figure 14.3(b).

The inverse DFT of the product of these two DFTs will be the circular convolution of the sequences $g(n)$ with the section of $W^{-\frac{n^2}{2}}$. We consider the overlap-save method of implementing a convolution, part of the circular convolution will correspond to a linear convolution and part will not. We can

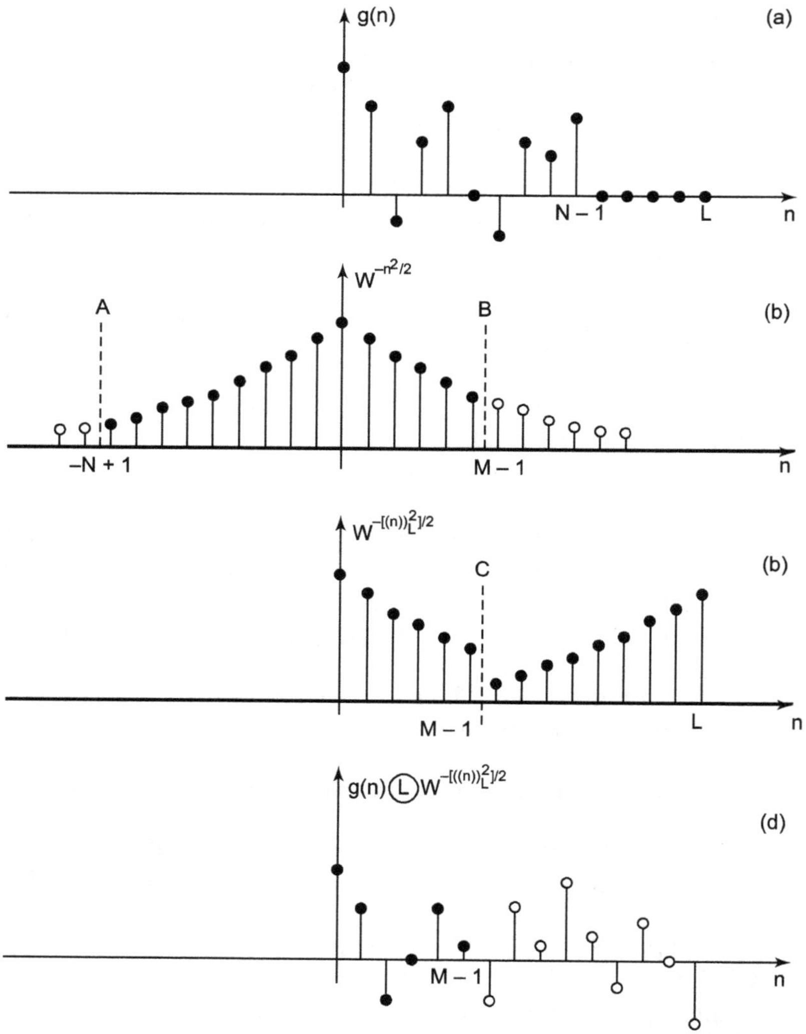

FIGURE 14.3 Illustration of sequences involved in the chirp *z*-transform (*CZT*) algorithm ($L = N + M - 1$).

arrange for the "good" or desired points to occur in the region $0 \leq n \leq M - 1$, by interpreting the index n-modulo $(N + M - 1)$. This means that we would compute the DFT of the sequence.

$$h(n) = \begin{cases} W^{-\frac{n^2}{2}}, & \text{for } 0 \leq n \leq M-1 \\ W^{-\frac{(N+M-1-n)^2}{2}}, & \text{for } m \leq n \leq N+-2 \end{cases} \tag{14.13}$$

$h(n)$ is shown in Figure 14.3(c).

If we multiply the DFTs of $g(n)$ and $h(n)$, the first M values of the corresponding inverse DFTs are the desired values of the convolution of $g(n)$ with $W^{-\frac{n^2}{2}}$. To obtain the desired M values of $S(k)$ as in $S(z_k) = W^{-\frac{k^2}{2}=\sum_{n=0}^{N-1} g(n)} W^{\frac{-(k-n)^2}{2}}$, we must multiply these values by $W^{\frac{k^2}{2}}$.

In the above discussion the size of the DFTs computed was $(M + N - 1)$. If we wish to compute the DFT using a power of 2 (2^r) algorithm. This can easily be accomplished by augmenting the $(M + N - 1)$-point sequences with a sufficient number of zeroes so that their total length is a power of 2. Since the number of complex multiplications required for the computation of each DFT is on the order of

$$(N + M - 1)\log_2(N + M - 1)$$

1. It is clear that the total computation required to implement the evaluation of $S(z_k) = \sum_{n=0}^{N-1} s(n)A^{-n}W^{nk}$ using the CZT algorithm is proportional $(N + M - 1).\ \log_2(N + M - 1)$.

   In contrast, the direct evaluation of $S(z_k) = \sum_{n=0}^{N-1} s(n)A^{-n}W^{nk}$ requires computation proportional to (N.M.). It is also clear that the direct method will be most efficient for small enough values of N or M, But, it is also true that for sufficiently large M and N (on the order of 50), the CZT algorithm will be most efficient.
2. In addition to increased efficiency, the CZT algorithm also offers added flexibility in the computation of samples of the z-transform of the finite length sequence. We do not require $N = M$ as m the FFT algorithms, and neither N *nor* M need be highly composite numbers; in fact, they may be prime numbers, if desired. The parameter ϕ_0 is required to be $\frac{2\pi}{N}$ in an FFT algorithm, whereas ϕ_0 is arbitrary in the CZT algorithm. Further, the samples of the z-transform are taken on a slightly more general contour that includes the unit circle as a special case.

The CZT algorithm is an example of the way that Fourier analysis can be performed using linear filtering. Similarly, the Goertzel algorithm is another example of the same.

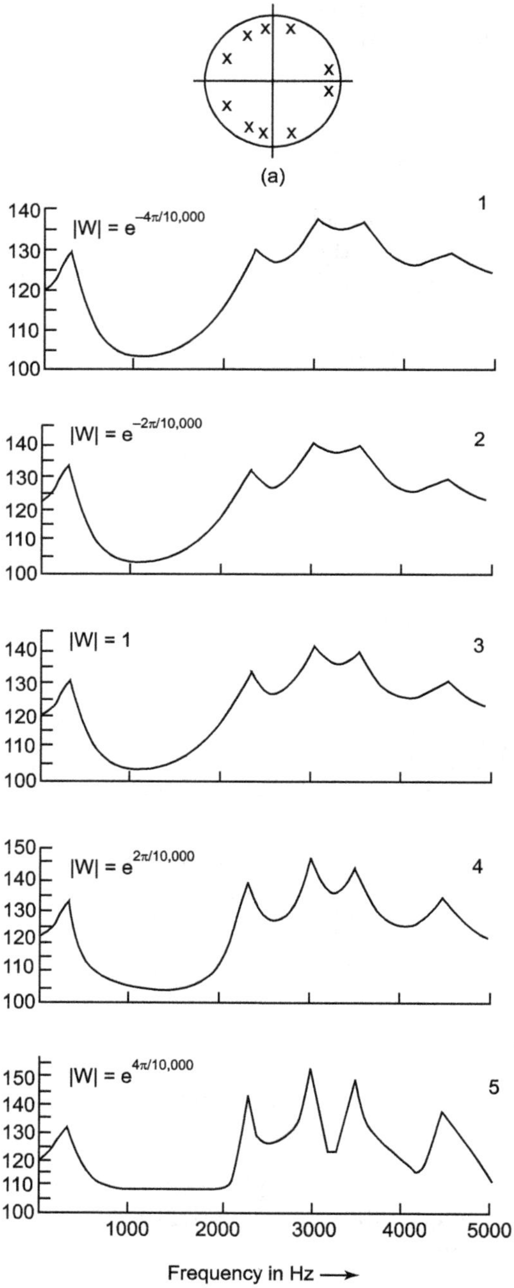

FIGURE 14.4 Use of chirp *z*-transform (CZT) algorithm, (a) *z*-plane pole locations for synthetic speech signal. (b) Evaluation of *z*-transform for several spiral contours.

EXAMPLE 14.1

An example of the use of the CZT algorithm to sharpen resonances by evaluating the z-transform of the unit circle, $|z| = 1$ is shown in Figure 14.4. The signal to be analyzed corresponds to a finite length segment of a synthetic speech signal.

The speech signal was generated by exciting a five-pole system with a periodic impulse train. The system was simulated to correspond to a 10 kHz sampling frequency. The poles were located at center frequencies of 270, 2290, 3010, 3500, and 4500 Hz with bandwidths of 30, 50, 60, 87, and 140 Hz, respectively.

Solution:

Figure 14.4(a) shows the z-plane plot indicating the location of the poles used to generate the signal. The CZT algorithm was applied to one period of the steady-state data for five different choices of $|W|$ with the results shown in Figure 14.4(b). The first two spectra correspond to spiral contours outside the unit circle with a resulting broadening of the resonance peaks.

$|W| = 1$ corresponds to evaluate the z-transform on the unit circle. As $|W|$ increases past unity the contour spirals inside the unit circle and closer to the pole locations resulting in a sharpening of the resonance peaks.

14.5 RADAR SYSTEM AND RADAR PARAMETERS

The major components of a radar are the antenna, the tracking computer, and the signal processor. Associated with the antenna are the transmitter and modulator and receiver hardware.

The tracking computer is the brain of the radar System which is used to schedule the appropriate antenna positions and transmit signals as a function of time, keep the track of important targets, and run the display system.

The major traditional functions of the signal processor are matched filtering and removal of useless information by threshold detection. A key element in the design of the overall radar is signal design. Transmitted radar signals may vary from simple pulse trains to high bandwidth chirp signals or low-frequency modulated (LFM) signals, bursts of pulses or chirps, nonuniform bursts, or polyphase circles.

The block diagram of a modern radar system is shown in Figure 14.5. Tracking computer controls all the functions. There is a control path from the tracking computer to the antenna system, if the antenna beam is electronically steerable, the tracking computer can control the beam position on a pulse-to-pulse basis and can determine whether monopulse information is called for.

Monopulse information is useful for better angle resolution. The tracking computer also controls the signal generator and coordinates the transmitted signals and the matched filter configuration of the tasks of the tracking computer becomes too great, and the jobs may be divided among several computers as shown in Figure 14.5.

14.5.1 Radar Parameters

We will consider the following radar parameters, which are discussed below:

1. Antenna aperture and wavelength.
2. Range and range resolution.
3. Doppler filtering.

1. **Antenna Aperture and Wavelength:** Antenna beamwidth is given by

$$B_0 \infty \frac{\lambda}{D} \tag{14.14}$$

where B_0 = beamwidth of the antenna, and λ = wavelength, D = antenna width.

If the antenna geometry is symmetric as in a parabolic reflector antenna, then antenna beamwidth B_0 is the same in both horizontal (azimuth) and vertical (elevation) dimensions, this corresponds to a pencil beam.

A spherical coordinate system is normally used for radar antennas. The radius of the sphere corresponds to range (distance from antenna), while azimuth is the angular dimensions parallel to the earth, elevation is the orthogonal angular dimensions perpendicular to the earth.

In many applications such as air traffic control (ATC), a fan beam is required to obtain full coverage in a reasonable tone. A fan-beam antenna is built with a large horizontal and small vertical aperture to obtain a beam that is narrow in azimuth and wide in elevation.

2. **Range and Range Resolution:** The maximum unambiguous range (R_{max}) of the Radar is given by

$$R_{max} = \frac{cT}{2} \tag{14.15}$$

where c = velocity of light = 3×10^8 meter/second and T = pulse repetition time in seconds.

For example, if pulse repetition time T is equal to 1 millisecond, then

$$\text{Radar Range}, R_{max} = \frac{cT}{2} = \frac{3\times10^8 \times 1\times10^{-3}}{2}$$

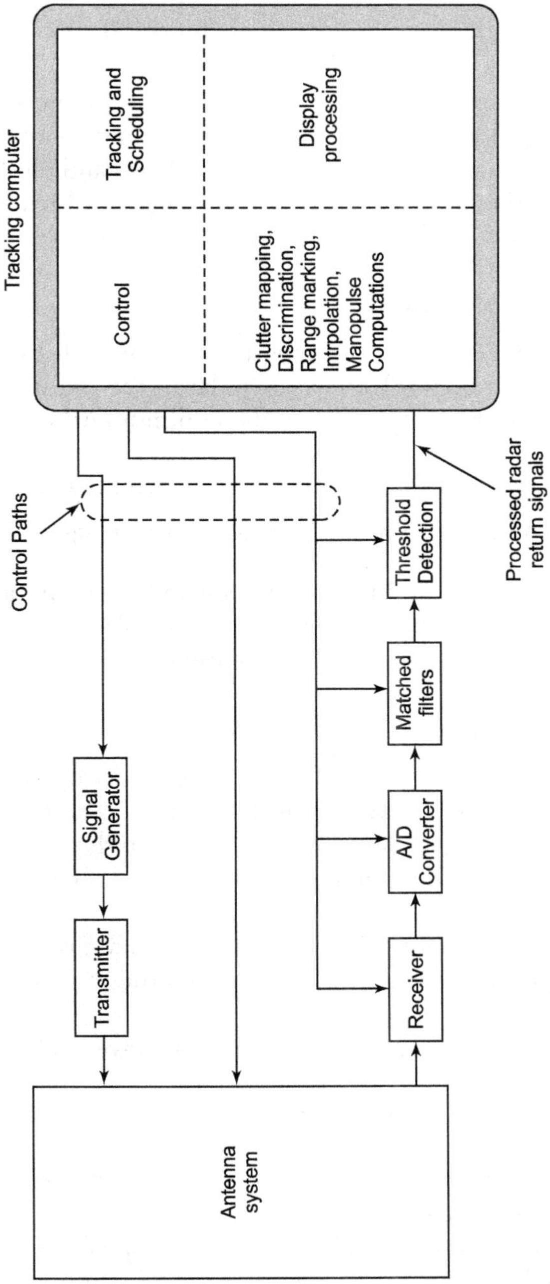

FIGURE 14.5 Block diagram of a modern radar system.

$$= 1.5 \times 10^5 \text{ meter} = 150 \times 10^3 \text{ meter}$$
$$= 150 \text{ kilometer}$$

If T is decreased, targets at ranges greater than R_{max} will appear after the next radar pulse has been transmitted, causing an ambiguity in the interpretation of the measured range. Receiver noise and clutter may clamp a much lower limit than R_{max} on the maximum range at which target detection probabilities are good.

14.5.2 Range Resolution

Range resolution ΔR is the measure of how well two targets that are near each other can be resolved by the radar. If the transmitted signal is a pulse of constant carrier frequency, then the minimum ΔR is determined by the pulse width p.

If the range difference ΔR between two targets in the same beam is less than $\frac{pc}{2}$, the two received radar signals (echo signals) will interfere and the two targets may be mistaken for a single target. By decreasing the width of the pulse, improvement in range resolution ΔR will take place but maximum range R_{max} will reduce due to a decrease in the average power of the signal.

14.5.3 Doppler Filtering

Moving targets can be identified by using the Doppler effect, When the target is moving relative to radar, then there will be a shift in the carrier frequency of the received signal, and this effect is called the Doppler effect. The shift in frequency is the Doppler shift and this is the measure of the velocity of the radar.

When a continuous sine wave of frequency f_0 is transmitted and the target is moving with a constant velocity n, then the received echo signal frequency is $f_0 + \Delta f$.

The resultant frequency shift in the received echo signal is given by

$$\Delta f = \frac{2\nu}{c} f_0 = \frac{2\nu}{\lambda} \tag{14.16}$$

where f_0 = carrier frequency; ν = target velocity, and λ = wavelength $= \frac{c}{f_0}$; $c = 3 \times 10^s$ m/second.

Continuous-time signals or practically speaking signals of long duration would result in poor range resolution. To obtain both range and velocity resolution it becomes necessary to deal with pulsed Doppler signals. We shall now show that detection of pulsed Doppler signals corresponds to sampling a sinusoid of frequency Δf at the pulse repetition period. Here, we assume that each pulse of carrier frequency f_0 begins at exactly the same phase, then at some given range, a return signal is received.

Since, during the pulse duration, the aircraft can be considered stationary, there is no measurable Doppler shift of the received signal. After T seconds, however, the aircraft has moved slightly and if the next return is sampled at the same range, a phase shift relative to the first return is discernible.

$$\text{Phase Shift } \phi = 2\pi \frac{vT}{\lambda} \tag{14.17}$$

If the aircraft maintains constant radial velocity, there will be an additional phase shift ϕ for each repetition interval T. Thus, any signal return from a given range can be represented as

$$s(n,t) = A(n)e^{j[2\pi f_0(t-T)+n\phi]} \tag{14.18}$$

where $A(n)$ = amplitude modulation caused by the motion of the antenna beam as it sweeps by the target.

For an electronically steerable antenna, it is possible to stop the antenna beam so that $A(n)$ can be unity.

In the radar receiver, we assume that the returned complex exponential is multiplied by a coherent oscillator source with an arbitrary but fixed phase y. This signal is denoted by $s_r(t)$.

The demodulated signal will be given as

$$f(n,t) = s(n,t)s_r(t) \tag{14.19}$$

Substituting Eq. (14.5) and $s_r(t) = e^{-j(2\pi f_0 t+\psi)}$ in Eq. (14.6), we get

$$\begin{aligned} f(n,t) &= s(n,t)s_r(t) \\ &= A(n)e^{j[2\pi f_0(t-T)+n\phi]}e^{-j(2\pi f_0 t+\psi)} \\ &= A(n)e^{0\, j\omega_0 T}e^{-j\psi}e^{jn\phi} \end{aligned} \tag{14.20}$$

Putting the value of ϕ from Eq. (14.4) in Eq. (14.7), we get

The exponentials $e^{-j\omega_0 T}$ and $e^{-j\psi}$ are constants of unity amplitude and can be ignored. The variable part is simply an oscillation of frequency $\frac{v}{\lambda}$, which is the Doppler frequency.

14.6 RADAR SIGNAL DESIGN AND AMBIGUITY FUNCTIONS

Radar signal design is directed toward achieving the best range and velocity measurement on one or more targets. We know that a transmitted narrow pulse results in good range but poor velocity measurement, a wide pulse of a single frequency yields good velocity but bad range information. From this, the signal design will consider a compromise between range and velocity measurement. We use the ambiguity function of two variables, range and velocity to achieve a compromise between range and velocity. The ambiguity function is at the center of analog radar signal design using analog matched filtered and the emphasis will be given to digital ambiguity functions. Here, we discuss and develop ambiguity functions entirely based on the assumption that the signal processor is digital.

The ambiguity function is an idealized mathematical model of the system shown in Figure 14.6. We assume that the signal is generated digitally but must pass through an analog filter on its way to the transmitter. Ideally the analog signal return $s\ (t - \tau)\ e^{j2\pi\ (t - \tau)}$ is a delayed and frequency shifted version of the transmitted signal $s(t)$. These effects, due to target displacement and velocity, are assumed to be carried undisturbed through the receiver analog filter and the A/D converter so that the input to the matched filter is the digital signal $s\left(nT_s - \tau\right)^{j2\pi f(nT_s - \tau)}$. Note that this signal is a function of two continuous parameters, τ and f (range and Doppler).

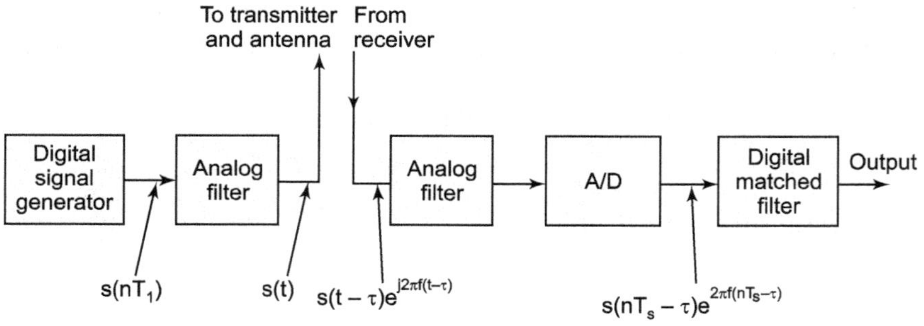

FIGURE 14.6 Block diagram of the radar model leading to the ambiguity function.

Now, we enquire about the nature of the digital matched filter used. To preserve radar power it is describable to make the signal duration large, but to preserve range resolution the output from a given range bin must be compact in time. This apparent contradiction is resolved by designing long-duration signals with short-duration correlation functions so that when the received signal passes through the appropriate matched filter, the output will be a very sharp pulse.

Hence, if we have a digital filter matched to the signal return for zero range and zero Doppler, then this filter must have the impulse response $s*(-nT_s)$. Therefore, the digital matched filter output will be defined as the digital ambiguity function. The ambiguity function is also in actuality the cross-correlation function between the signal and the matched filter impulse response.

Ambiguity function will be given as

$$A(\tau, f) = \sum_{n=-\infty}^{\infty} s(nT_s + \tau)s*(-nT_s)e^{j2\pi f(nT_s+\tau)} \tag{14.22}$$

where T_s is the sampling period at A/D converter.

If the matched filter is matched to a given range and Doppler, this corresponds to a two-dimensional displacement of $A(t, f)$. Therefore, Eq. (14.22) is a perfectly general formulation.

The major importance of the ambiguity function, that is, *the* magnitude of $A(\tau, f)$ is to bring into focus a basic constraint of radar signal design. The basic constraint of a radar signal design is that a signal cannot be designed that gives high performance everywhere in the range-velocity plane. The mathematical manifestation of this constraint is that the total volume under the squared magnitude of the ambiguity function is independent of the signal wave shape. Thus pushing $A(\tau, f)$ down anywhere in the (τ, f) plane will cause it to pop up elsewhere. To prove this, we use Eq. (14.22).

$$V = \int_{-\infty}^{\infty} \int_{\frac{-F_s}{2}}^{\frac{F_s}{2}} A(\tau, f)A*(\tau, f)dfd\tau \tag{14.23}$$

where F_s = sampling frequency = $\frac{1}{T_s}$ and T_s = sampling period.

Substituting Eq. (14.22) in Eq. (14.23), we get

$$V = \int_{-\infty}^{\infty} \int_{\frac{-F_s}{2}}^{\frac{F_s}{2}} A(\tau, f)A*(\tau, f)dfd\tau$$

$$= \int_{-\infty}^{\infty} \int_{\frac{-F_s}{2}}^{\frac{F_s}{2}} \left[\sum_{n=-\infty}^{\infty} s(nT_s + \tau)s*(nT_s)e^{j2\pi(nT_s+\tau)} \right.$$

$$\left. \times \sum_{m=-\infty}^{\infty} s(mT_s + \tau)s(mT_s)e^{-j2\pi(mT_s+\tau)} \right] dfd\tau$$

$$= \sum_{n=-\infty}^{\infty} \sum_{m=-\infty}^{\infty} \int_{\frac{-F_s}{2}}^{\frac{F_s}{2}} \left[\int_{-\infty}^{\infty} s(nT_s + \tau)s*(nT_s)s*(mT_s + \tau)s(mT_s) \right.$$

$$\left. \times \left\{ e^{j2\pi f(nT_s+\tau)} e^{-j2\pi f(mT_s+\tau)} \right\} \right] df\, d\tau$$

$$= \sum_{n=-\infty}^{\infty} \sum_{m=-\infty}^{\infty} \int_{-\infty}^{\infty} \int_{\frac{-F_s}{2}}^{\frac{F_s}{2}} s(nT_s + \tau)s*(nT_s)s*(mT_s + \tau)s(mT_s)$$

$$e^{j2\pi f(n-m)T_s} df\, d\tau$$

(on integration with respect to f)

$$= \sum_{n=-\infty}^{\infty} \sum_{m=-\infty}^{\infty} \int_{-\infty}^{\infty} s(nT_s + \tau)s*(nT_s)s*(mT_s + \tau)s(mT_s)$$

$$\frac{\sin[\pi F_s T_s (n-m)]}{\pi T_s (n-m)} d\tau \tag{14.24}$$

Since $R[(n-m)T_s] = \int_{-\infty}^{\infty} s(nT_s + \tau)s*(mT_s + \tau)d\tau$ (14.25)

Substituting Eq. (14.25) in Eq. (14.24), we get the autocorrelation function of the signal

$$V = \sum_{n=-\infty}^{\infty} \sum_{m=-\infty}^{\infty} s*(nT_s)R[(n-m)T_s] \frac{\sin[\pi F_s T_s (n-m)]}{\pi T_s (n-m)} \tag{14.26}$$

Since $F_s = \frac{1}{T_s}$ or $F_s T_s = 1.$

$$\frac{\sin[\pi F_s T_s (n-m)]}{\pi T_s (n-m)} = \begin{cases} 0, & \text{when } n \neq m \\ F_s & \text{when } n = m \end{cases} \tag{14.27}$$

Therefore, the double sum of Eq. (13) can be replaced with the single sum

$$V = F_s R(0) \sum_{n=-\infty}^{\infty} |s(nT_s)|^2 \tag{14.28}$$

We see from Eq. (14.28) that V (volume under the ambiguity function) is dependent only on the total signal energy and not in any way on the signal shape.

Thus, for a given amount of signal energy, we have demonstrated that any decrease of the ambiguity function must result in an increase somewhere else in the (τ, f) plane.

14.7 AMBIGUITY FUNCTIONS OF CHIRPS AND SINUSOIDAL PULSES

A chirp waveform is a linear frequency modulated (LFM) signal which combines some of the useful properties of both long and short pulses of a single sinusoidal carrier.

Now we first find the ambiguity function for the chirp and then obtain that of the continuous wave (CW) pulse as a special case.

From a practical point of view, we are interested in signals and systems, that is, filters with a finite duration. It means that we have to establish some convention on the limits in Eq. (14.9) that takes into account the precise way that $s(nT_s + \tau)$ and $s*(nT_s)$ overlap. We agree on the following convention as shown in Figure 14.7.

Let the width of the signal T be exactly equal to Mt_s where M is the number of samples in the matched filter. Thus for $-T_s < t \leq 0$, the overlap between signal and impulse response is perfect.

For $0 < t \leq T_s$, they are misaligned by one sample.

Now let us define $I(\tau)$ as the nearest rounded up integer of the ratio $|i|/T_s$. Thus, when $0 < t \leq T_s$, $T(\tau) = 1$, when $T_s < t \leq 2T_s$, $I(\tau) = 2$, etc. For t negative, Eq. (9) can be written as

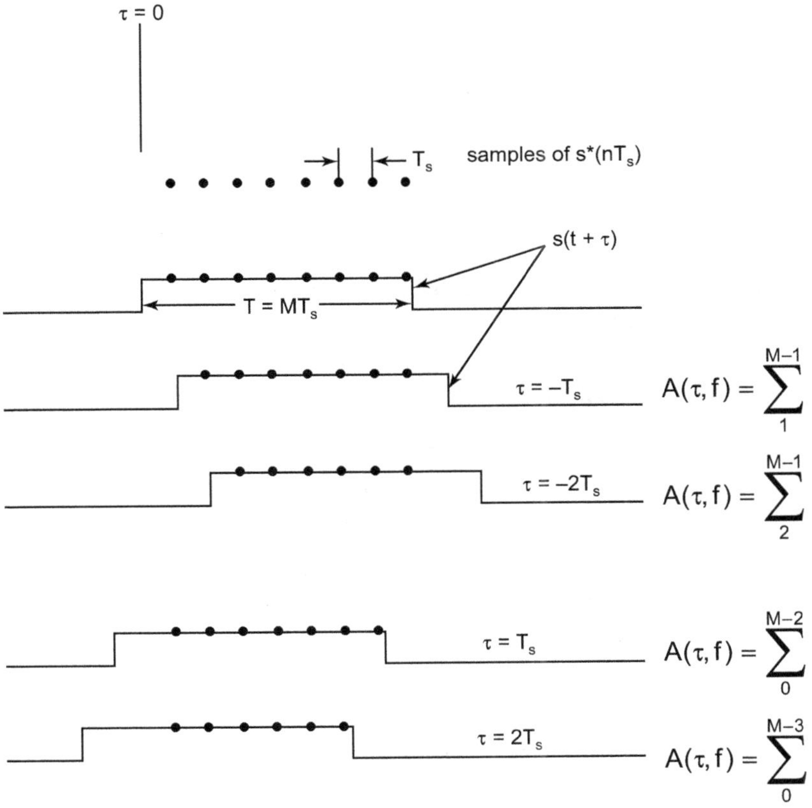

FIGURE 14.7 Synchronization conventions between digital matched filter and sampled signal.

$$A(\tau, f) = \sum_{n=I(\tau)}^{M-1} s\left(nT_s + \tau\right)s*\left(nT_s\right)e^{j2\pi fnT_s} \tag{14.29}$$

Similarly, for τ positive, Eq. (14.22) can be written as

$$A(\tau, f) = \sum_{n=0}^{M-1-I(\tau)} s\left(nT_s + \tau\right)s*\left(nT_s\right)e^{j2\pi fnT_s} \tag{14.30}$$

If $s(nT_s)$ is of the form of an exponential, then Eq. (14.29) can be changed into Eq. (14.30) by the simple change of variable $k = n - I(\tau)$, the inclusion of $I(\tau)$ in the arguments of the three functions inside the sum of Eq. (14.29) has been factored out to become a phase term that can be discarded. Thus, any computation can be done using Eq. (14.30) and $A(\tau, f)$ is really a function of $|I(\tau)|$, where $|I(\tau)|$ is the magnitude of $I(\tau)$. Now, we can develop Eq. (14.30) for the special case of a chirp.

$$s(t) = e^{j(\pi W/T)t^2} \tag{14.31}$$

where W = swept bandwidth of the chirp and T = total signal time duration.

We replace the parameter W by $\frac{N}{T}$ so that N is the time-bandwidth product of the signal.

For the case when T_s is the Nyquist sampling interval, that is, $\frac{1}{W}$, we find $N = \frac{T}{T_s} = M$. If we derive the result keeping M and N independent variables, then our answer will be valid for any sampling rate.

After neglecting phase terms from Eqs. (14.30) and (14.31), we get

$$|A(\tau,f)| = \sum_{n=0}^{M-1-I(\tau)} e^{j\left[2\pi\left(\frac{W}{T}\right)\tau nT_s + 2\pi fnT_s\right]} \tag{14.32}$$

Normalization of τ and f will be done by using the following relationships

$$\gamma = \frac{\tau}{T} \text{ and } \nu = fT \tag{14.33}$$

Substituting Eq. (14.33) in Eq. (14.32), we get

$$|A(\gamma,\nu)| = \sum_{n=0}^{M-1-I(\gamma T)} e^{j\left[w\pi W\gamma nT_s + 2\pi fn\frac{T}{M}\right]}$$

$$= \sum_{n=0}^{M-1-I(\gamma T)} e^{j\left[2\pi fn + \frac{2\pi\gamma n}{M}\right]}$$

or

$$|A(\gamma,\nu)| = \sum_{n=0}^{M-1-I(\gamma T)} e^{j2\pi n\left[\gamma + \frac{\nu}{M}\right]} \tag{14.34}$$

After solving Eq. (14.34) and further manipulating, we obtain

$$|A(\gamma,\nu)| = \frac{\sin[\pi(N\gamma+\nu)][1-(|(\gamma T)|/M)]}{\sin[(\pi/M)(N\gamma+\nu)]} \tag{14.35}$$

Eq. (14.35) is our fundamental result.

Now we check two conditions to see what happens when γ and ν*n* are zero.

If ν = 0, then Eq. (14.35) can be written as

$$|A(\gamma,0)| = \frac{\sin(\pi N\gamma)[1-(|I(\gamma T)|/M)]}{\sin[(\pi N/M)\gamma]} \tag{14.36}$$

If γ = 0, then Eq. (14.35) can be written as

$$|A(0,\nu)| = \frac{\sin(\pi\nu)[1-(|I(0)|/M)]}{\sin(\pi\nu/M)} \tag{14.37}$$

The important point to note is that both Eqs. (14.36) and (14.37) look like sharp pulses (about either γ = 0 and ν = 0) on about the same width, given that *M* is reasonably large integer. The term 1 – (|*I*(γ*T*)|/*M*) has the effect of reducing the frequency of the side lobe ripples at large range offsets.

Because the term *N*γ + ν appears as an entity in both arguments of Eq. (14.35). There is no way to separate the effects of range and velocity offsets. The ambiguity function of the chirp signal is shown in Figure 14.8 This figure shows two range cross-sections of the ambiguity function at different Doppler. Assume the chirp signal is used to track two targets in the same range bin traveling at different velocities. The result will be target returns apparently displayed in the range.

Since the chirp may be of long duration, it illuminates a target with substantial energy, thus increasing range. In addition, since the matched filter response is always a sharp pulse, the range resolution is also obtained. In this sense, the chirp has several good properties. Range–Doppler coupling makes it impossible to separate range from velocity measurements, thus the chirp is not a good signal for velocity measurements. Fortunately, in many practical problems, the range offset caused by Doppler is very small so that the signal return can also be used as a range measure.

Along the line *N*γ + ν = 0 in the (γ, ν) plane Eq. (14.35) reduces to

$$|A(\gamma,\nu)| = \left(1-\frac{|I(\gamma T)|}{M}\right)M \tag{14.38}$$

Thus, in three dimensions one can imagine a ridge along the line *N*γ + ν = 0 with a triangular decrease in the height of the ridge with increasing range offset. Due to the stepwise nature of |*I*(γ*T*)|, the top of the ridge

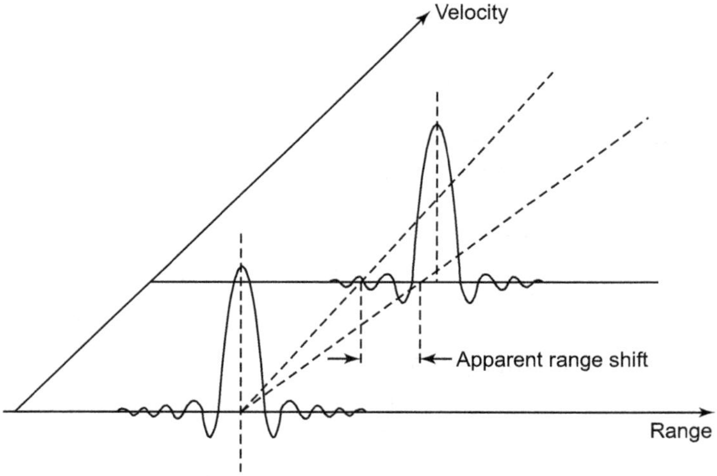

FIGURE 14.8 Ambiguity function of Chirp.

actually decreases in discrete steps. Figure 14.9(a) is a sketch of Eq. (14.35). This sketch shows $|A(\gamma, \nu)|$ (or $|A(\tau, f)|$) has substantial value (shaded region) and where it does not. For the case $M = N$ or $T_s = \frac{1}{W}$, which corresponds to Nyquist sampling. Figure 14.9(b) shows what happens when the sampling rate is halved so that $T_s = \frac{1}{2W}$, extra shaded regions due to leasing appear.

By comparing equation given below

$$A(\tau, f) = \sum_{n=-\infty}^{\infty} s(nT_s + \tau)s*(nT_s)e^{j2\pi f(nT_s+\tau)}$$

With the well-known formula for the analog ambiguity function

$$a(\tau, f) = \int_{-\infty}^{\infty} s(t+\tau)s*(t)e^{j2\pi ft}\,dt \tag{14.39}$$

We can derive the relationship between $a(\tau, f)$ and $A(\tau, f)$.

Relationship between the continuous-time Fourier transform of an analog signal $g(t)$ and the Fourier transform of the sampled version of the same analog signal are as follows:

if $$G_a(f) = \int_{-\infty}^{\infty} g(t)e^{-j2\pi ft}\,dt$$

and $$G\left(e^{j2\pi f}\right) = \sum_{n=-\infty}^{\infty} g(nT_s)e^{-j2\pi fnT_s}$$

then $$G\left(e^{j2\pi f}\right) = \sum_{n=-\infty}^{\infty} G_a\left(f + \frac{n}{T_s}\right) \tag{14.40}$$

By inspecting Eqs. (14.22) and (14.39), we conclude that the Eq. (14.39) is the Fourier transform of the "signal" $s(t + \tau)\, s*(t)$ and Eq. (14.22) is the Fourier transform of the sampled version, hence

$$A(\tau, f) = \sum_{n=-\infty}^{\infty} a\left(\tau, f + \frac{n}{T_s}\right) \tag{14.41}$$

Digital ambiguity function is periodic in frequency but not in time.

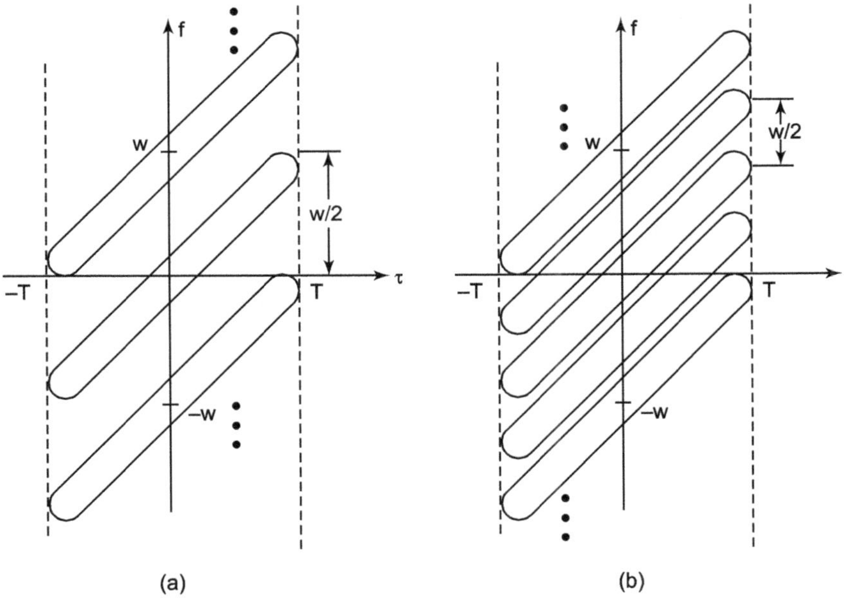

FIGURE 14.9 (a) Dense Portion of Chirp ambiguity function when sampling at the Nyquist rate. (b) Dense portion of Chirp Ambiguity function when sampling at half the Nyquist rate.

14.8 AMBIGUITY FUNCTION OF A CW PULSE

From Eq. (14.35), if we set $W = 0$ so that $N = 0$, this corresponds to a CW pulse and we find

$$|A(\gamma,\nu)| = \frac{\sin(\pi\nu)[1-(|I(\gamma T)|/M]}{\sin(\pi\nu/M)} \tag{14.42}$$

A few cross-sections versus n for fixed values of γ are shown in Figure 14.10 We can note that as range offsets increases not only does the main lobe decrease in size but it widens so that velocity resolution is lost.

If we take cross-section cuts versus γ for fixed values of n, then poor results would be obtained. For $\nu = 0$, $|A(\gamma, 0)|$ has a triangular shape, for other values of ν, the shape is sinusoidal with peaks occurring in arbitrary places.

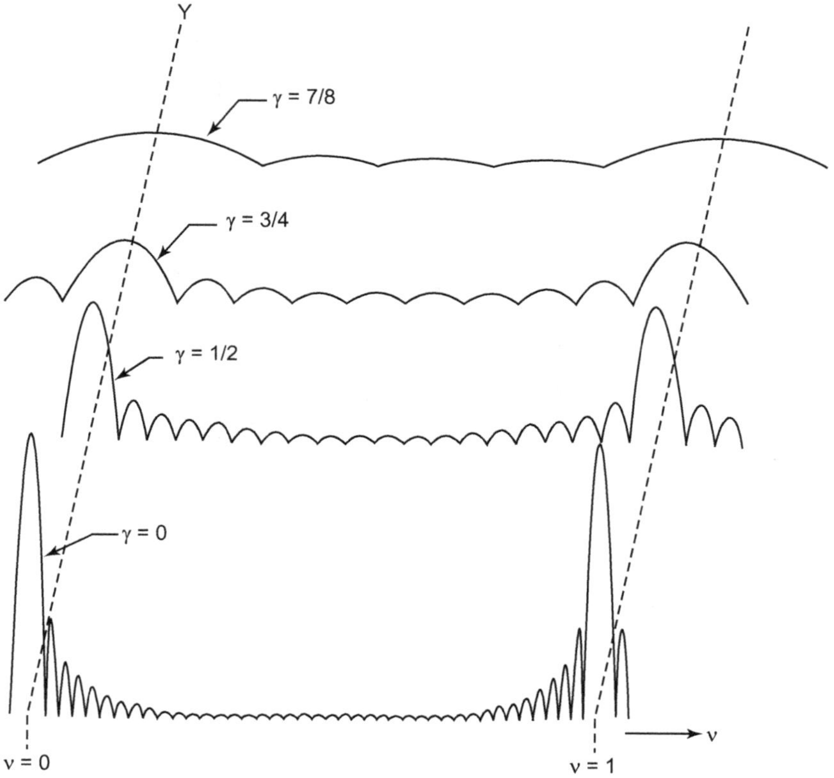

FIGURE 14.10 Cross-sections of the Arbitrary functions of a CW pulse.

14.9 AMBIGUITY FUNCTIONS OF A BURST

A chirp signal results in good signal detectability but it leaves unanswered the numerical values of velocity for a given target. The chirp is a popular radar signal to help increase range, Precise velocity measurements require additional radar signals.

An increased unambiguous range window is obtained through transmitting at a low repetition rate, while an increased unambiguous velocity window requires increasing the pulse repetition rate. In either case, the signal being transmitted is a succession of pulses, for which we now derive and discuss the ambiguity function. Since the burst is a pulse sequence, the analog and digital ambiguity exhibit the same basic features and there is no new property imparted to the signal by virtue of its digitization. A sketch of the ambiguity function of a 10-pulse burst with uniform spacing is shown in Figure 14.11.

This is also called "Bed of Nails." The parameter Δ is the spacing between pulses Figure 14.10 displays both range and velocity ambiguity. Since the periods in both τ and f are functions of only Δ, we see that increasing Δ lessens range ambiguity and increases velocity ambiguity and vice-versa. Thus, a burst most usefully yields range and velocity measurements. A burst provides sought but reliable information either range or velocity or both.

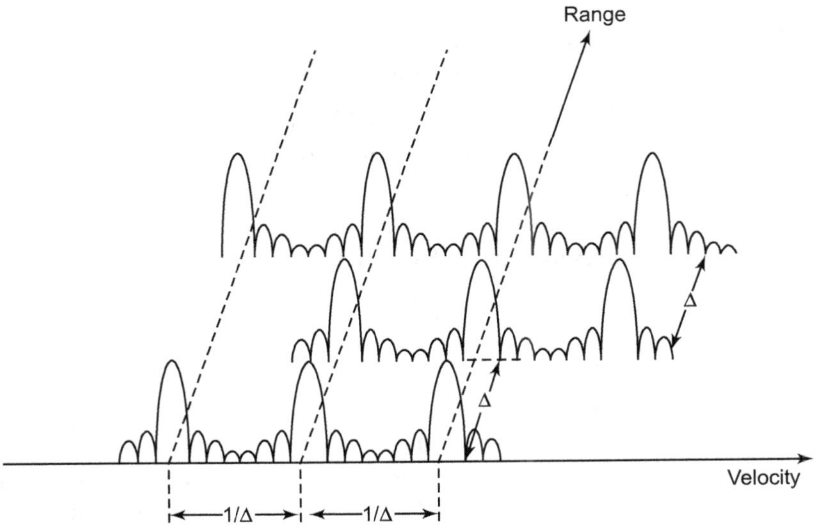

FIGURE 14.11 Ambiguity function cross-section of a burst.

14.10 OTHER SIGNALS

Sometimes, it may be useful during radar research to try to measure both range and velocity. This can be done by designing a signal with an ambiguity function that approximates a thumbtack, with a substantial peak of energy concentrated in a small section of the range–Doppler plane. Many such signals have been invented, such as up-down chirps, Barker and polyphase codes, and shift register codes.

There is a similarity between the design of filters and the design of signals having desired ambiguity functions.

There are several reasons why a formal signal synthesis procedure *is quite* difficult.

1. One is faced with a criterion for approximating a two-dimensional ideal function, it is more difficult than the filter design counterpart of approximating either a one-dimensional spectrum or impulse response.
2. Radar signal design is greatly influenced by many other radar design parameters, involving the antenna, transmitter, and receiver tracking computer, which is not necessarily under the control of the radar signal processor.
3. It is most important, clutter environments are often very difficult to model and the "right" ambiguity function depends on clutter properties.

It means that a radar system will not be optimized until the radar has undergone much field testing. Thus, extra flexibility inherent in DSP should be exploited as fully as possible.

14.10.1 Digital Matched Filters for Radar Signals

For any given range-angle cell, the signal return from the desired target may be marked by undesired background clutter. If the target-velocity is sufficiently different from clutter-velocity, Doppler filtering can be applied to extract the signal component. In a broad sense, the work of a radar signal processor is to perform matched filtering of the return signal for every range–Doppler (and angle) celt of interest. This results in very complex equipments and leads us to treat from the general case. Therefore, we require an appropriate matched filter for a variety of more specialized situations.

14.10.2 Filter Matched to a Long Pulse of Constant Frequency

Here, we consider the problem of obtaining a radar track on a satellite. We assume that the approximate angular position of the satellite relative to the radar is known but both range and velocity are relatively unknown.

We already know that a long pulse results in poor range resolution but the major problem is detection because of the large distance and small size of the target. In this case, it behooves us to construct a bank of filters. Here each filter is tuned to a different presumed.

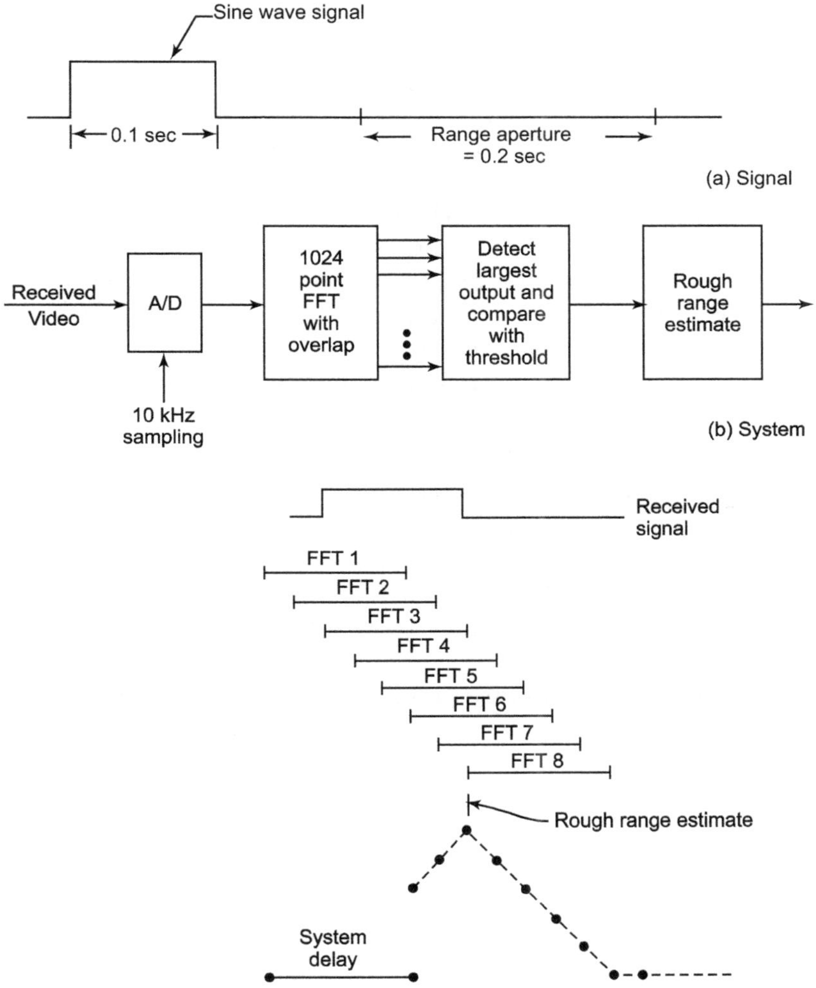

FIGURE 14.12 System for Detection and crude range estimation at very long ranges.

Doppler frequency: A pulse (assumed) is on for 0.1 s as shown in Figure 14.12 and the search interval encompasses 0.2 s, which corresponds to an unambiguous range aperture of 18,600 mi. Let the maximum satellite velocity be 1000 mph and choose an S-band radar, with $\lambda \approx 10$ cm. Then, maximum Doppler frequency $\Delta f = \frac{2v}{\lambda} = \frac{2 \times 1000}{0.10} = 20{,}000$ Hz. The velocity resolution is inversely proportional to the signal "on" time and is therefore 10 Hz. Hence, a bank of 1000 Doppler filters and a sampling rate of 10,000 Hz is called for.

We know that a sliding DFT is equivalent to a bank of filters. In fact, it is also true that the sliding unweighted FFT corresponds exactly to a bank of "matched" filters since the impulse responses of the filters comprising the sliding FFT correspond exactly to the received signal. A sliding 1024-Point FFT at a sampling rate of 10 kHz is a more expensive item than is really needed. However, because each FFT filter is only 10 Hz wide and there is no need to sample at 10 kHz. If we replace the sliding FFT by a hopping FFT, performing the FFT at (say) 50 times per second, this is equivalent to sampling the sliding FFT output 50 times per second. This results in a 200: 1 reduction in required computation speed.

As shown in Figure 14.12(*c*), the detected returns from the successive hopping FFTs can be used to yield a rough range marking.

14.10.3 Matched Filters for General Signals

In actuality, radar signal is often a function of the environment (which can be time-varying). It is useful to contemplate the design of digital matched filters with arbitrary impulse response and inquire as to how to design and implement them. Two different implementations of such a filter can be postulated. One is a straightforward direct-form FIR digital filter and the other an FFT realization of the same filter.

In the Direct-Form FIR Filter, the flexibility is attained by changing the impulse response, which means the changing of filter coefficients. In the FFT realization, the DFT of this impulse response is made flexible. The relative desirability of these two methods depends primarily on the time-bandwidth product (TW). The larger TW, the more one should favor the FFT. Time-Bandwidth (TW) product is an extremely important concept for matched filtering.

Bandwidth (W) determines the best range resolution possible with a given signal, and time (T) determines the best possible velocity resolution. In digital systems, the total "on" time T of a signal is equal to NT_s, where T_s is the sampling period or interval and N is the total number of signal samples.

But, in a digital system that samples the received analog signal at the Nyquist rate, T_s is precisely $\frac{1}{W}$. Thus

$$N = TW \tag{14.43}$$

Here, we can say that the time-Bandwidth product is equal to the number of signal samples. Since, in general, a filter matched to a given signal has the same number of samples in the impulse response as does the signal. N also determines the complexity of the digital matched filter.

We will see that the time-Bandwidth products vary over a very wide range, yet in most instances, the FFT version of digital matched filtering seems most appropriate.

In any given radar application. Time-Bandwidth product is determined by radar requirements. Hence, the number of samples in the matched filter impulse response is equal to the number of signal samples. It is a design condition more or less imposed by system considerations. The section size (i.e., FFT size) is an additional parameter, however, that can be chosen to minimize the hardware cost. Two possibilities are shown in Figure 14.13. In case (*a*) the section length is twice the signal length and also twice the impulse length.

But case (*b*) the section length is made four times as big as the signal and impulse length. The advantage of case (*a*) *is* the use of a smaller FFT size, while the advantage of case (*b*) as shown in Figure 14.13 is the throwing away of fewer points.

This trade-off efficiency can be given by the following formula

$$E = \underbrace{\left(\frac{L_s - L_h}{L_s}\right)}_{I}\underbrace{\left(\frac{1+\log L_h}{\log L_s}\right)}_{II}, L_s \geq 2L_h \tag{14.44}$$

where L_h is the impulse response length and is defined as the number of samples in the impulse response and L_s is the section length.

We are assuming equal signal and impulse response lengths so that the smallest section length must be at least twice the impulse response length. The first term of above Eq. (14.44) expresses the increased efficiency. Increased efficiency is determined in terms of the fraction of good samples obtained from the evaluation.

The second term of Eq. (14.44) expresses the decreased efficiency resulting from the logarithmic increase in multiplication power needed to perform a larger FFT. The trade-off efficiency (E) is therefore efficiency of computation and can be thought of as the multiplication power per processed datum.

As a norm, we take $L_s - 2L_h$, in this case, trade-off efficiency (E) will be $\frac{1}{2}$ for $L_H >> 1$. We can now plot E verses L_s with L_h as a parameter. This plot is shown in Figure 14.14. We see that improvements in multiplication efficiency are possible with larger sections, but that the curves are not monotonic, that is, there is a "best" section length for each value of L_h. In fact, the best length for the cases we have shown will be $L_s = 8L_h$. Here only the multiplication cost has been computed. For example, the memory size increases linearly with FFT size so that the gain in multiplication efficiency may be more than offset by the increased memory hardware.

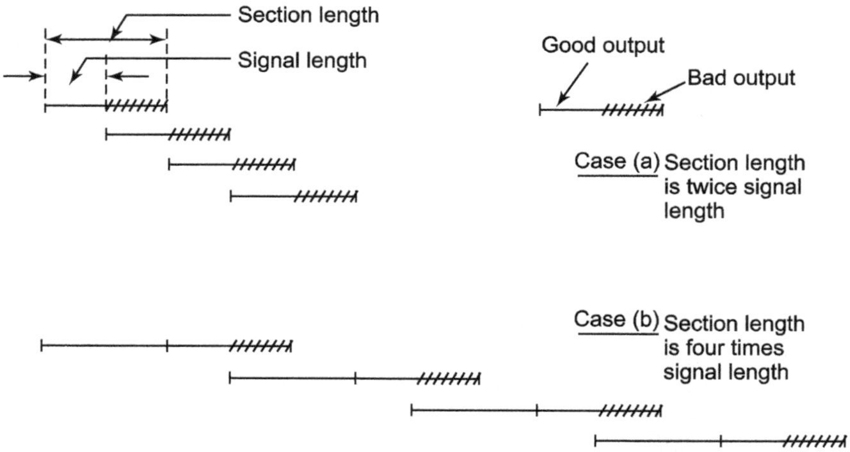

FIGURE 14.13 Comparison of Information thrown away for two different section lengths, with some signal lengths.

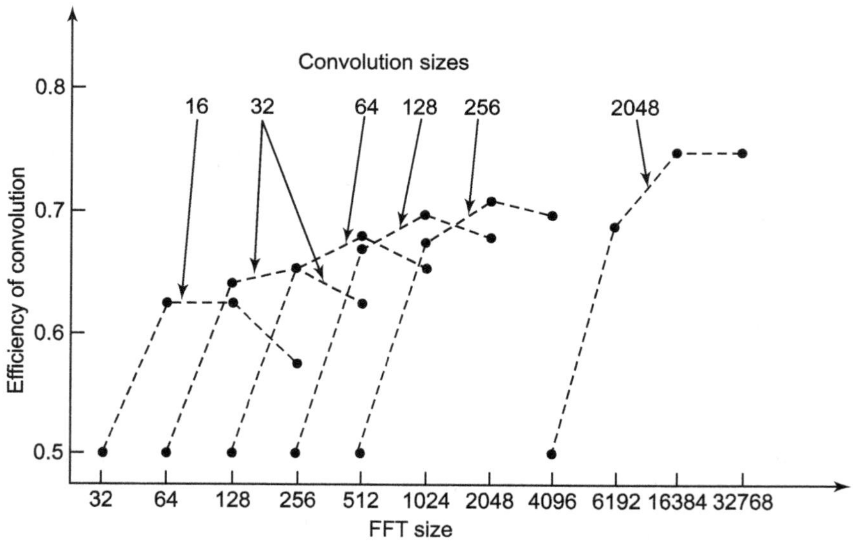

FIGURE 14.14 The efficiency of convolution versus FFT size.

14.10.4 Weighting to Reduce Matched Filter Side Lobes

The matched filter for a typical radar signal such as chirp is designed to yield the greatest signal-to-noise ratio (SNR) at a single instant of time. It means that the matched filter output will look like a sharp pulse. For most systems, this desirable main lobe of the filtered signal is accompanied by side lobes of fairly high amplitude. This can be decreasing when the radar is processing signals from several targets of differing cross-sections. This is because the main lobe of the smaller target can be marked by a side lobe of the large target.

The side lobes can be reduced by windowing or weighting at the cost of both range and SNR. This can be done in the time-domain by passing the matched filter output through another filter or in the frequency-domain by appropriate spectral weighting sandwiched between the forward and inverse FFTs. With reference to Figure 14.15, weighting can be accomplished by the appropriate design of the analog filter prior to the A/D converter. Three methods of weighting are shown in Figure 14.15.

Method I: In the method of Figure 14.15(a), extra computation is required if r is made too large where r is the filter length in the number of samples. For this reason, Hamming and Hann Windows are very useful since they correspond to $r = 3$ and still reduce the sidelobe levels to more than 40 dB below the main lobe.

Method II: In the second method of Figure 14.15(b) the situation is a bit more subtle. Since the output signal is always $2L_k$ long, where L_k is the length of both the signal as well as an impulse response, any spectral weighting

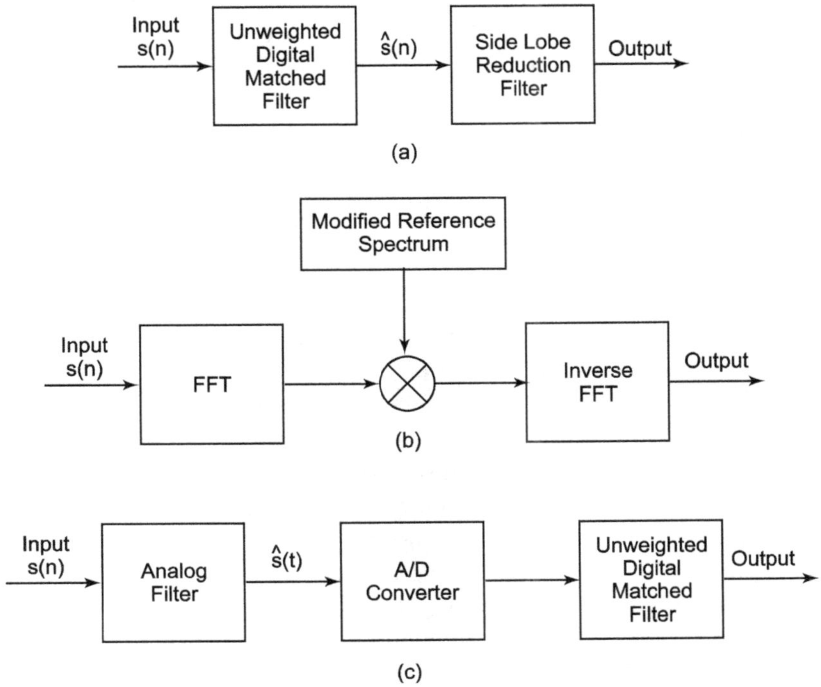

FIGURE 14.15 Various methods to reduce matched filter synthesis.

equivalent to an r-point FIR filter must be taken away from the matched filter duration. For example, let the signal consist of 256 samples, then we perform 512-point forward and inverse FFTs. With or without spectral weighting, the number of points generated by the inverse FFT is 512 but since r of these samples are associated with weighting, the effective time-bandwidth product shrinks slightly but this loss is minor compared to mismatch loss.

Method III: In the method of Figure 14.15(c), we note that if the A/D converter is ignored, we have a linear system and the weighting filter can be the presampling analog filter. The disadvantage in this form of weighting is the fact that this analog filter serves another rule, that is, removing noise mixed with the signal that could be aliased in with the signal due to sampling.

14.10.5 Matched Filter for a Burst

When trying first to detect a target at long range and then track the target as the range decreases, a standard procedure is first to transmit a chirp followed by a burst. The chirp is useful for detection but one cannot distinguish between range and velocity changes. To measure both velocity and range unambiguously, the radar signal generator will send out a burst. This signal may consist of a succession of short pulses or, to increase signal strength, it may consist of a succession of subpulses each of which is a chirp. To increase signal strength, we first pass the signal through a filter matched to these subpulses producing a succession of very narrow pulses with greater peak amplitude.

Thus, in either case, we are interested in the filter or filters matched to presumed target velocity or velocities for a burst of sharp pulses.

One method of matched filtering is shown in Figure 14.16. This method corresponds to a sliding FFT on N signal samples separated in time by the interpulse interval. Here we assume that one of the N FFT. Outputs are tuned to the correct Doppler. Now, we would expect that output to be a pulse train with a rectangular envelope.

This type of burst processing is inherently wasteful because k samples, later $(N - 1)$ of the original N outputs at the delay line taps, will be reprocessed when k is the interpulse spacing. If T_s is the sampling interval, then every T_s seconds a complete N-point FPT must be performed.

For example, in a 10^6-sample system processing 16 Doppler channels, a 16-point FFT must be performed every 100 ns.

The efficiency shortcoming of the scheme above can be overcome through the use of a permit memory structure and a band of digital filters, as shown in Figure 14.17. It is one of the rare cases for which a digital filter system is more efficient than an FFT.

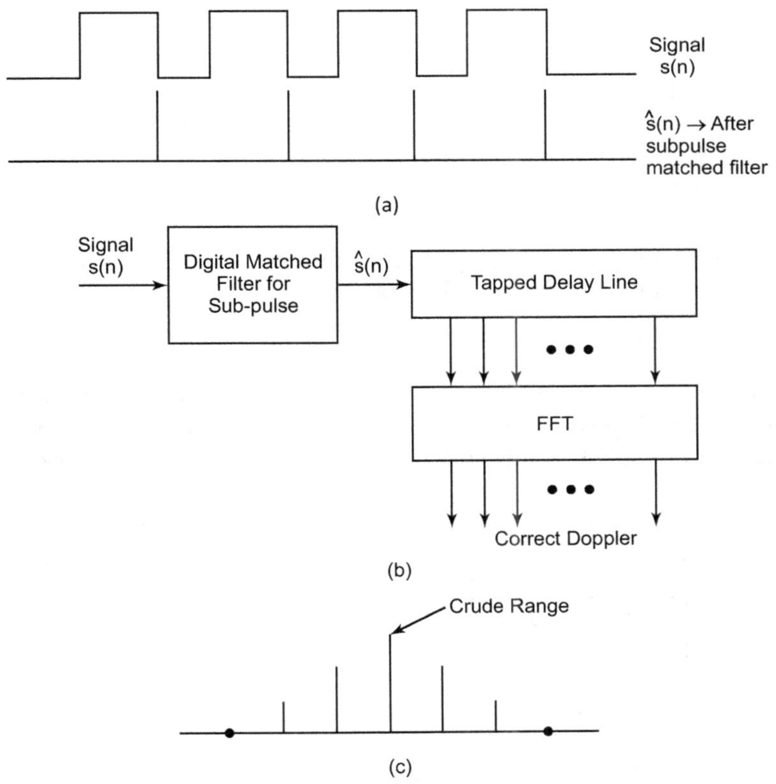

FIGURE 14.16 Burst processing by sliding FFT.

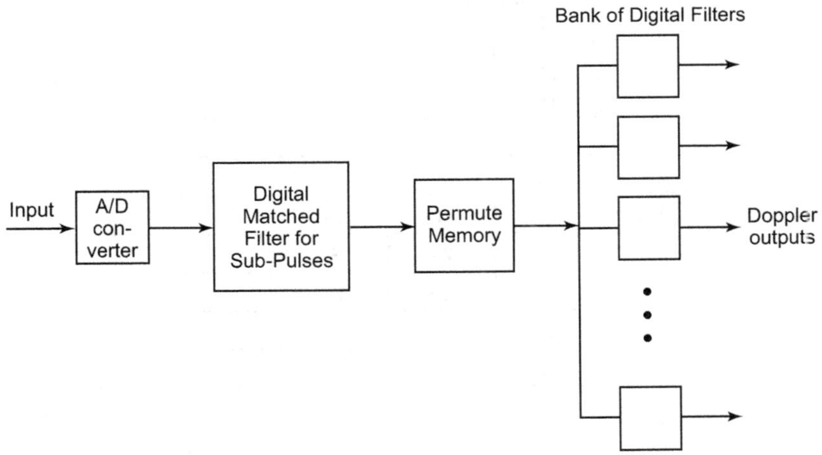

FIGURE 14.17 Digital realization of a burst processor using a permute memory.

The permute memory can be understood as follows:

We imagine a sequence of numbers labeled 0, 1, 2, 3, 4, ... as input to this memory. Also, we assume that the spacing between adjacent pulses is k samples. The function of the permute memory is to rearrange the ordering of the input sequence to be 0, k, $2k$, ...1, $k + 1$, $2k + 1$ ..., 2, $k + 2$, $2k + 2$. ....

With this reordering, Doppler processing now takes place on each range bin, beginning with range bin 0, then range 1, etc. For each new input sample, if we wish to process all N Doppler channels, we have to perform N complex multiplications, compared to $\frac{N}{2}\log_2(N)$ complex multiplications using burst processing by sliding FFT method. A further saving in the filter hank implementation is possible if not all N Doppler channels are needed. In FFT implementation, all channels are computed whether needed or not.

0	0	0	0	
1	5	25	8	
2	10	11	16	
3	15	36	24	
4	20	22	32	
5	25	8	1	
6	30	33	9	
7	35	19	17	
8	1	5	25	
9	6	30	33	
10	11	16	2	
11	16	2	10	
12	21	27	18	
13	26	13	26	
14	31	38	34	
15	36	24	3	
16	2	10	11	
17	7	35	19	
18	12	21	27	
19	17	7	35	Sequences Repeat →

20	22	32	3	
21	27	18	12	
22	32	4	20	
23	37	29	28	
24	3	15	36	
25	8	1	5	
26	13	26	13	
27	18	12	21	
28	23	37	29	
29	28	23	37	
30	33	9	6	
31	38	34	14	
32	4	20	22	
33	9	6	30	
34	14	31	38	
35	19	17	7	
36	24	3	15	
37	29	28	23	
38	34	14	31	
39	39	39	39	

FIGURE 14.18 Example of addressing sequence (for eight-pulse bursts and five range bins).

Implementation of the permute memory is done using a random access memory and a special addressing algorithm. By using this algorithm, real-time permutation of the data can be accomplished with no extra memory beyond that needed to hold the original database.

This addressing algorithm is shown in Figure 14.18. This algorithm is illustrated for a system where five range bins are to be processed for an eight-pulse burst. Each column represents the permitting of all 40 data samples. For the first 40 inputs, the memory addressing is sequential, 0, 1, 2, 3, etc. The Doppler Filter bank must receive samples in the order 0, 5, 10, 15, ..., 1, 6, 11, 16, etc., however, and this can be done by using the address sequence in the second column. To present extra buffering, each new input datum must replace the most recent output datum. Thus, for each new set of 40 data points, entry and exit from the memory follow a new addressing pattern, as shown in the succeeding columns.

14.11 AIRBORNE SURVEILLANCE RADAR FOR AIR TRAFFIC CONTROL (ATC)

In present-day ATC radars, the antenna rotates mechanically, sweeping out a full 360° every 4–12 s.

In this radar, azimuth beam resolution is about 1°–2°, and the vertical antenna pattern is a fan beam usually having a 30°–45° width. Thus, as the antenna sweeps by a target aircraft, there will be a succession of hits, that is, radar returns at the pulse repetition period after which no information is obtained from that particular target until the antenna has made a complete revolution. Based on this information, the system must track up to perhaps 50 aircraft within its field of view and display these tracks in a useful manner to the air traffic controller who can then correlate this information with his knowledge of the traffic schedule, planned flight paths of the various commercial and private aircraft, and his audio communication links with the aircraft and other control towers.

Present airport surveillance radars operate at S. band (wavelength approximately $\lambda = 10$ cm). At this wavelength, weather clutter, caused by radar reflections from rain, can be quite troublesome but perhaps the most important disturbing effect is ground clutter, which is picked by the antenna. Such effects can be alleviated through better signal processing, in particular, by Doppler filtering that permits the relatively fast aircraft target to be discriminated from the ground and weather returns.

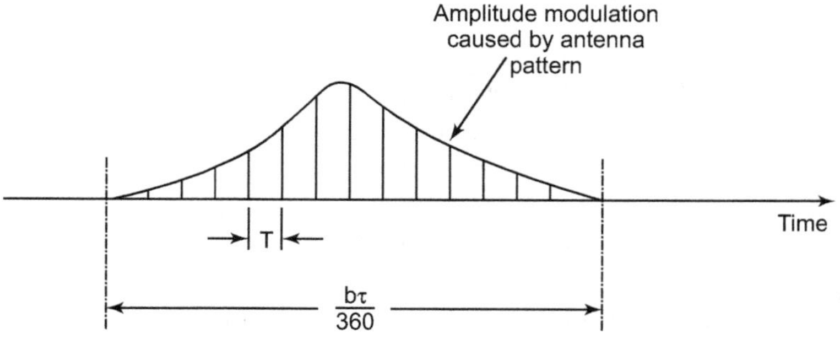

FIGURE 14.19 Radar return signal from a target at a given range as antenna sweeps by.

The ground clutters are having only a DC spectral component. These clutters actually have a spectral spread induced by the antenna motion. This makes Doppler processing less effective and could be avoided by the other antenna designs (such as a phased array).

A sketch of the experimental setup is shown in Figure 14.20.

The antenna and associated radar equipment were about $\frac{1}{2}$ miles away from the FDP facility. Two communication links were used. One link is used to relay the received radar signal to the FDP for signal processing and threshold detection. But, the other link is used to relay back the processed information to the radar display. The radar used was a coherent S-band radar with a 1° fan-beam antenna having a width of 5.25 m, a rotation rate of 1.36 radian/s, and a wavelength of 10.7 cm. The pulse repetition frequency (PRF) was 1000 pulses per second and these pulses were about 1 μ second wide, corresponding to a 1-MHz bandwidth. Given the antenna beamwidth and PRF, the number of hits as the antenna scanned by the target was 15. This corresponds to the burst signal for any range bin in the range aperture the problem would be to design a matched filter to the transmitted signal, taking into account some reasonable model of clutter.

Now, we discuss the system parameters to obtain an idea of the Range-velocity ambiguity problem.

First, the pulse repetition period of 1 ms is needed to give the required unambiguous range coverage of 60 miles. For an S-band radar, the range of Doppler frequencies. corresponding to the velocity range of $\frac{1}{3}$ km/s is about 3300 Hz. Since PRF is 1000 Hz, the Doppler spectrum will be periodic with a 1000-Hz period so that, for example, the system will not be able to distinguish among, say, 400-Hz, 1400-Hz, or 2400-Hz Doppler. Furthermore, since the Doppler frequency of ground clutter is close to DC, the 1000-Hz sampling will introduce large clutter components at 1000, 2000, 3000 Hz, etc., these frequencies correspond to "blind speeds." At blind speed, airplanes will be lost in the clutter.

Thus, an important aspect of a complete signal processing system is Disambiguation. Disambiguation may be accomplished by transmitting a signal at two different PRFs. The scheme devised for the present experiment was to transmit 8 pulses at one PRF, switch to another PRF for the next 8 pulses, and then switch back again, etc. This will allow a target whose radial velocity corresponds to a blind speed at one PRF to become detectable at a different PRF. Thus, the matched filter for our system becomes an 8-pulse processor. A bank of such processors is needed for many Doppler frequencies since there is no prior way of knowing the target's radial velocity. Optimum Processing from a computational point of view needs to be done for each range bin and perform a weighted sum of eight signal returns.

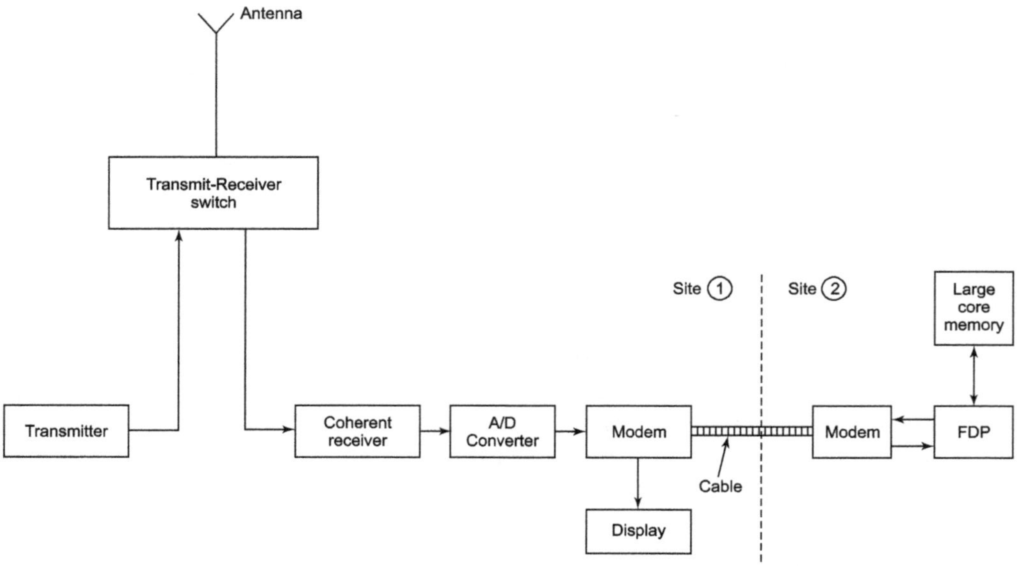

FIGURE 14.20 Experimental setup for airborne surveillance radar (ASR) signal processing.

Both weights and signals are assumed to be complex numbers. If N Doppler basis is to be examined, then $8N$ multiplications must be performed, for real-time operation, since each range gate is 1 μs wide (the width of the transmitted pulse), these multiplications must be performed in an 8-μs interval. For example, if $N = 8$, then $8 \times 8 = 64$ complex multiplications must be performed every 8 μs or one complex multiplication every 125 μs. One question arises in our mind that how the algorithm can be altered to maintain close to optimum performance while reducing the computation load. A filter greatly alternates the large DC clutter component followed by a bank of filters turned to the various Dopplers of interest should yield good results. Since an FFT resembles a filter bank and since FFTs tend to be computationally effort, a suboptimal signal processor using three pulse canceler, followed by an FFT was designed and simulated and compared with the simulated optimum processor.

Comparisons of "optimal 8-pulse processors" and "suboptimal processor three pulse canceler and unweighted DFT filter" are shown in Figures 14.21–14.23 for 0, 125, and 500 Hz, respectively. These plots are drawn between signal-to-interference ratio (SIR) improvement in dB and Target Doppler frequency in Hz for 0, 125, and 500 Hz frequency. In all cases, SIR improvement due to filtering is plotted as a function of the Doppler. There are three cases for this study:

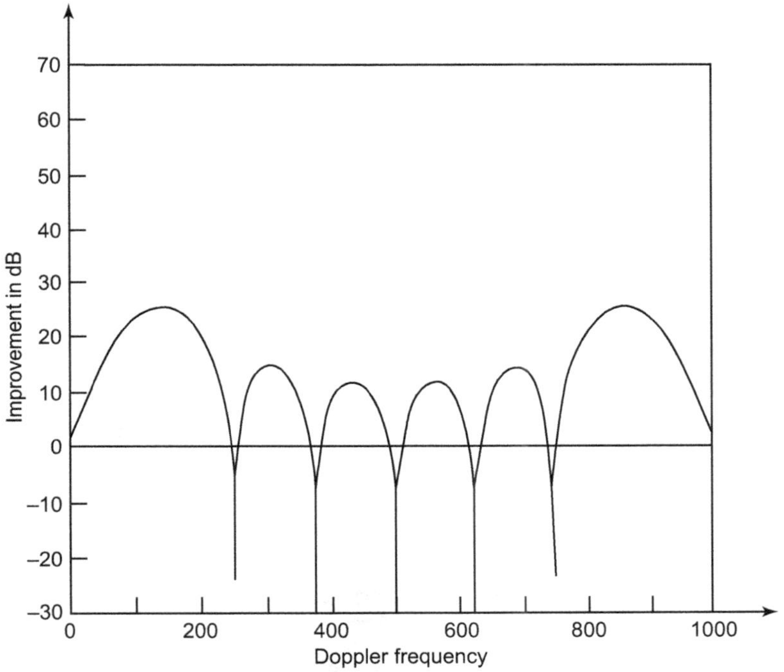

FIGURE 14.21 Signal-to-interference ratio (SIR) improvement of optimal S-pulse processor optimized at 0 Hz.

1. 0-Hz optimization,
2. 125-Hz optimization, and
3. 500-Hz optimization.

For 0-Hz optimization, we note that the best improvement for both cases is not at 0 Hz. The peak improvement for the optimum processor occurs at about 130 Hz and is about 26 dB, whereas the peak improvement for a suboptimal processor is about 80 Hz but is only about 18 dB.

For the 500-Hz case, both curves peak at 500 Hz. It appears that the suboptimal peak is actually bigger than the optimal but this is misleading because to perform three-point clutter cancelation filtering followed by an 8-point FFT really requires 10 input samples, whereas the optimum processor uses 8-samples.

It was shown that 2 multiplications are needed to perform an 8-point FFT compared to 64 for the optimal processor. Among the two designs, it can be argued that the clutter cancelations filter plus 8-point DFT are the correct filters to build. Actually, benefits result from the weighting of the signal prior to

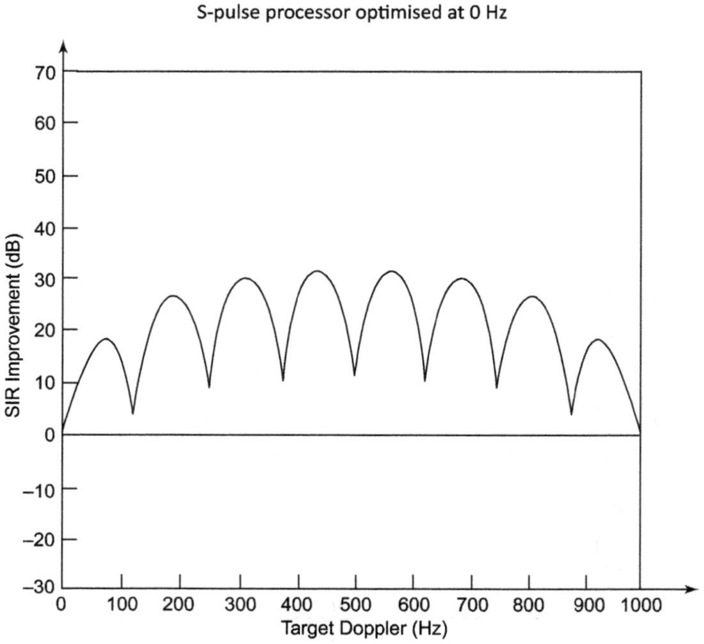

FIGURE 14.22 Signal-to-interference ratio (SIR) improvement of suboptimal processor 3-pulse canceler and unweighted DFT filter tuned to 0 Hz.

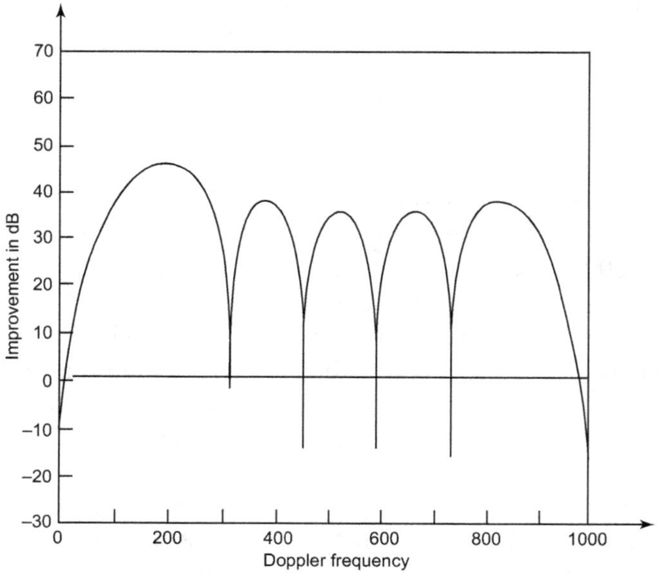

FIGURE 14.23 Signal-to-interference ratio (SIR) improvement of optimal 8-pulse processor optimized at 125 Hz.

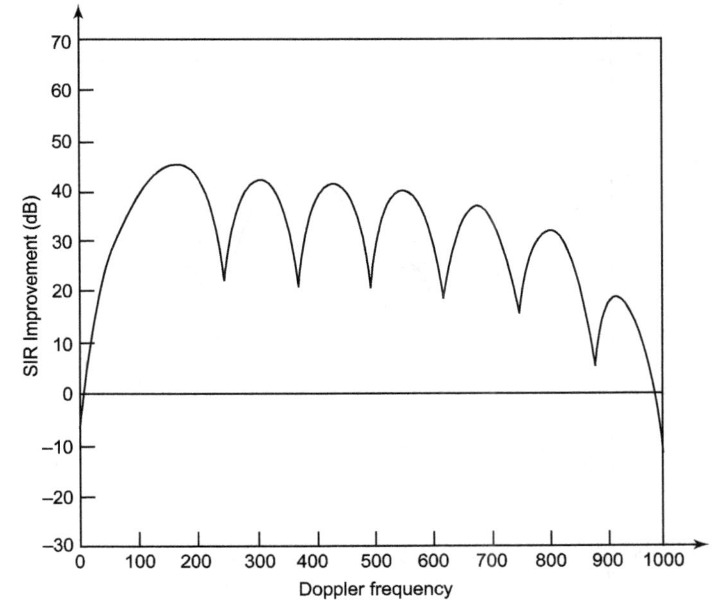

FIGURE 14.24 Signal-to-interference ratio (SIR) improvement of suboptimal processor 3-pulse canceler and unweighted DFT filter tuned to 125 Hz.

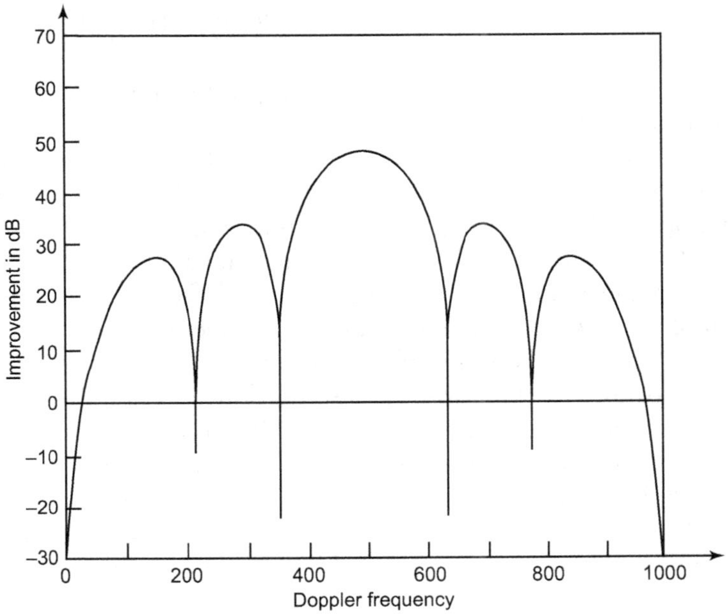

FIGURE 14.25 Signal-to-interference ratio (SIR) improvement of optimal 8-pulse processor optimized at 500 Hz.

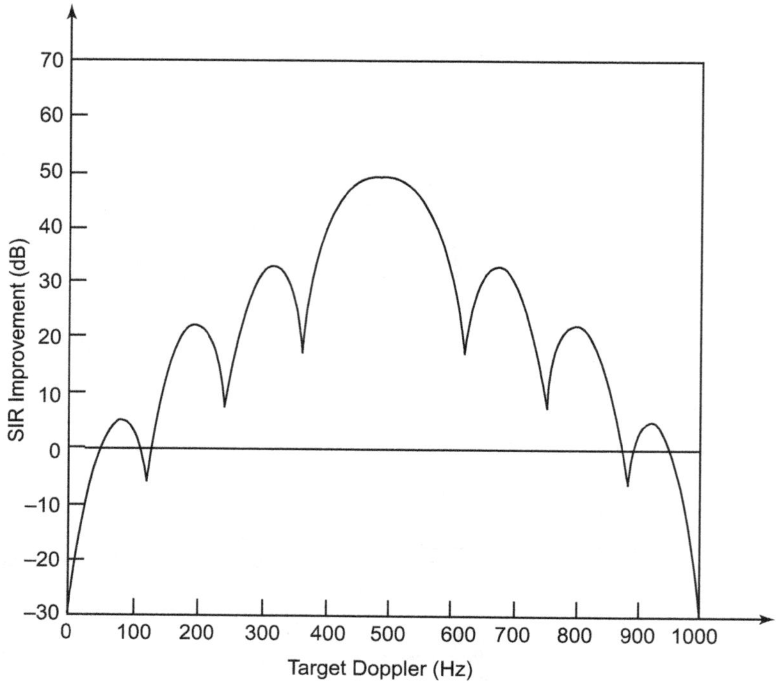

FIGURE 14.26 Signal-to-interference ratio (SIR) improvement of a suboptimal 3-pulse canceler and unweighted DFT filter tuned to 500 Hz.

the FFT, Weighting of the signal means passing the signal through the clutter cancelation filter.

This adds four complex multiplications, the benefits are reduction in the side lobes of curves of Figures 14.21–14.23. Figure 14.24 shows a complete suboptimal filter block diagram. This filter consists of a third-order FIR digital filter for pulse cancelation, weighting, and an 8-point FFT.

Given ground clutter suppression are still faced with the very severe problem of signal detection. SIR improvement factor is shown in Figures 14.21–14.23. Since the intensity of the clutter varies greatly with the terrain, the curves of these figures will be raised or lowered as a function of the range-angle sector. In order to obtain reasonably constant false alarm probabilities for all the illuminated space, we need a "clutter map" as a reference, that is, an averaged clutter intensity for each range-angle sector. This is accomplished by means of a scan-to-scan averaging of the *dC* component of the signal. Letting x_i be that component due to the *d*th scan, we can then prescribe the clutter map intensity to be

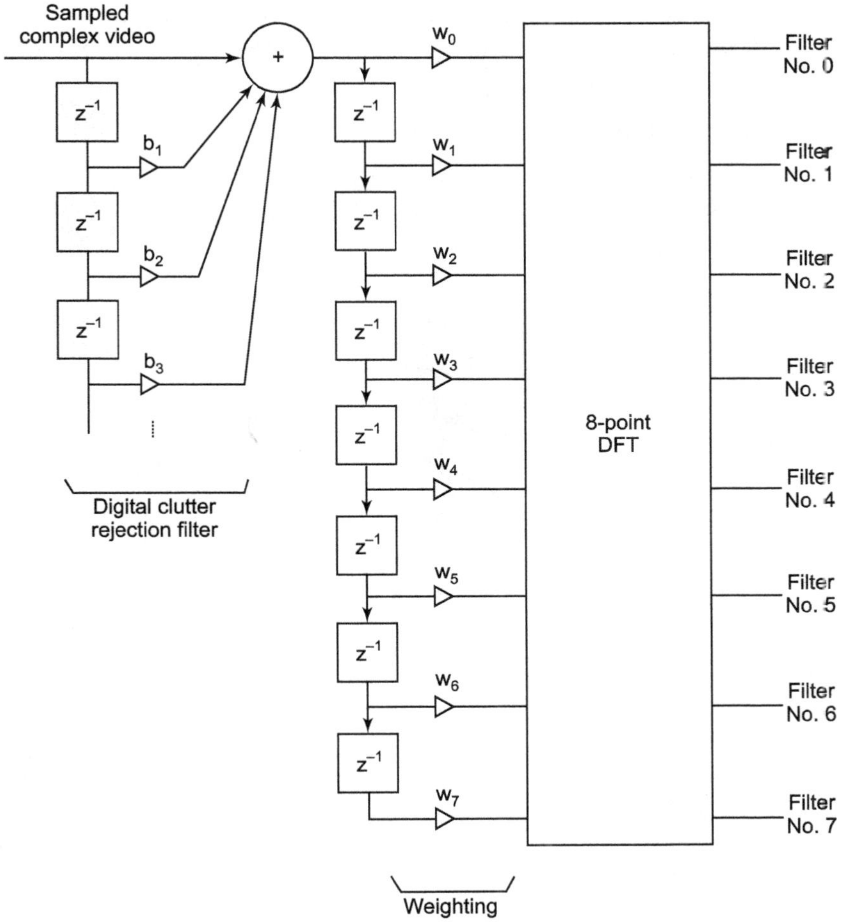

FIGURE 14.27 Block diagram of the suboptimal processor.

$$y = \left(y_{j-1} - x_j\right)a + x_i \tag{14.45}$$

For any range-angle sector, threshold detection is accomplished by comparing the largest output of the FFT with y_j. α will determine the length of time to build up a reliable clutter map indication.

Now we will discuss the FDP simulation of the signal processing and clutter mapping algorithms. The FDP is not a fast enough processor to process all information coming to the antenna in a 60 miles radius. This means that, in order to approximate real-time simulation, the data from a given section of space must be buffered and then processed while the information from the next scan is being accumulated. This situation is shown in Figure 14.28. The FDP facility proved to be useful for this processing for two reasons:

1. Its high speed meant that a large enough sector could be examined so that a useful number of targets are displayed.
2. The large core memory connected as an input–output device to the FDP (1,60,00018 Bit registers) allowed the buffing of raw video from a reasonably large sector.

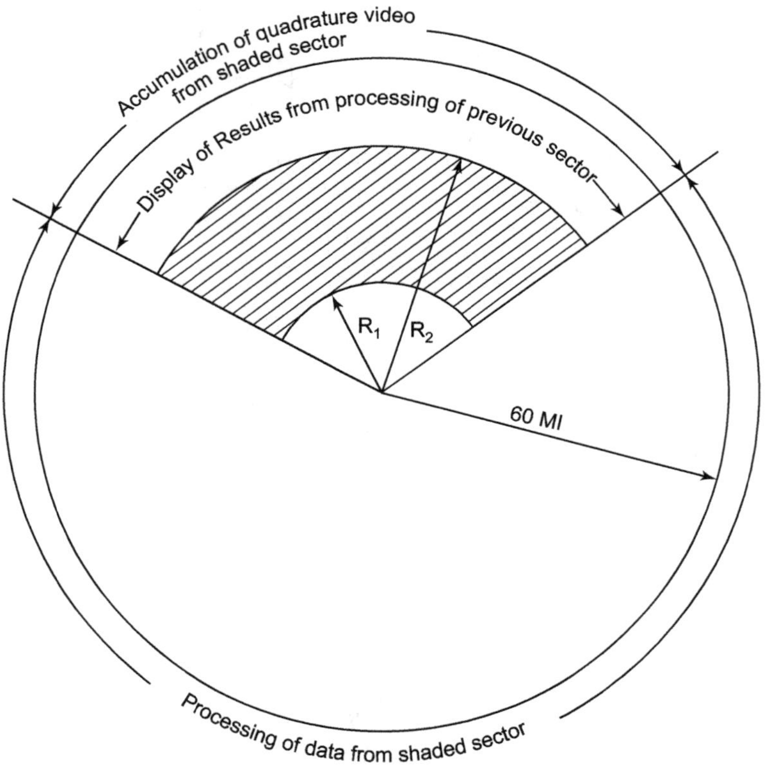

FIGURE 14.28 FDP real-time processing of a range-angle sector.

For determining the buffer storage requirements for the full 60 miles and 360°, we assume 600 range bins that are needed for 60 miles (at 1 μs per range bin and 10 μs per mile). It means that 600 numbers must be stored in 1 ms or 6,00,000 numbers per second and 2,700,000 numbers for a full 4.5-s scan. Given the buffer memory size round off to 1,50,000, this means that at most one-eighteenth of the total space can be processed in real time (with a one-scan delay before display). Thus, for example, the shaded sector could be 10 miles by 120° or 30 miles by 40°.

The buffer size is the real limitation in our case since the FDP can be real-time processing sectors 2 or 3 times these sizes. For display purposes, the sector was rectangularized.

Good Doppler processing and clutter mapping using special-purpose DSP hardware could be a useful addition to airport surveillance radar (ASR) system.

14.12 LONG-RANGE DEMONSTRATION RADAR (LRDR)

The purpose of a radar system using FDP (as a real-time simulator) was the detection of moving objects in large amounts of ground clutter background. The clutter spectrum consists of a very large DC return plus an AC component caused by foliage motion. The moving object, which could be either a vehicle, such as a car or truck or an airplane, or an animal or a human lost in the woods or an enemy soldier, returned a signal level about 80 dB below the steady-state clutter and 60 dB below the fluctuating background caused by foliage motion. The radar antenna was a phased array UHF antenna. With a 1.5° horizontal beamwidth and a vertical fan beam. This experimental antenna consisted of several thousand array elements arranged on half a cylinder, which permitted illumination of a 45° sector, a full cylindrical array would of course cover a full 360° sector.

Because of the very high. Clutter-to-Signal Ratio, a variety of signal processing tricks were necessary to obtain good signal detection. Following tricks were used in signal processing to obtain good detection:

1. pulse compression using complementary coding,
2. presuming,
3. Doppler processing, and
4. post-detection integration.

We describe the idea of implementation of each of these techniques briefly.

First of all, we present a diagram of the complete experimental setup.

The block diagram of LRDR is given in Figure 14.29. Prior to the A/D converter, the received radar signals are amplified and heterodyned.

Sensitivity time control (STC), which is a range-dependent gain control, takes place in a digitally controlled stepped alternator. Demodulation is performed in quadrature so that the video return signal is a coherent complex signal.

After A/D conversion at 10° samples and seven bits, the video is pulse compressed. This radar is assumed to survey 2048 range bins for each beam position, since there are 30 beamwidths in 45°, a total of 2048 × 30 = 61,440 range–azimuth bins are illuminated.

14.12.1 A Criterion of Choosing the Various Parameters of this Particular Radar

Since one of the main purposes of the project was to study the possibility of detecting moving targets in foliage, the wavelength had to be large enough to penetrate foliage. Here we choose UHF (a wavelength of about 5 m). Targets of interest moved as slowly as several tenths of a meter per second. In this case, the frequency resolution requirement is about 0.5 Hz. This meant that the total integration time before target detection was about 2 s.

If we assume that the maximum Doppler we desire to detect is about 32 Hz, then we need 64 velocity bins. This means that the information in the received signal is collected and put in a "bungle" 32 times per second and 64 of these bundles are eventually velocity filtered. These Doppler requirements determine the lowest permissible repetition period.

The highest permissible repetition period is determined by range ambiguity. The largest unambiguous range was assumed to be 22.4 mm (nautical miles), which led to a radar PRF of 3600 per second. Thus, approximately 113 radar return signals at the PRF rate can be combined to create a single bundle. The bandwidth of 10 MHz was chosen to give the required 50 ft range resolution.

The succession of signal processing devices also includes a range–azimuth gate selector, This selector was necessitated by the inability of the FDP to process the entire field of view in the basic 2-s coherent time interval. The FDP was able to process 2048 bins in that time, thus the function of the gate selector was the selection of 2048 out of the possible 64,440 bins for each 2-s interval.

Following gate selection the partially processed radar data is sent, *via* a Modem and appropriate digital interfaces, to the large core memory that resides nest to the FDP. The selected data enters the memory sequentially with respect to range bins, the memory addressing algorithm permutes the data so that 64 successive inputs from a single range bin are sequentially interested in the FDP. This reordered data can now be processed by the FDP that performs FFTs for Doppler discrimination.

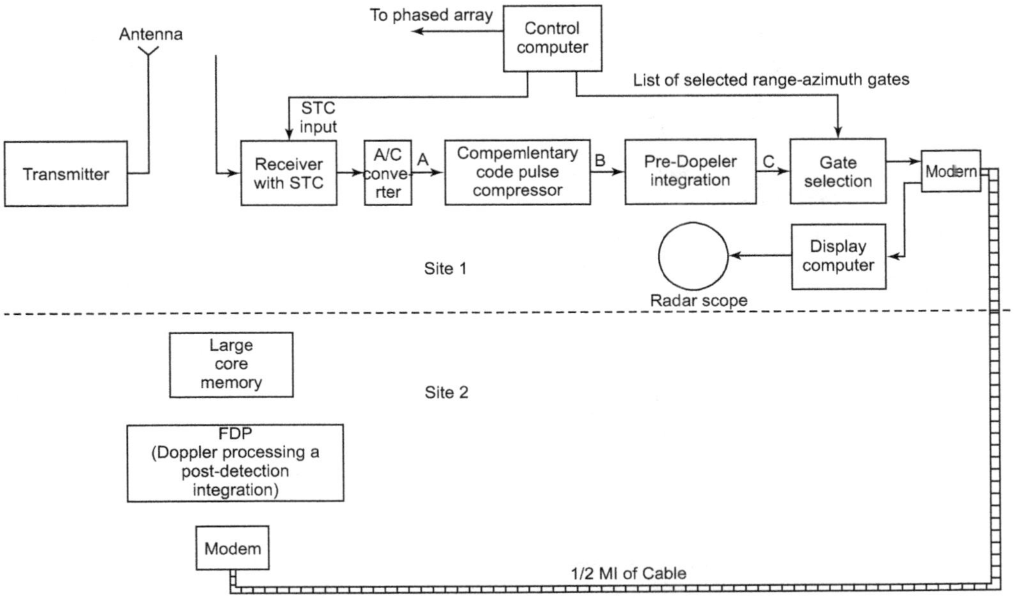

FIGURE 14.29 Block diagram of long-range demonstration radar (LRDR).

Then the EDP is in conjunction with the large memory to perform post-detection integration on each Doppler bin by adding fine successive FFT magnitudes at that Doppler frequency. Following this, various statistical decision algorithms are implemented fay FDP programs, leading to the detection of targets.

Finally, this processed information is sent back to the radar site, entering a general-purpose computer for formatting and display purposes. The other general-purpose computer at the radar site acts primarily as an input–output processor, controlling the antenna beam, the STC, the radar timing, and the selection pattern for use by the gate selection processor. Various time-epochs in the system are shown in Figure 14.30.

Here we will discuss some details of A/D conversion and implementation of the pulse compressor algorithm.

14.12.2 A/D Conversion

In the presence of large ground clutter, a minute target return rides within a huge clutter signal. Eventually, the target detection will depend on the Doppler discrimination properties of the radar signal processor, meanwhile, there is a concern list of the minuscule signal gets wiped out by non-linear and noise effects if quantization beginning with the A/D converter. It is really

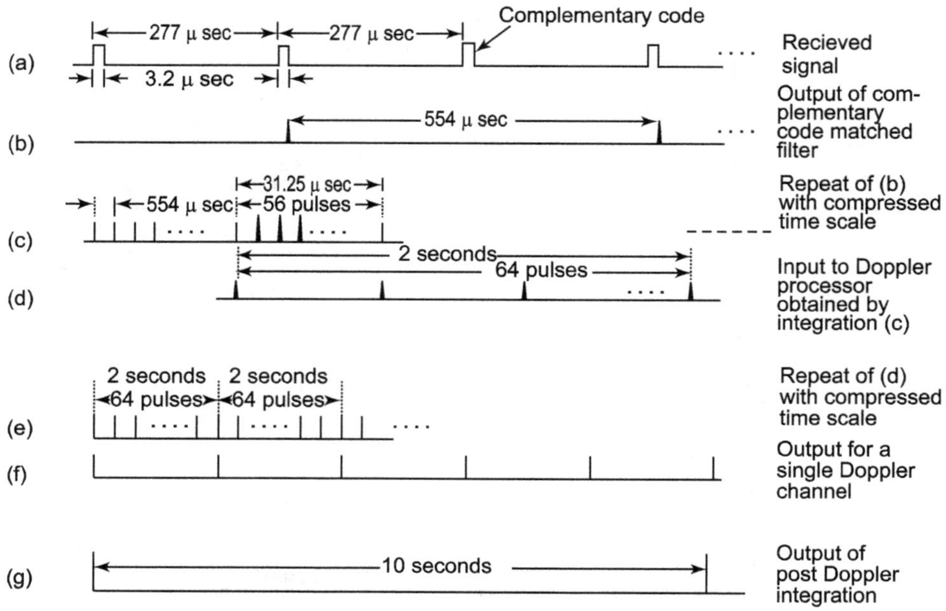

FIGURE 14.30 Illustration of various time-epochs in long-range demonstration radar (LRDR).

not necessary for the target return strength to be as big as a complete quantum step. As long as the clutter fluctuates and does not "get stuck" between two quantum levels at a precise value that would wipe out the target, one can expect that target presence will not be wiped out. In the present radar, 7 bits are used and the target is presumed to be about 12 bits below clutter level, yet enough signal strength is present to make it useful to perform further processing to raise SNR.

14.12.3 Pulse Processor

The transmitted signal consists of a 32 bits code and its complement sent out at alternative repetition intervals. The matched filter response to this pair of signals is a sharp pulse with no side lobes for zero Doppler. For this particular radar, if the requirement is imposed that only very slowly moving targets be detected but the theoretical result is quite valid for a noiseless signal.

Now we can define complementary codes, we begin with a very simple example, let $s_1(n)$ be the sequence +1, +1, and $s_2(n)$ be the sequence +1, −1. The matched filters at zero Doppler's will yield the autocorrelation sequences, for $s_1(n)$ the correlation sequence is +1, +2, +1, while for the $s_2(n)$ we find −1, +2, −1.

Adding the two matched filter outputs yields the final result 0, 4, and 0.

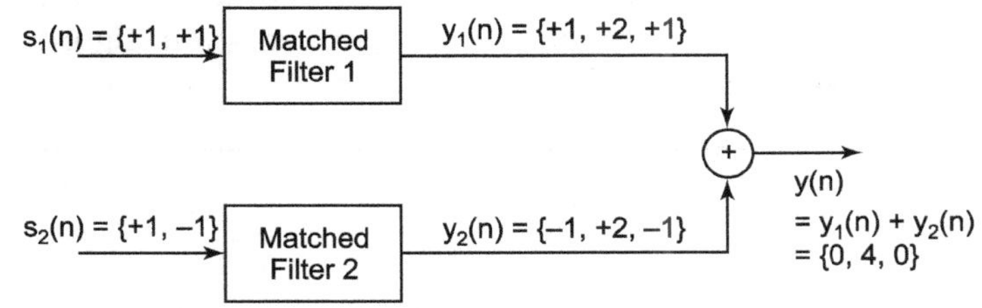

FIGURE 14.31 Illustration of the output of two matched filters. Here $y_1(n)$ and $y_2(n)$ are the autocorrelation sequences for two sequences $s_1(n)$ and $s_2(n)$, respectively.

Larger codes can be generated from $s_1(n)$ and $s_2(n)$ by the following algorithms:

1. Define a new signal $s_3(n)$ as $s_1(n)$ followed by $s_2(n)$, namely, +1, +1, +, −1. Define another new signal $S_4(n)$ as $S_1(n)$ followed by $\hat{s}_1(n)$ [$s_2(n)$ with all sign reversed], or +1, +1, −1, +1. These two new sequences form a complementary code pair of length 4 and have the matched filter outputs −1, 0, 1, 4, 1, 0, −1 and 1, 0, −1, 4, −1, 0, 1, which when added yield the sidelobeless sequences 0, 0, 0, 8, 0, 0, 0.

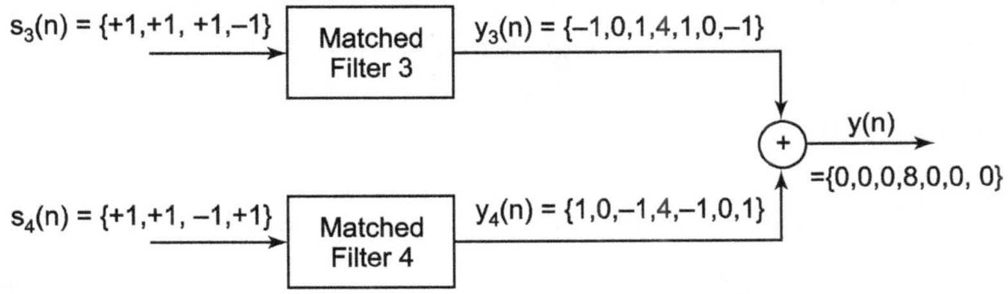

FIGURE 14.32 Illustration of the output of another two matched filters here $y_3(n)$ and $y_4(n)$ are the autocorrelation sequences for two sequences $s_3(n)$ and $s_4(n)$, respectively.

2. The procedure above can be iterated to double the length of succeeding complementary codes. Thus, the pair would be +1, +1, +1, −1, +1, +1, −1, −1 and +1, +1, +1, −1, −1, −1, +1, −1.

Figure 14.33 lists the complementary code pairs through length 3.2. The LRDR transmits a pair of length 32 (or 3.2 μs, since the bandwidth is 10 MHz and the subsequent sampling rate is 10^6 samples per second.

Implementation of the matched filter was a digital FIR filter with an interesting property that all the 32 multiplications per iteration were +1 or −1.

If not for this, 32 multiplications per 100 ns would have been necessary. Instead, 32 additions per 100 ns are required, a much less formidable task although certainly nontrivial. The implementation algorithm is shown in Figure 14.34 for a code of length 4.

+ + + –	2
+ + + – + + – +	4
+ + + – + + – + + + + – – – + –	8
+ + + – + + – + + + + – – – + + + + + – + + – + – – – + + + – +	16
+ + + – + + – + + + + – – – + – + + + – + + – + – – – + + + – + + + + – + + – + + + + – – – + – – – – + – – + – + + + – – – + –	32

FIGURE 14.33 List of complementary code pairs from 2 to 32.

The hardware was implemented with 32 hardware adders operating in parallel, using TTL adder logic and TTL shift register memory for the delays. Thus, one clock cycle corresponded to a complete iteration of the Digital matched filter.

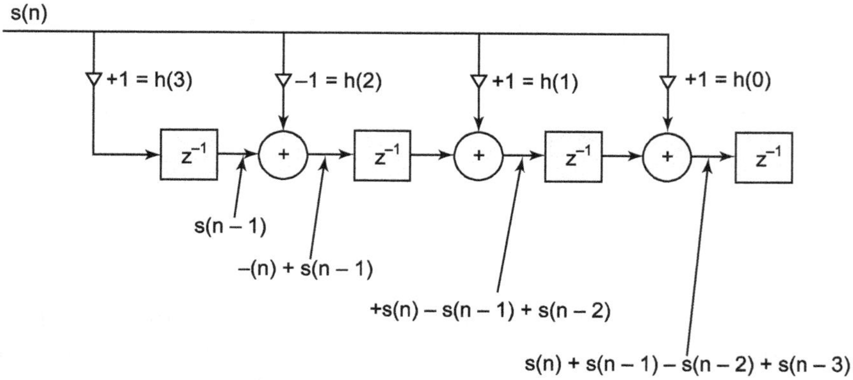

FIGURE 14.34 Filter matched to one of a length 4 complementary code pair.

14.13 DIGITAL MATCHED FILTER FOR A HIGH-PERFORMANCE RADAR (HPR)

The specification for this example included a phased array antenna with scheduling activities controlled by a large g.p. computer and a digital signal processor capable of handling bandwidths somewhat higher than 10 MHz with time-bandwidth (TW) products of about 2000. Chirps, bursts with chirp subpulses and nonuniform bursts signals will be handled. Here we treat one important part of this as digital matched filter.

First of all, a digital filter matched to a 10-MHz signal with TW = 2000 must be a two-thousandth-order FIR filter. If one attempted to implement such a filter by non-FFT methods, then 2000 complex multiplications would have to be performed every 100 ns.

A system with eight thousand 100-ns multipliers has a very good chance of never working. The use of the FFT cuts this requirement considerably. One would have to perform a 4096-point forward FFT followed by multiplication of the output by a reference spectrum (that of the matched filter) followed by a 4096-point universe FFT. Half of the processed result would be the correct matched filter output, while the other half would be worthless results of circularly convolved information. Thus the total number of multiplications for a radix-2 FFTs to process 2048 points is

$$\left(\frac{N}{2}\right)\log_2(N) + N + \left(\frac{N}{2}\right)\log_2(N) = N\left(1+\log_2 N\right) = 4096 \times 13$$

or a total of 26 complex multiplications every 100 ns. This is nearly two orders of magnitude reduction compared to non-FFT convolution. Radar specifications demand a pipeline convover.

The next step was to determine the most suitable radix number. Only the radix-2 and radix-4 cases were considered. For radix-4, we have six stages in the forward FFT and six stages in the inverse FFT. Each stage consists of three complex multipliers, for a total of 36 + 1 for the reference spectrum, making 37. This is greater than the 2fi attributed to radix-2 but, for the same clock rate, the radix-4 can process at twice the data rate of the radix-2 since it has four parallel paths compared to 2. This advantage can be used either by designing a more powerful system or by slowing the clock rate so that the multiplication speed requirement is eased. Radix-4 structure is complicated even quite valid. Therefore radix-4 structure is chosen. Since the speed requirements of this matched filter are so severe and since it was hoped that this radar would lead the way to even more powerful radars. We looked for the fastest logic family with sufficient versatility and this turned out to be 2-ns

ECL. Important design decisions dealing with the precision and the overall strategy of computation could only be answered by extensive computer simulation experiments.

The following problems will be studied:

1. register length of coefficient (W^{nk}),
2. register the length of samples of the reference spectrum,
3. register length of data,
4. type of arithmetic (fixed or floating-point or hybrid scheme),
5. scaling strategy, and
6. Koundmg or truncation.

Following simulation experiments were used to help the evaluation of problems given above.

1. Computation of the matched filter output for a single LFM or chirp signal with hamming weighting to determine whether any spurious side lobe peaks appear.
2. Computation of the matched filter output for the sum of a large LFM signal and a small LFM signal close in range to the large signal to see if the small signal is lost in a side lobe of the large one.
3. A low-level LFM signal is processed in the presence of simulated receiver noise. The purpose of this experiment is to ascertain whether the process has altered the statistics of the random signal component thereby affecting SNR.

The first simulation experiments were run with fixed-point arithmetic. For LFM signals, the signal amplitude as it passes through the successive FFT stages build up quite rapidly, essentially one bit per stage. Thus, to avoid using very long register lengths, the result after each butterfly was right-shifted by one bit. For this strategy in both forward and inverse FFT, it was found that coefficient register lengths of 8-bits and data register lengths of 16 bits were required.

Next, the system was altered to do floating-point arithmetic. Some decrease in the number of bits needed could be expected. The simulation showed this is true, it was found that a 9-bit mantissa and 4-bit exponent was adequate for all registers in the system. Despite the decrease in memory size, however, arithmetic complexity increase and arithmetic speed decrease made this system less attractive than the fixed-point case.

The system that was simulated most extensively used a hybrid arithmetic scheme that was cross between fixed and floating points. The hybrid scheme had the following features:

1. Coefficients were fixed-point 9-bit fractions.
2. The complex data word had two mantissa and a single exponent that served both mantissas. One mantissa is used for the real part and the other for the imaging part.
3. Mantissas were never left-justified (normalized). On overflow of either mantissa of the complex datum, however, both mantissas were right-shifted one bit and the exponent incremented by one.
4. The exponent is assumed to be a positive number, mantissas are signed, 2's-complement numbers.

An arithmetic system for one state of a radix-*r* pipeline FFT processor is shown in Figure 14.35. Here we have used decimation-in-time (DIT) algorithm.

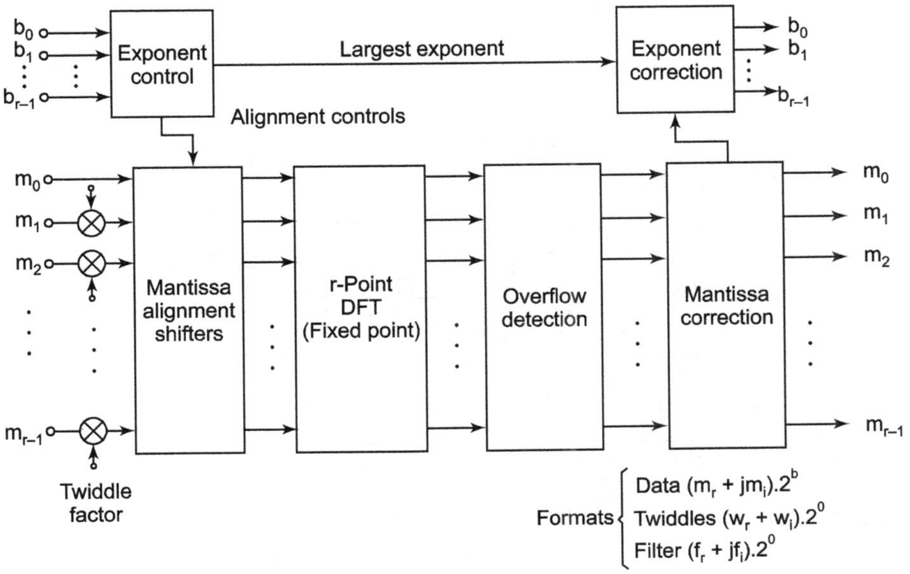

FIGURE 14.35 Hybrid arithmetic scheme for radix-*r* pipeline DIT-FFT algorithm.

Processing begins by immediately performing the $(r - 1)$ complex multiplications. These are simply fixed-point operations, and overflow is prevented in the cross-product combination by carrying an extended sign bit. In parallel with the multiplications, the largest of the r exponents is determined and transmitted to the output of the stage for later use. Before proceeding with the r-point DFT it is necessary to align the twiddled mantissa pairs to have the same exponents. The r-alignment shifter is controlled by the difference between the maximum exponent and the individual exponents of each mantissa pair. Following this alignment, all operations are fixed points. Notice that the twiddles and the search for the largest exponent, this was possible because, in the DIT-FFT algorithm, multiplication proceeds the r-point DFT,

Depending on the value of r, several further extended sign bits are carried through the fixed-point DFT computations to avoid possible overflow. At the DFT output, each complex mantissa pair is inspected to see if these extended bits have been used. Special logic examines these bits and determines the number of right shifts needed to bring the mantissa back into the range so that the length of the memory registers need not include these extra bits. A more detailed sketch of this overflow convection logic is shown in Figure 14.36.

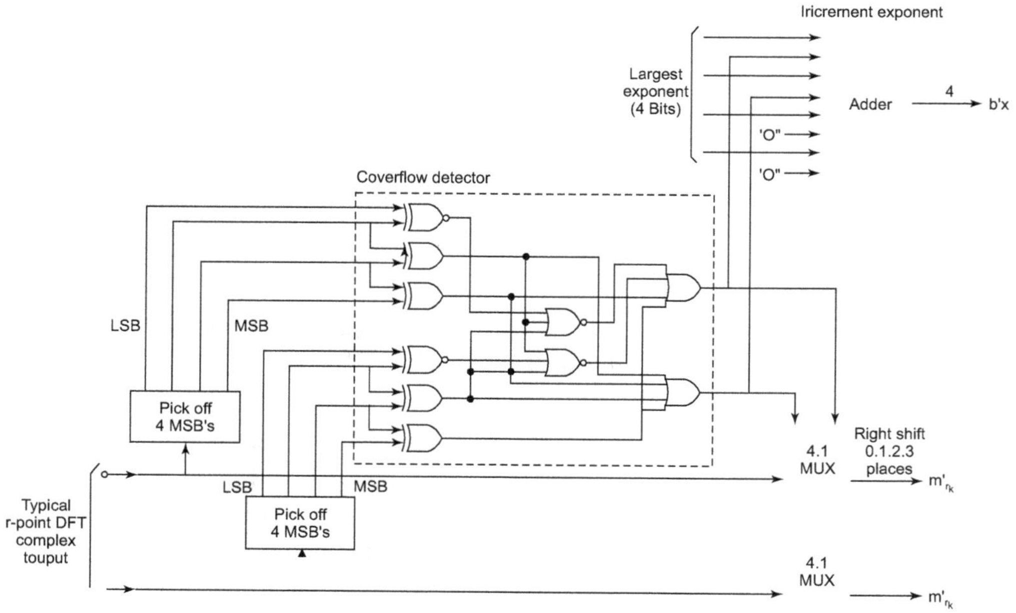

FIGURE 14.36 Details of overflow correction for the hybrid arithmetic scheme.

Simulation results convinced us that the scheme given above was highly competitive in performance with either the fixed or floating-point schemes. From a hardware point of view, this scheme was faster and cheaper than the full floating point and yet overcome the crucial problem of dynamic range that made the fixed-point system worrisome.

Finally, the question of truncation versus rounding was addressed. Various simulation experiments were performed to compare the two methods.

EXAMPLE 14.2

What is "Chirp Signal"? Why it is called so?

Solution:

Chirp: A short, high-pitched sound, such as that made by small bird or an insect is called a chirp.

A chirp is a signal in which the frequency increases (up-chirp) or decreases (down-chirp) with time. It is commonly used in sonar and radar but has other applications, such as in spread spectrum communications. In spread spectrum usage, SAW devices such as RACs are often used to generate and demodulate the chirped signals. In optics, ultrashort laser pulses also exhibit chirp due to the dispersion of the materials they propagate through.

In a *linear* chirp, the frequency varies linearly with time:

$$f(t) = f_0 + kt$$

where f_0 is the starting frequency (at time $t = 0$), and k is the rate of frequency increase. A corresponding time-domain function for a sinusoidal chirp is:

$$x(t) = \sin(2\pi f(t)t) = \sin\left(2\pi\left(f_0 + kt\right)t\right)$$

In a *geometric* or *exponential* chirp, the frequency of the signal varies with a geometric relationship over time. In other words, if two points in the waveform are chosen, t_1 and t_2, and the time interval between them $t_2 - t_1$ is kept constant, the frequency ratio $f(t_2) / f(t_1)$ will also be constant. The frequency varies exponentially as a function of time:

$$f(t) - f_0 k^t$$

In this case, f_0 is the frequency at $t = 0$, and k *is* the rate of exponential increase in frequency. A corresponding sinusoidal chirp waveform would be defined by:

$$x(t) = \sin(2\pi f(t)t) = \sin\left(2\pi\left(f_0 k^t t\right)\right.$$

Although somewhat harder to generate, the geometric type does not suffer from a reduction in correlation gain if Doppler is shifted by a moving target. This is because the Doppler shift actually *scales* the frequencies of a wave by a multiplier (shown below as the constant c).

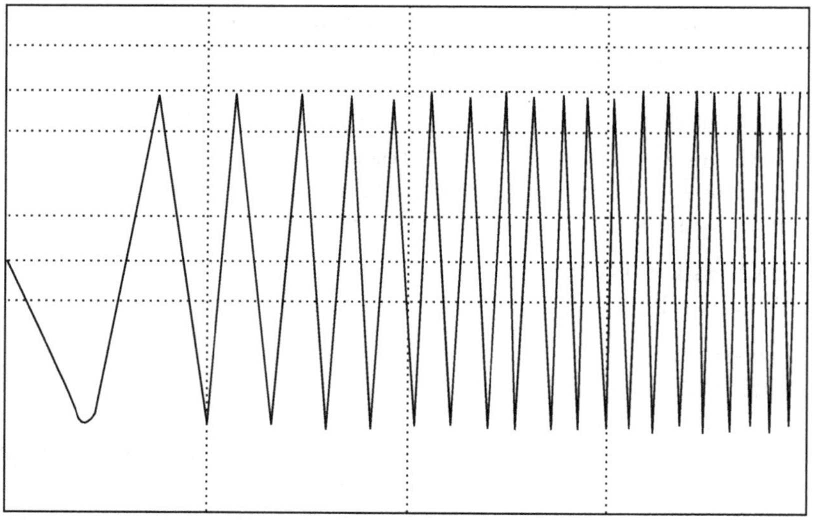

A linear chrip waveform

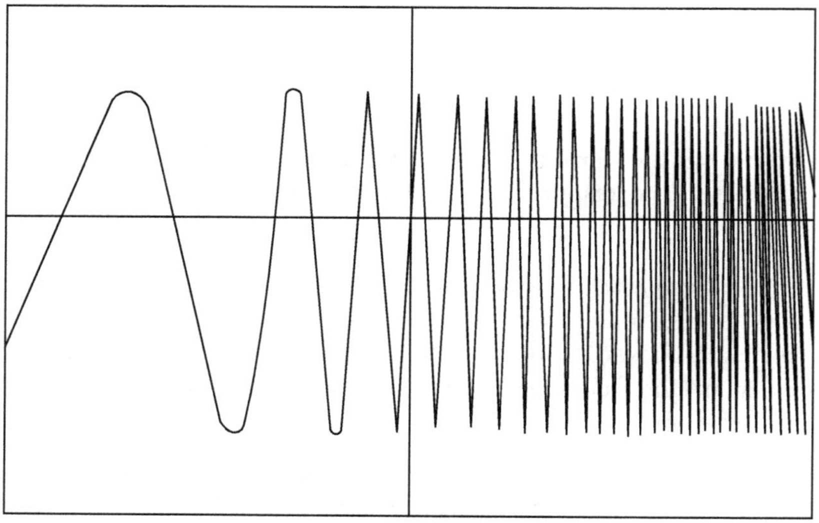

An exponential chirp waveform

$$f(t)_{\text{Doppler}} = cf(t)_{\text{Original}}$$

From the equations above, it can be seen that this actually changes the rate of frequency increase of a linear chirp (kt multiplied by a constant) so that the correlation of the original function with the reflected function is low.

Because of the geometric relationship, the Doppler-shifted geometric chirp will effectively start at a different frequency (f_0 multiplied by a constant), but follow the same pattern of exponential frequency increase, so the end of the original wave, for instance, will still overlap perfectly with the beginning of the reflected wave, and the magnitude of the correlation will be high for that section of the wave,

A chirp signal can be generated with analog circuitry via a VCO, and a linearly or exponentially ramping control voltage. It can also be generated digitally by a DSP and D/A converter (DAC), by varying the phase angle coefficient in the sinusoid generating function.

EXERCISES

1. What is radar? Give various applications of radar.
2. Write down various advantages and limitations of a radar.
3. Describe the modern radar system with a block diagram. Define various radar parameters.
4. What is Doppler filtering?
5. Write short notes on radar signal design.
6. What is the ambiguity function? Draw a block diagram of the radar model leading to the ambiguity function.
7. Discuss the followings:
   a. Ambiguity functions of chirps and sinusoidal pulses.
   b. Ambiguity function of a CW pulse.
   c. Ambiguity function of a burst.
   d. Other signals.
8. What are digital matched filters for radar signals?
9. Discuss filter matched to a long pulse of constant frequency.
10. What are Doppler effect and Doppler frequency?
11. Describe a system for detection and crude range estimation at very long ranges.

12. Describe a matched filter for general signals.
13. Discuss three methods of weighting to reduce matched filter side lobes.
14. Describe a matched filter for a burst.
15. Discuss burst processing by sliding FFT.
16. Discuss the digital realization of a burst processor using a permit memory.
17. Describe airborne surveillance radar (ASR) for air traffic control (ATC).
18. Compare optimal 8-pulse processors and suboptimal processor three pulse canceler and unweighted DFT filter for the following cases:
    a. 0-Hz optimization
    b. 125-Hz optimization
    c. 500-Hz optimization
19. Describe a suboptimal processor with the help of a block diagram.
20. Discuss FDP real-time processing of a range-angle section with help of a diagram.
21. Describe long-range-demonstration radar (LRDR) with the help of a diagram.
22. Illustrate various time-epochs in LRDR.
23. Discuss pulse processor.
24. Describe a digital matched filter for a high-performance radar (HPR).
25. Discuss hybrid arithmetic scheme for radix-r pipeline DTT-FFT algorithm with the help of block diagram.
26. Discuss details of overflow correction for the hybrid arithmetic scheme with the help of a diagram.
27. Define chirp z-transform. Explain with the help of suitable mathematical expressions, how to chirp z-transform can be used for radar signal spectrum analysis.
28. i. What is a "chirp signal"? and why it is called so?
    ii. What are clutter signals and how these can be alleviated?

29. Briefly explain the working of the following subsystems of a Digital signal generator used in a radar system.

i. Memory and recursive generator

ii. Memory for table lookup

iii. D/A converter (DAC)

iv. Analog filter

v. How mixer DSP can be applied in this digital signal generator

30. Why DSP hardware/algorithms are becoming popular in speech and radar signal processing? Explain how DSP hardware/algorithm can improve speech processing.

GLOSSARY

Adder: A device used to add two or more signals.

A/D conversion noise: The difference between the quantized signal and the sampled signal.

Aliasing: The phenomenon in which a high-frequency component in the frequency spectrum of a signal takes the identity of a lower frequency component in the spectrum of the sampled signal.

Amplitude spectrum: A plot of the amplitude of Fourier coefficients versus frequency.

Anti-aliasing filter: The low-pass filter is placed before the downsamples to prevent the effect of aliasing by band-limiting the input signal.

Anti-imaging filter: The low-pass filter is placed after the upsampler to remove the images created due to up-sampling.

Aperiodic signal: A signal which does not repeat at regular intervals of time.

Autocorrelation: A measure of similarity or match or relatedness or coherence between a signal and its time-delayed version.

Average power: The power dissipated by a voltage applied across a 1W resistor (or by a current flowing through a 1W resistor).

BIBO stable system: A system that produces a bounded output for a bounded input.

Bilinear transformation: It is a conformal mapping that transforms the s-plane into a z-plane on a one-to-one basis.

Bounded signal: A signal whose magnitude is always a finite value.

Butterworth filter: A filter designed by selecting an error function such that the magnitude is maximally flat in the passband and monotonically decreasing in the stopband.

Canonical structure: A form of realization in which the number of delay elements used is equal to the order of the difference equation.

Cascade form: A series interconnection of the sub-transfer functions.

Cascade form realization: Realization of a complex system as a cascade of subsystems.

Causal signal: A signal which does not exist for $n < 0$.

Causal system: A system in which the output at any instant depends only on the present and past values of the input and not on future inputs.

Circular convolution: Convolution that can be performed only when at least one of the two sequences is periodic.

Coefficient quantization error: The error that arises due to the quantization of the coefficients.

Constant multiplier: A device used to multiply the signal by a constant.

Continuous-time Fourier transform: Fourier transform of continuous-time signals.

Continuous-time signals: Signals are defined for all instants of time.

Continuous-time system: A system that transforms continuous-time input signals into continuous-time output signals.

Convolution: A mathematical operation that is used to express the input-output relationship of an LTT system.

Correlation theorem: A theorem that states that the cross-correlation of two energy signals corresponds to the multiplication of the Fourier transform of one signal by the complex conjugate of the Fourier transform of the second signal.

Correlation: An operation between two signals which gives the degree of similarity between those two signals.

Cross-correlation: A measure of similarity or match of relatedness or coherence between one signal and the time-delayed version of another signal.

Dead band of the filter: The range of values to which the output of a filter is confined during a limit cycle that occurs as a result of the quantization effect in multiplication.

Deconvolution: The process of finding the input once the output and impulse response is given.

Deterministic signal: A signal exhibiting no uncertainty of its magnitude and phase at any given instant of time. It can be represented by a mathematical equation.

DIF algorithm: An algorithm used to compute the DFT efficiently by decimating the frequency domain sequence into smaller sequences.

Digital signal processor: A large programmable digital computer or microprocessor programmed to perform the desired operations on the signal.

Digital signals: Signals which are discrete in time and quantized in amplitude.

Discrete Fourier series: The Fourier series representation of a periodic discrete-time sequence.

Discrete-time Fourier transform: A transformation technique that transforms signals from the discrete-time domain to the corresponding frequency domain.

Discrete-time signals: Signals are defined only at discrete instants of time.

Discrete-time system: A system that transforms discrete-time input signals into discrete-time output signals.

DIT algorithm: An algorithm used to compute the DFT efficiently by decimating the given time sequence and computing the DFTs of smaller sequences and combining them.

Downsampling: Reducing the sampling rate of a discrete-time signal.

Dynamic system: A system in which the output is due to past or future inputs also.

Energy signal: A signal whose total energy is finite and average power is zero.

Even signal: A symmetric signal with $x(-n) = x(n)$ for all it.

Even symmetry: Also called mirror symmetry which is said to exist if the signal $x(n)$ satisfies the condition $x(n) = x(-n)$.

Fast Fourier transform: An algorithm for computing the DFT efficiently.

Filter: A frequency selective network.

Finite word length effects: The effects due to finite precision representaticn of numbers in a digital system.

FIR systems: Systems whose impulse response has a finite number of signals.

Flipping a sequence: Time reversing a sequence.

Forced response: The response is due to input alone when the initial conditions are zero.

Fourier transform: A transformation technique that transforms signals from the continuous-time domain to the corresponding frequency domain.

Frequency response: The transfer function in the frequency domain.

Frequency spectrum: Amplitude spectrum and phase spectrum together.

Frequency warping: The distortion in the frequency axis introduced when s-plane is mapped into z-plane using the bilinear transformation.

Functional representation: A way of representing a discrete-time signal where the amplitude of the signal is written against the value of n.

Fundamental period: The smallest value of N that satisfies the condition $x\ (n + N) = x(n)$ for all values of n.

General-purpose DSPs: High-speed microprocessors with architecture and instruction sets optimized for DSP operations.

Graphical representation: A way of representing a discrete-time signal as a plot with value $o\ x(n)$ at the sampling instant n indicated.

Gibbs phenomenon: The ripples present at the point of discontinuity in signal approximation.

Host port: A special parallel port in P-DSPs that enables them to communicate with a microprocessor or PC.

ITR systems: S systems whose impulse response has infinite number of samples.

Imaging: The phenomenon of producing additional spectra by the upsampler.
Impulse invariant transformation: Transformation of an analog filter into a digital filter without modifying the impulse response of the filter.
Impulse response: The output of the system for a unit impulse input.
Input quantization error: The error that arises due to the representation of the input signal by a fixed number of digits.
Interpolation: The process of increasing the sampling rate by an integer factor I by interpolating I-1 new samples between successive sampling instants.
Interpolator: The anti-imaging filter and upsampler together.
Inverse discrete-time Fourier transform: The process of finding the discrete-time sequence from this frequency response.
Inverse Fourier transform: A transformation technique that transforms signals from the frequency domain to the corresponding continuous-time domain.
Inverse *z*-transform: A transformation technique that transforms signals from the z-domain to the corresponding discrete-time domain.
Invertible system: A system that has a unique relationship between its input and output.
Laplace transform: A transformation technique that transforms signals from the continuous-time domain to the corresponding complex frequency domain (s-domain)
Linear phase systems: Systems for which the phase drops linearly with an increase in frequency.
Linear system: A system that obeys the principle of superposition and homogeneity.
LPF: A frequency selective network which allows transmission of low-frequency signals.
LTI system: A system that is both linear and time-variant.
LTV system: A system that is linear but time-variant.
Memory system: Same as a dynamic system.
Memoryless system: Same as a static system.
Multirate systems: Discrete-time systems that process data at more than one sampling rate.
Narrowband low-pass filter: A low-pass filter characterized by a narrow passband and narrow transition band.
Noise transfer function: The transfer function from the noise source to the filter output.
Non-recursive filters: Filters that do not employ any kind of feedback connection.
Non-canonical structure: A form of realization in which the number of delay elements used is more than the order of the difference equation.
Non-causal signal: A signal which exists for $n < 0$ also.

Non-causal system: A system in which the output at any time depends on future inputs.
Non-invertible system: A system that does not have a unique relationship between its input and output.
Non-linear system: A system that does not obey the principle of superposition and homogeneity.
Non-recursive system: A system whose output depends only on the present and past inputs and not on past outputs.
Nyquist interval: The time interval between any two adjacent samples when the sampling rate is Nyquist rate.
Nyquist rate of sampling: The theoretical minimum sampling rate at which a signal can be sampled and still be recovered from its samples without any distortion.
Odd signal: An antisymmetric signal with $x(-n) = -x(n)$ all n.
One-dimensional signal: A signal which depends on only one independent variable.
One's–complement form: A type of representation in which a negative number is obtained by complementing each bit of the positive number.
Parallel form realization: Realization of a complex system as a parallel connection of subsystems.
Periodic signal: A signal which repeats itself at regular intervals of time.
Phase factor: A factor that is equal to *N*th root of unity. Exploiting its symmetry properties the DFT is computed efficiency.
Power signal: A signal whose average power is finite and total energy is infinite.
Prewarping: The conversion of the specified digital frequencies to analog equivalent frequencies to nullify the effect of warping in IIR filter design using the bilinear transformation.
Product quantization error: The error that arises due to the truncation of the multiplier output.
Product round-off noise: Same as product quantization error.
Quantization: Process of reducing the size of a binary word.
Quantization noise: The error introduced is due to the quantization of the signal.
Quantization step: The interval between successive quantization levels.
Radix: Number of independent symbols used in a number system.
Random signal: A signal characterized by uncertainty about its occurrence. It cannot be represented by a mathematical equation.
Recursive filters: Filters that make use of feedback connections to get the desired filter implementation.
Recursive system: A system whose output depends on the present and any number of past inputs and outputs.
Right-sided signal: A signal which is equal to zero for $n < N1$ for some finite $N1$.

ROC of z-transform: The range of values of $|z|$ for which $X(z)$ converges.
Rounding: The process of reducing the size of a binary number so that the rounded number is closest to the original unquantized number.
Sampling frequency: The reciprocal of the sampling period indicates the number of samples per second.
Sampling interval: Same as sampling period.
Sampling period: The time interval between two successive sampling instants.
Sampling rate conversion: The process of converting a sequence with one sampling rate into another sequence with a different sampling rate.
Sampling theorem: A condition to be satisfied by the sampling frequency for a band-limited signal to be recovered from its samples without distortion.
Sampling: The process of converting a continuous-time signal into a discrete-time signal.
Signal processing: A method of extracting information from the signal.
Signal to noise ratio: The ratio of signal power to noise power.
Signal: A single-valued function of one or more independent variables which contain some information.
Stable system: A system that produces a bounded output for a bounded input.
Standard signals: Signals like unit step, unit ramp, unit impulse, etc., in terms of which any given signal can be expressed.
Static system: A system in which the response is due to present input alone.
Steady-state response: The response due to the poles of the input function. It remains as $n \to \infty$.
Step response: The output of the system for a unit step input.
Step band: The band of frequencies that is rejected by the filter.
System: An entity that acts on an input signal and transformed it into an output signal.
Tabular representation: A way of representing a discrete-time signal where the magnitude of the signal at the sampling instants is represented in tabular form.
Time convolution theorem: A theorem that states that convolution in the time domain is equivalent to multiplication of their spectra in the frequency domain.
Time invariant system: A system whose input/output characteristics do not change with time.
Time-variant system: A system whose input/output characteristics change with time.
Total response: The sum of the natural and forced responses.
Transfer function: The ratio of the Fourier transform/Laplace transform/Z-transform of the output to the Fourier transform/Laplace transform of the input of the system when the initial conditions are neglected. It is also the Fourier transform/Laplace transform/DTFT/z-transform of the impulse response of the system.

Transient response: The response due to the poles of the system function. It vanishes after some time.

Truncation: The process of reducing the binary number by discarding all bits less significant than the least significant bit that is retained.

Two's complement form: A type of representation in which a negative number is obtained by complementing each bit of the positive number and adding 1 to MSB.

Two-sided signal: A signal which exists in both positive and negative times.

Unit impulse function: A function that exists only at $n = 0$ with a magnitude of unity.

Unit parabolic function: A function whose magnitude is zero for $r < 0$ and is $n2/2$ for $n \geq 0$.

Unit ramp function: A function whose magnitude is zero for $n < 0$, and rises linearly with a slope of unity for $n > 0$.

Unit step function: A function whose magnitude is zero for $n < 0$, suddenly jumps to 1 level at $n = 0$ and remains constant at that value for $n > 0$.

Unstable system: A system that produces an unbounded output for a bounded input.

Up-sampling: Increasing the sampling rate of a discrete-time signal.

Word length: The maximum size of binary information that can be stored in a register.

Zero padding: Appending zeros to a sequence in order to increase the size or length of the sequence.

***z*-transform:** A transform technique that transforms signals from the discrete-time domain to the corresponding *z*-domain.

INDEX

N

P

R

S

T

U

V

Z